General Psychotherapy

Lotte Hartmann-Kottek

General Psychotherapy

Principles and Common Theoretical Aspects - Rediscovering Humanity

Springer

Lotte Hartmann-Kottek
Kassel, Germany

ISBN 978-3-030-87465-0 ISBN 978-3-030-87466-7 (eBook)
https://doi.org/10.1007/978-3-030-87466-7

This Springer imprint is published by the registered company Springer Nature Switzerland AG
The registered company address is: Gewerbestrasse 11, 6330 Cham, Switzerland

Preface

Our field of psychotherapy can be compared to a conglomerate or kaleidoscope of many approaches and partial truths, each of which depicts something significant and valuable from the right perspective. However, the different basic stances, approaches, focuses and relatively unconnected relationship of each school of thought prevents them from forming a common, general framework. This fragmentation in the field of psychology mirrors the upheaval of our times, among other things, whereby, prior to now, even individual scientific fields were able to reach a consensus on an adequate shared worldview and concept of man.

This pertains to the complete integration of classical and quantum physics into "mainstream" science, which is yet to occur. Supporters of the former hold the basic assumption that the world is the sum of particles that can be measured, weighed, statistically evaluated and pinpointed. Others are of the opinion that the smallest physical entities are energetic quanta that join, pseudo-corpuscularly accumulate or exist in an intermediate state – and which can be influenced by human consciousness, among other things. This is a key notion of Einstein's legacy. $E = mc^2$ is a formula known the world over. One does not need to be a quantum physicist to discuss the principle of mass-energy equivalence.

Why is this relevant to psychotherapy? As, on the one hand, these assumptions presume a different underlying basis to the science of psychotherapy and, on the other hand, they are linked to a different scientific methodology. The first group solely focuses on the materially measurable and observable factual substrate, for instance in neurophysiology, biology, embryology, behavioural observations, etc., and employs statistical methods from classical physics. The second group faces the task of engaging with the energies and information radiated and absorbed at different levels in the interpersonal and intrapsychic interspace. Analogous to the smallest element, the quantum, more complex beings such as humans are understood as a whole entity in their simultaneously material and non-material constitution. The scientific treatment of the quantum amount linked in waves requires reference to quantum theory. (The methodology of classical physics and statistics is not applicable for this amount.) Both humans and brains, in particular, are a meeting place for both modes of being and their intermediate forms. We find constructive and quantum states in parallel, or, perhaps better said, intertwined. The "consciousness" subsystem is clearly present in the quantum state, whilst the materialised system of the constructive state seems to provide necessary shielding; on its simplest neuronal organisational level, it works – with limited processing capacity – using the on/off principle. There remains an evident great need for research on the non-material complementary perspective. The findings obtained would likely be greatly beneficial in this regard.

The dual mind-body problem is an issue of the concept in our minds, our own biological construction that pre-structures the way in which we think. As such, it is not a problem of reality. We outwardly project decoherence. Due to the structure of fragmented, sensory perception and the cognitive processing thereof in the dominant brain hemisphere according to the "either-or" principle, we struggle to grasp the entirety of the basic element or the substantially energetic identity of matter and its equivalent energy field.

Philosophers have pondered the dualistic mind-body problem for thousands of years. Over the centuries, proponents of Plato and Aristotle have fought fiercely as realists and idealists. Descartes paved the way for another milestone for the seemingly irreconcilable divide between matter (res extensa) and spirit (res cogitans). Even modern-day philosophers such as the mathematics enthusiast David Chalmers (born 1966), who opposes reductive materialism in its many forms and adopted the bold position that everything is spirit, are at risk of falling into a polarising either/or position. (Aristotelian theory only recognises "A or B" and explicitly not "Third".) Our place in this matter-spirit dimension is the shift zone, the "limbo" the "as well as" between materialised and connected energy, a fascinating energy that is capable of transforming and is simultaneously "spirit/consciousness" potential and "matter" potential. On the one hand, this understanding of our underlying basis leads to categorical confusion, but, on the other hand, it grants opportunities to psychotherapy that have been too little utilised to date.

If we are unable to find consensus on the aspects of the worldview, it is unlikely that we will be able to agree on the concept of man and psychotherapeutic approaches. As such, reflecting on the background aspects takes up a comparatively large part of this book.

The promising importance of these gaps for psychotherapy has not remained undetected and the natural sciences and humanities have each developed a number of interesting concepts to fill and bridge them. However, in this regard, a large number of unconnected elements coexist, such as Buber's "I-Thou" principle, Merleau-Ponty's notion of intercorporeality, Freud's focus on transference and countertransference, Roger's demand for empathy, Watlzawick's "Human Communication" and a range of other concepts from systemic therapy and communication sciences, the discovery of mirror neurons in neuroscience, Thomas Fuchs' explanatory model of the "The Brain – A Mediating Organ" or modern resonance theory from sociology (Hartmut Rosa).

The times in which we live are difficult and characterised by upheaval. Accordingly, there is an increasing need for people to act in a humane and generally responsible manner. Psychotherapy is one of the many fields of science that can strongly influence the contemporary concept of man. For this reason, a detailed chapter of this book is dedicated to the spiritual background trends that form our subject in the primarily hidden, inner dialogue and express, support and interact with our concept of man. In times of crisis, it is particularly pertinent to review the course we have pursued thus far. Is our field currently focused on developing well-functioning humans who are controlled by others or rather on facilitating the development of individuals who are strong enough to listen to their inner voice with their finely tuned antennas on the inside and out and integrate this into their peer community? The good news behind all of this is that there remains a significant amount to be achieved with regard to the scope for the development of human potential. Fittingly, this book is concerned with this scope.

Some may view the claim made in the subtitle as rather audacious. There is a strong likelihood that it may provoke resistance amongst colleagues who are unfamiliar with the term "Gestalt therapy", which is entirely understandable. As such, this perspective requires further explanation. Those familiar with "Gestalt therapy" will be aware of its foundations in the natural, human growth and maturation processes and consequent occurrence in the inner, unspoken knowledge that is intuitively available to each individual and likewise can and should be squared with his experiences. In reality, every individual is aware of their fundamental orientation, even if this is not described with used terminology.

Active members in the school of psychotherapy are also aware of Gestalt therapy's position as the natural intersection of psychotherapeutic aspects and traditions, which exhibits both the potential for a mutual, appreciative process of understanding between the current schools and also the opportunity to increase the psychotherapeutic effectiveness for patient care.

As a diverse range of personalities can be found amongst our patients and colleagues, we should ideally strive to provide a range of treatments to enable individual therapists to develop their own authentic style within the framework of the generally applicable standards.

The large group of colleagues who aim to improve efficacy have always been capable of seeing the bigger picture and working in a targeted manner. There is no doubt that this transformation poses a significant challenge to personal flexibility and the scientific understanding we all gain during our initial experiences in the field. Loyalty is commendable, but it should not become something to which we are bound. Rather, our initial knowledge should act as a sound foundation. The old "black box" has become increasingly translucent in recent years in any case. Behavioural therapy has come on in leaps and bounds at breathtaking speed and now aims to achieve an appreciative and stable therapeutic relationship, primarily in line with the client-centred approach. As opposed to the outcome of techniques that were previously overrated, this approach guarantees a high success rate for therapy.

A glance at the transformation of the psychodynamic group of methods that remain beholden to tradition may allude to a less impressive pace, certainly if the focus is only on shorter periods of time and mainstream practices. However, generally speaking, mainstream psychotherapy has also undergone an impressive period of development through the various stages of drive psychology, ego psychology, object relations psychology, self-psychology and, most recently, relational psychoanalysis. Many colleagues who have contributed to this revival demonstrated commendable courage to stand against threats of exclusion aimed at reformers by the conservative wing. An entire book could be written on personal experiences of the exclusion tradition of psychoanalysis (PA), the fate of its dissidents or tragic rebels and their constructive contributions to the development of our "general psychotherapy". The majority of today's PA representatives would have been named and shamed by their colleagues a century ago. Advancement is tough at times, but the tolerance gained is worthy of respect.

The more experimental colleagues with a fundamental psychodynamic concept, who aimed to integrate body therapy forms, staged and artistic means and expression, primarily convened under the more tolerant umbrella of depth psychology (DP), the "little sister" of PA.

From the outset, therapy forms outside the Psychodynamic Association were characterised by a more creative and experimental nature and fundamentally enriched the school of psychotherapy. This was and remains especially true for the humanist and systemic methods, including Gestalt therapy. They also include rarely mentioned derivatives of hypnosis, the (re)integration of which is otherwise primarily supported by systemic therapy forms.

The increasing joint adoption of methods and methodological aspects through the partitioned barriers of the otherwise politicly competing schools in recent decades has, of course, been met with mixed reception, especially when the adoption was coupled with contemporary political and social repression.

However, it may be precisely this difficult topic of the mutually uncoordinated, intellectual adoption and incorporation of vicinity aspects on a macro scale that may indicate that the time has come for an integrative reconciliation or fusion of the essence of psychotherapeutic approaches. As perhaps needs no mention, this path will require an extensive field of play.

I ask in advance for your understanding and would like to apologise to the restricted and incomplete nature of the recital of the various approaches. This study aims to provide an open system. Consistent amendments and extensions are always encouraged. I also kindly ask for your understanding that I shall not conceal my loyalty to the approach that I have previously gained the most experience with. We are all entitled to our own loyalties. Following the example of Lessing's "Nathen the Wise", a model of tolerance, or adherence to the principle "Ye shall know them by their fruits" (Matthew 7:16), the development of what is proven should be guaranteed, would oversee the recognition and removal of ominous, fundamentalist, professional politicians at an early enough stage.

In the case of the issue of intellectual expropriation and the lack of sources which has already been alluded to, we can all provide subsequent suggestions on how to categorise information. Therapies that may initially appear to be new often crop up, the majority of which contain universal principles. Usually, these therapies have simply been assigned other terms and given an up-to-date emphasis, in line with the times. The success thereof is then contingent on the initiator's charisma. He may be able to convince some that he has "reinvented the wheel".

However, that is not to say that the revival of the ancient cultural heritage of humanity is not welcomed in modern psychotherapy. This revival has been observed in the high regard that has been held for the Zen tradition since the mid-twentieth century, or the wave of mindfulness that swept through the world of psychotherapy in the last two to three decades, with the second wave acting as the "later child" of the first, after which it was enable to gain legitimacy as a scientifically effective method.

The natural, unofficial integration process has been observed both on a nationwide, in Germany, and global scale over the past three to four decades and has shone a light on the counteractive, association-based attempts at demarcation that prevailed then and now. This environment has given rise to the challenge for our generation and profession, embodied in the battle between the claims of supremacy asserted by individual groups against increasing the potential of patient care for all.

In the USA, where behavioural therapy has traditionally reigned supreme, a Society for the Exploration of Psychotherapy Integration (SEPI) was officially established as a sub-organisation of the APA (American Psychological Association). Researchers such as Louis Castonguay have examined partial aspects such as the "corrective experience" or the "insight" according to their relative importance, significance and variations throughout the various schools.

A number of adjustments and convergences have occurred over the last century, particularly in the following areas: (1) the importance of the therapeutic relationship, which has since become exonerated from suppositions and instead more authentic and personalised; (2) on the notion of the concept of man with its manifold, intrapsychic stages of development, its interwoven concepts, its need for mindfulness and its perception of resources, both internally and externally; and (3) in the area of specific, technical-methodological methods, regardless of whether these are derived from the learning theory-based humanistic-systemic or psychodynamic traditions. The notion that technical skills alone do not significantly impact the outcome of therapy has been discussed in research reports and textbooks over the past 20–30 years. However, this information is yet to be disseminated to all teachers and further passed on to students. Practising colleagues are primarily those behind the push for the integration process, i.e. those who are typically more likely to strongly identify with the recovery hopes of their patients than with the demands placed by professional politicians to define and demarcate approaches whilst simultaneously acknowledging the supremacy of individual associations.

From the outset, the concept for integration presented in this book can be viewed as particularly effective and consequently financially beneficial due to its consolidated potential. It is primarily concerned with the cultivation of human potential within our field, otherwise referred to as "psychotherapy with a human face".

Accordingly, in this regard, the field of psychotherapy has exhibited a rare and welcome occurrence whereby humanity, ethics and cost-effectiveness have positively complemented each other and shall hopefully continue to do so, provided this is not hindered by future developments.

In general, all methods employed in psychotherapy are based on advances in scientific research in neighbouring fields, which subsequently promotes the convergence process. The following areas in particular have helped to facilitate convergence:

(a) Published findings on the central nervous system, brain and neurophysiology.
(b) Systematic observations made through research on infants, mother-child interactions, attachment research and general developmental psychology.
(c) Findings from pedagogy, sociology and cultural anthropology.
(d) The fundamental, methodological development of science, including paradigm shifts, a number of which remain to be fully integrated, such as quantum theory which provides a key underlying approach for psychotherapy.

The field of psychotherapy is unfortunately not free from the deliberate misappropriation of research findings from other scientific fields, whereby the focus should, in an ideal world, be on using the findings to obtain insights in an honest way.

As can be presumed, the author includes traces of her own lifelong professional experience from working in inpatient and outpatient care covering all degrees of severity over the past six decades – in addition to extensive experience in education and training. Our actions should always be fundamentally aimed at helping patients to heal and get back on their feet in the best way possible.

Klaus Grawe was one of the pioneers with regard to the early phases of integration. His publications *Psychological Therapy* (1998, Hogrefe) and *Neuropsychotherapy* (2004, Hogrefe) were released following a period during which Grawe, the behavioural therapist and renowned psychotherapy researcher, established close collaborations with colleagues in the humanist and systemic branch, and he was astonished by the success achieved with their working methods.

I can recollect the strong conviction that strongest efficacy could be achieved in therapy through a combination of behavioural therapy and a humanistic-systemic approach he evinced during personal conversations I enjoyed with him around 1989–1992. At this time, he also became familiar with and held in high esteem the therapeutic relationship building method practised by Carl Rogers that placed an emphasis on interpersonality – the individual process-based approach spearheaded by the humanists, notably Gestalt therapists – as well as the research orientation of the humanistic and systemic branches.

The school of psychotherapy had already made a public call for resource orientation in the 1962 "Human Potential Movement" manifesto in the USA. Hitherto, the field of psychotherapy officially solely consisted of contemporary psychoanalysis, which was as limited in the sparseness of relationships and the libido-centeredness of the post-war generation shaken by the Holocaust as the behavioural therapy practised at that time, which negated emotions, experiences and relationships in favour of a scientific focus.

Around the same time, Virginia Satir, a charismatic pioneer of the humanist-systemic movement, incorporated resource orientation into systemic therapy. However, the latter was only lightly touched upon by the Palo Alto Group (G. Bateson, D. Jackson, J. Haley, J. Weakland, P. Watzlawick, J. Bell, V. Satir), which primarily focused on the system as a whole instead of the individual. Signatories of the 1962 humanist manifesto included: A. Maslow, Ch. Bühler, K. Goldstein, C. Rogers, A. Watts, J. Buggental, A. Köstler, R. May, A. Huxley, and A. Sutich. The extended circle of proponents also included Virginia Satir and Fritz Perls, who worked together at the time.

The integral visions presented by Grawe in his works *Psychological Therapy* (1998, Hogrefe) and *Neuropsychotherapy* (2004, Hogrefe) answered the question of what makes a good therapist. However, his answer, primarily from a humanistic-systemic perspective, failed to mention a source. According to Grawe, therapists should work in a resource-oriented manner and think, perceive and act in a process-oriented manner; become relationship experts but nevertheless remain experts on disorders; become experts in the motivational dynamics of psychological events; be able to intervene in a way that focuses on coping and clarification; learn to employ a range of interpersonal settings; learn to base their approach on insight gained from the case at hand; and develop as a therapist to perceive, think and act in a multidimensional manner. Later, Grawe condensed his points into five essential common factors: (1) therapeutic relationship, (2) activation of resources, (3) actualisation of the patient's problems, (4) motivational clarification and (5) (active help for) problem solving.

Summary: In contrast to the official professional policy pursued in Germany, on a content level, an expert demonstrated that perhaps the most significant potential for the future effectiveness of psychotherapy can be discovered in the approaches employed by the group of methods that have been officially excluded from Germany since 1998. His suggestion of combining notions from individual schools not only encouraged the assumption of these methods but also triggered interdisciplinary integration.

As such, this book aims to follow up on the series of contributions made to integration in recent decades, albeit with a slightly diverging emphasis.

At present, we are faced with the challenge of overcoming the fear of competition between the schools of psychotherapy and – at least in Germany – the fears and historical, mutual attacks between our colleagues from the fields of psychology and medicine. A wounded and often combative and bitter group of colleagues, a number of whom remain trapped by mutually devaluating bias, who continue to shape today's generation of psychotherapists. It is then perhaps no wonder that young talent in our field still struggles with impartial cooperation and obtaining their own findings in an unfeigned manner. I would like to see future psychotherapists achieve success in these endeavours. Psychotherapy requires all of its roots, those from psychology and medicine, and it also requires the findings of approaches that lead to policy changes and those that do not. First and foremost, it requires respectful cooperation between all colleagues.

Lotte Hartmann-Kottek
Kassel, Germany

Contents

Acknowledgements

No one can claim to have mapped their world of ideas without the input from others. The same applies to all "schools" of thought. We are all shaped by our culture, biased by the respective zeitgeist and have relied on the work of countless visionaries for generations. I would first like to direct my thanks towards these visionaries.

I would like to extend a big thank you to my many teachers from various fields and areas; my loyal friends, with whom I have enjoyed a lifelong intellectual exchange – and, last but not least, I would like to thank my patients. Whilst we were striving to achieve cooperation, they permitted me to enter their minds and world of experience and dared to make significant changes.

My thanks also go to the Springer editorial office, specifically Ms. M. Radecki and her team. They encouraged me to put pen to paper and write the introductory overview "General Psychotherapy" in light of my humanistic background amidst these peculiar times of upheaval. I have been provided with a great level of professional and emotional support at every point.

For readers in a hurry:

If the main reason you are consulting this book lies in the practical know-how of psychotherapy and gaining an understanding of disorders, I recommend turning directly to ▶ Chaps. 2, 5, and 6.

If you are more interested in and open to epistemic questions, you will find all the answers in ▶ Chaps. 3 and 4.

If the cultural and historical background on the development of our human potential are of particular interest to you, a look at ▶ Chap. 7 would be advisable.

Alternatively, turn to ▶ Chap. 8 to learn more about current research against the backdrop of contemporary professional politics and the history of psychotherapy.

Note on role and gender designations. This entire book pointedly encompasses all of the diversity that is exhibited in the field of psychotherapy: Patients, clients, subjects, protagonists. Therapists are alternately referred to as either male or female throughout.

General Psychotherapy

» An attempt to integrate the common factors and special contributions
of all current psychotherapy methods
with particular regard to the basic model of Gestalt therapy,
whereby key aspects are confronted from the outset
against the backdrop of an overdue scientific paradigm shift.

In lieu of an abstract notion:

» Behold the moon – and wonder
why half of her stands yonder
yet she is round and fair.
We follow empty visions
and artisans' ambitions
because our minds are unaware.
Matthias Claudius

Psychotherapy Between Tradition and Awakening and Between Multiplicity and Integrity

Contents

L. Hartmann-Kottek, *General Psychotherapy*,
https://doi.org/10.1007/978-3-030-87466-7_1

Overview

Psychotherapy unites the intellectual traditions of the natural and human sciences. It also reflects the current upheaval within the natural sciences.

Over the past two to three centuries, the notions and evaluation of objectivity and subjectivity have undergone significant transformations, a fact that has been reflected in the various schools of psychotherapy. Subsequent reflections on these shifts can help to build bridges.

Difficulties in understanding often arise when it comes to the equivalence of "field and structure", "wave and corpuscular theory" and subsequently Einstein's equation "$E = mc^2$", which touches on the equivalence of matter to spirit: two sides of the same coin. This is further linked to the "digital" and "analogue" use of our brain. It may perhaps surprise the reader to learn that this aspect is relevant to psychotherapy.

With regard to research, this book is geared towards the interdisciplinary spirit and international standard, as has been firmly established for half a century in Bergin & Garfield's Handbook (Lambert ed./ 6 ed., NY).

1.1 Psychotherapy as Science During a Period of Upheaval and the Search for Common Ground

1.1.1 What Is Psychotherapy and Where Does It Belong as a Field?

The title of this section likely gives rise to a number of questions: Can psychotherapy be regarded as a science? If so, to what school does it belong? Was psychotherapy not originally an appendage of philosophy, similarly to psychology? Is it not primarily based on opinions, models, subjectivism and speculation to the detriment of hard facts? Was psychotherapy not only first gradually switched to the natural sciences departments of universities in recent decades? Does it really belong there? Or does it simply want to use its position there to prove at all costs that it should be regarded as an "exact natural science" and has it since outpaced itself with statistic-methodical brilliance? Is the traditional division of natural and human sciences suitable for our school of thought, or are we already in the midst of an upheaval that challenges the conventional either/or way of thinking?

The methodical treatment problem: It may sometimes appear that psychological psychotherapy continues to emulate the standards set by Wilhelm Wundt (1832–1920), who established the first highly respected, psychophysiological, experimental psychology laboratory linked to a university in Leipzig in the mid-nineteenth century. He was a master of exact "objective" introspection, something that remained undisputed at the time. He influenced the work of his contemporaries around the world, including Vladimir Bechterew (1857–1936) and his student Iwan Pawlow (1849–1936) in St. Petersburg, Hugo Münsterberg (1863–1916), who emigrated to the USA, Freud (1856–1939) from Vienna, Husserl (1859–1938), who ended up teaching in Freiburg, amongst others.

However, as we all know, the history of the natural sciences did not end there. It has undergone an intellectual upheaval. And there is much yet to come. Although it may appear that theoretical physics – as paradoxical as that may sound – has found new, solid ground in the "indeterminacy of quantum theory" as far as this is possible, despite many open questions, and although "new physics" has long been fully implemented in the technical field and has boomed economically, other schools of "Applied Natural Sciences" have remained reluctant to incorporate this new body of thought. This includes psychology and medicine (to the exclusion of medical technology). For a long time, in order to appease, it was claimed that "new physics" only applied to the macro and micro levels, but not to the meso level, in which we humans live. However, this notion remains unchallenged and gravely affects the field of psychotherapy.

With his "Generalised Quantum Theory", quantum physicist Hartmann Römer (University of Freiburg) supported the stance that the principles of complementarity and entanglement, etc., also apply to the meso level of our existence. This also includes the phenomenon whereby the order of sequential measurements is not reversible (as postulated in classical physics) and their individual experimental results cannot be repeated.

When applied to psychotherapy, this subsequently entails that each human requires both experiences of belonging and dissociation for their own development – but the order thereof is not random, as each experience changes the existing situation. If an individual experiences equal numbers of disassociation or decoherence experiences at the beginning of their life as experiences of belonging later in life, this will not subsequently result in the same self-perception or same personality structure as would be the case if the order was reversed.

This also applies to effective psychotherapy, whereby a "natural" order must be maintained: we first need to establish trustful and therapeutic relationships, we need the most existential encounters with a high level of coherence and contact with the patient's/client's potential and resources prior to confronting patterns in need of correction without any perception of fear of defence by the patient. In this way, we can then begin to transform patterns.

Similarly, if we intend to use the medium of a healing, viable group matrix to work in group therapy, we first need to establish a trusting relationship network whereby all participants can experience being elevated from their existential being before any profound corrective processes can be successfully started – group concepts that begin with the promotion of confrontation, differentiation and individualisation from the outset primarily mobilise protective and defence mechanisms and rarely manage to reach deeper levels.

The laws of classical physics are based on statistical calculations. They are statements from the human perspective on the potential behaviour of the smallest possible hypothetical particles. If we take gas molecules as an example, for instance, there is a significant dimensional difference between man and molecule (eight mathematical dimensional spaces). The error range for the behaviour of an individual gas molecule is negligible, and, accordingly, the average calculating operations are fundamentally precise.

However, the situation is different if there is no dimensional difference, when it is a matter of one man confronting another with all his unique development in face-to-face interaction. In this case, the laws of classical physics and statistics are no longer meaningful. They can only provide a ranking sequence position for a tested dimension in relation to the population of an exceedingly large group, e.g. the contemporary population of the same age. This is comparatively poor information which is typically avoided in individual psychotherapy. It does not provide any information on the classification and significance of an event through the individual processing of the event within the subjective construction of the world by the subject.

Cave

If we were to believe that we can get by in psychotherapy with solely or primarily the ideological edifice of classical physics, we would be making an error in terms of category.

Similarly to the "uncertainty" principle in quantum mechanics, therapists need to develop a certain tolerance for uncertainty and use this carefully and encouragingly for the coherent development of unconscious, personal, dialogical and situational abilities to achieve intuitive self-management and creative reorganisation. This aspect remains absent from the curriculum of emerging, university-level psychotherapy, potentially due to the fact that it is scientifically harder to grasp. Blindness to uncertainty is thus likely to prevail amongst the following generation.

In the majority of psychotherapy and (Gestalt) pedagogical schools, however, uncertain fields have long and successfully been regarded as a development space for latent potential, e.g. in the various versions of play therapy, in "experiments" on group dynamics, in the systemic work of sociograms, in family constellations and scenario-based psychodrama therapy (Moreno), as a key element of Gestalt therapy (Perls), Emotion Focused Therapy (Greenberg) and in experiments to substantiate unconscious subject matters in the works of Jungian proponents A. Mindell, R. Assagioli and A. Pesso, amongst others.

This placement of significance on the scope for development has also led to the formulation and publication of the call for the individualisation of psychotherapy as one of the postulations from the research results of the APA (American Psychological Association) review summary by its authors, Norcross & Lambert in autumn 2018 – in addition to the principle demand for consideration of the therapeutic relationship and therapist's personalities.

Conclusion

If solely the logic of classical physics and all of its statistics applies to psychotherapy, all processes should result in the same outcome when the order is reversed. This is clearly not the case, as in quantum theory. Psychotherapy is one of the fields whereby the manifestations and calculation methods employed in classical physics meet and supplement quantum theory.

1.1.2 The Multiplicity of Our Psychotherapeutic Approaches – Confusion or Abundance?

Does this equate to a vast, primary multifaceted side of nature, or solely to the varying superimposition of a few principles?

A sensible approach would be to determine the therapeutically effective phenomenon behind the various, school-dependent viewpoints, descriptions and traditionally suppressed approaches.

Our psychotherapeutic "initial imprint" at first provides us all with a perception and allocation filter to internalise meanings, which seemingly objectively structures processes that are relevant to therapy. However, as we broaden our horizons, we typically discover that the former meanings are true to some extent but only encompass partial truths.

One group focuses on the conflictual processing that accompanies sexual development, the other on inadequately compensated inferiority conflicts, and another gains added insight into the primary lack of love and bonds at the beginning of life, another filters out mislearned behaviours through stabilising, beneficial consequences from life stories, another focuses on inadequate appraisals – yet another group focuses entirely on involuntary, physical expression-behaviour, etc.

On the one hand, the individual schools attach importance to the different stages of development; on the other hand, they focus on the different levels of understanding and integration: the intrapersonal, the psychosomatic, the interpersonal, the organisational or social. In turn, this results in different snippets of reality.

Each with its own justification. Wampold et al. [1], p. 336: ...

Important

"In the thousands of (RCT) studies conducted, we are still yet to find a specific treatment for a particular disorder, which was, on average, clinically and systematically superior to any other psychotherapy".

A tolerant scale was applied in this case.

The aim is to find a suitable coordination system that incorporates variety.

1.1.3 The Risk: When Research Turns into the Plaything of Competing Associations

If we intend to solely and uncritically base our search on school-based research, we will likewise find ourselves in the minefield of interest-led pluralism.

A classic example of a distortive influence is the impact of the allegiance effect. According to which, the dedication and conviction of the therapist in his methods is passed on to the patient seeking help, consequently increasing the patient's hope and trust in the therapist's skills and changing the final outcome from the point of view of the therapist with strong conviction in his methods up to an effect size of approximately 0.35. In the past, it has been commonly been observed for decades that this effect is falsely used to accredit the method or technique in question. According to recent calculations by Wampold, Norcross et al., the applied technique only made a 10–15% difference to the final outcome in psychotherapy. When combined, the therapeutic relationship and the therapist's personality, however, were twice to three times as likely to make a difference.

A futher risk: RCT studies (randomised controlled trials) have long been held in high regard amongst researchers. This method serves very well in the field of pharmacology. But "blinding" cannot be achieved in psychotherapy: the allegiance effect is typically observed. Ther German WBP (Scientific Advisory Board on Psychotherapy) continues to grasp onto these studies as though there is no alternative. The fact that no differences could be ascertained in meta-analyses on whether exclusively RCTs have been miscalculated or also normal controlled studies (Lambert's Handbook, [2], Chap. 13) has not yet reached all colleagues in the field.

Tip

The research group associated with the editor of "Bergin & Garfield's Handbook for Psychotherapy and Behavior Change "(Wiley, New York), M. Lambert, has adopted a neutral stance in this regard and consequently deserves full respect. It provides a retreat of relative balance and integrity amidst the competing battle amongst associations for superiority in the field of psychotherapy

1.1.4 Are There Any Interdisciplinary Common Factors?

The following factors arose in particular whilst looking for perspectives that exist and can be compared on an interdisciplinary level: the therapeutical relationship, the therapist's personality and the corrective experience. Thematically, the need for a close bond, security, development, self-efficacy, order and developing centring is widespread; however, each aspect is differently accentuated with regard to its significance.

The therapeutic relationship is the most effective factor in psychotherapy (APA Report 2018). Although it is referred to as a "common factor", marked differences can be found throughout the field. Carl Rogers has placed the therapeutic relationship at the heart of his therapist training since the 1950s and has trained candidates to show empathy, authenticity (congruence) and unconditional, positive attention (appreciation) in their therapeutic behaviour under the supervision of tape recordings. Klaus Grawe proposed the incorporation of this stance into behavioural therapy in the 1990s. However, systematic training along these lines remains woefully sparse. That being said, behavioural therapy, parallel to observing emotional impact, has achieved an above-average position since its mid-place in Klaus Grawe's ranking table from his meta-analysis (1934–1984)

(ES 0.93, Lamberts Handbook, [2], Chap. 13). The world average has been calculated at 0.83 ES. In mainstream and Gestalt therapy, the therapeutic relationship is multilayer and relatively complexly constructed according to the Martin Buber I-Thou relationship. This model shall be explained in more detail in ► Chap. 2. In all probability, this is likely the primary reason behind the extraordinarily high effect size value of Gestalt therapy, which lies well above 1.0.

The therapist's personality is also addressed in its own section in ► Chap. 2.

Psychotherapists must be both inwardly and outwardly capable of forming relationships – they need to be aware of their own areas that are yet to be healed, to ensure that if they are triggered, they are able to differentiate these from the problem areas of their counterpart and distance them from the therapy process. Training should also prepare psychotherapists to be capable of forming "selective, authentic, therapeutic relationships". As such, the therapist will care for his own needs in other relationships and not let them enter his relationship with patients. This is the minimum prerequisite to ensure therapy does not enter the realm of malpractice. Furthermore, therapists are required to be counterparts who can flexibly empathise with their patient's needs. This calls for a complete development ladder that features all key life issues. The therapist must be capable of adopting any attitude that helps the patient to take the next development step.

In certain cases, patients may become infatuated with their therapists. This can be a good indication. The ability to develop feelings of affection may touch upon the patient's potential, a potential that has become apparent primarily due to the trusting collaboration with the therapist, and may consequently entail that he will be most likely able to give other people a chance in the future. The primary aspects in this regard are appreciation and detachment.

A long-acknowledged common factor that can be found in different variations in all psychotherapy approaches is the "emotional corrective experience". This expression was first coined in the psychoanalytical work of Franz Alexander (1891–1964). A comparable occurrence may – in the case of behavioural therapeutic socialisation – also be described as a persuasive learning experience with a rewarding outcome, which the behavioural therapist may have motivated the patient to undergo. This appears to work most impressively at the therapy level, which can be proffered as real, symbolic or imaginary. The humanistic procedure, in particular Gestalt therapy, psychodrama, psychosynthesis, Pesso therapy, and, in part, physical therapy and systemic methods consciously orchestrate scenarios and role plays, whereby emotionally corrective experiences are easy to employ. The corrective experience is linked with the neurophysiological idea that the old, activated engrams are overwritten or additionally coupled with the engrams from the new experience.

This was previously studied by Louis G. Castonguay 2005, 2012 through interdisciplinary research. With studies on therapy behaviours and on the principle of insight, the interdisciplinary research reported an insight that corresponded to the viewpoint traditionally pursued by the school of psychoanalysis.

There are many variations on the emotional need for order in intrapsychic or interpersonal spaces. On the one hand, the level of maturity of the individual in question being must be determined. On the other hand, the chronification needs to be classified: does the case require chronified or fixed conflict resolution work that may disturb the patient's personality? In this regard, the analysis and simulation of the partial aspects, the concretisation thereof, the identification of the blocked issue, within the scope of the development context, the thus far avoided dialogue of confrontation and reintegration into a more harmonious whole have proven successful.

Various maturing approaches concern structurally weak personalities, people who have experienced poor bonding at an early age or coherence experiences and people who have been traumatised by crises and those susceptible to addiction. Sensitisation to states of equilibrium, or delicate internal and external regulation, stabilises individuals exhibiting behaviours similar to psychosis. In the initial stages, the key learning objective is to regulate, integrate and fine-tune. A variety of contrastive pairs that can be emotionally experienced such as value and non-value, closeness and distance, contact and withdrawal, love and hate, idealisation and invalidation, activity and inaction, outward and inward orientation etc. can be used as exercises in this regard.

The employment of awareness and mindfulness to achieve centring in variations of (secular) meditation is both a way in which structural weakness can be developed and the personality further developed. This approach comprises a number of methods based on the religious practices of the Far East, which travelled to the western world in several waves. It has triggered a strong response and maturing centring amongst a large portion of the population. By virtue of its physiologically stabilising, stress-reducing abilities, it has also gained popularity in hospital psychosomatic health programmes and pre- and post-treatment areas. Mindfulness approaches can be conceivably linked and adjoined to all schools of psychotherapy. Throughout the 1970s, 1980s and 1990s, they formed part of the integrated standard of humanistic therapy approaches, especially Gestalt therapy with its primary aim of increasing inner coherence. Subsequently, behavioural therapy increasingly adopted standardised mindfulness methods and has since been responsible for the impressive spread of mindfulness approaches, especially in the treatment of structurally weak patients.

1.2 "Objectivity" – The Uncontroversial Main Feature of "Scientific Character"

1.2.1 The Belief in Reason – A Gift from the Enlightenment

Now it is time to take a step back and look at the emergence of "new psychotherapy", which, for the purpose of this book, starts at the second half of the nineteenth century, despite the use of many, much earlier, illuminating sources at a later point in this chapter. As previously mentioned, the distinguished expert Wilhelm

Wundt spearheaded the fundamental scientific attitude of the era. A strong belief in reason and the spirit of the Enlightenment prevailed at the time. Following long periods of religious and emotional neglect, this was sorely needed.

As a special virtue, the objective of human development consequently arose as observation without emotion, the withdrawal from any personal needs and the ability to neutrally evaluate rational arguments. Objectivity became the name of the game, especially in scientific work.

As such, it was only logical that Freud, who exchanged letters with Wilhelm Wundt at the time, called on the therapist to declare his position on Freud's form of treatment at the end of the nineteenth century, whereby only close observation was required – without the inclusion of the practitioner's own, emotional feelings. This permitted the practitioner to become a neutral mirror of the patient, in a sense.

1.2.2 The Drawback of Bias

Pavlov encouraged his colleagues and their successors to adopt a similar attitude, based on the same spirit, when compiling the accurate records of these animal experiments. The primary notion lied in objective "scientific" observation, without the formation of any relationships. At that time, objectivity and science were understood as terms that were all but synonymous.

It is surmised that, by virtue of his own upbringing, Freud may have simply not been able to grasp the notion that people with fragile, early relationship experiences may quickly feel existential insecurity when placed under exclusively objective observation without any basic acceptance to grant them stability. As a child, he enjoyed a loving relationship and a secure attachment to his mother. Furthermore, as the first-born child in a Jewish family, he was raised with the special status of the future head of the family. As a result of Freud's early years, psychoanalysis particularly struggled to provide people with fragile, early relationship experiences with an adequate, beneficial therapeutic relationship structure.

Since the turn of the last century, the new, abstruse paradigm of "New Physics" slowly began to gain momentum and the notion that there could be no neutral, objective observation without observer effect, but that, instead, both the relationship and "non-relationship" have an effect, came to the fore.

At the same time as puzzling findings were emerging from New Physics, in 1923, Martin Buber (1878–1965) penned an entirely different, almost visionary perspective in his philosophical-literary text, which subsequently gained international acclaim: "I-and-Du" (I- -Thou), the existential significance of interpersonal (subject-to-subject) relationships – in comparison to encounters void of relationships (subject-to-object). He referred to the latter as "I--It".

Decades later (1967), Paul Watzlawik (1921–2007), in his slight modulation of the Buber approach, coined the famous saying: "One cannot not communicate!" Either one enters into a relationship or he refuses to do so, which, in itself, is a statement. Indeed, consciously or unconsciously ignoring another person is an expression of existential rescission.

The preserved and idealised notion of "objectivity" was used to decipher the problematic reverse side of the coin for the interpersonal area relevant to psychotherapy.

Therapy methods marked by a pronounced reliance on the aforementioned "scientific objectivity" such as behavioural therapy based on learning theory and traditional psychoanalysis typically struggle more to accept the findings of current research, primarily that the therapy outcome is primarily contingent on a positive therapeutic relationship. These methods only acquired their main character following the period of "belief in objectivity", i.e. the humanistic and also the systemic group of methods.

To avoid abandoning valuable ideas in an attempt to cast aside the unfavourable: the aim of objectivity is an important function. We cannot do without it, but we must nevertheless remain aware that objective awareness can act as an opposite or complementary pole to involvement in a relationship. Both skills are different in nature. They hinge on the situation and role, what is signified, and whether or when an "either-or" or a "both-and" is appropriate. In this regard, it is now sensible to divert our attention to the development of the therapeutic relationship in education and training.

1.2.3 The Temptation to Define "Science" Using Currently Available Measurement Tools

Undisputedly, the ability to present measurement data indicates a statement is likely based on objectively sound grounds. As we are all aware, this can put you at ease. However, if we fail to evaluate whether the measuring tool in question is sufficiently adequate at measuring what it is intended to measure and whether it has captured any secondary effects or distorted such effects etc. then we would instead be basing our observations on pseudo-objectivity. This bias has been long observed in the schools of science.

In his lectures on quantum-theoretical basic understanding for psychotherapists, quantum physicist Hans-Peter Dürr was known to enjoy recounting the following parable from Sir Arthur Eddington, an English astrophysicist (1882–1944) taken from the book "Philosophy of Physical Science". It is retold below from hearsay.

In his parable, Eddington compared a natural scientist to an ichthyologist, a marine biologist who studies fish and life in the sea. The ichthyologist casts his net, pulls it ashore and reviews his catch as is customary for a scientist. Following a number of fishing expeditions and conscientious inspections, he succeeds in establishing a principle of ichthyology: All fishes are larger than 5 cm!

He determines that this statement is a principle, as it has been confirmed with each subsequent catch with no exceptions. He tentatively assumes that this statement will apply for each future catch.

However, his friend, who we shall refer to as the metaphysician, questions the fundamental significance of this "principle". In his opinion, the upper size limit of fish found by the ichthyologist is simply the result of the 5 cm mesh size of the net. Unsurprisingly, the ichthyologist is not impressed by this objection and responds: anything that cannot be caught by my net is, as a matter of principle, outside the realm of ichthyological knowledge. As it exceeds the boundaries of the principles of ichthyology, the object is not defined as an object in ichthyology. For me, as an ichthyologist, what I cannot catch is not a fish.

This may be amusing at first. It would perhaps surprise you to learn that such narrow-mindedness is not simply a remnant of the nineteenth century, but it remains an element in the current field of psychotherapy. In the politics of the German professional field, a pseudo-scientific "net" called a "methodological paper" has been woven with the aid of behavioural therapy. This "net" claims to define: anything that sneaks through our meshes is not "scientifically recognised psychotherapy", regardless of how well it can be identified using other measuring tools. In contrast to Sir Eddington's ichthyologist, who appeared to hold conviction in his experiments, there is scant evidence that the aforesaid scientific politicians are aware that they are employing their own specially woven web of scientific pretence as a tool to achieve their objectives.

1.3 The "Necessary" Rediscovery of Subjectivity and Its Stake with Regard to Healing Therapy

1.3.1 The Lack of Emotion Creates a New Abundance

Here, it is worth taking a look back at the situation in the USA in the 1950s and 1960s, a time when mainstream behavioural therapy was strongly entrenched in science and primarily focused on trial-and-error and externally controlled learning and reward experiments. A time when, despite its theoretical interest in the spaces of interpersonal relationships, psychoanalysis surpassed itself in its professional approach with solidified abstinence (which led to Ralph R. Greenson (1911–1979) questioning and discussing whether a training analyst should be permitted to shake the hand of a candidate who had just come from his father's funeral as a one-time, exceptional gesture of condolence, for instance, in a psychoanalytic textbook in the 1950s). A time when exaggerated abstinence would cause one to freeze.

The somewhat delayed attempt made by Heinz Kohut to explore the significance of empathy for psychoanalysis almost cost him his affiliation to the field. His bold pioneering work is worthy of respect.

In contrast to the highly traditional, rationally outmanoeuvred, over-controlled, emotionally sparse world of contemporary, academically established mainstream psychotherapy, practising psychotherapists, a number of whom were themselves emigrants, were confronted, through their care for the traumatised Holocaust generation, with an intense hunger for solace, understanding, security, appreciative humanity and a search for the restoration of their dignity, without being officially equipped to handle the situation.

Shortly after, the same psychotherapists were faced with treating returnees from the Vietnam war (on their part, the youth reacted with opposition and used their "Flower Power" and "Make Love, Not War" movements to make waves in line with the mentality of large swathes of the contemporary post-war population).

The opposition also gained traction in the field of psychotherapy against the contemporary academic establishment. Starting in the 1950s, this counterculture branch of psychotherapists culminated in the Human Potential Movement. In its 1962 manifesto, the psychotherapists behind the movement (Maslow, CH. Bühler, Rogers, Köstler, May, Goldstein, Huxley, Mumford, Sutich, Bugental et al., also loosely connected with V. Satir and F. Perls) placed the subjectively experiencing individual who seeks meaning at the heart of their work. They focused on how much could be achieved by an individual plagued by bad luck in spite of everything and enriched the field of psychotherapy with a resource-oriented perspective, which was previously frowned upon as unprofessional and has since, around 80 years later, become a matter of course in all schools of psychotherapy. Their work permitted the opportunities presented by the therapeutic relationships to be recognised and advanced within the humanistic school into a consciously applied, healing instrument. The movement was pioneered by Martin Buber (1878–1965), Maurice Friedman (1921–2012), Lore Perls (1905–1990) and, particularly, Carl Rogers (1902–1987) et al. In the current generation, the avant-garde branch of psychoanalysis, "Relational Psychoanalysis" and the Gestalt therapy branch of "Relational Gestalt Therapists" have merged so closely that only insiders are capable of distinguishing between the two schools.

1.3.2 In Praise of Setting Limits: Crisis Intervention for Traumatic Inundation

Stabilisation and distancing techniques (in combination with contralateral self-touches whilst standing during exposure – at that time without EMDR) were established as the basis for later trauma therapy out of necessity as part of Gestalt therapy crisis intervention. These techniques were subsequently revisited by F. Shapiro in 1987. In Germany, Gestalt therapy students, including the author, were taught this know-how in the early 1970s during their basic training. It is arguably a good thing that the German psychoanalytical colleagues who had previously contacted F. Shapiro in the USA in order to introduce trauma therapy across the pond were unaware of the origin of the procedure. This made it easier for the procedure to gain recognition in Germany with regard to professional policy.

1.3.3 The Birth of Resource Perception, Encounters and Searching for Meaning

In a swift break from the gaps of contemporary, established psychotherapies, the Humanist Movement offered a new approach to therapy:

- Reliance on a personal, dialogical and healing therapy encounter culture between patients and therapists, whereby the therapist exhibits solidarity with the areas of the patient in need of treatment – (and thus is required to pay attention to detachment from any negative communications in sufficient time)
- A focus on forming a resource-oriented perspective on the patient in a departure from the previous approach that was solely based on acknowledging the patient's flaws
- Indicating appreciation for how emotions are handled and their means of expression, in addition to body-oriented resource work
- Contact with the unconscious and with its chains of association on all sensory channels
- The acceptance of consciously experienced subjectivity and inter-subjectivity in every real encounter as a valuable occurrence of its own accord
- The transformation of the therapist's self-perception into a compassionate counterpart who promotes life enhancement and development, which is most often required in a parent variation
- A focus on each individual's values and dignity, especially the patient's dignity
- The psychotherapeutically accepted significance of experienced coherence, sense experience and the search for meaning

1.3.4 The Acceptance of Creativity, Physical Comfort and Role Play

Not solely the approach but also the style marked a significant departure from previous methods: interpersonal, authentic and vibrant expression was promoted alongside the individual's resourcefulness, physical and artistic expression, whether this is expressed through dance and movement, music, poetry or creative media (painting, sculpting, clay work etc.), in role play, improvisational play or improvised theatre. A stage presentation of the patient's problem areas was no longer dismissed as simply "acting out" and physical expression was instead encouraged as one of the key insights into the patient's state of being.

Therapists were now expected to employ "selective authenticity" in interactions with patients. As such, the therapist was only required to address the experiences that would assist the patient's process, meaning he remained emotionally interactive in the interaction and an appreciable counterpart for the patient (in seminars aimed at enhancing the personality of resilient non-patients, the leader was permitted to participate with full authenticity).

1.3.5 Simultaneous Fascination and Anti-establishment Sentiment: The Humanist Wave Hits Germany

The body of thought behind the humanist wave first travelled from the USA to Europe through official invitations, unofficial liaisons and youth exchanges. I recall Charlotte Bühler and her husband Karl Bühler flew across the pond in 1961 to receive an official honour in Bonn and also hold a lecture at the Psychological Institute in Freiburg. Charlotte Bühler introduced her "live event research", discussed the new psychotherapy understanding of her emerging avant-garde reference group and touched on Maslow's research on the growth of self-actualisation and peak experiences. Around the same time, Gretel Leutz was coincidentally working as an "au pair" for the Moreno family and enthusiastically became involved in the psychodrama treatment method, which she later helped to spread throughout Germany.

In the mid to late 1960s and the start of the 1970s, American representatives from the Humanist Interactional Wave, otherwise known as the Encounter movement, were invited to lead seminars by the Group Dynamics Research Group in Bonn, Prof. A. Däumling, amongst others. This led to a number of inspirational incognito encounters and the additional training of psychoanalysts such as the couple Franz and Anneliese Heigl (1967–1969), who attempted to introduce the encounter culture of humanist therapy to psychoanalysis despite fierce opposition faced by the psychoanalysis mainstream (I can still vividly recall the appalling contributions to which F. Heigl was subjected at the annual DPG congress in 1979). Indeed, Franz Heigl once personally described to me how happy he was with his emerging group leader identity, which enabled him to act naturally instead of working towards transference neuroses and unleashing the aggressions thereof.

Not long afterwards, Heigl's "psychoanalytic-interactional psychotherapy" impressively triumphed at psychoanalytically-run hospitals in the face of opposition from the guild – however, not without deliberately withholding certain sources. The "response" principle coined by Heigl is purportedly derived from Fritz Perl's beloved wordplay: response ability – responsability. In line with this, the ability to respond is contingent on assuming responsibility for doing so. Put otherwise: Take responsibility for your responses (and non-responses).

In a similar vein, the following generation also sought to hide the Gestalt therapy source behind "Psychoanalytic Structure-Related Psychotherapy" (Rudolf [3] ff), as if the integrated reference would constitute an unreasonable slight to the psychoanalytic mainstream.

The young, contemporary generation thus far exhibited the strongest reception of the humanistic-systemic wave (which initially appeared only to have something in common with certain differences of emphasis) and linked it to their idealism and longing for an emotionally healed, more spontaneous and moral world. This accordingly went hand-in-hand with criticism of the somewhat emotionally frozen post-war generation. A provocative, anti-establishment attitude has since come to characterise certain sections of the young generation. Psychotherapy simultaneously captured and polarised the general zeitgeist. The surmounting tensions between the generations were partially responsible for providing the fuel behind the 1968 riots.

1.3.6 The Convincing Triumph of Empathy and Its Social Commitment

Buber's renowned I-Thou relationship (1923) has excelled to overtake the general concept of empathy in popularity; however, it should be regarded as a precursor to and formative element of modern Gestalt therapy, as well as a quality that without which Gestalt could not be. The correspondingly different emphasis placed on the therapeutic relationship marked the primary reason behind the separation of Gestalt therapy from psychoanalysis in the 1930s/40s.

It was predominately through the client-centred psychotherapy approach of Carl Rogers that empathy gained worldwide recognition as one of the key aspects of psychotherapy. Behavioural therapist Klaus Grawe (1943–2005) was amongst the first to acknowledge the significance of empathy in therapy outcomes within the scope of his comparative meta-analyses (from studies dated 1934–1984) and subsequently recommend the integration thereof into behavioural therapy. The author recollects conversations held with Klaus Grawe between 1989and 1991, during which time he was concerned with how psychotherapy could best be improved in the future. At that point in time, he was able to clearly state individual sources. Unfortunately, he did not do so at later points.

Around the time Carl Rogers (1902–1987) had highlighted the significance of empathy, the psychoanalyst Heinz Kohut was also discovering the importance thereof for healthy human development and placed it at the centre of his "self-psychology". This signified a clear deviation from Freud's preference for therapeutic neutrality, which was, at the time, perhaps more strongly advocated by Freud's successors than Freud himself. The tension that followed Kohut's assertions almost resulted in his exclusion from the Psychoanalytic Association, which would have significantly impacted his work. In the USA, it is claimed that only the shortcomings of the Psychoanalytic Association that began in the 1970s saved him from the fate of a dissident.

1.4 The Discovery of the Abundance of Interspaces and Interpersonal Multi-person Structures

1.4.1 The Apparent "Gaps" in Relationships

Now we shall take a look at the psychotherapeutically creative years of the 1960/70s: Hitherto, another "gap" remained, although its content had been theoretically foreshadowed in a number of ways:

1. Through the establishment of holistic and Gestalt psychology in the mid-nineteenth century
2. Through Alfred Adlers' (1870–1936) concept of meaningful life development, which included the individual's ability to integrate into a community
3. Through Jakob Moreno (1889–1974), who was one of the first to guide the investigative spirit to the interpersonal space through the sociogram and the interpersonal structures found in groups and families
4. Through Eric Berne's (1910–1970) transactional analysis concept based on the Freudian structural model
5. Further developments within the field of psychoanalysis, including the incorporation of the interpersonal space (in object relationship theory by Kernberg et al., for instance)

At that time, living beings had already been divided into three categories of complexity by biologists from the previous generation: (1) single-cell organisms, (2) multicellular organisms, (3) eusocial (ants, bees etc.). Humans were classified as between (2) and (3).

This alignment pertains to the "gap" between supra-individual forms of relationship and organisation. As such, the list could be extended.

1.4.2 The Pioneering Spirit of the First Generation of Couple, Family Therapy and Organisational Counsellors

I distinctly remember the growing interest in the field of couple and family therapy in the 1970s. At the time, there were three versions: the humanistic approach, which focused on the opportunities for couples and families to grow and develop (Maria Bosch Institute, Weinheim); the psychoanalytical approach, which employed a multi-generational perspective and the connective unconscious to enhance the psychotherapy concept (E. Sperling, 1925–2007, A. Massing, G. Reich / Göttingen); and the systemically oriented approach pioneered by H. Stierlin, which was based on work of the Palo Alto Group and the Milan Group. The latter were primarily involved in the pursuit of strategic factors and focusing on the system as a whole. Examples of strategic considerations included the separation of the generations and using suitable tasks to strengthen the bond between age-appropriate subgroups. Or, in certain cases, the system was completely "disrupted" by corresponding instructions and then its self-organisation was observed with curiosity (M. Selvini-Palazzoli, 1916–1999). In this case, the kind of relationship reverted back into the observer role; otherwise, it was adapted to achieve impartiality.

Supra-individual, multiperson forms of therapy constitute an equally significant complementary pole to the individual as psychosomatic medicine with its functional and organic "subsystems". Both schools of thought can be observed in a common complex series of relationship constellations. In each case, this concerns the relationship of the parts to their relational whole and their mutual interaction. At the centre of this complex series (in our approach) lies the individual human being, with his intrapsychic relationship constellation. Disorders – and also healing impulses – can easily be "mirrored" from one order of magnitude to another, as if connected by a resonance principle. For example, belittling affronts against and within the social group are often reflected in blood pressure crises and muscular misinnervations in the neck–shoulder area.

The highly interesting topic of social interspace and its various levels is addressed in its own separate chapter. Although the topic straddles and employs the aid of socio-psychological symbols and categories, a great need for research remains in the natural sciences.

The intrapsychic space and its ability to mirror the social and psychosomatic-functional space and vice versa has proven just as interesting.

1.5 The Unconscious – A Treasure Trove or an Unpredictable Sphinx? Discoveries in the Depths and Distance of Consciousness

1.5.1 The Multifaceted Nature of the Unconscious

The unconscious has previously been afforded a large number of ascriptions:

- The great unknown
- The scorn of science
- The personal landfill and storage room
- The helpful release in everyday life
- The extrapolator of probabilities
- The covert advance decision-maker
- The potential antagonist to the conscious agenda
- The intuitive artist

- The sleepwalker
- The wise seer
- The guardian of collective treasures
- The feedback amplifier or suppressor
- The mutual confidence in eloquent silence
- The expert in your melody and frequency
- The concealed or almost clear reflection of the "big picture", etc.

As perhaps is clear, the above list of ascriptions for the unconscious is rather exhaustive. Indeed, an entire book could be dedicated to the topic of the unconscious alone. The dynamic between the unconscious and the conscious is a central element in most variations of psychotherapy.

It would be almost inconceivable to imagine psychotherapy without the involvement of the unconscious. Would it be logical to introduce such a restriction on the unconscious? If this renunciation of the unconscious were to occur, would this not represent an expression of resignation, as the unconscious does not adhere to the rules that we use to manage in the realm of consciousness? Would this not evince a defiant "nevertheless" stance? Would any such renunciation be based on the fear of the relative unpredictability and incomprehensibility of the unconscious?

Traditionally, in the past, behavioural therapy attempted to evade any associations to the hard-to-grasp or inexplicable notion of the unconscious (as well as the therapeutic relationship) or to simply banish them into the black box. Whilst the sincerity required to openly navigate the understanding of the scientific limitations at the time is worthy of praise, the question of the extent to which therapists are able to engage in internal or external contact with patients if the unconscious being is simply ignored remains unanswered. Can research that focuses on the conscious partial aspects of the processes of intention and learning truly be considered relevant to everyday life? Or does it simply implicate an outdated practice? How valid are research findings that are interpreted solely as the result of technical methods with no regard towards the structure of the relationship?

The unconscious is traditionally the pivotal element of psychoanalysis and is promoted through free association. However, conventional criticism is often directed at how the psychological material is treated, predominantly against the hypothesis-led prerogative of interpretation supported by therapists, which is often experienced as invasive. This kind of criticism has gained traction in recent decades and has generally paved the way for a more sensitive and self-critical approach at any rate.

Other approaches such as artistic and body therapy, acting with theatre improvisation methods and an experimental scope have all provided opportunities for projections of the unconscious on many dimensions. The "bilingualism" of these therapeutic approaches has seemingly successfully landed on the shore of consciousness: one foot on the solid, dry ground of consciousness, the other swimming in the sea of the unconscious. As such, this has facilitated the development of understanding and translation skills between the two realms.

1.5.2 Historical Milestones

Sigmund Freud (1856–1938) described the difference between the primary and secondary levels of awareness with astonishing accuracy: Consciousness is defined as the secondary level, which develops at a later stage than the primary. It is characterised by a temporal "if-then" relationship, a local before and after structure, in addition to a causal relationship based on logic. All things have clear contours and identities. Precise observation is key. Aristotelian logic applies in this regard (A is not equal to B, tertium non datur, no third [possibility] is given). In contrast to this notion, the above causal and spatial-temporal structures cannot be found at the primary level. We are instead faced with a pars pro toto incident, lack of negation, consolidation, multiple interpretations of a thing, "both-and" implications and the resonance principle time and again.... His recognition of the qualitative differences between the conscious and unconscious clearly distinguish Freud as ahead of his time. However, he did not venture as far as to make any further classifications in the primary categories of natural science (classical physics and quantum theory). The tension between these two fields, which remained unresolved during Freud's lifetime, was simply not within his field of interest.

The first to significantly expand on the notion of the unconscious as a personal storage room for the repressed was C. G. Jung (1875–1961). It was his understanding that the unconscious was a vast, rich, spiritual space that intuitively connects us with the experiences and knowledge of other beings and people from the present and past. Jung gifted the field of psychotherapy with the notion of the "collective unconscious".

Jung's concepts were adopted by contemporaries such as Roberto Assagioli (1888–1974), amongst others, who made the distinction between the collective and individual unconscious. He differentiated between a lower, middle and upper unconscious, a relatively small zone of consciousness and an integrating higher self-function within the scope of the individual unconscious.

Physicist and Jungian psychoanalyst Arnold Mindell (born 1940), based dream, body and group work on unconscious communication and the higher states of consciousness.

In everyday psychotherapy work, we often experience the shared consciousness phenomenon in couple and family therapy settings as well as in regular and large study groups and especially in constellation work. This phenomenon can be visualised in a graded manner as the transitional phenomena to Jung's collective unconscious.

1.5.3 Hypnosis, Trance and Hypnotherapy

Important

No other method so clearly demonstrates that the brain constructs a version of reality that does not constitute a portrayal of reality as we naively assume. Accordingly, it can be said that we live in a "digital illusionary world"!

The construction puzzled together by our brain emerges as an imaginary everyday reality that is, for the most part, sufficiently coherent for us. We perceive the hypnotically constructed image as real predominately if sensual perceptions are processed by as many sensory channels as possible, which are associated with each other and integrated into the triggered emotions. During the initial stages of hypnosis, great care is taken to ensure that the brain constructs a realistic scenario.

Gestalt psychology perceptual research conducted during the nineteenth and twentieth centuries succeeded in illuminating the construction rules of our dominant brain hemisphere, at least partially.

Important

Our brain works with relational structures, i.e. frames of reference. Each detail receives its own identity through its reference horizon and learns its position in the whole – or potentially loses it.

The many confusing, deceitful sensory illusions illustrate this cerebral processing and construction method – and the limitations thereof! The individual objects themselves do not remain the same, but their relation to their surroundings does. As such, our "brain" is equipped to present a world of relatively stable relationships whilst nevertheless constantly being in the midst of the changing stream of stimuli to which it is exposed. Objects also occasionally change their appearance and significance. As the popular saying goes, "In the land of the blind, the one-eyed man is king". Our constructed reality focuses on the relationship between the part and its referential whole. The findings of Gestalt psychology appear to have fundamentally recognised the relation-centred construction rules of our brain (Gestalt psychologists would still have to posthumously award honorary membership to Einstein).

Cave

The illusionary objects alert us to the fact that this construction of reality is a portrayal of reality.

It then follows that our everyday world is constructed from illusions that we view as objective reality. Ancient Far Eastern traditions placed a strong emphasis on this viewpoint and juxtaposed the illusionary (Sanskrit Maya) with a corrective that is not formed through a cerebral construct but rather through direct participation in the interconnected "all-one", for which neither a linguistic concept nor any mental images suffice for the resonance thereof in us.

We are typically capable of handling illusionary objects in our daily lives. The struggle often first arises when our illusions are manipulated and we have no available external criteria to identify discrepancies and thus remain sceptical, in the case of rumours, tempting advertisements or political propaganda, for instance.

Important

Hypnosis, trance induction and hypnotherapy all work by influencing the reality construction process of our brains.

Cave
As such, no other processes are so clearly contingent on an ethically honest fundamental stance of the individual in question or quite as entirely susceptible to abusing power as the above.

They have accordingly been subjected to reservation, scepticism and mistrust, which has, in turn, delayed their dissemination, despite – or perhaps because of – the fact that they exhibit significant potential for the field of psychotherapy.

However, albeit in a partially concealed way, the aforementioned elements have nevertheless since been integrated into the basic approaches of all methods. This is a matter that requires clarity.

A resource-oriented approach, which was originally a humanistic concern, has been expressly pushed in systemic procedures, integrated into behavioural therapy a quarter of a century ago following Klaus Grawe's recommendation, has diffused into depth psychology due to its affinity for humanistic methods and also appears to have finally arrived in certain branches of psychoanalysis.

Important
A resource-oriented approach shifts the evaluation benchmark and works with reference system changes to achieve a more beneficial emotional evaluation.

Inner child regression therapy has been a popular Gestalt method since its first mention in 1969 (Finney). Likewise, it is also touched upon in Moreno's psychodrama and used in trauma therapy and depth psychology. The Gestalt therapy variant is also encountered in schema therapy. In a waking trance, the protagonist follows his biographical path backwards until he stops at a scenario or constellation that resonates with the topic that is currently being updated. Hypothesising the situation faced by the child at the time through the eyes of today's adult with the support of the therapeutic relationship and, if required, an imagined circle of helpers helps the patient to make a more coherent, convincing and effective contribution to the problematic situation faced at the time and allow it to end in a more coherent way. This process typically forces the patient to change their innermost identity and, in doing so, experience a "rebirth". Wishful fantasies to compensate for the damage inflicted by shortcomings at an early age (for instance, fantasies for a good mother who offers love and security or a protective family/big brother) can be scenically and symbolically fulfilled through regressive waking trances that put the defining power of the previous childhood identity into perspective. These trances provide a realistic way for patients to experience that their development could have taken this or various other paths (e.g. in Pesso and Gestalt therapy).

Trauma therapy employs similar approaches to link the engrams from experiences of helplessness with experiences of empowerment and enable the subjectively damaged identity to beneficially regenerate.

Imagination work. This work is associated with the guided daydreaming undertaken during trances, whereby, on the one hand, inner resonances come into contact with significant inner parts and, on the other hand, new challenges can potentially awaken latent abilities and potential. Symbolic as-if games framed in parallel worlds are also included in this regard. These games are particularly popular and at home in child therapy. Adults can often benefit from an imaginary visit to their own "old wise man" to receive a meaningful review of life prior to making upcoming decisions.

Hypnotic trances can also be used to successfully, imaginatively conclude remnants from past relationships that project phenomena of transference – as described above – and thus bring an end to distortive, involuntary self-hypnosis and pave the way for more genuine forms of relationships. This is particularly beneficial for personality disorders and the de-identification of superior survival concepts they require.

The improvement observed in patients as a result of hopeful expectations, even before therapy begins is well acknowledged in the field, as well as the fact that the scope of placebo and nocebo research overlaps with psychotherapy. It can, of course, be assumed that when a patient is informed that a certain approach is "world-class in its effectiveness", the patient is likely to internally respond to it differently than to an approach described as "unrecognised by science", even if the entire procedure follows the same path. Suggestions can become weapons of war, as, of course, can their opposite.

The above refers to a consolidation of the ubiquitous aspects in psychotherapy that can be attributed to the hypnotic condition.

What does the notion of "hypnosis" entail? In "classical hypnosis", the asymmetrical distribution of roles and activities is clearly defined and noted in advance. In this case, the subject is at least able to decide beforehand whether he would like to engage in the process of confiding in the hypnotist. The subject places his trust in the hypnotist that he will use his temporary power to help and for the well-being of the subject, and not in any detrimental way.

Hypnosis is a method used to potentially change an individual's construction of reality in an involuntary and evident manner. When under true hypnosis, the subject experiences the "hypnotic hallucination" as reality, not just as "mere mental images".

According to Milton Erickson, hypnotherapy is characterised by "indirectness" from the outset. Revenstorf writes ([4], p. 169): "Erickson would typically not openly explain his approach to the client – with the underlying purpose thereof being that the conscious perception of methods and the objectives can present an obstacle to achieving therapeutic change. That is not to say that Erickson intended to consciously deceive his clients. Rather, he viewed this approach as "passively leaving the client in the dark". Whilst transparency is often a desirable trait to therapy in many regards, this ambiguity can benefit the therapeutic process in a variety of ways. ...Indirectness is helpful when it comes to inducing trances, as it bypasses rational understanding and comprehension and triggers associative reactions". Whilst this is logical, it is nonetheless viewed by some as an invitation to manipulate the client.

This desired indirect attitude which results in permanent, concealed asymmetry, has been met with praise by colleagues, provided this approach is based on a concept of man that features equality and humanity. In practice, however, it is entirely feasible that sequences in an active dominating style may be included amongst others with more endorsing, receptive styles.

1.6 "Artificial Intelligence" and the Complexity of Natural Brain Activity

Machine learning processes are based on computer codes, or algorithms, which rely on feedback loops between artificial, reward-driven learning neurons and an externally controlled learning programme, which can also include a randomly controlled margin, such as in a speech or pattern recognition programme. Algorithms are the action rules aimed at solving problems that comprise defined individual steps to achieve predefined outcomes. At present, learning machines are being developed on this basis to take over and perform valuable support functions in a limited number of areas.

They are modelled on the basal, neural "language" of "excitation" or "non-excitation". The computer language widely used in this process to date uses a combination of "one" and "zero". In simplified terms, this process mimics the fractional processing principle of the dominant brain hemisphere, otherwise known as our logically thinking intelligence.

Freud's secondary level of awareness is characterised by this causal and spatial-temporal sense of order and is widely regarded as the culmination of development in this regard. This notion has helped to form the Western understanding of science and enabled it to seamlessly draw on the principle of learning psychology. When this principle is applied to humans as well as machines, individuals are assumed to be absent from preconceived ideas or predetermined goals; a clean slate. Impulses or needs that come from the system itself are not taken into account but rather considered to be a negligible factor. Philosophically, this stance correlates with empiricism and materialism and also draws on ancient visionaries (the Democritus school of thought).

It is not without hesitation that the notion that our brain physiology features two intertwined processing principles is referred to and discussed in this field. There are energetic holistic phenomena, starting with saltatory conduction, looping, re-entry circuits, synchronisation phenomena and circulating, energetic distribution patterns featuring an asymmetrical "observer position", which have decoupled themselves from neuronal single excitation and facilitated the understanding of an energetic, partially conscious, holistic flow equilibrium. This notion of the dual processing principle in the human brain is anatomically corroborated by the fact that there are parts of the brain that can be functionally localised and allocated to the periphery and other parts that appear to solely facilitate associative cross-linking and more holistic processing.

For almost a century, we have been tapping the electromagnetic field discovered by Hans Berger in research on brain waves as an extracranial, tangible vector sum through EEGs (electroencephalograms). We have been astonished both by the variety of individual electromagnetic events, area-to-area group activities, different frequencies and amplitude levels, rhythmic and synchronisation phenomena exhibited in the electromagnetic field – and also by its extensive influence, for which we do not have a direct sensory organ. The phenomena referred to above pertain to the changing function exhibited by the differently integrated activity patterns of areas of the brain. The incalculable degree of freedom exhibited by human brain activity appears to be at odds with the straightforward instructions of computer-affine programmes.

Artificial intelligence research, or "machine learning" and "deep learning", is almost entirely based on the side of neurophysiological activity characterised by binary code.

As such, highly cognitive functions are not touched upon or achieved. Intuition, as well as psychotherapeutically relevant patterns of coherence, appears to be more related to holistic processing abilities.

1.7 Mindfulness – The Borderline Impact of Transcendental Spaces

1.7.1 Historical Waves of Encounters Between the "West" and the "Far East"

The frontier to spirituality, which has historically often been pursued in a searching manner, relatively or completely free from religious restrictions, occupies a special chapter of this book. Increased contact between the Far East and Europe paves the way for waves of fascination, which was culturally expressed in this art. Of course, this topic could occupy an entire book in and of itself. In this regard, we shall narrow our focus to a few references, such as "western-eastern" themes found in the works of J.W.v.Goethe, Hermann Hesse (Siddhartha) and in Herrigel's cult classic "Zen in the Art of Archery" etc. In each case, an impressively strong intellectual mentality is always present.

In the 1950s to 1990s, the tragic fate of Tibet, amongst other factors, sparked a growing western interest in non-institutional, spiritual traditions. At first, this primarily concerned the Zen concept from the Far East. This concept notably had a fundamental influence on Gestalt therapy and also on wide swathes of the young population at the time. The Zen concept of awareness, which is today viewed as more akin to mindfulness, forms one of Gestalt's key approaches. Awareness of meditation's ability to influence the development of personalities spread during this time.

On the other hand, primarily in the "West", we have a tendency to naively permit our perception to be influenced by presuppositions. We allow our brain to piece together constructs and prompt us to rationalise through posing "why" questions. As a result, we often miss out on a deeper, important life experience. Deeper experiences require a more mindful approach in the many facets of the here and now; sometimes they reflect our painful oppressive confinement, although at other times they can lead to an uplifting, inner self and present the opportunity to achieve personal transformation, authority and development. As such, this path enables us to be more tolerant of our insecurities towards life. Influences from the Far East have gifted our culture with knowledge from the ancient world that has a vast amount of therapeutic potential and provides an effective counterbalance to the unilateral development of rationality.

(Fritz Perls, the main architect of the Gestalt movement made two trips to a monastery in Kyoto, Japan during the 1960s to receive training in Zen meditation)

1.7.2 Current Interest in "Mindfulness-Based" Therapy Approaches

At the end of the 1990s, the stabilising, regulation-promoting anti-stress effects meditation and yoga could achieve for the physiology and the general controllability of personalities was demonstrated by colleagues from the school of behavioural therapy (Jon Kabat-Zinn, born in 1994, et al.). These pioneers laid the foundations for a number of mindfulness-based approaches to the secular application of originally spiritual exercises of old that were used to aid human development, empathy, emotional regulation, centring and perceptiveness. Due to its overwhelming resonance amongst colleagues and the general public, this approach may not appear compatible with a behavioural hardliner concept and a purely cognitive approach. However, its success could also be interpreted as an indication of the needs and expertise presented in the field and perhaps also a welcome wish to depart from the unilateralism experienced thus far.

The turn of the last millennium signified a new chapter in the meeting place of cultures when a group of monks, with the Dalai Lama's blessing, were studied during meditation using EEG and imaging techniques by western neuroscientists (such as Prof. Dr. Wolf Singer and his team). The findings indicated special cerebral potential development as a result of meditation and facilitated the distinction between empathy and a meditation posture that provides strength and amicability. Furthermore, through synchronisation, the EEG strikingly captured the strong, radiated, energetic force field with its high degree of order.

The further investigations by Prof. Tanja Singer et al. have already roused curiosity and expectations.

1.8 The "New Science of Psychotherapy"

1.8.1 Diversity with Common Reference Sciences

The sheer breadth of psychotherapeutic approaches at present is impossible to overlook. On the one hand, this appears to represent an abundance, but on the other, it could be perceived as overload. It is simply no longer possible to be sufficiently knowledgeable on all variants. In this regard, the rough allocation into clusters of psychodynamic, behavioural-therapeutic, humanistic and systemic groups of methods can aid our understanding. This allocation harks back to Klaus Grawe's meta-analysis, which is based on his assessment of the studies spanning half a century (1934–1984). In addition to their common aspects, each of these clusters featured an array of different approaches and its own evolution. As can perhaps be expected, emphases have shifted over the past 150 years.

The findings from research on neurophysiology and developmental psychology, including findings from more recent mother-child and attachment research, have since been almost universally recognised and accepted as reference points against a common backdrop.

1.8.2 General Development Trends

Therapeutic relationship. Increased significance has been placed on the therapeutic relationship – and the recognition of its relevance to successful outcomes – in all schools, even those that differ greatly in their fundamental stance and even if this knowledge has unfortunately been subject to a wide range of implementation approaches.

Resource orientation. The resource-oriented basic attitude finds its origins in the humanistic-systemic wave experienced during the 1950s and 1960s. At first, it was cast aside as dubious by the established mainstream but has since gained increasing recognition in all groups of methods over the past two decades – once again upon Grawe's recommendation. This attitude forms a partial aspect of the creation of therapeutic relationships: am I, as a therapist, prepared to primarily spot and acknowledge the potential of my counterpart? Through this, the therapist is mainly or partially able to establish a relationship at eye level. Am I able to handle this, or potentially even look forward to it – or am I only capable of practising my role as a therapist with a clear constellation of differentials?

Part and whole work. Part and whole work has been subject to a similar adoption trend under a number of different names, as well as one of its variants, "inner-child work". Moreno and F. Perls, two theatre enthusiasts who delighted in mise-en-scène, first met in 1948/49. During this period, they exchanged their experiences. Moreno enjoyed entering the therapeutic process with biographically significant key scenes, whereby the role assignment of the governor contained potential that could be psychotherapeutically exploited.

The former psychoanalyst, Fritz Perls, was more interested in the inner dynamics of the conflicting forces; the subject was required to alternately change roles and debate the existence of both opponents, which was used to modulate the evaluations of both poles. Due to the change in location, the meta-level primarily gained significance and aided in the establishment of solutions – and also promoted the realisation that some wishes would remain unfulfilled and a certain level of grief would be required in this regard.

I came across the earliest published "inner-child work" in the first Handbook for Gestalt Therapy: an article by Ben C. Finney [5] "Let the Little Child Talk".

Behavioural therapists have also found stimuli in Moreno's roleplays and role-swapping interventions, a field in which semi-structured, exercise-oriented play scenarios are often employed. Moreno's work has also been welcomed by the young generation of depth psychologists with their open access to technical equipment.

When suffering a bout of illness, the founder of schema therapy, Jeffrey Young, was advised by a friend to consult a Gestalt therapist. After his session, Young reported that he had learned more about himself during those 10 hours than during the entire previous year of work with this CBT. Although he did acknowledge that Gestalt required more structural determination. As such, he then attempted to structure and manualise Gestalt therapy using methods he was familiar with from cognitive-behavioural therapy. To a certain extent, the depth of his work can be viewed as a win for the field, however, it was perhaps to the detriment of the individual-centred approach, spontaneity and the knowledge of how to open a dialogue with the deepest layer of dissatisfied basic needs and work from there to healingly build.

A very similar concept emerged in the 1980s with ego-state therapy. This form of therapy finds its roots in psychoanalysis and is primarily the sister of trauma therapy; "Family Therapy without Family" *Gunther Schmidt) was publicised as a novelty in systemic therapy. Friedemann Schulz von Thun's "inner team" is also based on the roots described by Moreno and Perls.

Paradigm research. The innovative Gestalt psychologist Kurt Lewin and his intensive research on paradigms in the 1930s/40s introduced crucial impulses on the relativisation of subjective experience. These have since been incorporated by the systemic school under the term "reframing". In Gestalt therapy, we refer to this notion as shifting the paradigm. It can prove to be a helpful intervention, especially in the case of changing an individual's identity as the individual models himself in his innermost self together with all of the relationships he has experienced, whether this be partnerships, family bonds, social, traditional, national or spiritual, etc.

Forms of body therapy. By and large, body language has been clearly integrated into the psychodynamic since the 1920s. Well-known body psychotherapists drew particular attention to it, and it has consequently developed into an essential component of the therapeutic process, especially in Gestalt therapy.

Body-oriented psychotherapy involves a number of facets. Although it experienced somewhat of a boom in the latter half of the last century, it is yet to be granted the recognition and level of integration that it is worth of in light of its effectiveness.

Trauma therapy. Trauma therapy, which finds its roots in crisis management practised originally in Gestalt (according to the author's experiences in Germany in the early 1970s) and humanistic therapy, has been defined as compatible with all schools. It has taken the fundamental stance of Gestalt and humanistic therapy and developed it into a generally implementable approach and further differentiated it. More specifically, trauma therapy is concerned with (1) security and stabilisation of self-esteem; (2) the permission and authorisation to establish boundaries as a protective mechanism (therapeutic distancing with containment) with the recovery of self-efficacy and defensiveness; (3) emotionally shaded differentiation and self-control with regard to confrontation capable of integration (screen method) and (4) body-oriented self-confidence (originally: contralaterial tapping whilst standing with any defensive reactions that were previously unconsciously maintained during exposure and permitting previously frozen, emotional reactions to defend an empathic and appreciative therapeutic relationship). Of course, there are a number of variations on the above. EMDR was first added at a later date by F. Shapiro (1984). Peter Levine also pioneered an impressive way of dealing with people referred to as "somatic experiencing". This approach draws on the individual's own experiences and is not rooted in the tradition described above yet follows a very similar approach, whereby it feels as though the natural path of healing restitution has been taken.

1.8.3 Abandoned Aspects of Therapy

Certain methodological paths have been abandoned over time due to their failure to achieve the required levels of success. For example, the fear-inducing emotional flooding employed by early BT has become a relic of the past. The practice of banishing feelings and relationships to the "black box" also ended long ago. With regard to psychodynamics, as of late, transference neurosis and (hypothesis-led) interpretation, including transference interpretations, which patients regularly found inappropriate and excessive and consequently triggered their need to distance themselves, has been met with increasing silence.

At the end of the 1970s, Gestalt therapists had already recognised the need to avoid excessive emotional expression in favour of increased authenticity and fine tuning. Furthermore, systematists have since abandoned their focus on the system, which had developed into their sole focus, in favour of a balance between the individual and system.

1.8.4 The Zeitgeist in Upheaval

Ultimately, all development trends are largely governed by the conscious default zeitgeist of mainstream's declining, reductive materialism concept of humanity on the one hand, and the initial teething pains regarding integrating the challenging findings of quantum theory with its incalculable degrees of freedom on the other hand. In this regard, the primary objective is to develop tolerance for uncertainty in a dynamic process to facilitate the development of and contact with the unconscious that was previously in resonance. Despite the fact that the principles of quantum theory have now been in circulation for over three generations, we are nevertheless progressing at an adequate pace – both in historical comparison and on the basis of the extent of the intellectual (r)evolutionary upheaval.

The two principles of natural scientific systems meet and interact in humans, as they do in each living being: (1) the system of the smallest particles (Newtonian psychics) and (2) the system of wholeness (quantum theory). Our brain is both constructed for and shaped by this interaction. In a more complex format, it mirrors the mode of existence of a single quantum, which can pseudo-corpuscularly fold in a number of ways – or connect as a wave as per Einstein's formula $E = mc^2$.

There is also evidence that this may apply to the cosmic macro world and can thus be regarded as the general principle of basic energetic substance.

As such, we have every reason to scrutinise and strive for the field of play between the structural fixation on time, on the one hand, and the great potential of possibilities, such as the fundamental freedom of development and decisions related thereto, for our concept of man and consequently also for adequate psychotherapy in the future.

The old body-mind dilemma of Western schools of thought emerges from this overarching perspective as an artefact. This dilemma principally arises if the intellectual constructive processing of the dominant brain hemisphere, which is based on the extrapolation of puzzle pieces that are recorded from reality through our sensory organs, is established as absolute as a concept of reality. This method adheres to the "either-or" principle and spatial-temporal notions of order. It lies in the domain of secondary awareness (according to Freud) – the holistic mode of consciousness that uses the wave identity of the being and follows the resonance principle. The majority of our psyche, as we perceive it, is formed by our unconscious. In any case, the unconscious is completely in control before the cerebral cortex matures and in situations where control is relinquished. It would appear that a special kind of extrapolation occurs during this process, which has the power to dictate the individual's own perceptions, ideas or intuitive decisions. This aspect has been recognised in the artistic world, but is not widely cultivated in the western world – likewise in education. As such, we only have rudimentary access to it.

In the ancient cultures of the Far East, the wave identity notion of our being is regarded as the most highly valued, true and realistic way to grasp reality. It enables us, through receptive inner hearing and seeing, to recognise the higher, spiritual connections in the connected being (mind) and absorb the information thereof. Whilst the complexity and possibilities of the whole cannot be ascertained by the parts, they can be reflected in a differentiated manner. The traditional, sacred syllable Aum or Om symbolically encompasses the entire oscillation spectrum of the individual. In contrast, the unfragmented puzzle concept of our human intellect (mind) that is projected onto reality was understood as a concept that distorts the reality of the artefact (illusion).

Seeing, or the Eye of Providence, has held special meaning in the western cultures, which are predominantly rooted in the traditions of ancient Egypt, Israel, Greece, Europe and America, from the outset. Vision favours perception based on focusing, analysing and differentiating more than any other sensory organ. In quantum mechanics, the observer's information, knowledge or conscious awareness always leads the collapse of the wave-function (Heisenberg cut). As such, our relationship to our environment helps to shape the world we live in.

Humans are naturally equipped for the interaction of both modes of being. To be bias means to be impoverished.

There is a current fascination surrounding the technical-digital progress towards "artificial intelligence" amongst our generation. To date, progress has been primarily based on large-scale statistically-based programming. However, these impressive achievements could potentially cause the concept of man to be reduced to the basic comprehensibility of machines. Similarly, there is also the risk that psychotherapy will be tailored to the machine model and its learning programmes and that the theoretical reduction and bias of qualification may be overlooked as the result of immense quantification.

Therapy relationship constellation. When assigning the groups of methods employed in behavioural therapy, psychoanalysis/depth psychology and humanistic approaches, into Jung's 4 quadrants, it becomes clear that;

"CBT", i.e. cognitive behavioural therapy, focuses on thinking, the rational, and the practical, sensual understanding of the outside world;

Freudian psychoanalysis aims to subordinate emotionality (and, if present, also intuition, if it is permitted to emerge from the unconscious) to thinking and conscious understanding, otherwise known as its "secondary level"; and the Humanistic methods, especially Gestalt therapy, prioritise feeling, intuition and a sensual experience of the world, only regarding reflective and categorising thinking as a secondary value.

Forms of therapy that involve more than one party (couple, family, group therapy, organisational counselling) require all four quadrants in equal measure. They are also found in relation to psychosomatic medicine in part and whole research in ascending and descending order. This results in a vertical grading of complexity, whereby mirror phenomena may occur as a disruption alongside salutary restructuring, depending on the experience.

Over the past few decades, it has become apparent that the main biases have at least partially been identified and efforts have been made to adjust this. Based on Klaus Grawe's (1934–2005) recommendations, CBT has introduced an emotional revolution and consequently benefited from the relationship culture of humanistic approaches, especially conversational therapy, despite the fact that the required skills for the therapeutic relationship have not yet been comparably introduced within the scope of BT training and will therefore likely not succeed in realising the true potential.

In certain aspects, depth psychology (DP) has seemingly come very close to achieving the humanistic way of forming relationships. However, with regard to the gradual orientation towards the traditional psychoanalysis mainstream as opposed to the psychoanalytical avant-garde, there seems to be a wide field of play depending on the institute. In recent years, DP has developed in a number of vastly different and often independent and creative ways and gained newfound confidence, not least as the result of the stronger research findings compared to its big sisters.

Psychoanalysis has also attempted to research new criteria over the past two decades. That being said, the fight to achieve meaning and legitimation for long-term psychoanalysis has been rather quiet in comparison.

With regard to research impetus, the humanist school has largely departed from the previous scepticism against theorising and scientific evaluation, to the extent that it now completely dismisses it. Gestalt therapy has since established a well-coordinated, intercontinental cooperation between Europe, North and South America, the Far East and Australia.

Mainstream scientists have taken it upon themselves to view all findings from psychotherapy research from a neutral angle and to juxtapose effective paths with the highest level of methodological competence and a balanced view. This refers to the contributors and, above all, the editor (Michael Lambert at present) of the "Handbooks of Psychotherapy Research and Behavior" (Wiley, New York), which has been published at regular intervals for over half a century. The neutral stance of the publications has helped to limit the egoism of associations and national pressures.

Unity and diversity. Even if the school of psychotherapy were to succeed in establishing a common framework, a certain amount of variation with regard to approach would nevertheless be required. In this respect, I am referring to the entire field of play between body-therapeutic interventions, scenic realisations, artistic expressions and more restrained verbal interactions. This is the result of the diversity of our human state. Only those who can internally fully accept their actions and way of being without having to bend over backwards will be able to fully develop their potential as a therapist and be a trustworthy counterpart. Of course, individual styles are also subject to change over the course of life. Usually, this change is coherent in nature.

1.8.5 Opportunities for Synopsis

Each school of psychotherapy offers a unique set of beneficial contributions, in addition to the ability to relativise and supplement existing knowledge in the field. At best, we will succeed in edging ever closer to the whole through the acknowledgement of partial truths. Psychotherapy is primarily concerned with establishing what makes a healthy, healed person, and what is missing otherwise.

The whole can be influenced through the change of one detail, just as a detail can be balanced through the optimisation of the whole. Both approaches can also work hand-in-hand.

As a part of nature, humans reflect the rules that apply and can be observed in nature. These rules are also reflected in psychotherapy, even though human knowledge may remain limited and maintaining an open attitude to expand our horizons would be advisable.

This outline attempts to integrate proven findings from practice and recent neurophysiological findings, as well as the current psychotherapeutic perspectives on effectiveness research and the two-faced nature of science, which has been overdue for over a century. Through this, psychotherapy shall gain access to the relationship of the part and the whole. The latter may promote new points of view that challenge old familiar taboos. The integration of both-and into quantum theory has proven to be the long-awaited gateway for brain research, and especially psychosomatics and psychotherapy, to depart from the centuries-old, self-propagating body-mind duality dilemma and Karl Popper's three worlds, as soon as the issue of separation is recognised as a real construct of our brain.

One of the key quantum theorists, Erwin Schrödinger (1887–1961) philosophised, "Multiplicity is only apparent, in truth, there is only one mind" (1985, p. 67). Amongst other things, Schrödinger explored the fundamental Vedantic concept of "Tat Tvam Asi": "Hence this life of yours which you are living is not merely a piece of the entire existence, but is in a certain sense the whole; only this whole is not so constituted that it can be surveyed in one single glance. This, as we know, is what the Brahmins express in that sacred, mystic formula which is yet really so simple and so clear: Tat tvam asi, this is you". ▶ Chapter 3 attempts to summarise the relevant aspects that form the background framework for our psychotherapy in an overview diagram.

In the fore of psychotherapy, research findings have spectacularly confirmed the significance of the emotion-modulating relationship axis. This has opened the door to new opportunities for psychotherapy in recent decades.

Literature

1. Wampold BE, Imel ZE, Flückinger C. Die Psychotherapie-Debatte. Hogrefe, Göttingen: Was Psychotherapie wirksam macht; 2018.
2. Lambert M. Bergin & Garfield's handbook of psychotherapy research and behavior change. New York: Wiley; 2013.
3. Rudolf G. Structure-related psychotherapy. Stuttgart: Schattauer; 2004.
4. Revenstorf D, Peter B. (Eds.) Hypnose in Psychotherapie, Psychosomatik und Medizin. Heidelberg: Springer; 2001.
5. Finney BC. Let the little child talk. In: Hatcher C, Himelstein P, editors. The handbook of gestalt therapy. New York: Jason Aronson, Inc.; 1969/1976.

2

Efficacies and Common Factors

Contents

L. Hartmann-Kottek, *General Psychotherapy*,
https://doi.org/10.1007/978-3-030-87466-7_2

Overview
The significance of common factors for the success of the therapeutic relationship and process remains generally unacknowledged, which has, in turn, led to a number of missed opportunities.

The therapeutic relationship, which is steeped in conscious appreciation and provides an adequate, life-enhancing new experience regardless of the method used, lies at the heart of efficacy.

The quality of the relationship, which was previously regarded as a "common factor" in the field correlates with therapy success more than any other factor (30%). However, said quality is vastly different across the various schools, which is most likely the reason for the high level of efficacy exhibited by Gestalt therapy.

The therapist's personality is responsible for 20% of the variance in the outcome, whereas disorder-specific methods only contributed 10 (−15%) to the variance in outcome. Proponents of these methods falsely tend to record their successes in previous research as method-specific. As such, certain methods could also be renounced. They may be abundant, but should only be used with sound judgement.

2.1 Mindful Awareness: As an Organ of Perception and as a Creative Force

Preliminary remark: For a large part, at present, mindfulness is not (yet) regarded as a concept that can be defined by science. It bears impressive structural parallelism to "energy" as the basic substance of reality which has thus far only been described in hypothetical and mathematical terms. That being said, both energy and mindfulness share a long tradition in the history of human culture.

The following chapter compiles observations, definitions and experiences from the fields of neurophysiology, psychology, psychotherapy, traumatology, hypnosis, sleep, dream, meditation and near-death research with a slightly different focus and subject-specific relativising background in each case.

In doing so, this chapter features a clash of two world views: the notion of the material definition of reality, whereby the spirit, if it indeed exists, is an epiphenomenon of its material basis, and the world view harboured by the "post-Einstein generations", which is based on the paradigm of changeable energy. We are currently living in a transitional period in this regard. This entire book is also characterised by a dual perspective: on the one hand, of the world of fragmenting distinction, the world of science and the Western mainstream. On the other hand, it deals with the world of connective resonances, without which effective psychotherapy would not be conceivable today. It is no coincidence that the remnants of the mindful Zen culture find balance in great resonance.

In the common "floating space" between the dualistic level that emerges as a result of fragmenting distinction, on the one hand, and the world that promotes the holistic view of the wave-like interconnected basic substance, on the other, there are seemingly – at least in comparison to light – different degrees of order and consequently different energy levels. The forms of order range from low-energy chaos and variously distinct synchronisations to high-energy neutralisation levels, for which the cosmic zero-point field is a notable example (according to Feynman, Nobel Prize winner in physics, the volume of a single light bulb contains enough energy to boil all the world's oceans).

In psychotherapy, a corresponding neutralisation of polarised forces would equate to the union of the inner, opposing poles.

The significance of mindfulness, mindful presence and traditions of awareness from the cultures of the ancient Far East travelled to the Western world in several waves, especially Zen traditions. These forms of meditation – in their secular form – have supplemented and helped to shape today's school of psychotherapy and reached large swathes of the population.

They have astonished the world of natural sciences in part through the previously unobserved synchronisation phenomena of the human energetic radiation with long-term mediators and the measurement thereof in the EEG field. This newfound curiosity has accordingly led to extensive research.

2.1.1 Two Consciousness-Modifying Forms of Being from One Basic Substance: The Focusing-Separating "Materialising-Locating" Pole – and the Interconnected-Oscillating, Non-local, Potentially Information-Receiving Pole with Floating Mixed States of Different Forms of Order: as Chaotic, Self-Organising Dynamics; as Categorically Assigned Pseudomatter; as Synchronised Resonance Patterns or in Neutralised-Interlocked Holistic Forms

Similarly to the way in which consciousness uses and shapes the human brain, it develops different cognitive paths and outcomes: if the sensory input encounters focused attention with an eye for detail, as is preferred, the information is directed to the intellectually extrapolating, dominant brain hemisphere. If this activity is rewarded with special interest and repeated time and again, this will result in the emergence of a more categorical, clearly structure world view based on Either-Or thinking and attuned to Aristotelian logic (A is not B).

If the sensory input encounters open, receptive attention that does not seek to influence anything, a phenomenon whereby oscillating fields appear to spread from the brain stem via the limbic system to the vertex and the prefrontal cortex occurs. This may also be reinforced by the special connective system of the vagal neurons between the heart and brain (i.e. the heart and limbic system), which can achieve joint synchronisation through slow breathing (once every 10 seconds) and an emotionally happy state (McCraty et al., Heart-Math-Institute). In this state, the regulative oscillation pattern is able to reach all cells. Feelings of happiness are experienced when this state is entered, and it also may grant the individual the feeling of higher insight.

During this process, we assume that epigenetic factors come into play with regard to the ability to achieve improved use of the brain in one way or another.

It then follows that the resonant, coherent wave fields constitute a special but nevertheless overlooked potential of our brain. They perform both a connective and relational role within the central nervous system and an interpersonal role between individuals, groups and even anonymous crowds. An equally high level of attention energy is a prerequisite in this regard, but not in a focused form, rather in a mindful, receptive and open-minded form. Wave fields foster a sense of belonging and override any feelings of isolation. They permit the individual to oscillate into other worlds, and thus facilitate a version of holistic, emotionally influenced perception, or "reading".

This principle also applies to our brain. Whilst the coherence principle primarily acts as a connective link, it also assists with differentiation in its least subdivided manner. This may at first seem paradoxical, but it is not. According to brain physiologist Wolf Singer [1], neuron groups that are involved in the coding of a certain object are locked in the same phase when they oscillate. This enables them to form a coherent object/shape/structure and separate this from the background activity of the remaining neuron ensembles that oscillate in other phases.

Note

Neuron groups that belong together, oscillate synchronously.

The oscillation that contrasts to the remaining background field subsequently becomes its own entity through interested (or unconscious) attention energy and thus the dividing line is drawn. In quantum physics, this is referred to as the Heisenberg cut, which first established the concept of delimited phenomena.

Note

Delimited things are created by attention energy.

We experience this right away when it comes to feasible plans and daydreams. The brain works according to the Gestalt law of figure-ground. According to this principle, synchronously oscillating elements are combined and separated from incoherent elements.

This fundamental principle of co-existence in the same phase of oscillation – or not, as the case may be – also applies within the organism, particularly within the brain, and also contextually between individuals (including the therapeutic relationship and in a couple, familial and group context) and between individuals and their – generally all-encompassing - higher relationship horizon. Individuals experience holistic relationships as healing and mismatches as stimulating and challenging to a certain extent. However, in excess, these mismatched relationships are perceived as threatening to their existence or perhaps as an attack or destruction.

2

Inspired by the notions presented by William James and Werner Heisenberg – the quantum physicist Henry Stapp [2, 3] established an integral model of consciousness. Subsequently, a conscious thought was regarded as a real object. In classical physics, it is not possible to create something that is more than the sum of its parts. However, that is precisely what a quantum state event would achieve. Previously separate states are accordingly combined into an ontological new whole.

> **Note**
>
> This whole-generating potential of the aspect of the basic substance that is capable of carrying waves has since been confirmed time and again.

We shall now return to the mind, which is understood as "conceptualising consciousness" and reflect on its significance in the ancient traditions of the Far East. In these traditions, the mind is understood solely as an extrapolation of the five senses and thus it forms a further sense, a sixth sense. The Far East places must less significance on its intellectual ability to construct than the West. The mind instead appears to prevent access to the advanced, non-conceptual state of consciousness that leads to higher insight. It is solely valued for its power as a coping mechanism for everyday life on the materially condensed level. At that point in time, there was evidently enough proof and consensus that higher knowledge and insight could be achieved through the non-conceptual state of consciousness.

Consciousness and the differentiation thereof have fascinated the minds of the Far East for millennia, notably since around 3000 BC in the Indus valley. Ancient Indian, or Vedic, ideology on this subject was predominately incorporated into Buddhism and has consequently been preserved, primarily in Tibetan translations that have been passed through the generations almost entirely unchanged (Tarab Tulku XI). One interesting aspect of "inner science" in this regard concerns the "conceptualising consciousness", which is understood as an abstracting and judgemental (intellectual) processing outcome of sensory experience and consequently as an additional, integrated 6th sense, which employs a selection of perceived details to create conceptual holistic images. This also includes the creation of concepts, values and everyone's "subjectively constructed world". The instrument that creates these images is the conscious, energising attention that brings elements to the fore. This attention extracts interesting or useful aspects from their contextual setting. As such, this imaging can be regarded as an objective, creative process in a mini format. The "object pole" exists as conceptual reality, but always only in an interactive connection with the "subject pole", as the conceptualising consciousness. Even with the addition of the qualities of the image and emotion-generating consciousness, it remains a construct, an "illusion" (Sanskrit: Maya), which can be coloured either positively or negatively. Constructs have traditionally been viewed with scepticism in the Far East; Daoism and Chan(Zen) Buddhism regard them as non-real generalisations that are imposed on things and prevent direct (transient oscillation) contact with them. The real "objective reality" of the whole remains unperceivable for the subject in any case.

The objective in this regard is to achieve a state of consciousness freed of concepts. Withdrawing from the self-referential system of evaluating and defining conceptualisation appears to be possible and something that can be practised through focusing on the real sensory reality in the here and now, which occurs in the preliminary stages of conceptual processing, and to become internally liberated from pre-determinations.

When achieved, this state results in a free, spontaneous life in the here and now. It facilitates broad access to the unconscious. We encounter this state in play, in playful interactions and social gatherings, in musical improvisation, improved theatre, in artistic pursuits and also in everyday encounters. It holds great potential for possibility and many degrees of freedom, and thus the chance to obtain a very special and coherent reaction. This state is encouraged and promoted through "free experimentation" in Gestalt therapy, provided the subject exhibits the required structural integrity.

The non-conceptual state of consciousness is cultivated and modified in meditation such that inner detachment or "non-attachment" is achieved. Parallel to making and breaking external contact, we encounter the same phenomenon within. This entails that we develop an inner relationship cosmos between all kinds of needs that constantly emerge and the "observer" or the "organiser", who forms changing alliances, temporarily places a number of them on the back burner and brings some of them to the fore with particular urgency and occasionally halts others. As such, a special kind of autonomy is displayed that is required to tolerate tangible needs. This, in turn, increases inner freedom, which facilitates decision-making and achieving a stance on the basis of certain values.

The potential development path of the consciousness, which seeks to block and switch off the conceptualising intellect in order to achieve higher levels of consciousness was first cultivated in ancient China and Japan, particularly in the traditional forms of "Koan" meditation. During "Koan" meditation, the intellect is tasked with achieving a rationally unsolvable feat, such as "hear the silent sound" until it ultimately gives up.

It is worth taking the time to compare the inner relationship between the conceptualising intellect and Western constructivism of recent decades. "The map is not the territory" warned Alfred Graf Korczybski (1879–1950), a pioneer of constructivism, whose work influenced Fritz and Lore Perls in the 1930s and whose school of thought was incorporated into Gestalt therapy.

There is perhaps no clearer juxtaposition of world views: the Western view, whereby the world is (inductively) constructed from the smallest possible particles and the spirit is understood as an epiphenomenon of matter, in contrast to the (deductive) view of the world and man, whereby an interconnected energy-spirit-consciousness entity is potentially, temporarily and primarily convoluted into the matter.

The thought processes of the ancient Far Eastern epistemological-theoretical-philosophical approach on the mutually interconnected nature (tib.: Tendrel) are tied to the unity of opposites of (1) Subject – Object, (2) Becoming – Cessation, (3)

Localisation – De-localisation and (4) Part – Whole. Each of these pairs of opposites is a separate, "interconnected" whole. The Tendrel approach similarly bridges spirit/energy and "matter" as two modes of the same substance.

It is perhaps notable how compatible this ancient school of thought is with "modern physics" (see ▶ Chap. 9). Conventional Western physics is based on the construction of the world from elementary particles, or material pseudo-particles. However, following Einstein's theory ($E = mc^2$) does not lead to a fruitless either-or state, but rather to a floating both-and state that encompasses both aspects of existential manifestation possibilities and a third floating state in between. This mixed floating state comprising contoured and interconnected basic substances can be found in a chaotic disorder whereby the materialised forms appear to dominate – or rather those of wave-like nature, such as in territories exposed to the play of wind and waves. A desired approach would be to strike a balance between the given poles in free-floating attention, which would be more than logical in a therapeutic sense.

2.1.1.1 The Integral Dual Nature of Mindfulness

The key concept of the cultural landscape of the Far East, mindfulness, appears to pose a challenge to translators. It combines two cognitive functions that seem incompatible: (1) object-related attentions, including memory, recollections (Pali: sati/ Sanskrit: smrti/ Tibetan: dran pa), and (2) the mental process by which one monitors one's own body and mind, otherwise known as introspection (Pali: sampajanna/ Sanskrit: samprajanya/ Tibetan: shes bzhin). The first aspect is conceptualising in nature, which perhaps facilitates external determination, whereas the second aspect is spontaneous, more improvisational, inward-listening in nature and arises from individual activity. The objective is to achieve a "middle path" between mindful spontaneity or spontaneous, refined mindfulness that unites these two poles.

This concept has caused many a Gestalt therapist to be astonished by the familiar basic structure of its theory, (1) for example, the attention energy that brings important things to the fore and thus creates structures (existere = to appear), (2) the aim of letting go of outdated structures in the death zone of the crisis cycle and to instead exist in the unstructured until new, more coherent contours emerge from within, (3) the appreciation of the here and now, (4) the "middle mode" which creates the ideal conditions for an attentive, responsible freedom of choice, amongst other aspects, and the (5) scepticism towards perceived superior intellectuality. Point 4 also applies to the basic notion of "free-floating attention" in psychoanalysis.

Point 5, which exhibits hostility against the intellect, was unfortunately just as exaggerated in Gestalt therapy in the 1960s/1970s as the overtly emotional demonstrations of expression at that time. These approaches initially hindered research efforts. However, an adjustive counter-current began to emerge in the mid-1970s (at least in Germany), which focused on the adequacy and authenticity of expression and supported small steps towards less spectacular but, for that, all the more coherent methods. The following intensification of the opportunities to achieve therapeutic change was both impressive and effortlessly natural at the same time.

As Fritz Perls wrote in his biography, "I did not invent Gestalt therapy, I simply rediscovered it" [4].

2.1.1.2 The Integrative Interplay of Modes of Being

The interaction of the two consciousness potentials that are anchored in the character of our brain has always been a key element of psychology. Here, we refer to the potential of differentiating and separating focus on the one hand and the potential of the connecting formation of the whole within the framework of common oscillating fields on the other.

As materialised reflections of wave fields that model matter, nature provides excellent examples of this phenomenon, especially in flora, in the structures of pinecones, seed pods, flowers and leaves, amongst others. Examples of an apparent constant dialogue between wave formations and material realisation on the one hand and between the part and whole on the other can also be found in the scale models of our own physical periphery (such as our ears, hands and feet).

The psychotherapeutic objective of achieving holistic "coherence" (one of the objectives of Gestalt therapy), or "congruence" (the conceptual equivalent in talk therapy) as a flexible regulator and expression of a healthy, healing balance of an organism makes complete sense in light of these observations.

2.1.2 Neurophysiological Hypothesis on Consciousness

Consciousness is understood to be a dynamic coherence centre for integral information. The neuroscientist Giulio Tononi [5], a proponent of the Western scientific school of thought, established the "Integrated Information Theory" (IIT) in an attempt to explain consciousness. According to this theory, the consciousness is based on a coherent, thalamocortical neuronal stimulation circuit with a dynamic re-entry mechanism (referred to as the "dynamic core" by Tononi), which may have the power to connect all areas of the brain – but which is interrupted in deep sleep in favour of the synapse-altering, post-processing of waking consciousness. This structure can be found in all mammals. Hence, consciousness requires a neuronal system that generates coherence. If the thalamocortical system sustains damage, consciousness will be lost.

Quantum physicist Carlo [6], p. 82) speaks highly of Tononi's "Integrated Information Theory" as an "attempt to characterise the structure that a system requires in order to possess consciousness in the quantitative form: a method to describe the physical changes that occur between our waking state (consciousness) and the dreamless sleeping state (unconsciousness), for example".

Tononi places a clear emphasis on the significance of a coherent energetic form or order as a prerequisite for consciousness. He does not address the phenomenon of conscious, incorporeal energy fields.

2.1.3 Corporeal Characteristics of Consciousness

Limited processing capacity. The central nervous system is the instrument that contains specific receptors and stimulation processes that enable us to come into contact with a part of the world. Our brain functions in part through local and functional allocation – and otherwise in an indefinable and free manner. It thus follows that certain sections of the brain are reserved for specific functions, particularly for integration work.

Our limited receptors and the limited abilities of our sensory organs render us blind to presumably a much larger part of reality.

The complexity of the reality in which we find ourselves overloads our brain with regard to stimulus processing, which appears to be configured to reduce stimuli: psychological experiments conducted in Gestalt therapy with the use of ambiguous images or stimuli that can be interpreted in a number of ways have demonstrated that our brain is only capable of accepting one interpretation of the respective stimulus at any one time. It then consequently manages this process by retracting the initial interpretation after a period of 3 seconds, letting it fade away and enabling another organised perspective to come to the fore. This process of letting go can be blocked by conflict. Certain individuals may experience fear and therefore unconsciously avoid the intermediate state of not knowing and not being able to recognise the image. Accordingly, the previously established understanding of the image gains increasing significance. In this case, an additional psychogenic restriction to capturing the multidimensional nature of the world emerges.

Despite all of the magnificence exhibited on the one hand: in order to compare a human brain to a radio, you would only require a simple model. Our neuronal saltatory conduction works according to the yes/no principle. The incomplete input of our sensory organs is pieced together by our brains on the basis of the over-simplified yes/no principle to form a rasterised map and impose this on reality. Anything to the contrary is unfamiliar to us, and we are only capable of participating in one programme at any one time.

Our intellect also works according to the yes/no principle in order to handle everyday life. Generally speaking, we view our perspective as without alternative. However, this is not the case. It represents a partial truth. This principle also applies to the parcellating differentiation between foreground and background, which enables us to concentrate and focus our attention: it is capable of bringing certain aspects of the entire stimulus to the fore and dimming down the rest. We are thus able to create our duality of the important and unimportant. We account for differentiated aspects with symbolic sounds such as language and thereby fix them. Particularly interesting is the way in which certain aspects are differentiated in language, such as the many variations on the word "snow" in the Eskimo–Aleut languages, whereas other languages only have one collective term for the concept. This is only a small example of the way in which our attention energy, which differentiates and categorises meaning, creates and maintains its own world. The differences can be endearing, quirky or disturbing, and sometimes quite subtle. As we have already established, every group of individuals and indeed individual lives in their own world.

How do we actually experience our own version of "being in the world"?

It would appear that there an inner meeting place with an observer can be found in our brains. According to American neurophysiologist Antonio Damasio, two different neurophysiological neural patterns meet within the most ancient parts of the brain (potentially in the paleocortex within the telencephalon, in the thalamus): one that rises from the inside of the body with all vegetative regulation parameters, which Damasio refers to as the protoself, and one that comes from integrated sensory perception together with the remote senses and represents the external world. As such, our brain seemingly features an internal meeting space that is simultaneously registered by an internal observer. Findings are then most likely communicated to an assessment body, who will then compare actual to target values, depending on the current level of motivation and on the largely unconscious paradigm. In the field of psychotherapy, it is preferable to both observe without judgement and to become aware of the unconscious readiness to judge and modify this if required.

2.1.4 Incorporeal Characteristics of Consciousness (Dissociations, Free Fields)

"The knowing field"

In research on the constellations of families, groups with conflicting configurations or large groups in political hotspots, therapists are required to work with representative perception, which implies a field of consciousness that connects us all. However, this has not yet been recorded in science and must be viewed as a hypothesis. Nevertheless, according to Albrecht Mahr and other experts in constellations, the way in which this field can be experienced is impressive. Doing so requires fearless sensory awareness on the one hand and an emotionally significant, energetic impulse on the other.

The expression "the knowing field", stems from constellation research and describes the experience whereby the "representatives" who are involved in the constellation and assume a role for which they have hardly any prior knowledge, sense clear impulses, predominately as body perceptions, which the protagonist (in most cases) recognises as helpful and coherent. The same applies to the head of the constellation. Findings made on "knowing fields" hint at a worldview connected by consciousness.

Fear and adversity repress the supra-individual sensory awareness and initially "ethnocentrically" (Jean Gebser) centre it on the clan that provides security, on the reference group that appears to provide a survival advantage or ultimately, when the level of fear increases, only on the individual himself.

Note

Fear isolates.

The phenomenon of the collective consciousness is non-local and transtemporal.

Critics of this method tend to pay close attention to the extent to which the collective impulses seem to correspond more to the unconscious self-reinforcing tendencies of the representative than accurately reflect the tendencies and problems of the protagonist.

This phenomenon remains to be scientifically and critically analysed.

2.1.4.1 Evidence of Targeted, Supra-individual Fields of Consciousness

A further example that we come across every day in psychotherapy is a phenomenon that is observed time and again parallel to awareness in supra-individual fields of consciousness by Balint group leaders and supervisors: If a problematic patient was intensively discussed in a group, in the next meeting he may appear to have gained insights or adopted a behaviour that would imply he sat in the previous session as quiet as a mouse and listened in.

This consciousness phenomenon does not appear to be bound by spatial-temporal restrictions. Jean Gebser (1949/[7], Ed. I, p. 234 ff) referred to the lack of space-time in the "magical phase" and to a connective experience of unity in a trance-like state of consciousness.

2.1.4.2 Dissociations Between the Body and Energy Field

Dissociation refers to the noticeable – if only slightly or significant – alleviation or temporary separation of the energy field from the body. However, it does not refer to complete separation.

As is clear, there a quite a few people who experience spontaneous dissociation in their sleep that they can still remember when they wake. Certain methods can be employed to encourage the occurrence of this phenomenon (e.g. hemi-sync). Robert Monroe (Monroe-Institute) et al. have dedicated their work to the research of out-of-body experiences that occur during sleep and released many publications on the topic.

Dissociation experienced by the victims of violence, abuse and torture is a sorrowful and macabre topic that regularly comes up in trauma therapy. In this regard, the ability to dissociate, to temporarily escape that agony to which the body is subjected, is considered a gift (Bessel van der Kolk, Peter Levine, M. Huber, etc.).

Over the past 30 years, near-death research has produced a vast amount of material on disassociation. (Moody, Sabom, Pim van Lommel, Eben Alexander et al.) Each near-death experience (NDE) begins with the separation of the energetic field of consciousness from the body and then generally follows typical stages. Interestingly, the detached energy field does not receive any input from the inferior near senses, i.e. it does not perceive pain but can still hear and see (!) in a 360-degree field. There appears to be an analogous self in the field that senses impulses for action but is nevertheless unable to express itself effectively. The sensory awareness seems to correspond to the higher frequency spectrum of the pineal gland, as encounters with entities are often reported in this detached state, which cannot be seen to the naked eye with its 400–800 nm frequency range.

2.1.5 Modifications of Consciousness in Psychotherapy

Freud's secondary and primary levels. Now we shall return to a time before high-tech studies were conducted on the physiology of the brain. Consciousness characterised by a logical structure was described by Freud at the turn of the previous century: he described "If-Then" relationships, clear yes/no statements, a front and back, a before and after, specifically causal, temporal and spatial order schemes. According to Freud, these are the characteristics of the desirable "secondary level", which is governed by Aristotelian "Either-Or" logic. Freud's therapy goal was to bring the "Id" (unconsciousness) into the "Ego" (consciousness). It was his understanding that conscious insight into the developmental history of a neurotic, conflict-related, unconscious disorder could hold the adequate healing potential to eliminate symptoms. The more alert and attentive an observation is at the conscious secondary level, the clearer and more defined the outcome.

In contrast, Freud also observed an entirely different structural quality for the unconscious field, which is governed by the resonance principle and does not exhibit any of the structure mentioned above. Freud's precise observation, which remains valid, is worthy of praise. However, it did not persuade him to undertake a contemporary, resolute revision of theories of equivalence. He did not address "new physics". Freud's worldview and assessment thereof was shaped by influences from the Enlightenment, which exhibited a clear preference for conscious processing over the unconscious.

In Freud's structural model of the mind (1923), in which instinctive impulses are attributed to the personality instance of "Ego", normative aspects to the instance of the "Superego" and reality-oriented and organisation aspects to the instance of the "Ego", each instance is granted a conscious (secondary level) and an unconscious (primary level) side. The Ego's defensive mechanisms to protect the self-image primarily succeed in a process-related manner.

The founding fathers of cognitive therapy, amongst others, regarded themselves as a conscious opposing force and alternative to Freud's position. Their approach was intentionally based on the rational attempt to correct pathogenic beliefs. This attempt to deliberately change negative concepts of life and eliminate them where possible received a warm response in the western world, however, has since seemingly retracted into the background. The primary aim of this approach was to fight, change, relativise or, ideally, replace one cognition with another.

Here, we refer to concepts of life that are either survival bias ("you can only rely on yourself"), internal equilibrium ("I do not derive any joy from this, but I do it for the joy of others") or adopting the definitions of others ("I'm good for nothing, which is something my father has already realised"). They are concrete patterns of thought derived from existentially significant chains of experience that are mentally fixed to a certain extent, or potentially "dismissed" as invalid and therefore cognitively distanced. Personal pain often lessens if it appears that it can be attributed to a higher principle. In order to dismiss these (unconsciously) emotion-based personal beliefs, the individual is required to first revisit the avoided problem area, which is a path worth taking.

The key minds behind cognitive therapy methods first emerged in the 1960s/1970s, initially Albert Ellis and Aaron T. Beck, followed by Lazarus,

Mahoney, Meichenbaum and Seligman at a later point. They developed a cognitive theory and method for the treatment of emotional disorders. Their primary aim was to change dysfunctional cognitions and beliefs, misinterpretations, negative opinions, distorted perceptions and discrepancies in reality. At first, it was assumed that these misinterpretations would trigger emotions. (However, this has not been confirmed by research on the physiology of the brain [8–10]. Emotional reactions typically precede cognitive ones, which can most likely be regulated by suitable cognitions.

The "cognitive revolution" of behavioural therapy was heavily influenced by the work of this group, a field which had previously based its approaches on a strict, behavioural understanding achieved solely through observable behaviour. At present, there is again impetus for a revolution through conscious willpower and acquired expertise. In the 1980s, both schools were combined and referred to as cognitive behavioural therapy (CBT).

The attempt to reach a compromise could give rise to the following: the image of the world we have in our minds also always represents a personal learning history of our encounters with it – and a personal history of conflict; the image is not objective in nature, whereby individual needs that arise and the changing responsiveness of our surroundings lead to their own problem constellations and potential solutions.

Journeys taken through life only mirror each other to a certain extent, and only if they are observed from afar: this may concern bringing life into the world, passing on life wherever possible and then releasing life once more (dying). A beneficial change process requires a compromise between the delicate combination of internal and external impulses.

If possible, CBT.

At this point, it is beneficial to determine whether we find examples of altered states of consciousness in everyday psychotherapy practice.

Transference and countertransference: in the case of the latter, therapists pay attention to the emotions they transfer to a specific patient on the client in therapy; transference, on the other hand, is concerned with determining in which role the patient's unconscious requires or would like to see the therapist, or which role it fears that therapist has assumed. In psychoanalysis and depth psychology – and also Gestalt therapy – these phenomena are regarded as standard phenomena which need to be registered. However, it is not possible to determine how they arise, even if partial aspects may be deduced through non-verbal signals. A certain receptive posture (which can be trained to some extent) towards the "counterpart" and mirroring physical awareness can facilitate obtaining and "reading" intuitive hunches, images, daydream factors. Or so it would seem. In this case, the ability to anticipate insight and good cooperation, or vice versa, is most likely attributable to consistency phenomena. The notion of consistency is so well known and evident amongst the population and colleagues that it now forms the foundation for each patient's right to three preliminary consultations, which are paid for by health insurance providers. Colloquially, people are either referred to as "on the right wavelength" or not.

Accordingly, in psychoanalysis, the therapist should ideally adopt "free-floating attention"; similarly referred to as a "middle mode" in Gestalt therapy. This pertains to an intermediate form between focused attention on observations and sensitive, receptive attention, which facilitates resonance.

Balint work and case supervision are already mentioned above. These are further examples from psychotherapeutic practice that demonstrate how we live in connected fields through which information can be exchanged. As is often observed, patients who are the subject of discussion in an expert group with an interest in the matter at hand may appear to have ideas, experiences or knowledge in the next session as if they had been present and unconsciously listened in on the discussions for which they were the subject.

The knowing field is recognised as an everyday experience in the aforementioned constellation works based on Jacob Moreno's psychodrama or the spatial-symbolic implementation of his "sociogram", a concept that has gained increasing popularity, especially amongst colleagues with a systemic background and major experts in the field of constellations (such as Albert Mahr, born 1943). Representatives appear to be able to grasp a certain level of information of the imagined, interactive system and verbalise information that the constellating person is not privy to, in a subconscious manner without any previous knowledge. This process is yet to be scientifically recognised.

Carl Gustav Jung (1875–1961), a former student of Freud, relied solely on "inner experience" to acquire knowledge. With the notions of synchronicity and the collective unconscious, he introduced an analogy to the connective understanding of reality to contemporary psychotherapy, which remained heavily influenced by the Enlightenment and used "intuition" to gain access to the spiritual dimension. This enhancement of the field was accepted by neither Freud nor the mainstream.

Jung's model distinguishes four main psychological functions: thinking, feeling, sensation (as sensual recognition of the outside world) and intuition. It is the task of each individual to achieve a balance between these functions and, above all, to develop and integrate their previously unexperienced aspects.

Roberto Assagioli (1888–1974), an Italian psychiatrist and psychotherapist who initially based his work on Jungian concepts, attempted to unite the psychoanalytically oriented currents of his time, which, to him, required expansion with regard to worldviews and concepts of man, with spiritual wisdom traditions. Assagioli established psychosynthesis, which is primarily concerned with the integration of partial aspects to facilitate personal growth.

Arnold Mindell (born 1940) started his career as an American physicist and then became a Jungian psychologist and teaching analyst at the C.G Jung Institute in Zurich, author and initiator of "process-oriented psychology". He combines quantum physics and systems theory, is convinced of the importance of "deep democracy" and works with the subtle aspects of experiences, primarily with dreams, physical symptoms, and with patients who have suffered comas or near-death experiences. To him, a solution can be found in the "dance of connectivity", which revolves around connected knowledge in which man is embedded

Now we return to current practices: Over the last century, meditation practices from the Far East have been embraced and adopted in a primarily secular form in the western world, especially in psychotherapy, in waves. Claudio Naranjo, Robert Ornstein, Hugo Enomiya-Lasalle, Francois-Albert Viallet, Charlotte Joko Beck,

Jack Kornfield, Karlfried Graf Dürckheim, Thich Nhat Hanh, the psychiatrist Christian Scharfetter and many others were pioneers in this regard.

Mindfulness beginners tend to benefit physiologically from stress regulation, an increase in psychovegetative, psychosomatic and mental strengthening and improved resilience. The pioneers mentioned above naturally aimed to achieve much more in this regard.

For instance, inner distancing and the training of inner observers, which increases the scope for inner decision-making, may be important for the subsequent development of beginners. There are a number of mindfulness practices, many that incorporate sensory grounding. The majority of modern therapists are most likely familiar with these techniques, especially those who work with structurally weak people. This positive, proximal basic attitude – when combined with a stabilising therapeutic relationship – promotes new, resource-oriented self-reference and identity development. During this process, the resource that emerges is the ability to take a pause, gain freedom and increase control over pressing wishes and needs.

Since the 1970s/1980s, meditation has formed an almost mandatory part of clinical treatment in the humanistic Gestalt therapeutic clinical concepts in the fields of psychotherapy, psychosomatics and psychiatry. Movement meditation is preferred over the silent form for patients that exhibit behaviours similar to psychosis as it prevents boundaries from dissipating whilst nevertheless enhancing feelings of self-efficacy.

Originally, meditation was a religious exercise used to induce a particularly awake state of consciousness beyond average daytime consciousness; this permits the individual to experience participation in a "one" or "whole" that infinitely transcends him for a short amount of time. As such, the centre of the world and the individual's own centre subjectively coincide. Unlike in religious communities, there are no textual interpretive definitions for this experience in the Zen tradition. Meditation itself is only a way in which to train consciousness. Generally speaking, in meditation, interest is initially fixed on rituals (looking at objects, counting breaths, changing mantras etc.). Those who practise meditation often benefit from tranquillity, the ability to distance themselves, calm, inner security and peace. Therapists who are experienced in meditation often encourage their patients to enter this safe calm, where it is easier to untangle emotions. They then act as a guide in this respect. The next step may be to invite patients who are adequately structurally developed to meditate.

In the USA, molecular biologist Jon Kabat Zinn succeeded in incorporating the enthusiasm he held for meditation in his earlier years into the health care system. He has established a "Stress Reduction Clinic", combined mindfulness exercises adopted from Hatha Yoga, Vipassana and Zen into an 8-week course in mindfulness meditation and gained scientific recognition for his Mindfulness-based Stress Reduction programme through accompany research with primarily physiological parameters. His work has been met with great response and gained loyal, continued support.

It was not initially clear that the cognitive-behavioural therapy community would engage in meditative methods, as they do not typically approach primarily reflexological or typical cognitive thinking approaches. At most, there is a link between the relaxation training to reduce anxiety (with the aid of the "......reflex") in therapy designed to confront anxiety and the vegetative stabilisation of the Mindfulness-based Stress Reduction programme.

However, the flexibility and practice-oriented nature of the CBT community to open the door to mindfulness is commendable. Firm inclusion of meditation throughout the different levels is a clear advantage for psychotherapy. Meditation increases inner security and autonomy and therefore the ability to resist stress. In the case of structural weaknesses, meditation provides the body with the opportunity to revert to its basic rhythm and consequently strengthen its internal connecting structures. In more developed personalities, meditation encourages inner regulation, order and differentiation of moral concepts.

2.1.6 Selected Individual Phenomena from the World of Consciousness

2.1.6.1 Daytime-Awake Consciousness

"Consciousness is consciously being", writes Scharfetter ([11], p. 49). In the case of Descartes, the substantiation of "consciousness", or conscientia or "res cogitans", does not exclude the possibility that consciousness is directed, intentional or listening. The conscious person experiences himself and the world to which he is directed. This is an activity.

Both common sense and the critical and demanding scientific intellect are primarily assigned to the undisturbed activity of the prefrontal cortex in terms of the brain's physiology.

Daytime consciousness with normal awareness and intelligence typically results in beta waves (13–21 Hz) in the EEG. As such, it appears to exhibit normal levels of self-confidence and abilities to perceive and act. When transitioning to light relaxation, the waves slow down into the alpha range (8–12 Hz). In the face of increased stress or alertness, they typically rise to 21–38 Hz.

2.1.6.2 "Superconsciousness"

The psychiatrist Scharfetter described and, through scientific research, became aware of the phenomenon of increased and expanded consciousness, not only through stimulation by hallucinogenic drugs but also in the scope of breathwork experiments pioneered by Stanislav Grof and pure meditation experiences.

In the 1970s, it quickly emerged that it was time for a categorical leap. At that time, in his book "General Psychopathology" (1st ed. 1976 ff) Christian Scharfetter presented a level of consciousness above daytime consciousness, a kind of "superconsciousness" described in the context of contemplative-meditative states. This concept was not supported by psychiatry professors or textbook writers, who dared not to accept the idea. It marked a decisive move against the mainstream. Perhaps there was simply no one with relevant experience in the field at the time. One key disadvantage to superconsciousness experiences is that they occur in silence and, as such, no one has the opportunity to witness them through another sense, as is often the case for certain other excellence achieved by others after years of training. For example, gymnastics are watched in awe and musicians enchant, etc. It is only in recent years that phenomena have been observed in the EEG for long-term mediators, which was otherwise deemed impossible: frontal, high-amplitude, coherent gamma frequency fields (40–100 Hz). This has also been corroborated by increases in cortical grey matter. The question of which experiences can be attributed to these recordings remains a sensitive matter as experience cannot be verified and is regarded as soft data.

The mere fact that a cooperation between representatives from different worlds has occurred is a notable scientific achievement in my opinion: the world of western neurophysiology on the one hand (Wolf Singer et al.) and the open-minded yet nevertheless tradition-oriented world of the teachings of the Dalai Lama on the other.

The above phenomenon has been linked to traditional Buddhist meditation with its strengthening goal of achieving happiness amongst all beings, a goal that is perhaps comparable to the strengthening basic attitude of Martin Buber's "I-Thou" and the Christian commandment "thou shalt love thy neighbour as thyself". The strengthening altruism found in imaging methods is not the same as a purely empathic basic attitude that incorporates empathic compassion and causes painful experiences to resonate in the brain. In a markedly differentiated approach, Matthieu Ricard has dealt with the difference between empathy and an all-embracing good attitude. Strengthening altruism fosters encouragement. It is of crucial importance that we are aware of this difference and consciously integrate it into our therapeutic approach.

Between 1982 and 1996, Elmer E. Green and his team at the Menninger Clinic in Topeka, Kansas discovered, amongst other things, increased electrostatic charges and increased, alpha wave discharge during the therapeutic process through their ground-breaking copper wall research with therapists. So called due to the fact that experiments were performed in a laboratory with copper-lined walls. The laboratory was also equipped with a bar magnet suspended over the head of the patient with the north pole facing upwards. The chair on which the patient was seated was also electrostatically insulated from the earth. Astonishingly high values were recorded. However, this research was not able to determine whether the ability to increase electrostatic and electromagnetic potential through occasional wave discharges was developed through training or simply innate. A number of the subjects who exhibited notably high potential were apparently known in advance for their "healer" characteristic. In another laboratory led by Popp, radiation that could be slightly increased in humans on purpose, as quanta or biophotons, was captured by a residual light amplifier.

Elmer Green's research was inspired by the "Tibetan system" described by a meditation teacher in the previously British newspaper publisher in India in 1882 (A.P. Sinnett): Each monastery features a dark room with a northside clad in polished copper disc. In front of this wall is an area for meditation, separated by glass and a free-floating bar magnet suspended above. Those seeking to become a monk find this arrangement helpful to reaching an experience of higher states of consciousness through their meditation than would otherwise be achieved without this external aid.

This research implies that the emergence and intensity of qualities of consciousness and the free availability of connectable (bio)quanta are directly related.

2.1.6.3 Preconsciousness

Freud first used the term "preconscious" to refer to all thoughts that can be called to mind at any time, but the individual is not actively thinking of in his first topographical model prior to 1923. These thoughts are (1) thoughts that were previously unconscious but have since been acknowledged due to a trigger, (2) processes that have been consciously learned or experienced, however, have since been performed routinely, such as counting, reading, writing, learning words in a foreign language, certain standard forms of dancing, swimming, riding a bike, driving a car, social greeting rituals, etc., (3) on the other hand, an individual may initially pick up something indirectly and only become aware of it at a later point, such as someone may appear to have been watching the front door from across the street when the "absent-minded" person left the house and greeted the person who was initially overlooked. (4) Atmospheric aspects may sometimes only be registered subliminally and incidentally.

2.1.6.4 Individual Unconscious

Behavioural patterns and ways of processing experiences that we learned consciously in the first place or primarily imitated unconsciously to cope with life and adjust to our cultural milieu, which resonated with us or provided great benefit and which have turned into a sure-fire success method in their entirety or partially form helpful, automated patterns that have since become unconscious (e.g. speaking, reading and writing, manual dexterity, driving a car, dancing, social behaviour). Unconscious, internalised encouragements and realistic, positive self-suggestion typically act as valuable resources. However, the same process also occurs with negative and destructive omens.

Depending on the environment, these programmes may appear to be inconsistent, dysfunctional and in need of correction in a different reference framework. Usually, these "programmes" can be modified and adapted in line with new experiences. Fixed automatisms in a changed environment can have a disruptive effect and lead to abnormal behaviour. As such, in therapy, it is key that the therapist determines which information the patient is unable or not allowed to let go of. Is the information linked to a fear of losing loyalties? Is it a symbol of a survival strategy that has been obtained through pain? Is it part of an unconscious conflict?

The frontier to the unconscious on top form: both our everyday life and the lively internal-external contact experienced during a therapy session occur at the "frontier" between consciousness and unconsciousness. Healthy "life on the frontier" is characterised by playful, creative, multimodal associations and experimentation. The more actively the frontier is passed, the easier it is to contact submerged automatisms in order to review their suitability and vary or adopt unsuitable concepts from previous experiences in a more intensive manner.

In the case of individuals who have erected symbolic barriers to safeguard their inner emotional world, the reason behind these barriers and the role they perform

2

must be questioned first and foremost. As such, it is crucial that the constellation that led to this emotional numbing is understood. This then needs to be empathically translated back into the experience, transformed into a relationship capable of dialogue, and what was previously unexpressed and unheard needs to be voiced.

It is precisely at this frontier between consciousness and unconsciousness that the elsewhere described phenomenon of experimental confrontation occurs. This refers to a highly effective, healing cooperation between the therapist and the patient, which requires the prior establishment of a special, existentially stable therapeutic relationship (as per "I-Thou").

2.1.6.5 Co-conscious and Subconscious

E. Pöppel (p. 169) postulated the existence of a co-conscious that unconsciously prepares conscious thought processes and consequently ensures mental links.

Accordingly, the unconscious helps the conscious.

Furthermore, Pöppel (p. 170) also refers to all of the processes of the neural machinery that we are unaware of despite the fact that they form the elementary building blocks of consciousness.

2.1.6.6 Collective Unconscious

The collective unconscious was first introduced to the field of psychotherapy by Carl Gustav Jung (1875–1961). It refers to an objective psyche based on previous findings. Jung's collective unconscious is an expansion of Freud's personal unconscious, filled with repressed images, with an all-encompassing connecting unconscious. According to Jung, the collective unconscious is populated by experiences that transcend time and culture and individuals live out its symbols through their own experiences. In other words, a shared treasure trove of ideas. Furthermore, it is populated by archetypes such as the Great Mother, the Anima/Animus, the Child, the Hero, the Wise Old Man, the Shadow, the Self etc., which, as structural principles, enable us to develop ideas about certain basic patterns.

Defined as a seemingly intuitive psyche, Jung's collective unconscious is closely linked to ancient concepts and attitudes from the Far East described above on the capability of the individual to read the information in the superior "bigger picture". However, a certain level of integration, centricity and development is required to achieve this higher sensory awareness.

2.1.6.7 Dreams, Dream Experience, Dream Content and Dream Interpretation

Medicine and science are yet to reach a consensus on the function of dreams. It would appear that aspects of our everyday experiences and additional physiological play a role in modulating our dreams. Dreams also seem to occur more often during rapid eye movement (REM) sleep than in non-REM sleep.

The contents of our dreams. Books on how to interpret dreams that allocate certain meaning to specific images have long been in circulation. They were common in antiquity, during the Middle Ages, and remain popular today. Generally speaking, these attempts by popular science to interpret dreams largely mirror the contemporary way of thinking and the way in which experiences were processed at the time of publication. As such, they cannot be regarded as generally applicable.

Freud's "The Interpretation of Dreams" (1900) was the pioneering masterpiece of the century. To Freud, dreams were the "via regia" to the unconscious. He exemplified the key points of his teachings on conflict and defence through his work on dreams. He viewed dreams as wish fulfilments. According to libido theory, Freud believed in fixed classifications with regard to the symbolism of male and female shaped objects. Since then, this concretist readiness to interpret has been abandoned and is now considered a remnant of the past. "External interpretation" also receded into the background at a much later date. However, this process may not still be fully complete.

The long-term development trend within and, especially, outside psychoanalysis, primarily in depth psychology, humanistic and Gestalt therapy, was (1) aimed at analysing aspects of relationships and (2) the use of self-interpreted chains of association within the scope of "subject-level dream work".

Practical Tip

As is often the case for Gestalt therapy, success was achieved through the recovery of all partial aspects as if they were living, independent, interacting partial identities of the dreamer through identification. Defensive mechanisms often clearly show that one mechanism is concealing, eliminating or allowing the dreamer to forget a certain aspect, for instance. Dream work is primarily concerned with images that include the strongest emotional or energetic objects. All details are examined in association with each other in the order of significance they are held by the dreamer. In the majority of cases, this associative exploration enables the individual to touch upon and unlock their dream Ego, followed by other people, animals, plants and finally the dead area. This vaguely corresponds to the order of inner distancing, or defamiliarisation, when partial aspects of the dreamer's personality have been cast into the dead area. The missing aspects without which there appears to be no solution are of particular note in this regard. The therapist may then request the following: Please take on the role of the missing aspect. Why has this aspect been cast aside? What does it hope to achieve? What does want the most? ...It is often beneficial if the dreamer assumes the role of the director (i.e. his unconscious) to explain the dream and starts each sentence with "I initiated this or that" or "I let this happen...". This enables the dreamer, in the role of the "unconscious director", to assume responsibility for the events and formulate them in a provoking, active way. Following this, the therapist can also ask about the daydream image that occurs once the dream appears to be interrupted and broken down. What would the next image have been? Is it the image you dread, or perhaps one you would like to see? Do you have the ability to influence it at all? The objective of dream experiences is to reintegrate the adequately transformed impulses into the inner workings of the personality.

2

2.1.6.8 Lucid Dreaming

Lucid dreaming is a type of dreaming whereby the dreamer becomes reflexively aware that they are dreaming and is thus able to control the further course of the dream. This can inspire creative actions. Aristotle is said to have had this ability. Furthermore, León d'Hervey de Saint-Denys (nineteenth century), Frederik van Eeden (twentieth century) and Paul Tholey [12] were also pioneers in the field. The control and internalisation of movement patterns during sleep is also something that many have tried to apply to sports training, based on the work of Tholey.

"Dream yoga" that developed within Buddhism is based on the ability to dream lucidly and aims to expand on this ability to control dreams.

2.1.6.9 Hypnoidal/Waking Trances, Hypnosis and Posthypnotic Suggestions, and Hypnotherapy

Hypnoidal or waking trances are states of limited consciousness and narrowed, possible inward-looking awareness. In psychotherapy, they are most commonly employed in regression work, which is part of Gestalt therapy (e.g. biography and inner child work, processing dreams and daydreaming).

Tip

It is important that the subject is able to particularly rely on a safe, trusting therapeutic relationship in these states. Therefore, the therapist is tasked with ensuring good rapport with the subject beforehand. This primarily takes place through an exchange of non-verbal signals as opposed to verbal requests. But both are conceivable. Accordingly, even during ongoing group work, if someone requires a regressive trip back into their childhood in order to reserve a conflict, I first look at him with an inquiring, encouraging glance to enable him to establish that he can trust me and that we have a bridge of understanding between us, and that he can trust me to guarantee the required level of protection.

In terms of brain physiology, a trance is characterised by reduced activity in the area that monitors whether a certain reaction fits our self-image and is appropriate, otherwise known as the prefrontal cortex, which is responsible for critical thinking. Whilst there is a risk of abuse occurring during a trance, it permits more freedom and penetrability for hidden traces of the past or representations of physical experiences for associative thinking. It enables repelled memories to emerge easier.

As such, the hypnotist's suggestions are no longer subject to critical examination. Provided they do not contradict fundamental beliefs on notions such as shooting someone, post-hypnotic suggestions are carried out even if they may seem completely nonsensical to the awake subject such as, for example, to approach someone on the street precisely when the clock strikes 12 to ask them what time it is. When asked this, the person in question will attempt to use rational thoughts to make excuses or shrug his shoulder and self-consciously move on with his daily business. In one experiment (from back when I was a student), when tasked with the post-hypnotic assignment of shooting someone with a rifle (which wasn't loaded, but the subject had no way to know this), the subject collapsed when he was handed the rifle. In this instance, the psychosomatic overreaction controlled by the unconscious was triggered by the existential, unconscious conflict. This test setting was also used during the post-war period to determine the soldier's fate between obedience, refusal of order and existential fear, amongst other aspects.

In the nineteenth century, hypnosis was regarded as the most promising method for treating mental illness. Charcot inspired his students with fascinating demonstrations that received high public acclaim during his psychiatry lectures in Paris. However, it was precisely this showmanship that brought the practice of hypnosis into disrepute and highlighted how it could be easily abused. It became a feature of carnivals. Those under hypnosis could be instructed to behave in a way that they would not permit of their own accord. Hypnotised people were accordingly often subject to public ridicule.

Milton Erickson (1901–1980) advocated a personalised, primarily resource-oriented approach with his outwardly unspectacular hypnotherapy, which was capable of using the unconscious as a "place" of creative self-healing during waking trances. (his methods influenced the Palo Alto group, systemic family therapy, neuro-linguistic programming (NLP) and system structural constellations, amongst others)

Self-hypnosis, autogenic training,

Autogenic training (1932) is an autosuggestive relaxation technique derived from hypnosis by J.H. Schulz (1884–1970) and was regarded as the reliable standard anti-stress method for decades.

"Graded active hypnosis" is an effective compromise solution between hetero-hypnosis and autosuggestion, with reduced potential for abuse, developed by E. Kretschmer (1888–1964) and further refined by D. Langen (1913–1980).

2.1.6.10 Inner Observer

The presence of an inner (ideally) non-judgemental observer is helpful in the inner relationship to supplement the establishment of an impulse which facilitates measured restraint with regard to the inner hierarchies of needs and the preconceived, polar inner reactions and enables the individual to no longer feel at the mercy thereof. Research has demonstrated that there is a hierarchy of inner observers who overtake each other during development. The observers belong to the internal structures that determine an individual's degree of will, control, freedom and decision-making. They can also prevent automatic judgement with regard to whether a process appears to be beneficial or unfavourable for the individual.

The instruction of inner observers is regarded as one of the characteristics of developing differentiation.

2.1.6.11 Flow Experiences

The fascinating "flow" experience, the focus of a large body of work by Mihály Csikszentmihály, has become a widely popular phenomenon: Charismatic, free speakers and, especially, improved artists, trustingly rely on this flow and deeply treasure their ability to have relevant, unconscious material flow to them. They feel part

2

of the energetic flow. It supports them and allows them to trustingly devote themselves to their speeches. The flow brings suitable ideas and intuitive solutions that fill and inspire with delight to the fore. Being in the "flow" is associated with a strong "I-Am" feeling, which is paradoxically a feeling of participation, of involvement in a larger flow of energy. The individual is faced with neither boredom nor fear of being overwhelmed. In this regard, the subject must possess a certain tolerance of uncertainty as he never knows what is coming next. He does not make the plans yet nevertheless still feels in control. He plays and gives himself over to the self-organising game, simultaneously assuming an active and passive role. At most, he is able to approximately guess the rough frame of reference for the next action. The brains of children and adults alike can enjoy their creative potential during free play.

Neurophysiologically speaking, this process is most likely a cardio-thalamocortical coherence phenomenon that features the typical increased parasympathetic tone, or a relatively slower heart rate and increased heart rate variability. Furthermore, the vagus may be stimulated by breathing slowly (once every 10 seconds). Heart-brain interactions occur in the area of the vagus nerve, which runs from the heart to the limbic system. The "subdominant" hemisphere of the brain appears to take the lead in this regard. It is typically associated with happiness and a cheerful, appreciative underlying mood.

Therapies with relationships that occur in a similar, common flow appear to be transparent, reliable, stabilising, relaxed, and at times even pleasant, coherent, meaningful and particularly effective for each party as a process.

The debilitating opposite pole: "Anti-Flow", when the free-floating quantum stream primed for connecting dries up. Could "fear" be the equivalent of a conscious experience?

What is the reverse of "Flow"? A high level of anxiety seemingly blocks the ability to spontaneously "read", pick up or verbalise suggestions from within. If observations are permanently characterised by anxiety and mistrust or fearful behaviour is exhibited, quantum networks are more susceptible to collapsing. As the popular saying goes: "Fear makes us weak". Having a blackout or only seeing a wall of white is typically viewed as a sign of the blockage of inner, spontaneous self-organisation.

Klaus Conrad, psychiatrist and Gestalt psychologist from Göttingen in the field of phenomenological anthropology accumulated an impressive, historical collection of patients with regard to anxiety-induced intrapsychic numbing. A major contribution was also provided by Karl Jaspers and the Heidelberg School's psychopathology and its investigations into the initial manifestations of schizophrenia amongst soldiers during World War II. The psychotic episode that featured delusions followed a stage of fearful observation with an increase in abnormal importance placed on particular individual events in each case. Following this, the loss of freedom enabled the individual to change the existing reference system and perspective at any time as he pleased in order to recognise that an unassignable sensation was caused elsewhere and that it is thus harmless and can be given the all-clear, for instance. A certain level of ability to constantly engage in an inner, flexible change of perspective is a prerequisite to a healthy mind. High levels of anxiety prevent (any further) changes in perspective.

Conrad's observation also corroborates the impression gained by other psychiatrists that the adolescent initial manifestation of schizophrenia is typically preceded by an "energetic loss of potential" (Glatzel).

We still do not entirely understand schizophrenia. It does not appear to be an object in terms of diagnostics, psychopathological condition, causes, neuropathology, the extent of genetic or environmental causes, the course, the responsiveness to drugs and other therapies, as well as the outcome. This is known as unspecified diagnostic vulnerability. Social psychiatry, and Luc Ciompi's Soteria approach, has taught us the importance of ensuring that patients exhibiting initial manifestations of psychosis are able to return to socially salutary networks of relationships to get the opportunity to be restored to their original condition (restitutio ad integrum). What can actually be healed in a social setting? Is it the same agent that resulted in the high infant death rates amongst babies observed by René Spitz due to anaclitic depression after weeks or no stimulation or eye contact? Is it excessive, stress-induced apoptosis (programmed cell death) caused by the underproduction of new cells and subsequent low availability of biophotons which are required to establish a dynamic, conscious ego field of awareness? (Hypothesis)

The association test that was popular at the end of the nineteenth century, which was partially interpreted in Freudian terms (Jung even wrote a thesis on the association test), may have already indicated the inadequate inner network but nevertheless permitted other perspectives on the altered chains of associations. Kraepelin obtained the test from Wundt. The way in which Kraepelin and Blueler understood schizophrenia was primarily based on a mental model of dissociation, division and inadequate connections. According to Scharfetter [13], all schizophrenias exhibit pathological limitation of the ego/self-awareness. This refers to the pathology of the identity, demarcation, coherence-consistency, activity and vitality, a functional unit that relies on a dynamic and, in particular, internally connected force field, which is understood and observed as an energetic whole according to H. Stapp and Tononi, and typically also experienced as natural wholeness.

2.1.7 The Variable Perspective of the Inner, Conscious Observer on the Hierarchy of Integral Vortex Formation

Our understanding of vortex formation is based on the observations of the brain researchers Edelman and Tononi, who likened the energy field that surrounds the brain to the large-scale weather patterns around the world that can be observed from satellites, in which atmospheric currents and vortexes are observed similarly to the circular connecting paths used to describe the limbic system. Of course, this image is only intended as a parable. However, it is capable of illustrating the notion that consciousness most likely requires a decentralised location and perceptual differentiation to the main event (a circular current or vortex). On the one hand, the inner flow patterns are subject to change during anatomical development. Yet, on the other hand, the location of the observer appears to be inherently variable, but nevertheless susceptible to being shaped by emotionally significant experiences. As such, there are preferred, inner perspectives and "locations". The consciousness responsible for determining the identity is focusing in different inner "bases". The main energetic location varies in our brain. It is modeling it by its organizational patterns. Someone, who likes sports or is playing the violin or is susceptible to addiction etc differs and evolves his brain in another way. The respective observation location or "seat of control" where the otherwise free-floating, energetic field is temporarily centred as consciousness, and relatively or semi-rooted, shapes the individual's identity and his perspective of the entire system.

Where are we most at home in our own body-mind-spirit cosmos? Each individual adopts their own different, preferred identification for each situation and stage of life. It is not crucial to determine whether this concerns a home for the consciousness or rather only a pattern of consciousness capabilities with unconscious experience and behavioural potential that has previously proven rewarding. This remains open to speculation.

Now we shall take a look at the various, prominent levels which can be found described in more detail in ► Sect. 3.7.

Firstly: Energetic, physical, functional desire, joy in being pure and identification with physical strength, functionality, dexterity and adopted, particular patterns of movement (such as running, swimming, dancing, etc.) are typical aspects of youth development that are intensified through every kind of sport. The ability to experience life in a body that has developed full functionality becomes the core of the self and self-consciousness, even more so if social acceptance is experienced as a reward. There are two sides to the delight experienced in knuckling down and the palpable effect – be it with constructive or destructive aggression (ad-greddi) – which can trigger fight or fight reactions in some.

The energetic negative side also includes equivalent painful experiences with regard to physical weakness and disabilities, such as the experience of a child who lost a leg in a minefield and can no longer run around and play with others. Or somewhat less spectacularly: experiences of weakness when suffering from a viral infection. Consequently, the decline of this level is often observed in the elderly, who attempt to gradually de-identify themselves from their ageing bodies and instead find their home in other, inner locations. The existence of other locations to choose from is, in itself, a good sign.

Secondly: The world of fulfilled sensual pleasures, including the need for trusting, social comfort offers a true, sensual and grounded paradise on earth. Satisfying basic needs such as eating and drinking, sleeping, clothing and having a safe roof over your head, as well as sexual satisfaction and the ability to have children if so desired is largely responsible for forming the level of expectations and the benchmark for a successful, happy life at this level. This level is characterised by a number of ups and downs between building and satisfying needs and requires balanced regulation. One of the strong regulating factors is the social surroundings of the group, which guarantee care but also set limits and encourage regulation. Conscious moderation becomes a skill and virtue. A lack thereof brings us to the negative side:

As children, we all learn that getting enough of something is a good thing, which is subsequently usually acknowledged with gratitude and can cause us to feel balanced. If this leads to us feeling just as good as when our needs were satisfied, then we have a good chance of being able to properly respond to the ups and downs of the changing deprivation and the satisfaction thereof. In this regard, there is a supreme inner authority that does not struggle to find the right balance and prevents the individual from falling into an emotional abyss once the satisfaction ends because nothing and nobody can make the satisfaction last and provide a sense of encouragement. Tender regulation is comparable to subtle solace; in successful cases, it is internalised. It clearly constructs a field between the regulating authority (the internalised mother) and the partial aspect that requires regulation, which never completely breaks away, similarly to how contact between an infant and mother should remain continuously palpable. Everything that applies to attachment theory in interpersonal external relationships also applies to internal ones; bonds can be secure, insecure and chaotic.

We are all aware of the many variants on the failure to develop this inner relationship: Alcoholism, binge eating, sex addiction, drug use, unwillingness to follow natural sleep cycles, the rhythm between exertion and relaxation, between proximity and distance, also the inability to control impulses against aggressive breakouts; more subtle, indirect rewards are observed in the case of workaholism. In the case of egotism, rewards are directly obtained from the social surroundings which can drown out the perceived lack of significance. Deriving joy from the devaluation of others appears to be related to this. Effective therapies are able to rectify strong internalisations through a captivating and dignified therapeutic relationship. In this regard, ex-users are able to convey hope in a more credible manner than others. They are more likely to subsequently internalise than others. Addiction therapies that rely solely on internalising external discipline and limitations only enjoy limited success.

Thirdly: Now we shall follow the inner location to the ego-enhancing level of experiences of power that provide structure and impulses. We shall adopt a neutral manner to fully explore this increase in functional desire. On the physical level, these experiences are linked to the pleasure of being able to control a well-functioning body, model it functionally and demand extreme and unusual performance from it such as extreme and competitive sports, gymnastics, ballet, martial arts, etc. Here, the focus is not primarily on a desire to live in the body, as in the basal level, but rather to have access to it. The body is regarded as a tool that can transform the individual's surroundings or a horse to be ridden. It turns into a performer worthy of pride. A power struggle is found both externally and internally. This struggle is also demonstrated when it comes to basic needs, which can be trained to achieve a different or higher goal.

The desire to accept and overcome challenges is also observed in the intellectual realm: Planning, calculating, establishing sets of rules, logical reasoning, erecting consistent edifices of ideas and inventive constructions up to artificial intelligence. The above realm constitutes the unfathomably large, rational domain. It is constructed and developed through pooled intelligence or focused attention energy. Intelligence (the mind) relies on the processing of the sensations that are captured by the neuronal filter according to the on/off principle and verbalised in the dominant hemisphere of the brain or precisely mathematically further processed according to the either/or principle. This form of working capacity may also be emotionally supported by willpower. In this regard, a man comes forth as the creative master of the world, who makes this world his subject. There is a strong risk that this world may be destroyed if he only develops awareness for his basic needs and the intellectual potential to satisfy these.

The will to rule can also be observed in social contexts, provided it is not softened through empathy and simultaneous identification with the concerns of others, an ability that should have been developed during the previous phase and thus transferred to this phase. If this is not the case, a grim, dissocial scenario can arise, featuring abuse of power, exploitation, incapacitation, coldness, cruelty etc., ultimately leading to sadism. If this destructive rage is directed inwards, there is also a risk of suicide. If this rage only sets out to destroy emotional parts, these will fall by the wayside and obsessive-compulsive symptoms may set in.

When paired with awakening sexuality, power plays may also be presented on the newly discovered stage of the reference groups with an attention-seeking attitude. They may also materialise through the desire to compete for power with rivals for the most beautiful partners. This combination of fighting spirit and sexual appetite is common throughout the animal kingdom, with many notable examples. Behind these fights for a higher rank is the idea that the most vigorous specimens have the highest likelihood of reproducing.

Fourthly: The ability to resonate with the different rhythm and frequency levels of others is something that develops and displays a readiness to respond from birth. Initially, it ensures that the carer provides adequate care for the child who requires protection. At the same time, the capacity for empathy and personal love represents a key opposite pole to the autonomous self-centredness touched upon in the previous phase. Empathy entails that an individual is able to put himself in another's shoes. The ability to love is an active, altruistic attitude based on emotions, which is generally oriented around supporting the life and development of beings, initially of others, but potentially also the individual himself. Love inhabits the lucid interspace of the I and Thou, ultimately the We. Everyone should live; everyone should succeed and be happy.

Now let's return to the very beginning. Hearing a mother's soothing, familiar heartbeat signifies security for a newborn baby. Breathing and the heartbeat form a link with each other and with those of the mother or carer. Lullabies further deepen this natural rhythm-finding process across a range of cultures. At the same time, a sensitive attunement process (D. Stern) takes place from the moment of birth between the mother and the "competent" infant [14]. Internal regulatory systems, trust in the world itself and the willingness to form bonds with people all develop to safeguard this common, coordinated oscillation pattern. As such, interactive games turn into a source of fun. The ability to become affirmative ultimately culminates in the ability to love, which has a beneficial effect both on inner psychological regulation and integration from birth, as well as on the acceptance of other beings to ensure that relationships can always be shaped from the basal level of the heart and decisions are always made from an integrated centre of the personality. This affirmative centre that is capable of love can be consciously experienced.

The typical consciousness experience of this level is commonly referred to as: "To be of one heart and soul"; "to open someone's heart"; "to steal someone's heart; "to hold someone in your heart"; "to love with all your heart"; "to lose your heart to someone; "to break someone's heart"; "to be cold hearted"; "to have a heart like a stone"; "to feel a stone fall from your heart"; "to feel your heart leap for you".

These phrases all pertain to the partial synchronisation of an individual's mental core with that or another. We are all aware that the heart possesses the strongest energy field of all organs. Its pulses can be felt in all cells, they coordinate the rhythms of other subsystems and are also tangible outside the body to other people. During synchronisation, parallels oscillate with each other and receive confirmation from this resonance. Aspects that are less suited are deemed less significant and pushed to the edge of consciousness. The oscillation results in the creation of a new whole, a couple, or perhaps even a community of destiny. Groups can also partially grow together through a shared experience.

Now we shall take a short foray into the principles of integration and separation: A similar process to which, albeit it on a much simpler level of complexity, can be observed when two neurons connect. When this occurs, they fuse to each other and are thus "wired", regardless of whether they will be assigned to Pavlov's dog-feeding example of the conditional and unconditional reflex, or whether (in line with Freudian thinking) a conflictual, excluded part of the experience, such as sexual desire, is combined with a situationally random object such as a dog that runs past and stored in a fused manner. In both instances, two primarily independent events that are inherently unrelated are combined to form one unit (creation of an experience) due to the simultaneous experience as per Gestalt laws (see W. [15]). The way in which these ties are broken down has been studied in depth through behavioural research. Despite the fact that both of the examples above were selected from the level of basic needs, the principle of becoming one (and separation, as required) also applies to a certain extent to Porges' "social system engagement (SSE)" and to making contact, as we shall discuss in more detail later in this book.

In this regard, the primary focus is on awareness of the need, the ability and willingness to be connected and actively able to shape connections, be it to intensify, enable, weaken or even eliminate them entirely.

Fifthly: On another level, the consciousness can also be centred in a beneficial way on the ability to consolidate personal, creative potential and to express it. As such, artists typically find themselves at home at this level. However, this level is less about pure products of art and more about the harmonious way of living, including the individual's take on purpose and moral concepts.

Sixthly: At a further level, the consciousness may be centred on the withdrawal of the individual's own needs and the receptive grasping of the power play of the environment, including all of its information. In this regard, the consciousness seemingly mimics an ear or satellite dish. In Zen, this form of active withdrawal to avoid disturbing the natural power play (wu wei) and also to learn how to "read" information is held in high regard.

Seventhly: Last but not least, the consciousness is also capable of centring itself on the relationship and the connective fields to the respective reference horizon, whatever this may be. Reference horizons shape and create identities or our relationship to them. They may grow with us in line with our own development, they may be lost or consciously eliminated and dismissed, they may also emerge from nowhere like a gift from the previous internal fog, or at least this is how it appears to us, as our inner eye is finally ready to perceive them.

2.2 The Therapist's Personality

2.2.1 "Good" and "Bad" Therapists as Research Findings

As we have yet not apprehended the structure, abilities and limitations of the individual psychotherapists themselves, it may be surmised that good or bad outcomes (with minimal variation) are generally always achieved by the same personalities, regardless of their methods [16, 17].

A large-scale study on the working relationship [18] first indicated that this phenomenon of therapeutic effectiveness was less contingent on the structural condition of the individual or the suitability of the therapist to the patient, but rather on the emotionally positive basic attitude of the therapist. Surprisingly, the original basic attitude of the patient was not found to play a role ([18], p. 125). "The patient relies on the positive attitude of his therapist to develop a strong working relationship, but the therapist does not depend on his patient's attitude...

This raises the question of how the therapist is able to offer his patient a positive relationship that convinces the patient to confidingly open up despite their difficult experiences to date, even if this may still be experienced as a gamble? This tacit invitation should not be confused with process-oriented friendly politeness, professionally polished generosity or comme il faut. The emotionally positive attitude emanates from the core of the therapist – or it does not exist. What should be done if there is no consolidated core? If secret or unsettled conflicts can be found within the therapist? Some life-long in nature, others currently on the cusp? As a professional, should you be capable of removing these inner aspects from therapy? This may be possible to a certain extent. However, in order to convey a credible, strong, calm and positive basic attitude towards patients, valance does not suffice. Patients have fine antennae to detect these aspects. In order to ensure an optimal relationship that enables the patient to feel they are in good hands, there must be scope for regression on the basis of trust, the scope to partially let go, to foster new development and reorganisation. Whilst this may, at first, appear to be a mammoth task. Others are naturally attuned to this concept of man.

We have now arrived at the issue – or perhaps opportunity – that the individual himself is the treatment instrument in the field of psychotherapy. This is a matter of course for the humanist methods; however, this was decidedly not that case in the early days of behavioural therapy. A major rethink appears to have occurred in the past two decades in this regard, which remains to be completely adopted in the mainstream and, in particular, by the older generation. All possible variations between classically promoted, conventional "therapeutic neutrality" on the one hand and more recent, relationship-oriented approaches on the other, whether in the form of Kohut's empathic psychology of the self, relational psychoanalysis or Fonagy's mentalisation concept, have also entered the school of psychodynamics. Although this necessary development trend has been comprehended by the Avant-Garde of both of the latter schools and an impressive commitment has been made to research, the education and training methods for the young generation of psychotherapists have been comparably neglected. Above all, the school of behavioural therapy is yet to fully grasp – or to halt the bashful resistance towards – the significance of adequate self-awareness for the development of the therapist's personality. Why should a psycho-technician who is well equipped with a range of "tools" require self-awareness? This is not to say that the psycho-technician is in need of therapy, but rather that a certain amount of "self-reflection" can be beneficial. However, this does not suffice to true psychotherapy.

What is the best approach in the field of psychoanalysis? Never-ending analysis over long periods of time on the couch still remains tainted by the inaccurate tinge of long-term regression and the typically unresolved relationship of dependency. The way power is used in psychoanalytical institutes and in professional politics has been the subject of complaint about well over half a century. In turn, this has shaped our relationship with power dynamics. Those who are oppressed easily slip into the role of the oppressor and simultaneously identify with the aggressor. Usually, retaining power appears to be the highest-ranking value above all others. This casts a shadow. At present, therapeutical practices lack a culture of equal esteem and appreciation, despite the attempts of the Avant-Garde schools to achieve a counterbalance to a certain extent.

Psychoanalysts have traditionally viewed the school of depth psychology (DP) as inadequately educated, less deep, solely adequate and structure-changing; a school that produces inferior work, the little sister of psychoanalysis. Depth psychology has always been primarily concerned with interactions and everyday events, geared towards crisis management and open to the inclusion of practical know-how from other schools. As such, the school developed a more tolerant style and spirit that places the patient above hypothesis. In contrast to its evaluation by its "big brother", the proven effectiveness of TP, which is the subject of far less research, outperforms that of psychoanalysis.

Humanistic therapists typically develop their therapeutic identity on the basis of a concept of man with a predominant emphasis on fellow humans. This identity is characterised by flat hierarchies, a willingness to experience and find a common understanding on a non-verbal and verbal level, and finally to conscious accept responsibility for one's decisions. If the therapist touches upon previous shortcomings and pain, they will only trigger temporary regression and then be harnessed by the potential of today's adult in the same session. Stress can be transformed into a resource. The mobilisation of resources has been embedded in all humanistic and systemic approaches from the outset. It is used to facilitate the provision of an emotionally positive relationship by the therapist, which is regarded as the top predictor of a successful working relationship [18]. That is not to say that exceptions are not observed even amongst the humanists. However, universal idealisation is not advisable.

Psychotherapists who are required to maintain an emotionally positive relationship throughout therapy first need opportunity to deal with their own shortcomings, pain, unresolved conflict and trauma, etc., in a professional sanctuary to find peace within themselves and gain natural self-acceptance. This simultaneously opens the door to the opportunity to learn by doing. Aside from being the best lesson to learn, it also gives the therapist hope that healing is possible, which can later be convincingly passed on to patients.

In my opinion, the political factions of psychotherapy – more than psychotherapists themselves – turn a blind eye to the fact that psychology/psychotherapy tends to attract a larger number of people who believe psychotherapy is required and beneficial by virtue of their own life experiences than other fields. In other words: people attracted to our field are more likely to have experienced stress in their life.

Tip

A question that is often posed in relevant seminars, especially those concerned with family, is, from the perspective of the former child, which family member needed help at the time, and from today's view, the help of a psychotherapist? At this point, most of the participants' eyes light up and they simultaneously agree that help would have been advantageous and saved a lot of pain.

Hardships, limits or disorders that afflict parents often overwhelm children. Many take on too much in their attempt to be good parents and fail to provide adequate light-heartedness. An entire, separate chapter could be written on experiences of abuse.

In principle, stress is not a contraindication for this approach. It can be used to increase the motivation to do something better and is generally a signal that the individual wants to actively or constructively deal with the stress or take action. All of those who venture down the path of facing their problem areas, whether this occurs at a point that appropriate in preceding or supportive official therapies or through self-awareness, many appear to benefit from the process and may consequently become particularly capable experts in the ailments they once suffered from. I can name a number of good examples in this regard.

However, this experience can also cause therapists to reach a deadlock. An acceptable solution can also be found when a therapist is aware of the problem areas he has not successfully solved for himself and, if these areas crop up in his therapy, the therapist can pass over to his patient without incurring any damage but nevertheless remain true to himself. This approach is likewise worthy of respect.

Cave

However, if the individual is unable to be honest with himself, he should perhaps not study psychotherapy.

The crux in this regard is that the individual is usually unable to recognise this himself. He likely does not seek supervision or intervision, rendering this potentially corrective measure ineffective. The community of insured persons pay a high price if psychotherapists are not adequately consolidated as personalities and continue to (and want to) conceal their need for therapy. Such therapists will subsequently squander chances for their patients. The lack of training does not explicitly pertain to theory or the knowledge of existing therapeutical components, but rather to the convincing personality emanated by the therapist, the ability to act as an adequate counterpart and provide patients with a personal relationship. Of course, self-awareness in individual and group settings (both are required) is a question of cost. There's no doubt in this regard. However, the price for corners cut during education and training is paid many times over as a result of inadequate patient care. Other fields are required to finance expensive equipment. However, in the field of psychotherapy, the psychotherapist himself is the instrument that requires tuning, refinement and cultivation.

Aptitude screenings are yet to be introduced for the intake of students into the field of psychotherapy.

(This was previously not the case in training institutions: in the early 1970s, the Gestalt institute that I trained at (the Fritz Perls Institute) put all applicants through 2 admission interviews, a group weekend during the selection phase and a final admission assessment by the group, both group leaders and the student at the end of the initial trial year of training.)

It is unavoidable that a certain number of individuals who themselves suffer from borderline disorders will register to study psychotherapy. Nevertheless, the establishment of crisis and counselling services and the ability to transfer to other fields and career paths if certain individuals come to the realisation that perhaps psychotherapy is no longer a viable career path for them can provide assistance in this regard.

2.2.2 Restrained or Active Psychotherapists?

In the most recent edition of the *Handbook for Psychotherapy Research and Behavior* (chap. 13), Elliot et al. addresses the question of whether psychotherapists are more effective when they conduct themselves in a restrained or active manner. Success outcomes cannot be achieved at either extreme end of the spectrum. The best results were observed at the mid-point between both approaches when the therapist delicately adjusts their approach to the patient. The groups compared all primarily used the emotion-focused therapy (EFT) method pioneered by Greenberg, which provides the therapist with a certain amount of leeway to act.

Other methods are either more restrained or actively structuring. It is beneficial to the outcome if the patient is already able to sense which approach he may be able to engage in and enjoy during preliminary discussions. In certain cases, the approach suggested by the therapist may be the perfect fit for one patient, yet too forceful or restrained for another. It is key that a relationship is established early on in order to commence with the practical interactional "dance" of the relationship.

2
If this does not occur, this failure will be deliberated at length by the therapist and potentially become a topic of discussion. Does the patient harbour any experiences from previous therapies with other methods? Has the patient mentioned any fears or aspirations from their personal anamnesis? Is it possible to identify any barriers caused by transference? What does countertransference indicate? If I were to reverse my role as a therapist, how would I feel in light of how everything has gone so far?

2.2.3 Relationship-Centred Versus Process-Oriented Therapies

Examination of the contextual research reviews compared to process-oriented ones (2015 Wampold et al., 2018 APA Report) reveals a different understanding of therapeutic competence in each case and the use of different criteria in the evaluation thereof. The process-oriented approaches favoured in behavioural therapy define competence according to loyalty or adherence to the process, i.e. the ability to comply with the rules. Therapists are regarded as well-informed, disciplined psychotechnicians. Other possible aspects of therapeutic interactions are side-lined as they may jeopardise therapy. Instead, the objective is to achieve standardised, technical implementation of the approach. This procedure has been met with positive (salvation) expectations. With this stance, the standardisation of conduct that was previously scientifically researched ensures that the approach will be taken seriously by the target scientific audience and acknowledged as scientifically proven. Accordingly, the understanding that loyalty to the process negatively impacts the final outcome is not compatible with this concept. An understanding that has, if anything, led the mainstream to turn a blind eye.

It has long been understood amongst the Avant-Garde of BT that the therapeutic relationship constitutes a missing element in the psychotechnical approach. They primarily have adopted the approach based on the model of conversational therapy in a similar vein to Klaus Grawe's recommendation from two decades prior. The variety of integral therapeutic relationship approaches towards the patient appears to be very mixed in this regard. If therapists are not able to experience relationship-oriented professional treatment during their education or training (including role reversal expectations), they are rarely in a position to offer this to others. They have not been exposed to learning to doing. Students of the Rogerian approach were and remain required to test, reflect upon and supervise their therapeutic attitude with regard to characteristics of empathy, authenticity and congruence over the course of many years. This practice refines their antennae. Many colleagues (but not all) who only first incorporate the therapeutic relationship into their inner repertoire post hoc impart the impression of a "light form" on their patients. In other words: this is better than no relationship at all

2.2.4 Self-Care, Mental Hygiene, Resource Care Amongst Therapists

Let us start with a proven measure: a sympathetic collegial supervisory group can provide strong support. A constructive atmosphere is key in this regard. However, this does not entail that criticism is not welcomed. A lack of criticism would be a setback that would challenge the purpose of the working group. That being said, easing and therapeutic factors must outweigh the ones that cause stress. Every established colleague should be a member of a supervisory group in which he feels comfortable. Changing groups is often a delicate issue, but must nevertheless remain a possibility. Consensus needs to be reached on the matter of whether therapists should be informed by the same stylistic attitude in order to promote affirmation or, instead, whether a stylistic mix is acceptable due to the ability thereof to inspire new ideas. Those at the beginning of their careers tend to gravitate towards to the first approach, whereas learned practitioners are more likely to lean towards to the second approach.

Over time, supervisory groups may metamorphosise into intervisory groups, whereby the "old timers" replace a supervisor with their skills. Once the "ranking" has been re-established and deemed acceptable, collegiality typically increases. If a relatively high level of solidarity is achieved, colleagues may use these groups to manage personal crises such as how to approach ageing, illness or the loss of an important person in their life. In certain cases, groups may provide more stability than familial relationships.

Self-care is a popular aspect in training groups that feature younger colleagues and the "middle generation". The topic of self-care permits participants to take a deep breath. However, finding the correct criteria for assessing balance is a bigger task than may at first be apparent. The assessment scale is not one-size-fits-all. It varies according to the internal and external context.

The main questions in this regard are: what can I use as a resource, as a source of energy? Do I have any hobbies? What requires energy? Where can I get energy? To what extent can the fact that I fail to adequately acknowledge rewarding experiences be attributed to my internal assessment method? How many positive interactions do I instigate in the world around me? Or how many do I block and consequently wonder why I am surrounded by gloom? How much do I permit myself to complain and to whom?

Young parents are often required to test the limits of their strength and learn to cope with everyday life on little to no sleep due to their children becoming sick every now and then and crying at night. The notion of a day off is simply not conceivable. Ten minutes of peace and quiet becomes a luxury. Others suffer unbearable loneliness, especially colleagues with a migrant background who face constant uncertainty about the safety of the family left behind in war zones. Where can these people find solidarity and empathic interest? How and with whom can they experience regular relief? Can they detach from their situation through writing?

Some people may experience shortcomings in relationships and feel as though their world is turning grey. In this case, it may be beneficial for the individual to take stock of their situation by putting their inner team of helpers into concrete terms in a collage or drawing. This can be compared to assessing a valuable asset, especially when it concerns all of the constructive, life-enhancing and loving experiences of relationships (and parts thereof) in life. What can the individual in question trigger within himself in this regard? Maybe the first time?

A key aspect of everyday life is the rediscovery and potential revival of viable possibilities for relationships. Many things are overlooked in this regard and not on the radar of the assessing awareness, such as the friendly newspaper boy, the dog owner you often come across when walking your dog who you have only shyly greeted thus far, the bus driver who likes to receive a nod but rarely does, the classmate who has been waiting to be invited over for a long time and, last but not least, the housewife, who has long been viewed by the individual in question more as a domestic fixture than a living being with her own world of interests that produces impulses if they perceive the opportunity to resonate. Of course, this begs the cardinal question: what caused the individual to withdraw? Where and to whom was this withdrawal originally directed? What does he most want to express?

In conclusion: colleagues should not feel any shame in admitting that they are overloaded under difficult circumstances and consequently organising time off to concentrate on their own therapy. To simply ignore the deficiency, continue, fail to feel any success, lose passion for work and ultimately end up in a chronically depressive state would be a far worse outcome.

I would advise all self-employed independent workers to get involved in a positive, stabilising and competent supervisory or intervisory group to support their professional endeavours.

2.3 The Therapeutic Relationship as a "Common Factor" – Despite an Abundance of Variants from Different Schools

The therapeutic relationship has thus far been regarded as a "common factor" in research without any further differentiation, as if there are no variations thereon.

2.3.1 Research Statement on the Therapeutic Relationship as a "Common Factor"

The influence of the therapeutic relationship has traditionally been averaged out across all methods as a "common factor" that accounts for 30% of variance in outcomes on average [19].

Note

The Evidence-based Psychotherapy Report compiled by Norcross & Lambert [20] for the APA, (American Psychological Association), concludes: "The psychotherapy relationship makes substantial and consistent contributions to the outcome independent of the type of treatment. Decades of research evidence and clinical experience converge. The relationship works! …"

In light of the vastly different emphasis placed on the therapeutic relationship and the varied introspection on its influence on the therapy process, this mainstream message is both puzzling and leads us to question its significance. On the other hand, this approach does appear to be legitimised by the fact that the variability between the therapeutic relationship models that focus on historic and conservative approaches and the modern Avant-Garde developments is greater than the variability between the individual schools. Almost in parallel, these trends converge in one shared objective, at least with regard to the therapeutic relationship. This is precisely the aspect used to legitimise the postulation of an integral platform for "general psychotherapy".

Furthermore: Wampold and Norcross [21] likewise attributed 30% of the variance in outcomes to the patient. This comprises influences from the patent's structure, his structural weaknesses as well as the impacts of different bonds he has experienced, his motivations, his interactional characteristics and his current crisis condition.

2.3.2 Psychoanalysis: The Developmental Sketch of the Therapeutic Relationship

In Freud's time, the notion of thought was embedded with the Enlightenment ideal of rationality and consequently regarded as a solely scientific aspect. As such, Freud demanded that his successors should act as mirrors to their patients; there was no room for the therapist's personality to enter into the equation. In any case, the work alliance formed with a relatively non-neurotic part of the patient is rational in nature. According to this stance, the therapeutic alliance is skewed on the patient-side through recurring transference reactions such as relationship expectations of a libidinous or aggressive nature as the result of previous, significant relationships that have not (yet) concluded, and thwarted by resistance impulses that favour maintaining the status quote. Countertransference, the overall emotional response to the patient, is touched upon on the therapist-side. After years of training analysis, no neurotic transference should take place on the therapist's side (anymore). Certain psychoanalysts were in favour of analysts adopting increasingly transference-oriented behaviour in order to encourage or accommodate the experience of transference neurosis. A patient requires a personality with an adequately strong ego to achieve the objective of transference neurosis, the clarifying differentiation and initially often pent-up, the emotional potential behind the transference. Those who do not exhibit a strong enough ego may end up coming to a standstill.

Working with transference and resistance processes despite a range of other forms is the core element and connecting commonality in psychoanalysis.

This applies to the psychoanalytical understanding of physical gestures, sequences and other non-verbal processes that have since been gained respect, which were initially (up to almost the end of the twentieth century) understood as avoidance or resistance mechanisms or disparagingly as "acting out" by psychoanalysis, based on the notion of the talking cure. Colleagues who defended the integration of these expressive and action levels were met with fierce opposition from the mainstream for the span of an entire generation (Müller-Braunschweig, Tilman Moser, Heisterkamp, …).

The most tolerated aspect at the time was the further, non-verbal relationship to one's self and others in addition to the direct interaction in special versions of therapy for structurally weak patients (F. Heigl, A. Heigl-Evers: Interactional therapy; G. Rudolf, Structure-oriented psychotherapy). The mainstream was most likely able to tolerate these compromises as they permitted the preservation of the intrinsic, historical understanding of therapy to a certain extent. Whether this tolerance would have been extended by the mainstream if the authors were open to their community about the fact that decisive impulses for their work were obtained from the humanistic group dynamics or from Gestalt therapy, remains open to interpretation. This shines a light on the fact that the interest in mutual, practical know-how that exceeds the boundaries of individual schools indicates significantly greater appreciation and integration possibilities than the proponents of policies based on boundaries would like to delineate.

Much attention has been paid to Thea Bauriedl's concept of relationship analysis as a differentiated bridge between psychoanalysis and systemic therapies: "Auch ohne Couch" [it also works without a couch], Psychoanalysis as relationship theory [22].

The understanding of the implicit relationship knowledge that leads to moments of meeting was developed through mother-infant research. ([23, 24]; Stern erected a spiritual bridge between modern psychoanalysis and Gestalt therapy as a result of his close involvement in the Italian school of Gestalt therapy.)

With regard to integration, the development of the therapeutic relationship was generally required to contend with similar teething issues to body therapy, as described above. Freud had not initially stipulated an intersubjective perspective. On the contrary, he believed that conscious failure to gratify instincts, or emotional abstinence and neutrality exhibited by the analyst would counter the repression experienced by the failure. Intersubjective therapists have identified the risk of triggering artefacts that are hardly related to the patient's issues as a result of this adoption of a frustrated attitude. The forerunners of the intersubjective revolution in psychoanalysis include Ferenczi and the entire Hungarian school, as well as Sullivan, Guntrip, Fairbairn et al., and, in particular, it drew heavily on the work of Stolorow and Atwood. Both of them were highly critical of fact that the constructs of psychoanalytical theorists (Freud, Adler, Jung) were interpreted as personal attempts to solve their own existential crises that could not be generalised. The therapist may attempt to understand his patient from the patient's point of view and thus facilitate an exchange of subjective experience and develop empathy. This attempt bridges into self-psychology and, in equal measure, to the humanistic understanding of the self, which developed separately to psychoanalysis, as well as to the mentalisation wave that emerged at a later date. As such, empathic therapeutic relationships have also become an integral element of psychoanalysis alongside the traditional mainstream. Only insiders are capable of distinguishing between relational psychoanalysis and relational Gestalt therapy. Indeed, the difference between the relational psychoanalysis and Freudian-conservative school of psychoanalysis is much greater.

2.3.3 The Tradition of the Therapeutic Relationship in Behavioural Therapy

Placing the psychotherapeutic method at the heart of therapy and regarding the therapist as replaceable lies at the heart of the basic principle of traditional behavioural therapy. Relationship factors were relegated to the "black box" for behaviouralists [25]. Nevertheless, the social relationship appears in the behavioural theory model as a prerequisite for the instigation of learning processes – at least as a marginal variable. After Klaus Grawe came into contact with talk therapy (TT) through his meta-analysis, he recommended the integration of its relationship concept into behaviour therapy as of the 1990s. (However, talk therapy practice requires years of training, which cannot be skipped without compromising therapy quality!)

The inclusion of a significant relationship concept in behavioural therapy would require a shift in identity, which is yet to occur, at least in the mainstream.

Kanfer and his colleagues [26] are proponents of the following approach in behavioural therapy: behaviour and solution-oriented, positive, small steps, flexible and forward-looking. For Schulte [27], competent expert status is regarded as the most crucial element. This was confirmed and expanded on by Lammers [28]: "The job of the expert is to provide the patient with a compelling and credible explanatory model for his issues, which also includes methods in which these can be changed. The expert is also required to possess expert knowledge in the specific strategies and methods and assume responsibility for the proper application thereof. The therapist's expertise permits the patient to feel hope and trust as he is able to understand his issues better and now has the chance to rise above his vulnerabilities and regain control over his issues".

On the subject of therapeutic talks, Heßler and Fiedler [29] describe how cognitive behavioural therapy attempts to improve mental disorders "in a goal-oriented manner through the direct practice of alternative behavioural and interaction patterns, which are primarily legitimised on the basis of knowledge on the functioning of diagnosable mental disorders and presentations of symptoms". This is further explained on page 31: "The aim of each school of psychotherapy is not to adjust the patient, per se. However, in many cases, therapy inevitably leads to temporary adjustment. …"

In his book on the subject of the therapeutic relationship, conducting talks and behavioural therapy methods, [28], p. 40/41) summarises the formation of the therapeutic relationship under the following aspects, provided the relationship in question does not involve patients with problems based on motivation or interactions, or personality disorders: The establishment of trust and security, provoking a sense of hope (an attitude that promotes positive expectations) and a willingness to actively participate. Working relationships in behavioural therapy can be characterised by the following: An orientation on tasks and objectives (solution orientation) through the cooperative, transparent and open-minded agreement on goals, approaches and methods, and through an active, guiding and structure-promoting attitude; a positive and resource-oriented basic attitude achieved through self-help, empathy and appreciation. In this regard, the focus is placed on the active and self-determined participation of the patient, whereby the therapist is required to act as a model and multiplier.

Accordingly, the therapist is primarily concerned with the content of therapy from a change-oriented viewpoint. He should only divert more attention to the relationship if the patient appears to be disgruntled or exhibits problems with regard to motivation or interactions. This attitude corresponds to Schulte & Eifert's dual model (2002).

Thomas Fydrich, a long-standing member of the German Scientific Advisory Board for Psychotherapy (WBP) who strongly advocated the traditional viewpoint of BT on the board and in public hearings, and defended the rejection of humanistic methods in 2018, rejected the separate research of methods and relationship factors (p. 274) as found in contextual research conducted by Wampold et al. [16] in an article published in 2019. Research conducted by Wampold et al. determined that relationship variables (30%) and variables in therapist personality (20%) were jointly estimated to be responsible for approximately 50% of the variance in the outcome, with the method used only accounting for 0.2%! However, Fydrich continued to press his creed in his article: "A separation of the therapist, therapeutic relationship and therapeutic method (is) artificial and consequently not objective". He clearly was unable to accept that this point was a fact. The very notion would jeopardise the inflated regard in which the tool-based approach is held in behavioural therapy and evidently touches on a central, faith-like fundamental belief. Paradoxically, in this case, the ground of scientific knowledge is abandoned in favour of solidified behaviour-therapeutic conceptual thinking.

(The "methodological paper", the WBP's screening tool is a product of the scientifically inadequate transfer of approaches employed in a laboratory setting to psychotherapy, which do not take the impact of the therapeutic relationship and the therapist's personality into account. As such, the WBP's methodology was and remains scientifically untenable. Rulings made on the basis thereof should be reflected and re-assessed under the microscope of the international scientific community. See also ► Sect. 1.2.3)

2.3.4 The Therapeutic Relationship in Humanistic Client-Centered-Therapy (CCT)

In Client-Centered-Therapy, the therapeutic relationship is based around a set of keywords: Empathy, appreciation, congruence, authenticity, self-actualisation, lifelong development. Client-Centered-Therapy placed the therapeutic relationship at the heart of the therapeutic healing process from the outset.

Carl Roger's [30] definition of his understanding of empathy is summarised below according to Schmidt-Traub ([31], p. 114 f):

Tip

Empathy is the therapist's sensitive ability and willingness to observe the patient's thoughts, feelings and inner conflicts from the patient's perspective and to understand them within this frame of reference. As a result, he may discover certain motives and implications of the patient that the patient is often not aware of himself.

The variable of empathy is highly complex, all but immeasurable and hard to separate from the other therapist variables such as authenticity, warmth and positive awareness. They are all connected. The extent to which the patient feels understood by the therapist is of the utmost importance for the therapeutic outcome and empirically particularly well established [32].

The Client-Centered-Therapy objectives of therapeutic relationship establishment were gradually adopted by other approaches, methods and procedures. This adoption appeared to occur most naturally in the humanistic approaches the emerged from the same zeitgeist. Apart from the various body therapy approaches, the adoption of these objectives also occurred in Frankl's logotherapy, Berne's transaction analysis, which continued to maintain psychoanalytical characterisations, Gestalt-based psychotherapy (from Walter), Moreno's psychodrama, which assigned relatively large significance to roles in his understanding of relationships, and, to a certain extent, also Petzold's integrative therapy and Greenberg's EFT (emotion-focused therapy). With regard to the therapeutic relationship, Gestalt theory is focused on Buber's I-Thou relationship, which can partially be observed as parallel lines but nevertheless also appear to be more differentiated and capable of confrontation.

This life-enhancing basic attitude was introduced to a number of variants of systemic family therapy by the charismatic Virginia Satir, who spent a period of time working in groups with Fritz Perls in Esalen in the 1960s.

Once Klaus Grawe completed his meta-analysis in the 1980s, he believed that, if combined, the know-how gained from BT and CCT (client centered therapy) would result in an unbeatable culmination and subsequently encouraged his colleagues in the field of BT to adopt the Rogerian relationship culture in the years that followed. (Oral statement).

It would appear that psychoanalysis has come the furthest in this regard. The field has finally succeeded in departing from the professional frustration that marked its initial generations and instead adopted a more affectionate therapeutic basic attitude from mother-infant research, attachment research (which was also initially briefly excluded from the mainstream), Kohut's psychology of the self (which was almost excluded), relational psychoanalysis and Fonagy's mentalisation concept. Ferenczi, whose active analysis also fell out of favour during his lifetime is now cited as a principle concept in this regard. Any references to the adoption of concepts from other schools were avoided, such as the inclusion of the "response" principle from the humanist group dynamics in interactional therapy (through the additional appointment of F. Heigl and Heigl-Evers to the chair of

2

group dynamics in Bonn, 1969–1971), or in structural psychotherapy (2004) following several years of liaison between G. Rudolf and the school of Gestalt therapy (1978–2003). As is evident, the field of psychotherapy is much more interlinked than some people who derive their perceived superior status and worth depending on the school to which they belong would like to admit.

With regard to the therapeutic relationship, the majority of concepts practised in body therapy are derived from the traditions of humanistic and/or depth psychology. Conscious, current awareness of physical sensations and/or impulses opens the door to the unconscious. Innervation processes are unconsciously influenced by old convictions that shape our identities as well as threshold shifts within the sensorium. Fright and humiliation are incarnated just as lightheartedness and dignified self-perception. Perceived, experience-based changes capture our attention both internally and externally. Body psychotherapy employs exercises to work on mindfulness, touch and/or experiential stimulation.

The reform movement of Art Nouveau, in combination with Elsa Gindler and her "Seminar for harmonious body training" in Berlin played a key role in the development of body psychotherapy. Expressive dance and theatre also provided a number of influences. This practice can also be linked to the psychoanalyst and physician Wilhelm Reich, who Freud ensured was excluded from the psychoanalytical association. As a result of this, a long shadow was cast over body psychotherapy, which prevented it from receiving the recognition and place it deserved in light of its effectiveness within the overall concept of psychotherapy. It found itself most at home in the group of humanistic approaches characterised by experience-oriented work. There are at least 30 recognised variations of body psychotherapy methods.

The therapeutic understanding of relationships needs to account for the relatively close physical proximity through the particularly mindful reflection and professionalism of the therapeutic attitude. Sensitive regulation of proximity and distance is just as crucial as a refined understanding of transference. Movement and body therapy methods build access routes to the previously obtained and unconscious identity-shaping traces. Physical approaches are particularly effective when it comes to a lasting change of identity. Its involvement in the main therapy (in contrast to its usual delegation to specialists) is underappreciated and advantageous.

2.3.5 The Relationship Variant in Gestalt Therapy as a Precursor to the "Special Therapeutic Relationship" (▶ Sect. 2.4)

Despite the fact that Gestalt therapy is correctly regarded as a humanistic approach, it also includes principles that are characteristic of behaviourism and psychoanalysis. With its inclusion of systemic therapies, it is further linked to the holistic approach. From the outset, the concept of Gestalt therapy combined common principles from all schools as well as schools of thought that span time and cultures thanks to the critical open-mindedness and searching stance of its founding members. Gestalt has succeeded in finding new balances between analytical and assimilating tendencies, between unconscious and conscious thoughts, between non-verbal and verbal as well as proactive emotional and physical experiences and reflective post-processing, between patient and context-related aspects, self-determination and coercion and many other facets.

The therapeutic relationship variation based on Buber's "I-Thou" relationship (1923) was the primary reason many later Gestalt therapists abandoned psychoanalysis of the day and its call for neutrality. The ability to enter as the subject into an I-Thou relationship denotes the ability to resonate and communicate with the self, which can be observed in the I-It relation. This recently discovered and welcomed relationship variant does not hold true to the black box concept that was popular amongst the behaviouralists at the time (John Watson) in the least. A century later, we are able to delight in witnessing the convergence of various approaches to the therapeutic relationship.

The priority of the I-Thou field lies in the strong, interpersonal acceptance that provides security and eliminates that fear of devaluation and rejection to such an extent the confrontations, transformations and "rebirths" can be achieved in a gratifying manner.

2.3.6 Therapeutic Relationship Concepts in Systemic Therapies

Systemic therapies comprise couples therapy, family therapy, organisational work and therapy for large groups as well as political peace work. The intellectual precursors to these forms of therapy can be found in holistic and Gestalt psychology, in biological closed-loop thinking and in the technical understanding of cybernetics. In the 1950s, the technical term "Gestalt" was defined as equivalent to system and wholeness.

We are therefore faced with a complex scope of duties and a historical process which has and remains to subject therapist and leadership behaviours to a number of different requirements As the name would suggest, the focus of interest is not primarily on the individual but rather on the system and its internal relationship dynamics. Intellectual understanding and not empathy is required to observe systems. Nevertheless, the family therapy advocated by Virginia Satir is a particularly exemplary historical example of charismatic, resource-oriented empathy that each family member knew to incorporate in a positive manner. As such, we are presented with an unclear view with regard to the therapeutic relationship and the overall self-perception in systemic therapy.

A book written by Gunthard Weber, Gunther Schmidt and Fritz B. Simon [33] explores the topic of the conflict between the various development trends in systemic therapy. It describes a systemic-constructivist line of development that represents Fritz Simon and a systemic-phenomenological line attributed to the other authors. The content of the book concerns the methods of constellation work. Simon repeatedly calls on his colleagues to provide a system-theoretical explanation for the phenomena observed in constellation work, which overextends the conventional categories of explanation without filling in the gaps of their own accord. These gaps pertain to the archaic "knowing fields" which cannot be placed by conventional natural science. System therapists who work with them are often labelled as shamans by other colleagues.

Willemse and von Ameln [34] also highlighted another controversial trend to seek therapy-relevant unity due to its importance to behaviour within the therapeutic relationship. In the eyes of Watzlawick and the Palo Alto Group, this unity remains in the individual, although the group is taking steps towards viewing the system as a whole; Luhmann instead describes a social system consisting of voiced or expressed communications. From a critical perspective, this entails, in the eyes of Luhmann, that unfulfilled dissatisfaction (fear, hatred), which reduces the perspective on reality that is relevant to psychotherapy does not yet belong to the system.

Willemse and von Ameln (p. 56) provide the following suggestions with regard to the systemic approach:

Tip

"Never consider your way of looking at things the only possible way and never trust simple explanations". "Help your clients to develop a positive view of the world. Work on resources and solutions, not on problems and shortcomings". "Pay attention to your language, as it creates a reality". "Concentrate on changing patterns of communication, not on the changing members of the system". "Consider the context if you aim to understand the system". "Think in a circular manner". "Treat your clients with appreciation and respect; they – not you – are the experts in their situation and you are the expert in finding effective solutions" (p. 59) "Those who have internalised systemic thinking will refrain from judging their clients' values, downplaying their problems and incorporating claims to their own superiority" ...

This attitude can be found almost 1:1 in Gestalt therapy, especially when the objective lies in encouraging clients with a deeply Socratic attitude to interpret non-verbal phenomena or thoughts through association.

A number of emotionally reserved basic attitudes can also be found alongside this considerate, nodding attitude, especially amongst colleagues who are more drawn to the constructivist and regulated understanding of the system. This does not produce an altogether uniform concept for the therapeutic relationship.

The Palo Alto school (rather officially) adopted the term supersummativity (emergence) to describe the added value of a system (Gestalt) recognising that the whole is different and more than the sum of its parts from Gestalt psychology. A system (Gestalt) is an internal relationship structure. Everything is connecting to everything. Any changes made to a part affect the whole. Of course, this acknowledgement also applies to the research on reference systems conducted by Kurt Lewin based on experiments on reframing.

2.3.7 Establishing Objectives in Psychotherapy as Part of an Extended Work Alliance

The objectives established within therapy are based on its context. They are achieved within a certain understanding of relationships. In this regard, the most natural approach is to initially try to reduce the symptoms that led to the patient seeking therapy. However, only a small number of patients restrict their field of vision to this approach and find satisfaction therein, as they would prefer not to touch on underlying problem areas that they can sense. In this scenario, it is often beneficial for the therapist to widen the field of vision to the point that he can point out indications that may lead to an understanding of the concealed conflicts, shortcomings or trauma and, although it does not currently appear to be the correct point in time to address these issues, that it would nevertheless be sensible for the patient to listen to his inner thoughts to determine when he wants to take the next step and grant himself further relief and alleviation. In this case, eliminating symptoms may provide motivation for further, more profound psychotherapy.

From the outset, the majority of patients only experience a symptom as an overflow mechanism and not as the actual problem itself, such as sleep disorders in the context of relationship crises, for instance. They are able to recognise the causal link, albeit it only through subjective bias or a narrowed point of view, and are motivated to engage with it.

Certain patients may mask their life situation with a practised phrase for their symptom that first needs to be deciphered, for example, "I have a circulatory problem" – (then, when inquired about) "I get dizzy when my husband comes home from the pub late at night on Fridays and makes a racket. It makes me angry and I worry that he might hit me or do something to the children and we will once again run out of money". In this case, the establishment of an objective may go beyond the field of psychotherapy and enter the realm of social work, family welfare, drug counselling etc.

Now we return to the establishment of objectives in psychotherapy. In practice, we rely on the capabilities of an inner observer to both draw comparisons to earlier mental states and to anticipate the yearned for, healing state. It asks us:

Tip

What does the patient need? What does he experience as a shortcoming that needs to be remedied? Which objectives is he capable of establishing and formulating for himself on the edge of consciousness? How does he feel he will experience himself afterwards? How does he think others will perceive him? Are there any underlying objectives that he has hidden from himself or that he does not feel comfortable having or expressing? Perhaps due to the fact that he views them as unrealistic?

The therapist must concede that it may not be possible to observe the entire whole, but that focusing on a short-term objective can benefit the therapeutic cooperation.

Does a patient feel overexerted when it comes to specifying objectives? Does the patient require assistance with perceiving and formulating specific objectives? This is often the case in structurally weak individuals. However, it does not indicate a lack of goodwill. The patient is simply unable to put the aspects that make his life so difficult into words. Patients in a state close to psychosis will struggle to understand this question. In this case, they are not necessarily capable of the required internal differentiation. Structural development is required in this regard.

2.3.8 The Therapeutic Alliance in the Strict Sense

In terms of content, the work alliance concerns agreements on space and time, or regularity, duration, frequency, setting, financial remuneration, cancellation rules and also agreements pertaining to confidentiality and formalities with insurance policies and health insurance providers.

By definition, the work alliance is linked to the "developed" part of the patient, the part that is, or at least appears to be, capable of agreement. It constitutes the section that is interested in improvement and change, and which can therefore be regarded as reasonable and insightful. This part is required to have sufficient trust in the person who offers this alliance. Which subsequently places a large number of requirements on relatively healthy parts.

Unfortunately, some patients do not possess the required "grown-up" part. This part only occasionally comes out on top in the patient. It is crucial that we take the time to assess our patient's capabilities with regard to forming an alliance. Simple wishful thinking on our part as therapists does not cause the part capable or making agreements to magically appear. In certain cases, this can be precisely the problem, such as with patients who are addicts, or who exhibit of acute or subacute suicidal tendencies and also in the case of many people going through a major crisis.

A therapist is somewhat more likely to succeed if he is able to reach the emotional core of the patient in a compassionate, emotional way and if the patient demonstrates a plausible interest in himself and improving his condition that through the "grown-up" part.

2.3.9 External Framework and Setting Variations

Individual setting. Psychotherapy predominantly takes place in an individual setting whereby the therapist typically sits opposite or diagonally opposite to the patient.

In classical psychoanalysis, the analysand lies on a couch and the analyst sits diagonally behind him so that he is shielded from the analysand's view. As such, contact purely takes place through conversation. The lying position encourages the patient to regress into the childlike mental state of dependency. Restricting the level of interaction promotes the analysand to reference himself and his mind.

In contrast, when the patient is placed in the seated position he is afforded much more control over the situation and enables the therapist to understand him through both non-verbal and verbal capabilities, remains open to interaction with the therapist and exhibits a potentially equal adult identity. The first to practice the seated position a century ago were the followers of neo-analysis such as Schultz-Hencke and his proponents.

In children's play therapy, the child is usually permitted to select how he would like to playfully develop within the provided framework of the playroom. The therapist establishes a gentle and supportive relationship with the child in order to support, understand, clarify, and respond to the child's process and to enable the latent aspects to emerge on the surface in order to process emotions and release them.

The setting referred to as "experiment" in Gestalt therapy, offers the patient a similar framework, whereby the patient is able to playfully and creatively develop in contact with his unconscious. The media used in this regard are interchangeable. Certain patients may find it easier to express their state of mind through a dance-like sequence of movements, or thorough acoustic improvisation on synthesisers, in fantastical storytelling, through artistic collages made from objects, or by painting with a certain colour schemes and styles, which is the most commonly observed form. This setting is consequently associated with art therapy. Internally, it is important the contact with the unconscious is intensified whilst in a kind of waking trance and at the same time the patient inquisitively registers the process with his non-judgemental observer, which is then simultaneously or subsequently discussed with the therapist. This ability to make free associations is well suited to more developed structures but may prove too much for structurally weak patients, who tend to feel more secure in semi-structured settings.

In order to achieve targeted, emotionally corrective experiences, both behavioural therapy and Gestalt therapy facilitate jointly seeking out situations that the patient/client has seemingly struggled to cope with thus far. A classic example in this regard can be found in lifting journeys which are employed in behavioural therapy to assist anxious patients after other relaxation training. This process begins with shared journeys and increases in a graded programme until the patient is able to successfully cope with their problem independently. A number of beneficial experiences have been observed with anxiety-mitigating and experience-based corrective programmes for a range of symptoms, especially for anxiety and phobias. Paradoxically, the most effective factor in this experience (the therapeutic relationship) remains the element that has received the least attention in scientific research.

In Gestalt therapy, the corresponding approaches are not standardised. They are employed after an individual, clarifying the processing of the history of origin and any potential shift in symptoms in order to facilitate the deidentification of the object to which the anxiety or phobia had secondarily attached itself through anxiety-free new experiences within the protective space of the therapeutic relationship. The fear or its equivalent is then retraced to its source of origin. Contact with the formerly frustrated basic need, such as the need for security which has now been accepted and understood within the scope of therapy and received solace for its former lack, is particularly effective in this regard. The fate of the detached fear which had fused to a random object is only of secondary interest and is relatively simple to supplant through corrective experiences.

Pair Setting: In Couple Therapy, the Therapist Must Remain Impartial at All Times

The ability of the therapist to make both divided and hurt parties feel able to understand and show goodwill in front of the therapist and offer common ground that enables both parties to feel adequately secure to show themselves at their most vulnerable can be regarded as a great achievement. This is precisely the situation in which the therapist is required to emanate a strong, sympathetic and confident "I-Thou" relationship, without which he will not succeed. This situation enables us to obtain an overview of the common ground that once existed and led to the establishment of the relationship. Can this common ground still be found in thoughts or emotions? Is there (still) any common ground that provides mutual "nourishment", a sense of home, encouragement or support? How is the mutual "balance" achieved? Who gives and takes at which levels? Has consensus been reached in this regard or does each party view and assess the same circumstances in an entirely different manner? At what levels of maturity does communication occur between the parties? Which part of one party speaks to which part of the other? Is one party acting the role of a child in order to achieve care? Or does the other party assume sole responsibility in excess to dominate the other? Are some parts concealed from the other party for the sake of the relationship and thus secretly turn into monsters that threaten the relationship? Have the couple successfully merged into the role of parents, or did this lead to an eruption of rivalry for mutual attention? Have the couple succeeded in rebooting their relationship once their adult children have moved out? etc.

The supportive therapeutic relationship and the therapist's ability to empathise is faced with twice the challenge. Both sides are home to a frustrated, hurt partial identity, which has either existed since childhood and now been a trigger or emerged as a result of frustration within the partnership. The first step can be regarded as complete once the two disappointed and injured partial identities are successfully confronted and invited to enter into a dialogue in a sympathetic environment. Following this, it then needs to be determined if the complete or partial absence of one of the other from the couple's relationship is conceivable, or from the outside or not at all, which may potentially trigger a process of dismissal that may need to be further addressed through individual support. In the end, it is important to determine whether the couple is able to see any opportunity for development for their future through the above steps.

Despite the fact that the examples of contextual aspects have been framed around the couple, the supporting background influence that provides both partners with integrity and dignity remains within the therapist's area of control. This setting determines whether couples will succeed in "getting along with each other" or "separating in a respectful manner".

2.3.9.1 Multi-person Setting in Family Therapy and Team Counselling

As can be expected, the therapeutic relationship takes on another level of complexity in family and team therapy. Family members are weaved into their familial network with all of the risks and opportunities it offers them, as well as its natural developmental tendencies. Members of teams are governed by the internal group dynamics as well as common, company-specific and socio-political external contextual factors.

The preferable methodological principle of multidirected partiality was proposed by Boszormenyi-Nagy in 1985 and Stierlin et al. in 1977. This method requires the therapist to consistently maintain her interest, empathy and identification with each individual participant in the system throughout the entire session. This ideal borders on a superhuman demand, yet it remains the goal in most cases.

Selvini-Palazzoli et al. [35] broke down this notion to the concept of neutrality, which in this regard refers to constantly changing partiality whilst nevertheless maintaining balanced care overall. Neutrality comprises several facets such as social neutrality, outcome neutrality and problem neutrality. The matter of whether the system therapist/consultant should or wants to address problems in an entirely neutral manner at all times should be addressed as soon as dissocial problem areas are broached.

In reference to Williams [36], Schwing and Fryszer [37] highlighted four roles for the systemic leadership figure, which may or should alternatively take the lead to a greater or lesser extent:

1. The teacher: in this role, the therapist is tasked with passed on information to the clients, providing suggestions and initiating learning processes.
2. The facilitator/empathic listener and conversation partner: this role borrows aspects from talk therapy in order to create an atmosphere of understanding and being understood.
3. The consultant/expert in systemic methods: in this role, the therapist integrates suitable methods into the process such as circular, future and solution-oriented questioning, work on genograms, a timeline or sculptures for instance. He adopts a neutral attitude with regard to the outcome and does not provide any advice.
4. The evaluator: When he assumes this role, the therapist is able to provide his own insights and observations, with every precaution, and his own personal statement, subjective feedback and therefore evaluate.

This approach results in a role-puzzle that cannot be achieved without conflict and inner reflection, yet nevertheless gives the individual personal wiggle room.

For the therapist, the skill in the beginning lies in first transforming the expectations born out of necessity that are held by individual family members into a certain level of assurance by making them feel understood and appreciated with regard to their concerns in general and their individual action. To some extent, the way in which contact is established can indicate the common system of values which is regarded as capable of reaching consensus is also significant in this regard.

2

2.4 The "Special Therapeutic Relationship": "Therapeutic Agape" or "Therapeutic Solidarity" – Development Assistance with the Integral Activation of Resources and with Integrated Confrontation in the Sense of a Targeted "Psychological Operation"

Achieving "therapeutic agape" is a high objective which is the target in Gestalt therapy and is sometimes possible to reach, but always only in close proximity. However, even this represents a vast achievement. Unconditional love facilitates and stabilises life. Constructive confrontation, transformation and healing can all be achieved within its sanctuary. In psychotherapy, unconditional love can be used as a compass needle for every form of decision.

This chapter provides a detailed overview of the therapeutic relationship that simultaneously works on different levels on the part of the therapist. As such, different relationships are achieved at different levels, Which delivers the background required for targeted, corrective internal and external relationship experiences. The "special therapeutic relationship" varies from the general one through its special structure and emphasis on solidarity and development. This difference likely forms the main reason for the differences between the effect sizes observed for the average value for psychotherapy (ES = 0.83) and Gestalt therapy (ES = 1.12–1.42), which has emerged as the most effective humanistic method but can nevertheless simultaneously be regarded as a connecting element to all other groups of methods. Furthermore, this difference in effect size may also be attributable to the method used to recover excluded partial aspects and to part-whole research that includes a dialogue.

As such, in this regard, Gestalt provides an example of an extended understanding of the therapeutic relationship.

2.4.1 The Preliminary Phase – A Hidden Gift

Prior to each therapy session, the therapist undertakes short and calm yet nevertheless conscious preparation in order to first establish a centred, balanced state of mind, which is free from basic personal needs that may make demands of the patient, in order to act as a receptive vessel for his patient to a certain extent. Centring also enables the therapist to enter the relationship with an attitude configured to sensitive, empathic perception (Rogerian approach) and ultimately to concentrate on fostering a caring and encourage perspective towards the patient that focuses on potential and resources. The patient directs an imaginative flow that imparts power, meaning and connections towards the patient. (A concept derived by Gestalt therapists from Martin Buber's I-Thou principle) This attitude is most comparable to the suprareligious fundamental belief of Buddhism, which concentrates on the aspiration: "May all living beings be happy!" (the western world of research has been astonished at the high-amplitude, synchronously ordered EEG fields the monk Matthieu Ricard capable of radiating onto his surroundings and his counterpart during meditation when he adopts this attitude". We can therefore surmise that the patient is touched by this form of radiation from that therapist, which, in turn, contributes to the therapeutic relationship as an atmospherically stabilising setting that conveys trust and security.

Whilst it would be desirable to meet all patients with this basic attitude. It remains important that each person is viewed as an individual and thus needs to feel individual appreciation. This approach combines a supraindividual and individual viewpoint.

The achievement thereof may initially appear too demanding, but it can turn into the effortless permanent attitude of those who foster a basic attitude that includes solidarity. Living this affirmative solidarity in a non "moral" but rather humorous manner is a key prerequisite for supporting, inspiring and even making open-mindedness, trust and commitment to common work fun and light-hearted. Many patients have never experienced this existential affirmation. They long to experience it and often relatively quickly demonstrate the ability to transfer through their communications and consequently experience positive resonance. As it would seem, an affirmation can be contagious.

2.4.2 The Therapist as a Development-Enhancing Counterpart

Adoption of a development-enhancing role by the therapist can often lead to controlled "therapeutic parenting". We learn from nature, or our nature if we are parents who love their children. This is a path that permits a certain amount of freedom. Furthermore, we all express love differently at different phases of our lives. And it will also be required to set limits in a range of ways. A loving attitude makes it easier for our children to accept frustration, providing we are consistent with setting limits that meet them halfway whilst ensuring we palpably maintain the affirmative and connective field. In certain cases, love needs to be prepared to face power struggles and disputes all whilst maintaining a sense of humour, if possible. Love does not evade or employ crooked ways during the conflict. It does not insist on authority and perfection, but it may place values and beliefs under scrutiny in order to concede its own limits and be prepared to let go when the time comes. It always upholds the dignity of its counterpart and remains a model for it. To a certain extent, this chapter is linked to 1 Corinthians 13, without any commitment to religion. The principle of love does not belong to an individual or group. It belongs to them all at the same time.

As relationship-oriented psychotherapists, we represent the principle of unconditional love to the best of our abilities. Of course, we are not able to achieve perfection in this regard. Unconditional love is an expansive superhuman concept. In order to prevent us from biting off more than we can chew, love is responsible for our internal balance – the internal open dialogue with our own limits – between pain that is yet to heal, our unfulfilled needs and adequate self-care. In this way, love provides relief. It addressed but the internal and external.

Following his introduction to Gestalt therapy as part of his own crisis intervention (upon recommendation from a friend), Jeffrey Young broke down Buber's "I-Thou" principle from Gestalt therapy to the concept of "limited reparenting" for his schema therapy. Although a reasonable compromise, Young's concept fails to encompass all of the facets included in the Gestalt approach. After his session, Young reported that he had learned more about himself during those 10 hours than during the entire previous year of work with this CBT. Although, he did acknowledge that Gestalt required more structural determination in order to facilitate increased feelings of security.

2.4.3 Therapeutic Intentionality and Resource Orientation

To the best of our knowledge, unconditional love exhibits the greatest healing potential available. We encounter it in a number of shades. As a "therapeutic agent", it strengthens accepting energy in a targeted and palpable manner to

1. reach the existential core of the counterpart in a non-verbal and verbal manner in order to stabilise it, encourage it to regenerate "It was lovely to get a first impression of you", "I'm glad that you found your way here and have courage to want to change something"). This only works if it is perceived as honest and not simply an empty phrase. Otherwise, the approach of less is more can also be effective. The objective is to establish an increasingly resilient field of coherence that summons hidden remnants of basic trust into resonance and thus facilitate the development of an alliance characterised by a concept of "we". Even minor success in this endeavour can make an impact.
2. The acceptance of the therapist should enable empathic solidarity to reach the part of the personality that has remained stuck in distress during the life cycle. Accordingly, part of us, as therapists, takes on the role of the child through empathy in order to help these parts. In certain cases, the patient may exhibit several parts that were held back from developing. Now they are no longer alone and no longer need to feel any shame for their miserable situation. Someone else shows an understanding of how "one" can end up in this situation. This other observer may be the "inner child" with his frustrated basic needs and longing to be at least subsequently seen and accepted – and consequent capacity to be integrated (see paragraph "Empathy and solidarity"). The good news is that this can be achieved and often leads to phenomenal satisfaction and inner peace.
3. Unconditional love has developed a positive accentuating approach to communications in the "here and now", whereby it transforms its frame of reference in an unconventional manner and delights in finding appreciative sides to its counterpart in the here and now, such as "I'm impressed by the caring way in which you talk about your children and husband, when I look back on everything you mentioned about your own emotionally deprived childhood. You appear to have managed to avoid continuing 1:1 with the coldness and hardship

you were subjected to, and instead – at least somewhat – toned it down or adopted an entirely opposite attitude. That alone is a significant life achievement that you can be proud of. Or: "You looked at me for a little longer and more intensively, but also more openly than before. This helped me and has not only given me hope but also the peace of mind that despite all that has happened, you have the courage at heart to give the individual people and the world another chance – following the necessary critical evaluation – and to again pass judgement on whether any different encounters or contact can be achieved, or not, as the case may be". This resource-oriented mindful awareness of the therapist to the patient's non-verbal (unconscious) manifestations that indicate that it may be possible to master this situation in a life-enhancing way provides incentive and reward. Indications of a credible witnessed identity shift in the desired direction can instigate an automatic process whereby the phoenix will soon arise from the ashes.

4. A common resource-oriented interest in the healing and capable part of the patient is also a child of unconditional love. What does the patient feel she has achieved in life? What is he truly proud of? Where and how does he recharge his batteries, find happiness and perhaps even gratitude? Which past and present relationships – at least to a certain extent or period of time – has he found to be nurturing, affirmative, encouraging, stimulating, reliable and, above all, caring. Was or is there a certain part of the patient that is behind the success of these relationships? Has the patient ever experienced a caring relationship? With another person, animals, plants, things or something even less tangible such as moral concepts or something abstract?
5. When interest is displayed and requests made, the "inner child" who has experienced hardship and a lack of protection and care, is often capable of looking into and increasing compensatory wishful fantasies. This reflects the individual's desire for life-enhancing and acceptable relationships. As such, this experiential longing can be harnessed as a personal affinity towards a caring, positive shift in identity. The patient is aware of the existential value of adopting this stance. If the patient concedes that he has suffered serious deprivation in this regard and subsequently experiences real and symbolic protection and consolation, this experience prevents the patient from drifting towards identification with the aggressor and instead helps him to increasingly experience the principle of caring and loving attention.

Application of point 5 is well suited to small workgroup in a group setting, whereby each individual in the small groups takes turns envisioning the ideal scene that would have liked to experience as a child. The other group members then assume the role of the perfect parents, if so desired, etc. These situations often include a number of unspectacular aspects: playing games with each other, sitting together in peace at the dinner table, going on a trip, receiving a birthday cake with a candle for each of life, taking the child by each hand and swinging them between both parents, etc.

The resource perspectives found in 1–5 form partial aspects of the "special therapeutic relationship", which, in itself, is not a separate common factor as its perspective is based on the life-enhancing basic attitude. However, if desired, the fourth aspect can be developed into the "team of inner helpers" as a separate method. This is often a beneficial undertaking prior to an intrapsychic or interpersonal encounter in order to strengthen the patient's position and enable the corrective experience to achieve the desired effect. During the process, the patient reflects on all successful, life-enhancing past and present relationships that did or continue to exist, symbolises them and then brings them together into a protective semicircle behind him, enabling their energy to transform into a reliable, strengthening tailwind. When equipped with this inner encouragement in addition to the therapist's empathy and, in a group setting, strengthening from the solidarity of a number of or all group members, the patient may feel capable of approaching or enduring confrontation that has been avoided to date, potentially for the patient's entire life with completely different conditions than those experienced without the above. This new experience is marked by understanding empathy and solidarity for the part of the personality which has remained stuck in misery (usually a childhood ego) that the patient was previously ashamed of. Supportive empathy liberates this part from its painful, internal, disconcerting exclusion and enables it to heal.

With regard to point 1: The inner encouragement is not formulated, nor is it exaggerated or implausible. Put in simple and secular terms, the supportive message of love may be: You, like me, are a being who needs and wants to cope with the world in this current period marked by (particularly) difficult circumstances; This is currently harder for you than me. You have my full respect and I also see hidden potential in you and would like to help, etc.

(In religious traditions, the message of unconditional love may be: May you be happy like all other beings! (Buddhist). Or, in the case of Buber's I-Thou (Hasidic Judaism): We are connected as god is equally reflected in both you and me. I cherish him in you, even if you are not (yet) able to perceive him. Or Christian: For where two or three gather in my name, there am I with them. Or Hindu: "Namaste" (Sanskrit) which is roughly translated as "I bow to the God within you". These messages are all closely linked.)

Note

Psychotherapy defines itself as a secular field.

The benefit of touching on or instead eliminating a connection to the spiritual varies depending on the patient's background. Any form of missionary work is a contraindication in this regard. Unconditional love is not bound to any religion, nor is it possessed or governed by anyone. However, we are all capable of giving it to everyone.

In recent years, even the most rational scientists have been astonished by the EEG findings achieved by monks versed in meditation such as Matthieu Ricard.

During classical "loving-kindness meditation", he radiated a high-energy, high-amplitude and synchronised ordered field, the kind of which had not previously been observed by EEG experts.

I believe that centred psychotherapists, despite the fact that this concerns an entirely different field, also emit palpable radiation and are thus capable of subtly coming into contact with their counterparts. Extensive, systematic research is required in this field.

The trick is to consciously establish fields of coherence between the therapist and patient on the levels where harmony is achievable and desirable. This enables the patient to experience being seen, strengthened and accepted and simultaneously endure an objective confrontation (I-It level) with aspects that require correction. The I-Thou field must be strong enough to withstand the frustration of partial correction. This approach may be viewed as a reflection of emotionally regulating comfort provided by adequately good parents: this supportive, accepting solidarity prevails over the current frustration and helps the individual to endure it.

In other words: the energy-giving and synchronised ordered field is often perceived by the patient as an underlying safety net despite his doubts that he cannot be accepted due to his shortcoming and it provides him with basic trust and hope.

6. The "special work alliance" constitutes an additional stabilising and progressive component of Gestalt therapy and forms a kind of partnership. The therapist is not regarded as above the patient in any sense. Instead, the therapist accepts with Socratic humility that he knows nothing about the patient and is instead reliant on the open internal and external with him. Partnerships entail appreciative, mutual exploration, loyalty in solidarity when either party encounters difficulties and also trust in the mutual stability when required. This form of partnership tends to benefit for interchangeable leadership roles. Sometimes the therapist is the one doing the listening, being tentative and seeking guidance through the world of the other party, and at other times he channels care. Once he gains a certain level of strength from the partnership, the patient allows himself to enter the flow of his associations without fear in a waking trance, takes the lead with this activity and, in response to the therapist's suggestions, translates images or non-verbal manifestations from his world and places them in the context of his chains of association linked with his life story. Fearless, partnership solidarity can cause defence mechanisms to melt away, as they are also "translated", and their original meaning is deciphered.

Tip

Paradoxically, meaning can be found in everything, whether in the skewed excluded aspects or in vetoing defence mechanisms. If the original existential fears and biases behind the dynamics that created them are successfully recognised, they can be eliminated and release the strength bound within.

2.4.4 Mental Operations – Separation and Emotional Reconnection

Mental operations require preparatory regression work during a waking trance; the objective thereof is to correct detrimental, identity-shaping experiences and damage.

> **Note**
>
> In order to perform regression work in a waking trance, a strong relationship of trust needs to have been established between the therapist and patient. This positive alliance is regarded as "therapeutic wholeness", a "we" field, that possesses its own truths, for example, it may provide the patient with the assurance that he is a respected and valuable being, regardless of what happened to him in the past or how often he has failed. The above can be regarded as the initial activation.

The therapist is responsible for ensuring that this field does not encourage the patient to passively regressively rest on his laurels but rather inspires him to tread down previously avoided paths. He rewards the patient with an appreciation for each display of initiative in the right direction and allows him to refuel in each case in the mutual "we" field. (Behavioural therapists are well versed in activating interventions.)

Biographical regression work entails travelling back in time to an earlier stage of life and earlier social reference systems. It is also an association with the unconscious retransformation into former identities (also referred to as Ego states by certain colleagues) which can be mentally retrieved and are unconsciously capable of resonating at all times. In certain cases, retransfomation may lead to the unintentional triggering of an "activating situation" through a problematic identity.

For example: Each time she is approached by her elderly neighbour who she hardly knows and has not had any bad experiences with, the patient reacts with inappropriate freezing and a blackout. In therapy, she associates her neighbour with her mother and her destructive and hurtful side.

Therapist: "We both know what a wonderful person you have developed into despite your difficult start in life. I admire you for your journey through life. Perhaps we should reach out to your inner child and integrate her more into your current life. How would you feel about that? (The patient nods) Then allow us both to meet her. I will be at your side the entire time on this journey. We will still be a strong team and listen to each other. Now, travel back in time and become your younger, smaller and childlike self. You find yourself in an atmosphere and scene that vaguely reminds you of your experience with your neighbour. At the same time, you can feel my presence. I can also place my hand on your shoulder, if you would prefer. Two people can withstand a lot more than one alone. You are already looking into mind. Where are we now? How old are you? Who else is there?"

Patient: "We're in my childhood bedroom. My mother is angry because I haven't tidied my room again. I'm stubbornly screaming as I'm only 4 or 5 years old. The window is open. We're on the 4th floor. She doesn't want the neighbours to hear the cries. She threatens to throw my toys out of the window if I don't stop. I continue to whine. She fires her favourite teddy bear out of the window whilst explaining to me that I'll "be next if I don't stop!" – My heart stopped. She really meant it this time. Then she went even further, this time with a piercing voice: "and if you survive that, I'm putting you in a home. I'm done with you!"

Towards the end of recalling this experience, the patient appears to lose her voice and involuntarily turns towards the therapist with raised shoulders, who then comfortingly takes her in her arms and gently sways her. This triggers the child ego to start sobbing. The tension in the shoulders disappears. After a while, the patient turns an eye to the therapist: "That wasn't the only time. I've experienced several other instances that were just as terrifying", says the patient whilst lightly stomping the ground with her right foot.

Therapist: "Your foot appears to have an important contribution to make". She imitates the patient's movement somewhat stronger and glances at the patient in an inquisitive and encouraging manner. The patient takes her up on this and bravely turns towards her imagined mother in the room.

Therapist: "What should you have finally told your mother in the past?"

Patient: "You scared me to death. I truly believed that you wanted to kill me and give me away. You seemed intent on it. For years I wasn't sure whether I would survive this hell with you. You were a nightmare. Every time I did anything at all that you didn't like. You took away all of my security and frightened me. I hold that against you!" (At this point, the patient stretches out her hand with her fingers apart and makes a defensive movement.)

Therapist: "Your hand seems to have something else to add on the topic of your mother and her parenting!"

Patient: "Oh yes, I'll leave that with you. I don't want to take that along with me. I don't want to be that horrible. Granted, sometimes I get the urge to flip out, but then I remember how devastating it was for me" The patient's shoulders once again become tense. Therapist: "What does the child need? What does she need today?" The patient looks the therapist in the eye: "Someone who will leave me be, who understands me and is reliable".

Therapist: "That I can understand. What can we do together to help your inner child at least symbolically feel and grasp this? How about if we were to both took this pillow into our arms, especially yours, and gently comforted the child? "Can you think of any comforting phrases to say?"

Patient: "You're safe now. Be calm and sleep. We are both there for you. Everything will be ok. ....Is it ok if I take the pillow with me until the next therapy session? The therapist nods.

Commentary: The threatened identity has actively freed itself from the previous internal alliance and has entered into a protective alliance which has now shifted into the realm of its own control. Sustainable correction is contingent on the depth and authenticity of the experience. Here, the term "operation" refers to conscious separation, to the denouement of inconsistent "wholes" (I am an unsuccessful, powerless being) and conscious acceptance of coherent, life-enhancing alliances as the establishment of inner wholes (I am likeable even with my own impulses). Both the separation and merging are acts of experience.

2.4.5 Holistically Embedded Confrontation

Objective observation corresponds to Buber's "I-It" relationship. It is a necessary prerequisite for confrontation work. Dissonant constellations offer opportunities for objective observation. Typically, they feature a therapeutic talk. The therapist often notices something inconsistent or eye-catching in the body language. The therapist may mirror a gesture with an inquisitive facial expression, he may repeat a notable term with deliberately equal emphasis, he may comment neutrally on how eye contact was just broken off, he might imitate a sigh (what exactly is the patient trying to say with this?), he may repeat a sentence from the beginning, the message of which has since been negated, he may repeat a sentence that was broken off with a slightly inquisitive emphasis (how should this question originally have ended?) etc. It may be that the beliefs represented do not correspond to the current framework. In this approach, discrepancies are highlighted not to judge or embarrass, but rather to pose the question: what does each party believe? Do different parties exist internally? It could be that both are right to some extent. (?) What is valuable about the suppressed part? Who won't let it be? Generally, this is followed by a clarifying discussion in the hot seat. This is a sensible approach when two opposite poles can be perceived and there is adequate consensus that they express different things.

For example: Therapist: "Is it ok with you if we give the two sides we have just observed a little more space? What do you feel is right? Where is the part that you just mentioned sat? Here on this chair? And, in this case, where does the part that you just allowed to emanate through your body expression sit or stand, as you prefer? There? Is that the right distance? Please arrange it in such a way that it looks correct to you. Thank you. Which of these parts is currently closer to you? Are you able to adopt its perspective and express what you stand for, what your worries are, especially when it appears on the surface? "Now the one over there has heard a lot. How do you think it is doing now? This is most likely not easy. It may have another view of the entire story. Are you able to imagine entering it in order to empathise and express what it all looks like from the other side? A sequence of several role transformations is often helpful in this regard.

It can be interesting to observe how both opposing parties change in this regard and how the positive reference to their core concerns, enabling new solutions or compromises to emerge. This methodology requires finesse. There is a risk that supportive observations may interrupt the process too often or too early, or be attributable to insignificant aspects. This approach is completely process oriented and if you are particularly talented, the safest shortcut to reach the updated core conflict, referred to as "experiential confrontation". The process may be completed with an imaginary meta or bird's eye perspective. Some may find it harrowing or calming to see two injured, unloved and needy beings facing each other from above or two beings, that were not granted the right to exist and passed this on to each other.

Changing roles and perspectives facilitates deidentification, letting go of fixations and identity shifts. It is therefore important that the role change takes place consciously and with the depth of feeling. The therapist is responsible for ensuring that the subject receives adequate time to immerse himself in his new role according to his internal rhythm. Changing perspectives facilitates the development and strengthening of the inner, connective field and the inner observer. This approach does involve passing judgement, but rather careful consideration, establishing understanding, letting each other live and finding compromises.

2.4.6 Overview of the Concurrent Components of the Complex Special Therapeutic Relationship

0. Mindfulness-centred preliminary phase with the therapist.
1. Existential level: I-Thou is directed at the essence of the counterpart as restorative basic acceptance: Therapeutic agape field
2. Real level:
    1. The therapist identifies dormant potential which he recognises and appreciatively designates as a resource for the developed abilities, successes and solution experiences thus far.
    2. He simultaneously registers the shortcomings, conflicts and defence mechanisms, etc., and assesses the patient's psychosocial development phase in order to be a development-promoting counterpart, a side he knows to activate within himself. Objective I-It.
    3. Partnership of "experiential confrontation", the solidarity in exploring the unconscious signals that arise during the therapeutic talks/narration.

4. In the case of "focused regression" and "inner child work", the therapist's role lies in empathy and solidarity for the part of the personality that remains stuck in misery.

2

3. Projection and transference levels (potentially transference detachment, provided this would prevent the therapist from assisting the patient if required, be it as a development-promoting guardian, a partner or as an expert);

   The therapist registers both the transference that is directed at him and his own countertransference in the strict sense (if provided).
4. Mindful "reading" of the countertransference (in a broad sense)
5. Expert level: The therapist provides instructions for exercises, body awareness exercises or roleplays, provided this is reasonable; he lends his expertise, sensitively supports and guides the process and provides feedback.

Cave

Here, it must be explicitly stated that the therapist, in the context of the "special therapeutic relationship" is not a psycho-educational teacher. His expertise and (hopefully) more stable situation are not valid reasons for him to be regarded as existentially superior to the patient. He is not qualified nor asked to interpret, explain or provide advice on the patient's world from his own perspective. Providing suggestions is already more than enough in this regard. However, he can significantly contribute to helping the patient find a suitable solution.

2.5 Learning and Practising – Between Coercion and Self-Determination

2.5.1 Spontaneous Learning by Experience

This principle forms a natural aspect in the development of all humans and animals. Ideally, it is motivated by curiosity. Healthy children delight in discovering new things – as do healthy adults. We learn through success. Spontaneous recoveries may occur when beneficial external and internal circumstances make it possible to achieve something that was hitherto not possible. We also learn to establish a harmonious relationship with our surroundings through involuntary imitative learning.

2.5.2 Guided Learning by Experience

This principle is otherwise referred to as upbringing when we are young, e.g., learning to get on well with others, learning to get dressed, learning to tie our shoes, to brush our teeth, to eat with a knife and fork, how to greet people, how to cross the street, etc. We also learn to refrain from doing certain things. Role models and our relationships with them are crucial in this regard. We learn a great deal from these models, both unconsciously and consciously.

Schools also incorporate learning by experience when they don't just impart theory but also facilitate and integrate practical and sensory experiences.

2.5.3 Therapeutic Learning by Experience

This principle has been adopted in many schools of behavioural therapy. It provides the framework for emotionally corrective new experiences in a targeted manner, such as taking an elevator, initially accompanied by a companion if the patient has a fear of elevators, with the end goal being to take an elevator alone.

Learning through an emotionally corrective experience is an old observation in psychotherapy derived from the traditional field of psychodynamics (Franz Alexander). It is seldom predictable, but for this, it is all the more valuable and usually leads to a positive turnaround in therapy. In behavioural therapy, the new experience is precisely planned and tailored as closely as possible to the problem in light of the disorder. A large amount of research has been conducted on the range of reward effects that stabilise the learning experience. Traditionally, the focus is not on the therapeutic relationship; however, it still plays a crucial role if the patient requires encouragement to take steps that hitherto were the subject of fear. The new experience may succeed in eliminating the symptom, but this is not a given. If it does not succeed, in the future, a therapist should then be in the position to pick up on the blocking counter impulse, decipher its meaning with the help of the patient and continue working with him. This would be useful in combination with the psychodynamic approach.

Negative, punishing consequences have proven to be less effective in behavioural therapy and have been side-lined for therapeutic application.

The critical response to this approach has drawn attention to the fact that strictly guided experiential learning, especially when it is offered as part of a fixed process, is exclusively coercive in nature, that it does not include the impulses of the patient but rather devalues and suppresses them as disruptions and thus does not contribute to a healthy integration of new experiences with previous ones, but instead promotes over-adapted, feeble-minded behaviour.

Effective pedagogy and psychotherapy are capable of allowing the old and new, and internal and external to encounter each other and adequately combine.

2.5.4 Learning and Motivation

When there is no inclusion of experiential components, learning requires the learner to exhibit a high level of willingness to learn or motivation. Learning can be arduous without additional, emotionally supportive motivation, as is observed on a daily basis in pedagogy. This is also reflected in psychotherapy: Patients who are sent to therapy by another individual or institution and not of their own accord typically fail to make any progress. Therapy without self-motivation is a contraindication. The patient's resistance needs to be taken seriously and mutually understood. Following this, the motivation to participate in therapy can once again be assessed.

Completely different results emerge from learning that is met with a high level of motivation due to an additional emotional motivation, such as learning a language to be better understood by loved ones or when learning a skill is particularly important in the patient's environment or it may help to save the patient's own existence or the existence of his family.

In psychotherapy, motivation is regarded as, for example, natural improvement of symptoms of any kind, a reduction in psychological strain, the ability to relate and integrate, security through increased autonomy and self-efficacy, security through material advantages, security through recognition and position within a social scope, security through being loved, an increase in the acceptance of others and oneself, consistency and conformity with one's moral concepts, etc.

2.6 General Aspects of Psychotherapeutic Methods

This concept is the least effective of all the common factors. Depending on the examiner, it varies between accounting for 0.2% and 10 (15%) of the variance in outcomes. For the most part, this goes against the opinions of different schools who each display varying levels of identification with their specific methods and techniques. In behavioural therapy, it is widely believed that the effectiveness of psychotherapy is achieved with a standardised sequence of determined, repeatable and therefore scientifically verifiable and tested methods. The inadequate fit of the scientific paradigm, which employs statistics to apply meaningful approximations to the masses, to the 1:1 ratio in psychotherapy has been explored in detail elsewhere.

Note

Psychotherapy can also achieve success without methods. In this case, attention is solely directed towards to establishment of a relationship and the modulation thereof.

Methods are additional elements to the therapeutic relationship. They are capable of stimulating and enriching the relationship, deepening the focus of awareness and triggering associative thoughts and impulses. However, they can also be used to block, to resist and as an obstacle to a relationship or a way in which the individual can distance himself. Both the patient and therapist are capable of employing methods. Methods can be used to unite both parties and to facilitate the closeness; however, they can also erect walls and prevent encounters. In certain cases, they can also be meaningfully employed to meet halfway with regard to closeness and distance, facilitating secure leeway that provides a good starting point for a process, the course of which remains unpredictable at first. This is also a logical approach.

Methods can open the door to spaces of opportunity, especially if they are only partially structured and provide the scope for spontaneity and projections. Then the methods may gain a playful character. However, methods can also restrict self-initiative and instead adopt a scholastic, challenging performance character. This may also be indicated. That being said, it remains important that the therapist is aware of which aspects he would like to promote in the patient/client (bias) and the intrapsychic and interpersonal "price" thereof.

These individual methods and techniques that have emerged from the various schools have gifted psychotherapy with broad and stimulating diversity. On the other hand, the most popular methods have since been imitated by almost all of the "schools", often under intentionally different names (to mask the fact that they were adopted). In this regard, we can highlight that everything that has proved helpful to many in therapy has gained the most popularity and we regard this as a common good. We shall leave the last word on this matter to our largely quiet (in the sense of professional politics) colleagues who provide care on the front lines to patients who remain largely untreated, Perhaps best summed up with the famous statement "We are the people".

According to Tschuschke [38], the work undertaken by all schools is must more similar than practitioners would like to believe and certainly more similar than association politicians believe. Every now and again, it may seem as though functionaries fight over illusionary front lines, where the practising colleagues have long since established the common area for play.

The flipside to this coin is that, through this unofficial system of information, we risk inadequate skill as a result of the "light" forms of know-how that spread through introductory courses and that the style of the individual may develop into a disjointed hotchpotch. As such, it may be advisable to revisit the topic of "general psychotherapy", which remains a whole despite its vast diversity and provide corresponding advanced courses.

The advantage of technical diversity is that it allows each colleague to select and follow the methodical paths that best suit them. This fosters the establishment of a natural sense of security in working with methods, which can consequently be passed on to the patient.

Literature

1. Singer W. Der Beobachter im Gehirn. Essays zur Hirnforschung. Frankfurt/M: Suhrkamp; 2002.
2. Stapp HP. Quantum propensities and the brain-mind connection. Found Phys. 1991;21(12):1451. Reprinted in Stapp (1993)
3. Stapp HP. Mind, matter, and quantum mechanics. Berlin, Heidelberg: Springer; 1993.
4. Perls F. In and out the garbage pail. Moab, Utah: Real People Press, Bantam Book; 1969.
5. Tononi G. Phi: a voyage from the brain to the soul. Pantheon Books; 2012. isbn:978-0-307-90721-9.
6. Rovelli C. Sieben kurze Lektionen über Physik (Sette brevi lezioni di fisica). Reinbek b. Hamburg: Rowohlt; 2016. /2014.

2

7. Gebser J. Ursprung und Gegenwart. Die Fundamente (Teil 1) und Manifestation (Teil 2) der aperspektivischen Welt. & Kommentarband. dtv, Munich. 1973.
8. Damasio A. Descartes Irrtum. Stuttgart: Schattauer-Verlag; 2001.
9. LeDoux JE. Das Netz der Gefühle. Munich: Dtv; 2001.
10. Zajonc RB. Feeling and thinking. In: Forgas JP, editor. Feeling and thinking. Pairs Cambridge: University Press; 2000.
11. Scharfetter C. Allgemeine Psychopathologie. Thieme,Stuttgart; 1991, 3.Aufl.
12. Tholey P, Utecht K. Schöpferisch Träumen. Falkenverlag, Niedernhausen: Der Klartraum als Lebenshilfe; 1987.
13. Scharfetter C. Psychopathologie. Sternenfels: Sinn/ Ernte/ Aufgabe. Verl. Wissenschaft und Praxis; 2008.
14. Dornes M. Die frühe Kindheit. Entwicklugspsychologie der ersten Lebensjahre. Reihe Geist und Psyche. Fischer. F/M (Erstveröffentlichung 1997); 1993.
15. Metzger W. Gesetze des Sehens. Frankfurt/M: Waldemar Kramer; 1953.
16. Wampold BE, Imel ZE, Flückinger C. Die Psychotherapie-Debatte. Was Psychotherapie wirksam macht. Bern: Hogrefe-Verlag; 2018.
17. Wampold BE, Imel ZE. The great psychotherapy debate. 2nd ed. Routledge/Taylor and Francis Group; 2015.
18. Rudolf G. Die therapeutische Arbeitsbeziehung. Untersuchungen zum Zustandekommen, Verlaufund ergebnis analytischer Psychotherapien - unter Mitarbeit von T. Grande und U. Porsch. Springer, Heidelberg; 1991.
19. Wampold BE, Imel ZE, Flückiger C. Die Psychotherapie-Debatte. Was Psychotherapie wirksam macht. Hogrefe, Bern; 2018.
20. Norcross JC, Lambert M. Evidence-based psychotherapy relationships III. 2018. http://rightslink.apa.org/index.cfm?uid=2018-51673-001&docType=j.
21. Norcross JC, Wampold BE. Evidence-based relationships: Research conclusions and clinical practice. Psychotherapy 2011 – psycnet.apa.or; 2011.
22. Bauriedel T. Auch ohne Couch – Psychoanalyse als Beziehungstheorie und ihre Anwendungen. Stuttgart: Klett-Cotta; 1994.
23. Stern D, et al. Nicht-deutende Mechanismen in der psychoanalytischen Therapie. Das "Etwas-Mehr" als Deutung. Psyche – Zeitschrift für Psychoanalyse. 2002;56(9–10):974–1006.
24. Wimmer B. Daniel Sterns Liebeserklärung an die Gestalttherapie – ein Kampf der Liebenden. In: Hartmann-Kottek L, editor. Gestalttherapie – Faszination und Wirksamkeit. Eine Bestandsaufnahme; 2014.
25. Watson JB. Behaviorism. Revised ed. Chicago: University of Chicago Press; 1913. /1930.
26. Kanfer FH, Reinecker H, Schmelzer D. Selbstmanagement- Therapie. Ein Lehrbuch für die klinische Praxis. Heidelberg: Springer; 2006.
27. Schulte D. Therapieplanung. Göttingen: Hogrefe; 1996.
28. Lammers C-H. Therapeutische Beziehung und Gesprächsführung. Beltz, Weinheim und Basel: Techniken der Verhaltenstherapie; 2017.
29. Heßler JB, Fiedler P. Transdiagnostische Interventionen in der Psychotherapie. Stuttgart: Schattauer; 2019.
30. Rogers C. Encounter-Gruppen. Das Erlebnis menschlicher Begegnung. Munich: Kindler-Verlag; 1979.
31. Schmidt-Traub S Therapeutische Beziehung - ein Überblick. Psychotherapeutische Praxis. 2003;3(3):111–29.
32. Bohart AC, Elliot R, Greenberg L, Watson JC. Empathy. In: Norcross JC, editor. Psychotherapy relationships that work. New York: Oxford University Press; 2002. p. 89–108.
33. Weber G, Schmidt G, Simon F. Aufstellungsarbeit revisted … nach Hellinger? Heidelberg: Carl Auer; 2005.
34. Willemse J, von Ameln F. Theorie und Praxis des Systemischen Ansatzes. Die Systemtheorie Watzlawicks und Luhmanns verständlich erklärt. Heidelberg: Springer; 2018.

35. Selvini-Palazzoli M, et al. Paradoxon und Gegen-Paradoxon. Klett-Cotta, Stuttgart; 1977, 1981.
36. Williams A. Visual and active supervision. London: Norton & Company; 1995.
37. Schwing R, Fryszer A. Systemisches Handwerk. Werkzeug für die Praxis. Göttingen: Vandenhoek & Ruprecht; 2013, 2012, 2010, 206.
38. Tschuschke V, Koemeda-Lutz M, Schlegel M. PAP-S Rating Manual (PAP-S-RM). Rating-Manual zur objektiven Einschätzung therapeutischer Interventionen von Psychotherapeuten unterschiedlicher schultheoretischer Konzepte. Zurich: Schweizer Charta für Psychotherapie; 2014.

Sequences of Order, Reflection and Development Dynamics: The Interplay of Multiplicity Within the Whole

Contents

L. Hartmann-Kottek, *General Psychotherapy*,
https://doi.org/10.1007/978-3-030-87466-7_3

3.1 Introduction and Survey

In this chapter, we shall address the familiar school of thought on the development of our material and neurophysiological structures, as well as the stages of developmental psychology up to death. In both fields – especially in recent years – a wealth of details has been compiled, which can only be summarised with a brief overview in this book.

From a biological perspective, humans, as "precocial animals", are born too early (Adolf Portmann, 1897–1982). As such, the developing baby is exposed to a higher level of vulnerability and gains a particular capability to learn from its surroundings and adjust to its environment and sociological context.

In an ultrafast time lapse, the ontological development mirrors the development of the phylogenetic tree. The delayed growth spurts are marked by development phases and transitional periods for the individual, which also modify social behaviours and provide grounds for conflict, identity shifts, initiation rites, changes in function and role, in addition to a chain of developmental peaks. Overall, this results in the need for resonance across all developmental levels in psychotherapeutic treatment.

The morphologically and functionally different branches of the vagus nerve, which serve the various phylogenetic defence mechanisms, were first discovered only a few decades ago by Stephen Porges. These branches are both of psychotherapeutic and, especially, of structural relevance. On the one hand, they confirm the traditional psychodynamic structural model, and on the other, they contribute to adequate accentuation.

The question of whether the contemporary cultural framework is in the position to exploit the developmental potential granted to us by nature, or which aspects of this potential can be promoted and ignored – and whether this will lead to correction in general and, in particular, in psychotherapy, remains unanswered. Cultural comparisons, such as those explored in ▶ Chap. 7 (and provided they are adequately corroborated by historical information), can be helpful with regard to putting our perspective and benchmarks into context.

Which natural dynamics keep things working from a contemporary perspective? Development spawns needs. Internal or external imbalance seeks balance. Physiological needs seek gratification. Curiosity yearns to discover and explore. The pairing of desire for harmony, peace, joy and balance with healing potential is rarely a lasting event, unless expressly nurtured. However, the mere fact that this is possible is one of the truly exciting developments from our current, scientific knowledge, as the field of tension between imbalance and balance in internal and external relationships is one of the life-long challenging regulatory tasks with which we are faced. As is perhaps evident, it is also one of the primary concerns of psychotherapy.

Interestingly, the concept of structural resemblance is characterised by a sequence of increasing or decreasing complexity. Are the complex structures mirrored in simple ones? And vice-versa: can the smallest basic structures such as one quantum, be found in the largest? This question belongs to the ancient cultural heritage of humanity. If we assume that we, as humans, are a reflection of the basic structure and look at ourselves binocularly, both as a being with a materialised form and an energetic being, which we are ill equipped to sensually perceive – what is the resulting worldview and concept of man?

The majority of philosophers feel alienated by the sciences. Scientists are not often interested or oriented towards philosophy. They are instead encouraged to take superficial phenomena at face value, or rather regard them as reality. As such, the dismay is typically rather significant when a quantum physicist such as Hans-Peter Dürr states, with reference to Einstein, that "there is no matter", matter is made out of energy, or deployed spirit. This unsettling dismay is often averted as everyday sensory experiences seem to confirm the existence of a material world.

There seems to be an analogic principle functioning in our brain and also functioning in the world as a whole: western science-oriented socialized colleagues are used to focus on differences, on details and to puzzle their "reality" by the reducing means of the small slides of sensory perception. These puzzles are also the raw material for our rational power, that is functioning as an on-off-process, as well as our neurons do. This principle is helping to localize, to count, to weight and find causalities. – But at the same time our brain is an universe of waves, floating in a greater universe of waves– as well as our private world and as well as our smallest cell does. This principle of connectedness, synchronicity and resonance is a nonlocal wholeness. Both principles are melting in the world, as well as functioning in our brains, – in ourselves and in between.

The development of the phylogenetically younger, neocortical, dominant brain hemisphere with its cognitive-structuring abilities facilitates the perception of contoured, delimited individual things. It first develops step by step during childhood. In early infancy, neither brain hemisphere is evidently dominant, the right "subdominant" hemisphere does appear to be the somewhat favoured one. The "other side" of the brain represents the ability to resonate, or rather to react to the wave character of reality and go along with it. In this regard, the notion of lateralisation should be treated with a certain level of scepticism; there are functional areas that, in an ideal scenario, work complementary with each other with changing dominance – a state that we can compare or view as analogous to the floating state of a quantum between wave and particle identity. This transformable floating state appears to be our true home in this dimension.

Despite impressive developments with regard to material form, neurophysiology still struggles to grasp its wave character, although has recently started to approach the notion with curiosity and interest.

Previously, it was almost entirely overlooked that an adult brain – all of the brains of researchers – and a child's brain inherently capture a different excerpt of reality and, above all, process what is captured differently as a result of different dominances. The regularly reported, different perception of children by adults is usually dismissed as deficient, not to be taken seriously, scientifically unreliable, or inferior variants. The privilege of being able to select the topic of research, the method and the interpretation thereof is solely possible in the scope of insight found in adult brains.

A truly scientific stance is characterised by registering the observed and reported, taking note thereof (potentially accumulating knowledge) even if it cannot (immediately) be placed in the existing categories – and to tolerate the not-knowing for the time being, even when this proves difficult. Many crises solved in this manner have led to the expansion of the existing systems of categories. As generally know, not-knowing can trigger anxiety and cognitive identity crises in, predominantly, insecure people and lead to denial with destructive tendencies. This occurs both as an individual and collective phenomenon.

It is always interesting to listen to empathic parents, child psychotherapist's and pedagogists, etc., to hear what they find out about the worldview held by children and also what adults are capable of retroactively recalling, despite the fact that as a child they were unable to confide in adults due to a fear of being laughed at (again). In systematic surveys taken by adults, up to 90% of those surveyed indicated memories of situations from very early childhood that seemed to point to "spiritual" aspects, predominantly in the sense of all-embracing experiences of a bond with nature. Here, the term "spiritual" explicitly does not have any religious connotations (see [1–5]).

This chapter shall end with an attempt to provide an overview of the various aspects of development categorically based on materialisation, or rather on the density of the oscillating frequencies which transform the essence of Einstein's formula (energy = the mass of a particle at rest multiplied by the speed of light squared) into a vertical position. The degree of order of realities, regardless of which mode of being they are located in, also appears to be of particular significance.

Due to our construction, we, as people, feel at home precisely in the threshold between the pseudo-corpuscular and energy-connected way of being, as tightrope walkers performing a balancing act on the "Heisenberg cut". Ideally, we walk this line consciously and not half asleep. This increases the chance of waking our latent connective potential from its deep sleep and reintegrating it.

3.2 Neurophysiological Developmental Stages

3.2.1 Epigenetic Influences

Knowledge on epigenetic processes bridges notions on the "Either-Or", with regard to predisposition, and the environment as reasons behind certain qualities solidified at the start of the previous century. Epigenesis facilitates the interplay of both. People who live under extreme circumstances (e.g. stress) may pass the corresponding gen-expression or disposition to the following generations. In this regard, epigenetics helps individuals to adapt to their environment, or, in a sense, to meet the demands of the zeitgeist if this presents any challenges. As such, in psychotherapy, it is thus sensible to enquire as to the attitude of life that the individual's ancestors held or were forced to hold.

The extensive scientific specialised knowledge on epigenetics could fill volumes and will not be explored further in this book.

3.2.2 Embryological Milestones in Development

Embryogenesis begins as soon as the sperm cell penetrates the egg (oocysts). The diploid chromosome sets then split in two into haploid chromosome sets. One female and one male haploid chromosome set then unite to form the nucleus for the new being, which then passes through the blastomere and blastocyst stages as a result of cell division. Finally, on the seventh day, the embryo attaches to the wall of the uterus. In the fourth week of development, the neural tube closes; following this, the nervous system dominates embryonic development. The eye and auditory vesicles first become visible in the fifth week of development, in addition to the hemisphere vesicles. At this point, the head and trunk are roughly the same size. This is followed by the differentiation of the internal organs, the spine, the upper limbs and the face, which is already unique. At the end of the eighth week, the embryo is 3 cm in diameter.

A foetus demonstrates motor, respiration and drinking motions, it drinks amniotic fluid, urinates and becomes conditioned, i.e. it learns to link noises and vibrations (= experiments performed during the 32nd week of pregnancy). It possesses a short-term (30th week of pregnancy) and long-term memory (34th week of pregnancy). Babies become familiar with pieces of music and their mother's voice in the womb.

The prenatal period and birth process have been verified with many hypotheses from psychoanalysts (O. Rank, D.W. Winnicott. N. Fodor, F.J. Mott, G.H. Graber and, above all, L. Janus). Behavioural therapists successfully proved prenatal conditionality.

3.2.3 Lower Limbic Developmental Level

According to Gerhard Roth [6, p. 62 ff] the following areas can be counted amongst the lower limbic developmental level, without attempting to go into any deep anatomical detail: The hypothalamus, the preoptic area of the hypothalamus, the central nucleus of the amygdala, periaqueductal grey, locus coeruleus, raphe nuclei and the vegetative brainstem centres.

G. Roth [6, p. 92] claims that the following affective-emotional circumstances and behavioural tendencies are located here: Flight, paralysis, defence, stress regulation, aggression, anger, fury, joy, sadness. Somatic functions that are key to survival such as temperature and hydration regulation are vegetatively and hormonally regulated at this lower, limbic developmental level, which is closely linked in function with the pituitary gland (hypophysis). Two nuclei (preopticus medialis nucleus and anterior hypothalamic nucleus) which react to testosterone generated in male gonads and trigger typical sexual, aggressive, dominance and territorial behaviour are emphasised in this regard. "The preopticus medialis nucleus is more pronounced in men than women – one of the less obvious anatomical differences between both sexes in the human brain" (Roth, p. 66) – In women, the ventromedialis hypothalami nucleus responds to the oestrogen secreted into the blood by the gonads to stimulate female sexual behaviour.

Digression I – sexuality: Fractionally delayed development of sex drive.

Since the paediatrician Heubner first observed the phenomenon of precocious puberty, i.e. puberty occurring at an unusually early age linked to tumour-related deterioration of the pineal gland (also known as the conarium or epiphysis cerebri, which selectively inhibit pituitary activity) in a 4-year-old in 1898, the puberty-blocking effect thereof has been confirmed time and again. Furthermore, during the night, the pineal gland produces melatonin from serotonin, in increasingly higher quantities during childhood than adulthood. During the day, melatonin levels sink considerably. Accordingly, the pineal gland has a regulating effect on the pituitary gland by unblocking the production of sexual hormones through the accrued melatonin levels. After puberty and increasingly until 20 years of age, radiological calcifications have been observed in the pineal gland. Further increases in the concentrations of corpora arenacea advance at a much slower pace.

Puberty typically occurs between 11 and 14 years. Yet those who have excessively partaken in competitive sports typically enter puberty at a much later stage. It is reported time and again that women who are forced to flee from mortal danger experience a reduction in hormonally regulated cycles as a result of the body reacting to fear and deprivation. Anorexic girls also stop having periods, if they had previously started. On the other hand, puberty often starts earlier in overweight children, with many starting as young as at 9 years of age.

The digression of sexuality, which is allocated to the lower limbic developmental level due to the pituitary gland, which nevertheless leads to the pineal gland, demonstrates how problematic the exclusive layer thinking in our brains truly is.

Sigmund Freud (1856–1939) owes much of his psychoanalysis to the difficulties faced in integrating the sex drive – and also his appreciation thereof as a timelessly significant, superior life instinct, the libido. Even in an attempt to view and understand it in the context of its time and under the particular circumstances of the founders of the school, Sigmund's Freud cognitive achievement remains a milestone for psychotherapy.

3

Note

Digression II: Stephen Porges' (Chicago) polyvagal theory is the result of over three decades of research on the autonomic nervous system (ANS), which has made a profound impact on its functional understanding. The autonomic nervous system connectively spreads over all levels of the limbic system and beyond, down into the periphery, for instance. Nevertheless, it is included in the area of the lower limbic system as it is rooted there with its danger-averting function. It consists of three hierarchically structured sub-systems:

1. The phylogenetically young, myelinated, ventral parasympathetic branch of the vagus nerve; this system is the system of social engagement (SES), which integrates the cranial nerves that are responsible for innervation of heart, facial expressions, head posture, vocalisation, vocal inflection and precise hearing of human voices, thus facilitating nuanced mutual understanding among people.
2. The entire sympathetic system, which triggers fight-or-flight responses when required and extensively prepares for this with the inclusion of the hormonal stress axis, can also be found in the ANS. It puts the organism into alert mode, ready for action and movement. An intuitive assessment of the balance of power in the overall situation is responsible for deciding whether the individual should fight or flee.
3. The phylogenetically ancient, non-boundary-afflicted, dorsal branch of the vagus nerve is capable of triggering immobilisation, i.e. paralysis, as tonic immobility in extreme cases.

These neurophysiologically substantiated findings are primarily helpful for gaining an understanding of traumatic events and also significant for the general psychotherapeutic understanding, especially for manifestations of fear, phobias, psychogenic paralysis, manifestations of overexcitation, depression and compulsion, and for the development of psychosomatic manifestations. It is beneficial to observe the pathogenic conflict from the perspective of the polyvagal theory.

In a gentle way, the archaic vagus branch demonstrably reduces all vital functions (breathing, heart activity, temperature regulation, intestinal activity, metabolism) to a minimum in hibernating animals and carefully revitalises the animal in spring. As such, it is clear that the archaic vagus branch is associated with the pineal gland, where there appears to be a receptor for high-frequency radiation that cannot be constrained by overlaying soil or the cranial bone. Lasering away the pineal gland found in hibernating animals prevents the animal from reawakening. Zoologists also regard the standstill behaviour often exhibited by females in the animal world during the copulation process as a species-preserving derivative of this archaic blocking mode.

In the last century, a comparable, deliberately triggered, temporary lowering of the metabolism in humans is said to have been observed several times by scientists studying exceptional Indian yogis (reported by physicist Elmer Green, 1917–2017).

Neurophysiologically, the above phenomenon concerns the function of the non-myelinated, dorsal branch of the nucleus ambiguous, which is regarded as the oldest part of the vagus nerve.

During the mid-1900s, the psychiatrist Ernst Kretschmer (1888–1964) arrived at the hypothesis that that phylogenetically ancient tonic immobility mechanism must also be activated in the severe form of schizophrenic waxy flexibility, a tendency to remain in a frozen state.

When subjected to sudden, life-threatening stress, the vagus nerve paralyses motor functions, also blocking sensory and cognitive functions. As a result of bradycardia and paralysis of the vasoconstriction of the blood vessels, blood pressure drops and the blood supply to the brain breaks down, leading to a collapse of the vagus nerve due to vasovagal syncope. It remains unclear whether this variant only occurs in its true form in the case of pre-existing damage, and further, the extent to which unconscious learning processes may be beneficially involved.

One form of syncope that was widespread amongst upper-class women in the eighteenth and nineteenth centuries with hypotonically low blood pressure, and primarily occurred when they were shocked, was classified as a histrionic condition whereby the power of imagination played a formative role. The above phenomenon was a somatic symptom disorder shaped by the contemporary zeitgeist that was not detrimental to the view of women at the time, but rather shed a positive light thereon. Similarly, this may also be an indication that the activation of the dorsal vagus has at least a small scope for learning processes based thereon. This social bonus has not existed for a long time. Vasovagal syncopes are a medical topic that is taken much more seriously today.

Blackouts experienced due to fear may be viewed as an extremely mild but nevertheless widespread form of tonic immobility in structurally fully developed individuals. Social-communicative abilities are restricted with all reactions that block stress and a withdrawal from the hostile world is initiated. The polyvagal theory suggests that our understanding of anxiety and panic disorders needs to be reviewed through a new lens.

Manifestations thereof are altered by the developmental stage the organism is at the time of the trauma or the anxiety-inducing event and the intensity thereof. Furthermore, a number of aspects play a role, such as whether the patient is able to cope with the event as a one-off or in a sequence, whether the event can be attributed to a force of nature, or instead to the whim and responsibility of an individual, and which relationship quality is thereby used or damaged. Notably, early and significantly threatening experiences are not stored in the declarative memory that can be consciously accessed, but rather recorded in sensory fragments in the amygdala, initially remaining susceptible to being triggered in isolated forms.

The tendency to socially withdraw and concentrate on oneself in situations characterised by stress and fear also touches on the psychiatric understanding of the aetiology of schizophrenia in part, despite the fact that the origin thereof remains unclear. Social fragmentation or the withdrawal of interpersonal referential fields has a distinctive, inner counterpart in this regard. Renowned psychiatrists from the old school of thought, such as Binswanger and Conrad, observed that this self-transcendence needs to be blocked upon the onset of psychotic manifestations. In other words, the ability to change a position internally, to behold oneself and the current situation externally, from above, from the eyes of one's counterpart, etc. Conrad [7] referred to a "prisoner within" and links this state with the notion of a loss of energetic potential. In the viscous cycle of extreme withdrawal, the loss of familiarity with the world leads to a paranoid phase followed by catatonia, whereby the body also becomes foreign and uncontrollable. When this process is reversed, provided all goes well, the individual travels back through a paranoid phase into the world in which he feels secure to once against establish relationships, on the basis of a trusting relationship with the outside world. Each shift in position that occurs in the inner and outer reference system appears to require an adequately strong, relationship-forming force field, which is weakened by fearful observation.

A small mental pathway is permitted in this regard: If we view ourselves as beings in an energetic in-between state with structurally concretised (materialised) energy on the one hand and free energy fields on the other, and we are aware of our dual-sided nature, then we are more capable of assuming responsibility for this balance. How many accepting, loving references that trigger positive feedback do I need in order to live freely, in balance and securely in my environment? What level of concretising objectivity, or critical or protective differentiation is sensible, appropriate for the situation and also tolerable for the entire system? The stance of the second notion corresponds to an observer's Heisenberg cut, which causes the connected wave formations to collapse in experiments on the quantum level. It would appear that we require a certain level of connectable, free quantum (such as biophotons) emissions for healthy internal and external functioning. (The average value for external emissions has been calculated as 10–27 quantum per second per human.)

The above paragraphs primarily pertain to psychiatry and dream therapy, whereby the focus is generally on the severe stress experienced during the processing of memories. However, the polyvagal theory may also provide a reaction matrix for everyday stress and life.

3.2.4 The Mid Limbic Developmental Level

As with the lower limbic level, the activity of the mid limbic developmental level remains largely inaccessible to the consciousness. It is the area in which emotions are formed, and unconscious feelings and attachment qualities, emotional assessments, reward experiences and corresponding expectations, unconscious conditioning and non-verbal communication occur.

▪▪ Neuroanatomy

In a functional sense, the mid limbic developmental level includes the following core areas: the septal region, the amygdala complex, the basal ganglia with dorsal and ventral striatum, the thalamus and the hippocampus with the surrounding cortex (Roth, p. 68).

An intact septum is a prerequisite for mindfulness, learning and the formation of memories. It plays a role both in cases of strong aggression and in the absorption of water and salt for fever and temperature regulation.

The central nucleus of the amygdala integrates visceral, vegetative, olfactory, gustatory, auditory and affective information and is correspondingly reconnected. The amygdala's basolateral nuclei group is connected to the hippocampus in addition to other areas and informs it of the emotional significance and character of the current situation. According to G. Roth, a number of different ideas exist with regard to the paths of conditioned reactions of fear. These nuclei group also send information to the sensory and limbic cortices as the basis for emotions that are capable of consciousness.

The various basal ganglia (Roth, p. 74) initialise the regulation of arbitrary actions on the one hand and influence preparations for action in an emotional and motivational sense on the other. The dorsomedial striatum area, which is connected to the prefrontal cortex among other areas, is responsible for unconscious regulation and unconscious leaning of targeted courses of action. Inhibitory feedback striata facilitate fine-tuning. The dorsolateral striatum appears to be primarily responsible for highly automated patterns of movement. The nucleus accumbens comes into play with regard to increasing sensory motivation, i.e. rewards. The substantia nigra and the ventral tegmental area jointly form the mesolimbic, dopaminergic reward pathway, which contains a high number of opioid receptors. The pathway's activity increases in the case of unexpected rewards, and conversely reduces in the case of negative experiences. Deviation from the standard is not observed if the outcome meets expectations.

Conscious preparations for behaviour are made for the entire brain in the basal ganglia, where they are compared with movements that have already been stored and released for materialisation in the dorsal striatum. Motivation, emotion, flaws, judgement and adjustment to social situations are further contributed by the ventral striatum. The neural patterns of both striata encounter each other in the substantia nigra, where they are influenced by the other basal ganglia and either inhibited or released, depending on the balance.

The thalamus is the large, articulated neural area located just above the brainstem between the cerebral cortex and the midbrain. It forms part of the interbrain and acts as a modulating relaying and switching centre between the cortex and the lower subcentres. Further detail is not required in this regard. The hippocampus is all that needs to be emphasised. The access codes for retrievals from episodic long-term memory appear to be recorded in chronological order with their individual context in the surrounding cortex. Eric Kandel even succeeded in identifying "place cells" for spatial memory. The transition into long-term memory is notably encouraged by emotional contrast levels that may come from the amygdala or the mesolimbic area.

3.2.5 Upper Limbic Level

At this level, we encounter feelings that are capable of consciousness. They can be modified by environmental influences, upbringing, social motives, ideals and also through pondered consequences. The desire to avoid negative consequences and empathy can lead to the inhibition of impulses, postponed rewards and to an increasing tolerance for frustration in general. This level develops from pre-school to adult age.

Neuroanatomy: This level comprises the limbic areas of the cerebral cortex, the orbitofrontal, ventromedial, the anterior cingulate cortex and the insular cortex. In the former (OFC), olfactory and gustatory sensations are phylogenetically assessed. In primates and humans, this process primarily involves emotions, motivation and assessing consequences. The anterior cingulate cortex is a particularly densely connected centre of integration within and outside the limbic system. It is connected to the motor cortex fields, and is further responsible for the following functions: Awareness, fault recognition, fault correction, the perception of pain as a warning signal and assessing risks. The upper limbic level is often also responsible for assessing the reward and punishment ramifications of an action with the help of access to the memory from the hippocampus.

3.2.6 Cognitive-Linguistic Functional Level

The cerebral cortex of the endbrain is more differentiated still, namely with six layers compared to the three to five layers of the limbic system. Various sensory stimuli are stimulated in the primary areas, and this stimulation is passed on to a cascade of several association areas. In this process, these areas gain awareness in addition to the ability to vocalise and compare the current situation with previously memorised knowledge and experiences.

The three-dimensional world is first constructed in the endbrain from the sensory stimuli, especially though the body and sense of balance. Furthermore, the endbrain exhibits functional differences between both sides: The right parietal lobe facilitates spatial orientation, including changes in perspective, whilst the left is employed for calculating, writing and reading, in other words, with handling language and symbolic and analytical information.

The temporal lobe identifies and categorises objects and understands situations and chains of events through hearing and sight. The speech centre situated on the left side helps to establish the meanings of words and simple sentences. On the right side, the temporal lobe specialises in understanding the more emotional aspects of language such as prosody and the meaning behind facial expressions. As such, this corresponds to the more phylogenetically recent acquisition of the system of social engagement (SES), according to Porges, on the basis of the neuronal afference of the ventral vagus nerve.

The centre that stores and instinctively aligns the grammatical rules of human language can be found in the prefrontal cortex of the cerebral cortex. Emotional influences sent by the limbic system play a ventrolateral role. Furthermore, this centre is also responsible for weighing up assessments, planning actions, conscious decision-making and the ability to think in a creative and original manner.

3.2.7 Learning and Forgetting as a Basic Neurophysiological Phenomenon

(a) Eric Kandel, winner of the Nobel Prize in Physiology or Medicine in 2000, was predominately celebrated for his outstanding research on temporally ordered, declarative memory. This memory stores our knowledge and memories for associated situations and life contexts that consciousness is able to access and the material of which provides reference for the growing personality in its identity.

That being said, it is also widely acknowledged that the memory is not as static as it may appear, but that every memory is a new, creative, easily transformable process to a certain extent. The human memory does not correspond to a precisely captured photographic record.

Learning theory, which can be traced back to Ivan Pavlov (St. Petersburg), strives to obtain exact knowledge on the conditions under which neuronal connections are created through learning, and then once again destroyed. As such, it has subsequently compiled extensive, substantially vetted, detailed knowledge in this regard.

Ultimately, it succeeded in confirming all of the pedagogical efforts to work in a relationship and experience-oriented manner and place active learning at the fore.

(b) Under traumatic conditions, if the hippocampus, the gateway to the declarative memory, fails to work due to sensory overload, partial aspects of the memory will be saved in the neuronal nuclei of the amygdala. These aspects can then be subsequently triggered through similar stimulation without the patient needing to understand the connection.

(c) Parallel to the central memory, decentralised memorisation of traces of experiences is saved in body cells. This was unintentionally discovered in transplantation medicine. It has been proven time and again that preferences such as eating habits, taste in music and other experiential aspects are transferred from the organ donors to recipients.

However, somewhat less spectacularly, touching, certain postures and movements can evoke an individual's own vivid memories. Body therapists have gained extensive experience with the manifold body memory and have developed a range of approaches in this regard [8, 9].

3.2.8 The Intellect as Formed or Structured Energy

In our culture, many view the intellect as the pinnacle of human development. However, others regard it as a primarily cognitive tool. Its function is linked to an intact brain, especially to the phylogenetically youngest part, the neocortex. The brain develops over a number of intermediate steps and spurts.

The developmental psychologist Jean Piaget (1896–1980) is considered one of the most significant scientists who closely, creatively and pioneeringly researched the development of human cognitive faculty, the intellect and, primarily, the development of a child's thinking from birth to adult age, and developed this research in a differentiated manner. He often struggled with regard to criteria of adequacy due to a lack of empirical comparative studies. His interests primarily lay in the question of how human cognitive processes (in the sense of the intellect, the mind) are structured, not in relation to a higher insight in the mind.

Piaget's schema has been widely revisited by a number of subsequent researchers since his time. The sensorimotor stage is the first phase of life, followed by the preoperation stage at age 2, then the stage of clear thinking as of ages 4–7/8, which is subsequently followed by the concrete operational, which ends at 11/12 years. According to Piaget, the formal operational stage is the highest stage of logical thinking, whereby the ability to think about abstract and theoretical concepts and use logic to come up with creative solutions to problems is developed. This rough division into categories also includes a number of sublevels. Subsequent researchers primarily focussed on higher competency, a greater ability to be influenced by external influences and faster stages of development for a child's cognitive development.

The sixth year of life marks a significant milestone in intellectual development, a notion that has been granted increasingly more attention in recent years with the "Theory of Mind". As such, the change of perspective is achieved in such a way that the observer understands the reasons behind why the observed individual has obtained a mistaken view of the true situation. During this process, the viewpoint of another person is absorbed, and the category of a perspective schema is formed. This is primarily an intellectual capability and not an emotional-empathic one. Children are capable of the latter at a much earlier point.

The "Theory of Mind" acts as a basis for mentalising, the primary focus of interest for Peter Fonagy, Anthony W. Bateman and their group of fellow proponents; the ability to mentalise is understood as an entire developmental psychological package linked with the ability to perceive and emotionally empathise developed up to that point, the quality of which depends on the various types of bonding and interaction experiences. A. Bateman and P. Fonagy ([10], p. 14) summarise: "Mentalisation-based therapy is, in itself, not a new method. The tenacity with which MBT therapists work to improve mentalisation is the only new aspect in this regard. In our opinion, all therapies are more successful if they pay consistent attention to the thoughts and feelings of the patient. This notion is nothing radically new; If patients feel that the therapist wants to understand their subjective states, they are likely to respond better to his interventions". The authors define mentalisation as a developmental-psychological construct. Deficits therein do not only likely form personality disorders, but may also be involved in psychopathological process in a milder form. Page 16 addresses the notion of taking the patient's subjective worlds seriously: "We believe that every practitioner, whether or not he explicitly employs a mentalisation method, must see the world from the perspective of his patient; whenever the consistent focus on her inner, mental process, i.e. her subjectivity, is in the foreground, it is intrinsically valuable. It is precisely this resolute concentration on the subjective reality of our patients that characterises all subsequent chapters".

Indeed, the psychoanalytical authors are correct in their statement that this (helpful) therapeutic attitude is not radically new: Half a century prior to their work, in 1962, the humanistic manifesto also demanded that the subjective worlds of patients are taken seriously and appreciated. This stance was adopted in the various humanistic methods of the day. In his experience-oriented, scenic approach, Jacob Moreno began to systemically experiment with changing perspectives and roles in the 1930s and 1920s, followed by Fritz Perls in the 1940s. Moreno allowed relationship problems to be resurrected and varied in distinctive, real-life scenarios, whilst Perls focused on the intrapsychic, dialogical resolution of conflict between partial aspects. Carl Roger's demand for empathy that was established in the 1940s and Heinz Kohut's empathy concept from the second half of the twentieth century further contributed to taking the patient's subjective world seriously.

Psychoanalysis, which initially based its approach to the therapeutic relationship on the spirit of the enlightenment and upheld the fictitious notion of objectively observing therapeutic neutrality for a long time, struggled to fall into line with the subjective reality of patients. This change process still requires time. However, it is likely precisely this withdrawn relationship quality that granted psychoanalysis its comparatively modest efficacy (effect strengths) in the past and now allows it to predict great efficacy in the future, following the transfer of other therapeutic relationship qualities into the perspective. Furthermore, in the second half of the twentieth century, it received a vehement proponent in the form of Heinz Kohut's empathy concept through self-psychology. At the turn of the millennium, it propelled the group of relational psychoanalysts on the edge of the avant garde to the fore, and those surrounding Fonagy, as just mentioned, accepted an effort to include the empathic and intellectual role reversal as a key element of psychoanalysis under the term mentalisation.

An increase in effectiveness was also experienced in cognitive behavioural therapy, which is ranked in Grawe's meta-analyses (from the studies dated 1934–1984) in middle place in terms of quality (however, it led in terms of quantity), following its emotional turnaround with the intentional adoption of conversational therapeutic relationship qualities, leading to the inclusion thereof in Lambers "Bergin & Garfields Handbook" in 2012 as above average yet equally effective (ES 0.93), as the humanistic group of methods, which had previously been superior to cognitive behavioural therapy. Since then, behavioural therapy has continued to cultivate the previously neglected therapeutic relationship.

All groups of methods benefit (1) from interest in the patient's subjective world and (2) from the differentiation and intensification of empathic therapeutic relationship work. As such, they are increasingly merging with one another and adapting.

Note

What exactly is the intellect? The question of the secret behind the process on which intellectual achievement is based constantly preoccupies brain researchers. Neurophysiologist Wolf Singer (born in 1943) states that the underlying strategy of our intellect is the tendency and ability to categorise (Singer [11], p. 171). Coherence has been observed in groups of self-organising, neural activities. The way in which we perceive the world is determined by the organisation of our cognitive tools. One of the basic prerequisites for cognitive recognition is the ability to distinguish figures from the background. This often occurs on the basis of oscillatory activity in stages of coherence that stand out from the surroundings. The process of recognition requires that the coherent systems, the ones that oscillate with each other, spontaneously come together, separate themselves from their neighbours as holistic entities and seek the most likely solution for the interpretation of an image, which initially prevails over slightly less distinct further solutions. As such, these possible solutions or interpretations often conflict with each other, something that we do not usually become aware off. A decisive entity behind this process remains to be discovered. The laws of perception in Gestalt psychology are of clear significance for understanding the brain's physiological processes [12]. It is clear that the oscillation behaviour organised by the cells themselves represents one of the transitions between matter-bound and matter-free energy.

3.2.9 Heart-Brain Coherence as Wave Identity

The existence and significance of our concurrent wave identity has hardly been broached by the mainstream of our scientific community, as western cerebral physiology and neurophysiology traditionally solely focus on the material substrate. Nevertheless, scientists have observed interesting phenomena with regard to connectivity, innervation and the potential coherent electromagnetic synchronisation between the brain and heart. In this regard, a recently discovered neural network that surrounds that heart works independently and belongs to the cardiac vagus branch that can be found in the ventral vagus system plays a key role. When placed under safe and calm conditions, it exhibits variability, or rather oscillation, that is synchronous to breathing. The cardiac vagus branch sends a large number of afferent nerve pathways to the brain, especially to the limbic system. This process may also lead to the formation of common, or rather synchronous oscillation processes. These higher energetic forms of order appear to promote physiological health, provide psychological uplifting and normalise with regard to pathological gene expressions amongst patients. Accordingly, it appears to aid the organism in reverting to its healthy starting position through this ordered, energetic pattern down to its last cell; a kind of reset button.

Publications on these natural, resilient oscillation processes between the heart and brain (especially the brainstem and limbic system) have only been in circulation for the past two to three decades. The neurophysiological prerequisites for these observations have since been recorded, the neuron pathways have been discovered, the electromagnetic processes have been recorded and the accompanying hormonal processes have been described. It is perhaps notable that the heart features its own neural network that appears to be capable of enabling it to perceive and process things independently (Antoine de St. Exupéry surmised this prior to research on the notion: "It is only with the heart that one can see rightly").

The research described above was primarily pioneered by the HeartMath-Institute, USA [13].

All of our oscillation processes can be psychologically influenced, similarly to how we can consciously strengthen them and bring them to the fore. They can also be corrodible and potential disruptive due to their capability to deorganise the entire system and make us sick in the long term.

The notion that inner coherence or congruence is beneficial is a basic conviction held by psychotherapists with a humanistic background, especially Gestalt therapists (and speech therapists with regard to the term congruence). The fact that this has since been scientifically proven in detail is staggering.

There is no doubt that the coherent field does not cease to exist at the material physical boundaries.

A number of studies have been conducted to research interpersonal synchronisation and therapeutic relationships (such as Stéphane [14]) or the correlations between couples and fields within groups.

This section emphasises the opportunity and need to take our wave identity just as seriously as the structural aspects of identity.

3.2.10 Inner Observer

Inner observers correspond to decentralised perspectives with a simultaneously increased inner centring of the consciousness. The hierarchy of inner observers, which is postulated to correlate to experiences, enables us to suspend judgements and tolerate increasing complexity. This predominately also includes the unification of opposites, which is experienced as a sign of maturity.

The amply substantiated, strong findings on the increase of vegetative regulation that is proven to stabilise following quantitatively manageable meditation have succeeded in convincing a wide swathe of the contemporary population of the significance of mindfulness and the secular approaches to the wave of mindfulness.

Patients that we regard as "structurally weak", whereby the primary objective is to catch up on emotional regulation, have found the meditative pause approach to be a blessing. It facilitates the establishment of a differentiated, internal contact behaviour for internal and external relationships to impulses or desires that arise, which enables acceptance and refusal to be graded. The subtle grading of internal and external contact behaviours is the main feature of a matured structure.

Neurophysiological classification can only be hypothetically attempted at this point in time. It is most likely an analogue process or a variation on the process described in the previous section on heart-brain coherence, which facilitates the stabilisation of the developing structure of the personality. Calm (with an emphasis on the vagus nerve), breathing-synchronised mediation enables the substrate of the brainstem and limbic system to oscillate, or rather synchronise, in its original inner orders without disruption and to establish a modulating, centring observer position. The discovery that the emergence of conflicting impulses and emotions can be observed internally from a certain distance and that automatism must exist to immediately realise or assess them signifies a significant step towards personal freedom. It creates a space for decision criteria, for balancing the originally polled desires and for balancing contradictions and personal values.

What is the point in having naturally paired poles for reconciling our emotions if we need to learn how to bring the poles closer together with great difficulty? At the beginning of life, the polarised distribution of neurophysiological islands reacting with pleasure or unhappiness creates polarised reactions of the baby. So the baby can call for care. And it is also able to reward its carers with radiant happiness, therby maintaining their motivations to continue their loving manner. The expressive appeal with a quite great bias makes sense in this situation.

As adolescents, we learn that we are less likely to damage relationships, but rather benefit them by increasingly self-balancing our emotions and desires and directing them to others, if at all, in an adequately subtle form and thereby taking the situation of the other person into account.

3.2.11 The Self as a Field

Through its actions, the self has proved itself to be a partially conscious, self-organising and sociable field, both internally and externally.

Our imagination forces us to quickly reach our limits when we attempt to imagine a living field which can neither be locally assigned nor defined by its limits, but rather can be understood as changing, energetic activity through its current internal and/or external contact behaviour. These qualities can be found in quantum theory.

As our imagination finds it easier to grasp concrete notions, we sympathise with models and are thus susceptible to regarding certain models as reality.

Perhaps the most widespread model is S. Freud's [15] structural model, which established the comprehensive envelope of the self at the time of the object relations theory. According to Freud, this additional space is the location in which significant relationship experiences are stored. The central Ego assumes the role of the organiser in the structural model, whereby the Id encompasses all driving impulses and basic needs that are often in conflict with each other and the Super Ego embraces internalised needs and rules, even if they are not in tune with each other. On the other hand, its positive shadow, the Ego Ideal, if present, beckons the Ego in the direction that rewards an increase in desire and subjective value. Although the model may obtain an additional level of complexity through tension attributable to dissonant internalisations within entities, in general, it provides a simplified, straightforward and practical scope of opportunity for conflict resolution work.

In the case that all tension is released, balance can be attained, even if only temporarily, whereby no new entities and discrepancies are discernible in the self. This potential capability was also acknowledged by Freud.

Current materialisation, i.e. the current resonance in the external and internal field of perception, has always been taken into consideration in the notion of the self harboured by behavioural therapists. In this respect, the self is regarded as contact in action.

Recent psychoanalysis-friendly neurophysiology [16, p. 41 ff] has encouraged us to view the entire right (subdominant) brain hemisphere as our self system, as it stores the image of the self and objects and their relationships, predominately to caregivers and reference persons. It is also responsible for monitoring vital functions and non-verbal communications, including the modulated emotions, stress management and expressions of empathy thereof.

3.2.12 The Right Prefrontal Neocortex

The right, coherent, prefrontal area sits at the apex of the limbic system, which is responsible for regulating the autonomic nervous system and assumes control for the entire right cortex. The right prefrontal cortex matures at 9–10 months of age and begins to code relationships as of 1 year. This leads to intersubjectivity, the sharing of the emotional state. Accordingly, the baby expects to be able to participate in the emotional state of its counterpart. Emotions are modulated in the right prefrontal cortex in addition to regulation of social and emotional behaviours; it is capable of interpreting the slightest indications of emotional changes to facial expressions that generally escape consciousness within a matter of seconds.

That being said, the activity of the prefrontal area is intensified by precisely this conscious attitude attuned to the well-being of fellow humans over an extended period of time. The neurophysiological research group led by Prof. Wolf Singer, Frankfurt am Main. studied monks with extensive experience in meditation with the Dalai Lama's blessing to establish extremely high amplitudes accentuated in the prefrontal region and synchronisation phenomena in the EEG (electroencephalogram) to the astonishment of western scientists. Values this high had never previously been recorded. Meditation changes the brain and increases the grey matter and radiative power in the stimulated regions. One of the monks studied using the EEG, Matthieu Ricard, introduced the general meditation theme of all-embracing, loving kindness, which clearly goes beyond empathy, and provided further insight into his stance in his publications. Empathy alone enables us to dive into the perception of pain experienced by the sufferer. The inherently Buddhist loving kindness and all variations on blessings permit us to radiate encouraging power to another individual, particularly to sufferers. When visualised through imaging techniques, the patterns found in the regions of the brain that are activated in each case exhibit significant differences. Similarly to sensitive mothers (as strived for by Daniel Stern), loving-kindness meditation or the I-Thou field is concerned with the combination of primarily sensitive interpretation, whereby the interpreter completely withdraws, with the ability to impart encouraging power to the patient, adjusted to his individual patterns.

Experienced psychotherapists who consciously direct their approaches towards the well-being of their patients in a genuine manner also develop a suitably accentuated field, at least rudimentarily. Their genuineness and coherent sustainability, etc., can be unconsciously interpreted by patients and other counterparts who are at least reasonably sensitive. Perhaps this is the key secret to the success of the "good therapist personality".

As is tradition, psychodynamically oriented therapists tend to look for transference and countertransference. However, this only represents one element of this potential communication. The patient also unconsciously studies his therapist in terms of countertransference to establish whether his inner attitude and presence feels sufficiently healing or not. Health insurance providers have consequently established trial sessions in a wise pre-emptive move. To this extent, it is psychotherapists who are called on to assume responsibility for our state or order, the chaos and any potential collapse of values or lack of energy radiated from our field.

3.2.13 The Scope of the Imagination

It is at this point that it is important to note that the somewhat vague perspective abilities referred to need to be distinguished from the rich world of imagination and fantasy exhibited in a child's brain towards the end of the infant phase. Accordingly, at this point, the brain is capable of creating several parallel worlds alongside the real one, constructing these and leading a compensatory life therein. The evaluation thereof is varied. These worlds provide certain children with the ability to survive desperate situations of hardship. Shortcomings in reality are often satisfied by imaginary, dream figures. Others become lost in their parallel world and fail to sufficiently find their way back to a realistic self-image in the given reality.

Adults often react with scepticism to the neurophysiological finding that the world perceived in their minds is not only perceived by the child's brain as (almost) real, be this for the good or bad. The above also applies to memories about our past, which may appear to me more or less transformed new creations. As such, broaching the topic of imagination is comparable to walking on a tightrope and the ability to control reality is a tremendous asset.

However, we can use this ability in psychotherapy throughout the course of our life, as will be demonstrated in the practical section. Compensatory wishful thinking in times of need can point us in the right direction by highlighting what is required for good personal development. It helps us develop and participate in the aspects that enhance our lives through a longing for what is missing. Otherwise, we are susceptible to passing on and reproducing the suffered negativity by identifying with the culprits and aggressors.

The consciously applied, beneficial use of compensatory wishful thinking and its symbolic substantiation as an alternative option to relativise the factual reality in its identity-forming abilities in therapy shall be addressed in more detail in the practical section.

3.2.14 Perceptive Capability of Higher Frequencies in the Pineal Gland

The epiphysis (pineal gland) is phylogenetically linked to a third eye on the roof of the calvaria; certain lizards have a third eye with a lens, cornea and retina. The majority of fish, reptiles and amphibians have photoreceptors on the calvaria with neural connections to the epiphysis. Humans also have a floating, neural connection between the retina and the pineal gland. Perhaps surprisingly, the pineal gland is not initially part of the brain: it develops in the foetal mouth, developing from the roof of the mouth and travelling towards the pineal gland. It can first be detected in an ultrasound 49 days after fertilisation.

The discovery that hibernating animals with a pineal gland that has been switched off with a laser no longer reawaken could indicate that the pineal gland predominately reacts to the higher-frequency radiation components of light, which are not hindered by, but rather pass through layers of soil and the calvaria.

Melatonin found in the pineal gland prevents the onset of puberty and regulates the biorhythm of sleeping and waking. It is primarily released by the pineal gland during sleep between 3 and 4 a.m. Animals habitual to regions with long hours of lightness and darkness, such as Arctic walruses are purported to have developed significantly larger pineal glands compared to those found in animals that live near the equator, such as elephants.

Humans who are habitually subjected to extended periods of darkness, such as Icelanders, are reported to be of the opinion that real entities inhabit another dimension (frequency level), e.g. elves. These beings exist in a higher, less dense frequency and are capable of living in and passing through materials that humans perceive as solid, such as rocks and trees. Only true Icelanders who naturally received darkness training during the dark winter months of their childhoods are capable of perceiving these beings. These beings should be respected and remain undisturbed in their traditional settlements, particularly in their traditional meeting points. Such locations can be found marked in current maps of Reykjavik. As the previous generation of Icelanders were not able to simply escape the long, dark winter by emigrating south (which is commonly practised by the current generation) and as such, the entire population was subjected to the natural "darkness experiment", they were required to train their eyesight for higher frequencies from childhood, whether they wanted to or not. Notably, no one who speaks of an encounter with a representative of the elves is regarded as mentally unwell. Quite the opposite: When the island ring road was constructed in Iceland during the previous century, it was diverted by official resolution of the Government committee, despite the extra costs, in order to bypass the land of the elves to avoid disturbing the elves, which could lead to the failure of electrical devices or cause car windows to break, as I was told in Iceland. Other people who live close to the earth's poles also report contact with trolls and dwarves, etc. In Germany, and other similar regions in more moderate climates, these characters can solely be found in fairy tales and children's fantasies. They cannot be found in the perceivable reality of the brains of adults who have not participated in extended darkness training. Traditionally, in pre-Buddhist times, systematic experiences of darkness were cultivated and used as an initiation ritual and path to inner clarity and the development of consciousness in regions such as Japan, the Himalayas and Tibet. However, this does solely entail friendly encounters or positive resonance. In certain cases, these experiences led to projections and resonances with one's own negative aspects, which has long been regarded as a pitfall in meditation. Individual darkness against one's will is regarded as one of the cruellest punishments.

The floating yet nevertheless interesting and receptive potential of the pineal gland with regard to higher light frequencies remains woefully under researched in science.

Alongside melatonin, the pineal gland also produces dimethyltryptamine (DMT) in deep sleep, which is regarded as the consciousness molecule. DMT is the active substance in the South American shamanic drink ayahuasca, which has been traditionally consumed for approx. 1500 years. It is inherently produced in essentially all plants and animals. In humans, it is released in large quantities when we are born, during sleep (in REM phases) and when we die. Potentially also in deep meditation. Mere seconds after oral administration by humans, DMT is broken down by MAO (mono amino oxidase), provided the latter is not blocked by an MAO inhibitor.

Earlier experiences of primordial interconnectedness, regarded as background radiation across time and cultures, may be attributable to the predominant dominance of the right hemisphere of the brain in the initial 3 years of life [16] and/or the activity of the not yet calcified pineal gland. The calcification of the pineal gland starts during childhood and significantly advances during puberty.

The magical, pre-cognitive phase between birth and school age can be found across all cultures, yet remains relatively unresearched with regard to psychological aspects. The pre-cognitive brain appears to exhibit abilities or qualities that are primarily, increasingly superimposed by the functions of the left brain hemisphere once the child is at school age and may potentially even partially regress.

As earlier mentioned, from a neurophysiological standpoint, it is not possible to assert that the hemisphere that becomes the dominant hemisphere at a later point is dominant at this phase. In the first 2–3 years of life, the right hemisphere is clearly the dominant hemisphere. Following this (according to Schore), this dominance switches multiple times until the left brain hemisphere finally assumes its dominant position during adulthood. Indeed, this could be the key behind the different qualities of childhood experiences and perception.

The precognitive phase is often characterised by developmental psychologists as deficient in comparison to the logical thinking that occurs at a later point from the adult's perspective. Certain cultures focus on precisely these resonance forming abilities exhibited in the precognitive phase and the dominance of the right hemisphere as the path to knowledge and fulfilment.

One such culture was the Far East, several thousands of years ago. The Chinese concept of the physical reality as a physical, continual whole was explored by Joseph Needham (cited as per Capra [17], p. 213). The omnipresent chi was perceived as a thin field of undetectable matter that could be densified into solid objects. Individual objects, including humans and their bodies, interacted with all other beings in a wave-like and oscillating manner according to their inner rhythm, in which the fundamental powers of Yin and Yang were also expressed. All beings were connected to each other through their oscillations and simultaneously integrated into higher-level harmonious world events. This notion is strikingly similar to the viewpoint adopted by new physics, which deals with the notion of different formations of a universal, energetic basic substance.

As it would seem, the experience of time is not objective, but rather a function of cerebral experience processing. Children and artistic individuals are more likely to exhibit a fascination with momentary experiences and lose themselves in the here and now in game situations or in creative actions, which cause time to stand still and seem to consequently impart a sense of deep connectivity with nature in most cases. It often requires tremendous effort on the part of adults to truly experience living in the here and now like naive children. In these moments, children perceive everything as animated: Plants, animals and objects. Everything is worthy of empathy and respect. Everything appears to communicate with everything.

Note

Nostalgically, an impressive number of adults indicate memories of a kind of spiritual involvement that may retain their positive radiant power through the course of the individual's life.

Toddlers often hazily perceive a world beyond the visible world, sometimes with protective entities, and other times with more threatening entities. It would seem that these phenomena are dependent on resonance. That being said, it is not a simple task to distinguish this level from projections with adequate certainty.

3.2.15 Vertex Waves and Presence

In certain experiments, such as the lighting of a randomly controlled lamp when the subject presses a button, it has been demonstrated that in instances whereby there is a highly concentrative expectation of an event that is about to occur, EEG potential is registered in the back of the head above the vertex. (In order to visualise this phenomenon, the EEG technology is required to add up a large quantity of EEG from the same person and create an average image starting from the press of the button and moving backwards.) Depending on the testing method, this process either involves anticipatory waves or readiness potential. In terms of energy, this corresponds to forms of intensive attention or conscious presence. This potential is attributed to increased activity of the ARAS (ascending reticular activating system from the brainstem to the neocortex). These waves exhibit a high level of awareness of the present. Whether this potential is also used for the focused mode of consciousness or the receptive remains to be determined.

ARAS activity is supported by the classical meditation posture during ZEN meditation, whereby the straightened posture of the back, neck and shoulders increases muscular tone and leads to a state of pre-tension ready to be released, which can further be intensified by special, symbolic hand and finger gestures (mudras). In this regard, conscious, deep breathing counteracts an excessively sympathicotonic state, instead promoting physiological balance in the vegetative system with parasympathetic guidance and a breath-inducing, ordered heart-brain synchronisation pattern can be initiated that floods the entire order. This state may act as a resonant antenna.

With regard to cultural history, it is notable that almost all depictions of meditators from the Far East, particularly from Vedic and Tibetan traditions, optically emphasise precisely the location of the vertex wave above the crown with a head of hair. (The counterpole of higher receptivity cultivated through meditation is also symbolised in these depictions from the Far East with long, large ears.)

When Native Americans are not adorned with an entire feathered headdress, they often tie a single eagle feather in their hair directly over the vertex. This feather represents the strong, energetic radiance of the wearer, as the down of an eagle's feather can be set in motion by even the finest electromagnetic oscillations, thus displaying the radiation of the wearer (verbally conveyed by Ref. R. Bruyere, Arizona). Achieving higher spiritual development marked by a deep respect for omnipresent nature to which every human being belongs as a tiny part thereof is seemingly a meaningful life goal for the majority of Native Americans.

3.2.16 Biological Radiation of Living Behinds

It is scarcely worth mentioning that measurable radiation exists. Indeed, EEG and ECGs are an everyday part of medicine nowadays. However, the extent thereof is astonishing, considering the available data. Needless to say, the scope of human radiation ([18] / Th. Görnitz) is a question of the technical equipment and the sensitivity thereof.

At present, whether, and if so, what impact scientifically verified electromagnetic radiation from one individual may have on another has not been officially researched and categorised. People are often colloquially referred to as "on the right wavelength", or indeed not. This phenomenon of seeking closeness to certain people due to the fact that it is good for you or, in the exact opposite, trying to distance yourself from others is a general experience that is reflected in our vocabulary.

Cerebral radiation

After all, technically sensitive devices can be used to locate different EEG wave factions, even if they are energetically weak, at great distances from their original location.

- Gamma waves (38–70 Hz) travel 3000–10,000 km (mental concentration),
- Beta waves travel over 10,000 km, (strong attention),
- Alpha waves (8–12 Hz) travel over 30,000 km (relaxation, attention directed inwards)
- Theta (4–7 Hz) 30,000–100,000 km (hypnosis, meditation) and
- Delta waves (lower than 3 Hz), 100,000 to 300,000 km (deep sleep), i.e. almost to the moon. The EEG recordings correspond to the total potential of primarily large, currently synchronised cell groups in the brain, which are found on the outer side of the skull. These figures have been taken from the book written by Frido Mann and Christine Mann [19, p. 152], the latter of whom is the daughter of Werner Heisenberg and ensured that all physical data in the book was checked by Thomas Görnitz, quantum physicist and previous colleague of F. von Weizsäcker (see Mann & Mann, note p. 235).

If the average brain radiates as impressively as described above, what would these figures look like for charismatic individuals who are experienced meditators? In this regard, it is perhaps worth mentioning the impressive, high amplitude, synchronised, frontally accentuated gamma fields observed in meditating monks close to the Dalai Lama, particularly Matthieu Ricard, which were introduced to experts around two decades ago in studies conducted by Wolf and Tanja Singer and their teams. Matthieu Ricard is responsible for opening our eyes to the precise description of the associated inner attitude of active loving kindness, which differs from conventional, primarily caring empathy through the establishment of a strong field of accepting encouragement, solidarity and loving security. Certainly a potentially interesting topic for psychotherapists.

But where does this radiation come from? The majority of living tissue radiates light. F.A. Popp [20–23] successfully used a residual light amplifier to detect ultra-weak cell radiation, i.e. the light of bio-photons (= quantum). Light originates in the cell nuclei, or rather the interspaces between DNA (deoxyribonucleic acid), where the DNA bases guanine, cytosine and thymine encounter each other as slightly pulsating cavity resonators. They bring the emerging photons to a higher level with the help of reverberatory surrounding tissue. An anticipated drop from a higher to lower energy level causes them to emit laser light energy. This ultra-weak light is not directly visible to the human eye. It is approximately as intense as a burning candle over 20 km in the distance. However, technical amplification can be used to objectivise this radiation. Despite the initial scepticism from the academic mainstream, these findings from biophoton research have been verified several times since their original discover.

Whilst we may not possess a direct sensory organ capable of detecting these electromagnetic oscillation processes, we can nevertheless assume that we encounter and react to each other with our radiation.

3.3 Stages of Psychological Development Along the Axis of Time

According to our understanding, development constitutes the overlapping of potential-oriented regulation phases. Our thinking in this regard is, of course, shaped by historically psychotherapeutic familiar development sequences from psychoanalytical tradition, for example, and from ancient, experience-led cultural traditions. Finally, current findings from neurophysiological science have both added to previous knowledge and set new waves, building bridges of understanding to the empirical knowledge passed down through the millennia.

All neurophysiological development is modulated by the development of the "social system" (according to Stephen Porges); in the ideal scenario, this is characterised by potential heart-brain coherence, but when faced with danger, reverts to fight-or-flight mode and, in cases of life or death, activates tonic immobility. A breadth of phylogenetic and neurophysiological research has been conducted on this topic. Ideas on the abortive impact of the polyvagal system may contribute to our psychopathological understanding.

The therapeutic consequences of integrating the above concept are (1) that therapy requires an inherently comparable social sanctuary for each newly arrived being, provided traumatisation or negative impacts similar to trauma have occurred, (2) and that the reversal of symptoms occurs in the reverse order to the onset thereof, namely out of the frozen state through potentially temporarily exuberant stimulation into normal balance. This is the same in both trauma and Gestalt therapy.

The phenomenological profile of the development sequence is shaped by the overlapping of relatively similar maturation (e.g. of the social system), gradual development (e.g. sexuality) and the subsequent and late-onset stages of development (e.g. the myelinogenesis of the corticospinal tracts, the corticothalamic neurons and cortical development that takes place during and after the 20th year of life). The various paths of development are not closely linked to one another, leading to the emergence of various combinations of growth spurts and periods of latency.

Psychotherapists can seek reassurance in the fact that research-based, meaningful interventions have taken place at the start of development in a current project on prophylaxis: In an extensive project on parent-infant psychotherapy to improve parent sensitivity in the development of emotional interactions with their child (PTIP) headed by Christiane Ludwig-Körner, IPU Berlin, therapeutic parental care has been successfully used to create improved conditions for a secure bond quality, thus ensuring that the parents pass on as little of their own pathology, if any, to the next generation as possible.

3.3.1 Existential Level: To Be or Not to Be

(a) Between Martin Bubers "Urdistanz und Beziehung", Primary distance and connectedness.

The referential whole is an existentially all-encompassing, pre- and perinatal cosmic mother-child entity which provides required shielding against sensory overload; the baby's pineal gland is fully active but not yet calcified (what it absorbs and uses has not been extensively researched). At first, the baby's life primarily consists of sleep. During birth, the pineal gland releases DMT (dimethyltryptamine) in order to protect the young organism against excessive stress. It is beneficial to the baby if its waking phases are filled with eye contact and warm vocal sounds. After birth, an "ideal", sensitive mother is able to use her sixth sense to intuitively ascertain why her baby is crying and find a solution. If the initial phase goes well, interconnectedness is joined by a buoyant groundedness. The latter may entirely or periodically overshadow the former, as if it is not possible to achieve dual connectedness, despite that fact that we are all dual beings in terms of the energetic pseudo-matter-spirit identity.

(b) Humans are born one year too early. Our babies are not mature. They cannot run away soon after birth as (other) precocial animals can do (A. Portmann), we are extremely reliant on care and the ability of our caregivers to intuitively understand our expressions. As such, an infant's brain is shaped by its new environment more than if it had remained in the womb for another year.

If an infant's basic needs are not met, the sleeping-waking rhythm is disturbed, shielding against sensory overload is not adequate and contact lacks in existential attention, the infant may be subject to irreversible delays in development or, ultimately, even death. When an infant experiences frustration due to a temporary interruption to its bond with its mother, initially, it is possible to observe an agitated, vital attempt to restore the relationship. After further frustration, this stage is followed by resignation, which though temporary at first, may ultimately become irreversible. This is an example of "anaclictic depression", a term coined by René Spitz, which ultimately leads to death. The example of the loss of a relationship that provides security highlights the observations of concurrent states of attempts to manage trauma, as described in polyvagal theory.

(c) Embryonic development.

Organs that first exhibit activity in the early phases of pregnancy are the heart, which is the first organ that can be detected through its autonomous rhythm (third week of pregnancy) and the pineal gland (seventh week of pregnancy). The physiological development of functional centres structured according to hierarchy, in particular the vegetative integration levels in the brainstem and interbrain takes place during the pre- and post-natal period. If the regulation mechanisms become overstrained and deflect, such as during digestive issues and teething experienced in the prenatal phase, the introduction of external impulses that intercept extreme emotions, provide a mitigating response and support calm regulation is often be beneficial. These processes are internalised once a resilient bond can be formed and subsequently used as a model to calm oneself.

(d) Certain toxic, infectious and traumatic influences due to malnutrition and emotional neglect, etc., may cause the development to come to a standstill. As such, this stage may lead to lasting reduced output for all organs and structures that were in the vulnerable phase of their development when the standstill occurred, e.g. vegetative function centres that fail to provide adequate regulation for the cardiovascular system. The earlier the age at which this kind of damage is incurred, the likelier it is that the damage will be compensated for by neighbouring structures that assume the affected functions.

(e) Regulatory survival mechanisms

In the ideal case, if the baby and infant is lovingly accepted and cared for, it shall consequently experience a feeling of security, connectedness and safety and can focus on its further developmental tasks with an emphasis on the parasympathetic system in its individual sleep-wake rhythm without any disruption. A healthy baby becomes increasingly confident in the world around it and develops a sense of purposeful curiosity. Above all, this curiosity is primarily interested in human eyes. In addition to breastfeeding, the familiar heartbeat, rocking in synchronisation with breathing and the sound of the soft mother's voice all have a calming effect, as the baby was first introduced to the mother's voice through the amniotic fluid.

As has long been acknowledged, from a neurophysiological standpoint, a paired hypothalamic, polarising system of islands of pleasure and unhappiness inherently support the baby with regard to regulation in everyday life [24]. Crying and whimpering are ways in which the baby communicates its distress or lack thereof to its caregivers, whom it rewards and greets, on the other hand, with a relaxed, joyful smile, which corresponds to the entering of the inner island of pleasure.

This phylogenetically ancient reward system functionally progresses in stages in brain structures, which develop at a later point, are younger in terms of phylogeny and travel to the frontal cortex. Its paired structures provide the system with motivation to change. However, these systems also include certain potential for addiction and may accordingly prove a challenging task for integration and inner order.

When an adequate secure bond is formed, a baby typically intrapsychically develops the following structures:

1. An inner-outer boundary, which helps the baby to distinguish between internal and external stimuli.
2. It merges various qualities and sensitivities that come from the same counterpart (object) and also the baby itself with the same original entity. Accordingly, the same person can be both the good and bad mother, who is not always there when needed, for example. Despite these differences, she is still recognised by the baby as the same person (entity). These formative merging processes ideally require a low-stress atmosphere and that the behaviour poles are not extremely far apart from one another, as may be the case with a borderline mother in the worst case.

3. It has been demonstrated that babies connect polar zones of representation, such as islands of pleasure and unhappiness, all basic needs, such as hunger and thirst, satiety receptors, and temperature sensing zones for hot and cold temperatures. in the brainstem and limbic system, and establishes connective fields, partially according to hierarchy, on the basis of integrative physiological regulation systems, which may serve the baby's perception of its state of being as well as the development of the subsequent, centred ego/self system, including its potential, decentralised, inner observers. On the other hand, this process also concerns the formation of a functional whole with an increasingly regulated and balanced centre, and later also with the potential to enter consciousness.

During this development stage, the most developed aspect is the self-worth axis. When this axis is not successfully centred, narcissistic tension between brilliance and plaintiveness shall continue to exist into adulthood, which can cause significant issues for the individual concerned and their environment. This lack of integration is echoed by the outside world, which is threatened by devaluation vs. idealisation and also by division. On the inside, fears of falling into plaintiveness emerge if the ideal of brilliance is not fulfilled (OPD 1.4 and the undeveloped form of 1.7). Mildly insufficient centring is a normal phenomenon for a child's development.

These functionally networked structures should first be able to withstand mild frustration as of the age of 3 and no longer instead retract into functional fragmentation or internal paralysis.

In psychoanalysis, mental overload in the case of structural weaknesses is not referred to as a defence mechanism, but rather as a survival or splitting mechanism. If the structure is able to withstand the frustration, this leads to the emergence of defence mechanisms that indicate the affliction of the inner-outer boundary: Projection and introjection.

(f) We now revert to the beginning of life: It makes sense that the entry into the world should be looked at from a trauma-specific viewpoint, or rather, from the viewpoint of the polyvagal theory [25]. If the discords to the rhythmic needs for adequate care are too large or if stimulations are overstraining, disturbing or even traumatising, or if stimulation is missing entirely, or the needs of the baby remain unsatisfied too often or for too long, or there is no stable, emotionally reliable caregiver, or even in cases of neglect or abuse, this will lead to the overreaction of the sympathetic nervous system and mobilise the archaic fight-or-flight response. As such, establishing a certain level of mistrust in the world and people can be regarded and sensible and reasonable. Anger and fear tend to initially dominate with regard to emotions, and resignation increasingly sets in after long periods of syncope.

When an individual's life is endangered, potentially due to abuse – also in the figurative sense, in situations involving social humiliation and shame – phylogenetically ancient tonic immobility invokes a motoric and emotionally frozen state. When faced with death, the body places the individual into a state of shock, often also with variations of dissociative experiences. This dissociation constitutes a merciful function, as it allows us to not feel any pain, similarly to near-death experiences. It is only in recent years that the neurophysiological principles behind this phylogenetically ancient reflex have been researched (see also ▶ Sect. 3.4).

It is only reasonable to investigate the impact of fear and existential threats not just in extreme cases, but also in a microanalysis. In principle, when under threat, the body responds with delimitative, contact-avoiding retraction as a protective mechanism. However, when exposed to subjective security, the mind is opened with intensified and integrative contact possibilities. This microanalysis helps us to understanding the onset of symptoms. If fear and danger are not intercepted by the intensification of trusting relationships from caregivers and the endangered individual instead allows himself to cling onto the fear and danger, they can instead be separated and lead towards isolation. As such, the trust in social security disintegrates.

(g) The very early phase, referred to by psychodynamicists as schizoid was not defined by Freud, yet nevertheless can be found in neoanalyses from Freud's time on the standard concepts of psychoanalysis. The term "schizoid" was coined by the psychiatrist Eugen Bleuler in 1908 and introduced by him to psychoanalysis.

According to the psychodynamic understanding, the schizoid element, which establishes a subjectively perceived glass wall between the sufferer and the world, is the result of earlier suboptimal, structure-forming experiences of the young baby or infant that led to a frustrated interaction with the outside world and, as such, now causes it to be met with distrust. It may indicate that as a child, the sufferer was not subject to any secure attachment experiences.

There are no correlations to the OPD in this regard.

(h) The corresponding healthy and desirable aspect of this development stage has been recorded as an intentional phase. It is characterised by inner security and a willingness to explore and interact with the baby's surroundings. In this case, the baby is primarily driven by curiosity based on a sufficiently balanced inner equilibrium and the experience of a reliable bond that provides security. Lotte Köhler, who has devoted herself to research on psychoanalytical development, was the first to observe this constellation.

(i) A fragile axis of self-worth that lacks a stable centre may emerge under unfavourable circumstances that lead to structural weaknesses. A strong need for recognition and increased morbidity can alternate with experiences of humiliating insignificance. Idealisation and devaluation alternate in precisely the same manner. As such, the value of the self and others is in constant conflict between being over or underestimated. This conflict is listed in the OPD under ▶ Sect. 1.4.

(j) Regulatory disorders: Pathological forms that we encounter at the beginning of life as patterns of subpar mother-child reconciliation: interpersonal regulatory disorders primarily characterised by overreactions of the sympathetic nervous system on both sides, in both mother and child, e.g. crying babies.

3.3.2 The Basic-Needs Level

This section is primarily concerned with the following physiological basic needs: Nourishment, security/belonging – and secondarily also with development that occurs later, such as sexuality.

(a) The magically animated world is discovered with curiosity and empathy.
(b) Satisfaction and frustration of basic needs, which generally emerge as rhythmically recurring motivations between polar opposites (wake vs. sleep, contact vs. no contact, tactile vs. vocal stimulation, happiness vs. annoyance or unhappiness, hunger and thirst vs. being full, activity vs. passiveness, interest in the outside world vs. interest directed inwards, I and Thou, inner calm vs. adapted resonance (with or without modification). Periods of frustration should ideally be as short as possible and not exceed the boundary of what is tolerable.
(c) Attachment experiences ("Man becomes an I through a You [Thou]" Buber)
This stage is characterised by interdependency, spontaneous, social interactions and modulating communication games. The referential whole is the personal I-Thou relationship.
(d) Emotional regulation between the preservation of a positive self-image with any partial emotional splits (repressions) and somatisation tendencies, and potentially with the polarising loss of self-image if the structure has not yet developed enough or in the case of structural weakness.
(e) Conventional psychoanalytic classification: depressive stage.

Again, this definition finds its roots in a pathological perspective.

Positive attributes: Ability to change and a willingness to communicate. Infants explore the personal relationships they have established with their trusted caregivers. They enjoy interactive games and the initial experiences of self-efficacy achieved by allowing themselves to make their own modifications to the trusted caregiver. If the level of security in the relationship decreases as a result of multiple experiences of frustration, certain impulses are suppressed in order to ensure that the strived-for interaction with the individual who may potentially provide care is not endangered. In this constellation, it may be the case that this pattern of (over) adjustment is included in the baby's personality as an early survival strategy.

Correlations to the OPD: This basic-needs level is included in multiple conflict constellations specified by the OPD.

- Interdependency vs. autonomy (OPD 1.1) provided the objective in this regard is to not permit or trust oneself to leave an interdependency behind at the end of this stage;
- Care vs. autarchy (OPD 1.3) provided the sufferer views the benefits of being cared for in a material sense as more important than the ability to support oneself. However, certain colleagues view a longing for material care as a replacement for love, security and emotional security.

Preferred defence mechanisms: Turning against oneself, somatisation.

- Oedipal sexual conflicts (OPD 1.6) are only included here in a qualitative sense as conflicts in basic needs, not from their initial onset. In this regard, this topic has a delayed unique role (see ▶ Sect. 3.3.6).

(f) The impact of the polyvagal theory on the basic-needs level: Fear leads to a loss of balance, or the right measure.

When we experience fear or enter fight-or-flight mode, our appetite is effectively blocked. Conversely, when we finally escape from periods of fear or distress that have shaped us, we begin to hoard greedily, and eat and drink excessively. As such, once things once again revert to normally, albeit perhaps still relatively meagre, we still continue to reward ourselves on the basis of the survival mechanism applied when we were in distress, namely to take care of ourselves and feel rewarded for doing so, as this mechanism has not been deactivated and replaced. It is perhaps therefore not hard to fathom how addictive tendencies can emerge.

Gestures of peace: Many cultures offer strangers food and drink to signify that the host has automatically or wants to switch from potential friend or foe mode and instead give the all-clear. (It is interesting to examine this phenomenon through the contrasting meanings in different languages: hostis (Latin for foe), host in English, Gast and Hostie (German for host) are all derived from the same word; Similarly, to how the word gift in English and German (meaning poison), is derived from the German word "geben" (to give). This is a cultural example of the dimensions of polyvagal theory experienced over millennia.)

Sexuality is also blocked as a result of fear. When subjected to permanent fear, overwork or extremely high performance, the body typically suppresses the production of sexual hormones in men and women. Conversely: at the end of a war, the accumulated sexual needs of the victorious soldiers are justifiably feared by many women amongst the conquered population.

3.3.3 Positions of the Early Autonomous Level

Discovery of self-centered autonomy, will and relishing in control

(Referential whole: "Me against the rest of the world" in conflicts within the family environment)

Development of voluntary motor functions and sphincteric activity (myelinisation of corticothalamic tracts with connective neurons in the peripheral systems)

Experiences of power and helplessness due to boundaries set by the surroundings

Regulation that benefits voluntary concepts, some of which may be magically glorified;

Regulatory motivation: The ability to block emotions, to distance oneself both internally and externally, self-realisation that aids empathy and interconnectedness that had remained latent prior to this stage.

Conventional, psychoanalytic classification: compulsive (anal) stage

Apportionment of blame both internally and externally (OPD, no. 1.5)

Additional aspects: The autonomous level constitutes one of the most important conflict constellations: The transition from interdependency is often tough.

Forced autonomy: Development from interdependency to autonomy may be accompanied by "taking the bull by the horns" or excessively defiant behaviour. It can also be forced to occur at a certain point if the previous stage was rather meagre, emotionally unfulfilled, conflicting or even abusive.

Banned autonomy: Autonomy can also be subtly prevented by a needy mother figure. "If you don't what to do what I need you to but would rather do your own thing, then I won't love you anymore. Stay small, don't grow up to become your own person and leave our cosy relationship!" Or: "You're horrible if you leave me, I will kill myself if you do as you are all I have". When internalised, this can trigger an obstructive fear in dependents. This stage is certainly multi-facetted in nature.

Correlations to OPD: 1.2 Submission vs. control. What motivates and equips the controlling and powerful subjugator to enter this role? Is it perhaps an additional ability that remains well balanced through social and empathic capabilities that have already been internalised, or is it instead something to fill a gap left behind by a lack of previous love and attention?

The first case provides the foundation for the ability to distance oneself, equality, neutrality, reliability, a sense of duty, orderliness, the ability to plan, sincerity, etc. However, in the second case, one of the following pathological distortions can easily develop: obsessive-compulsivity, the compulsive need to be in control, striving for dominance, pedantry, perfectionism, the urge to clean, hostility towards feelings, precision, a dogmatic claim to power, punctuality, greed and possessiveness, in the case of scant emotionalism.

Preferred defence mechanisms: The isolation of emotions, rationality and intellectualism

General information on the OPD differentiation between active and passive: As per modern trauma theory, it is beneficial to take a deeper look into the parallelism of active to fight mode and of passive to flight mode.

The OPD refers to a form (2.0) of limited awareness of the conflict and feelings with regard to oneself and others with the evasion of objectivity. In this context, this limitation can be understood as a pathological form.

3.3.4 The Early Childhood Integration of Various Strands of Development at Pre-school Age

Change of perspectives, social integration and competition, gender curiosity, fantasy, etc., triggered by a growth spurt in the prefrontal area (cited as per Th. Fuchs)

Gender curiosity as the result of a sex hormone development spurt in early childhood (OPD 1.6) This concerns the integration of gender roles.

Joy in creating imaginative parallel worlds during play and As-If realities as opposed to reality checks. This may be attributable to a developed form of identity conflicts (OPD 1.7).

Interaction of fine motor skills (development of the motor integration centres)

Ability to change intellectual perspectives (theory of mind) and focus attention. Referred to as mentalisation in combination with the empathy that has developed thus far.

Social skills on the one hand: Rivalry, power struggles in social settings, the forerunner to ranking conflicts. Finding one's place in a group.

On the other hand, there are regulatory motives: Sticking to the status quo with a sense of belonging; The desire to achieve self-actualisation in familial and extended social settings.

Moral concepts as intrapsychic criteria of development

Conventional, psychoanalytic classification (a) the histrionic (formerly: hysterical) phase that tests the incentives of the developing sexual identity, (b) the urethral phase, which was assigned to competitive urges by Freud (Who can last the longest...?).

3.3.5 Intellectual Development, Social Development (Curiosity for Expanded Knowledge): The Level of School Children's Developmental Incentives and Discoveries

The joy in discovering motor-sport performance potential

Self-worth stabiliser: Material interests, performance at school or in the child's surroundings, assumption of responsibilities suitable for a child (with the risk of overload in the case of parentification)

Popularity obtained through social efforts as a source of reward

A functional desire for creative-cognitive potential (eagerness for technology)

Regulatory motivation: Acknowledgement of oneself and others, rivalry

Conventional, psychoanalytic classification: Latency stage

There are no correlations to the OPD in this regard

3.3.6 The Integration of Sex Hormones During Puberty

This transitional period attributable to a spurt in hormonal development is a notoriously, perilous challenge that can often land in stormy waters.

Puberty is an extended transition period experienced on all levels: bodily, mentally, sociopsychologically and also in terms of values and ethics. As such, it destabilises the entire personality with regard to its previous references from the ground up. Previous consequences of deficiencies and precipices from earlier self-development and subsequent problem periods once again come to the fore in certain circumstances. Perhaps most commonly, disconcerting issues with self-worth featuring depressive and suicidal tendencies tend to re-emerge if these consequences are directed inwards, or with all variations of dissociation spanning from asocial behaviour to criminality, if they are offloaded externally. Certain individuals may find themselves seeking to make up for a lack of familial attachment and satisfy their significant need for age-appropriate attention and recognition by engaging in asocial clans, groups or gangs.

In psychiatry, we are often faced with a number of adolescents who experience a form of loss of energetic potential during this transition, combined with structural regression of the personality culminating in the initial manifestations of psychosis. An emotionally constructive, manageable, sustainable and age-appropriate relationship can help them to stabilise once again by facilitating the reorganisation and stabilisation of their interpersonal "I-Thou" field. The significance of a social-psychiatric therapeutic healing relationship after the initial manifestations of psychosis in order to prevent further relapses was pointed out by Luc Ciompi back in the 1980s within the scope of his Soteria project. Accordingly, an ordered, manageable, acceptable social relationship cosmos with limited stimulation is used as a therapy and medicine.

Those going through puberty often experience considerable inner chaos. They require clarity, order, values yet also flexibility and understanding, both for encouragement and rejecting.

This transition phase also may cause experienced values, and values that are only prescribed to, to be brought into question, thus reaching new depths. Furthermore, this time is also characterised by great idealism and the search for meaning that may spark new direction in life. This idealism often decreases later in life, yet it can still impart a new accent on the newly born identity. Puberty can be regarded as the rebirth of forms of the personality.

3.3.7 Adolescent Identity

In a typical scenario, adolescents try to find new security and recognition in their new gender role, at least externally, and, at a slightly later point, their own path in the world and place in society. Internally, adolescents are required to integrate their sex drives with their emotions and values and accept their new identity. Initial partnership relationships contribute a great deal to the improvement of self-acceptance. Painful experiences often cause adolescents to fall into existential crises regarding self-worth.

Adolescents who are about to enter a long period of training and studying are often exposed to the particular burden caused by the conflict between the continued, financial dependence on their families and aspirations to lead an autonomous life.

Loyalty conflicts often arise among adolescents who feel guilty for leaving the family that supported them during their training or initial social ascent. This is particularly the case if the family in question has experienced and passed on resentment and social envy. On the other hand, the adolescent may instead become the pride and hope of the family, which provides encouragement. Although, many adolescents feel burdened by their families' expectations for social careers, in the knowledge or feeling that they will never be able to meet them, for whichever reason. As such, they are plagued by a feeling of perpetual failure; the family is turned into a source of negative reflections. The only solution in this regard is to encourage the adolescent to change their reference system. What benchmark is satisfactory and realistic for him, for good reasons? What are his strengths and qualities that his family have failed (or are unable) to see, because they are perhaps beyond their scope of reference? Is a form of grieving required if it becomes apparent that the parents – for whichever reasons – are not capable of ever giving the desired recognition as they are unable to grasp or appreciate the achievements or development of the adolescent? Conversely: Is the older adolescent able to appreciate the lifetime achievements of his forefathers?

3.3.8 Young Adult Identity

Young adults are also subject to the same struggle with similar dissonances to adolescents, primarily still with partial financial dependency. On the one hand, they experience pride in finally being able to stand on their own two feet and master life on their own. They have determined which direction they are headed in for the most part. Professional responsibilities have been defined. The path thereto may be steep, but it has nevertheless been outlined. The challenges posed by an institutional pressure to adapt in the face of fierce competition can trigger borderline states of tension.

In their private lives, young adults typically decide on a partner. If this is reasonably successful, a holistic ability to love will develop and – despite certain necessary and even crises-ridden relationship work – the young adult attains an essentially constant source of happiness and security. They may have even already welcomed a baby to their world.

Women after required to initially implement a balance between work and/or family, which can lead to markedly different conflict potential, depending on the individual. The recent popularity in taking paternity leave, which was certainly not perceived as the norm in previous generations, can help to modulate conflict. Despite successful cooperation, young parents may, after an extended period in the new three-person constellation and demanding care for the baby, exhaust all of their power and no longer have any energy left to maintain the potential source of happiness that the partnership brings to life. The couple may only first mention which different relationship and partnership aspects the individuals take for granted and unnecessarily get caught up in the details after the initial crisis.

3.3.9 Finding Satisfaction in Careers, Family and Society as an Adult

Let us first explore the case whereby everything goes well in the imagination: We gain a certain positive balance and satisfaction with life through the development of our own self-determined lifestyle, the realisation of our own concepts and values, potentially also with the corresponding assumption of social responsibility. Even when everything doesn't go quite as planned, we develop sufficient tolerance for handling frustration with any failures. Through prior experience, we learn that these failures do not ultimately constitute a serious threat to us, and consequently increase our own seasoned stamina. However, internally, the situation is not always so ideal. Each time we hit a new decade, we are reminding of our mortality. Our children fly the nest. Marriage becomes bleak and functional. Particularly successful people with above-average differentiation are susceptible to a form of midlife crisis. "Is that all there is?", is often asked. Whilst this crisis is extremely unpleasant for many, it can also fill life and relationships with a new and previously unimaginable quality and depth.

Life experience can also vary greatly. Strategies from childhood may not work out, similarly to the most important personal relationships, for an array of reasons. There is seemingly no demonstrable external balance in life. However, internally, precisely these fundamental life crises can help to turn a new leaf, whereby the entire personality and its life plan is put into question, which leads to a stock-take under different standards of values, thus dismissing the ineffectual and granting a new stage of life a chance.

3.3.10 Leaving Work, Entering Retirement, Departures and New Beginnings

The end of a career is often perceived and feared as social death. Someone who was viewed as a senior expert in his role and whose opinion was highly valued is now, in the best-case scenario, bid farewell in high esteem. However, in certain cases, this experience is more akin to getting rid of those sitting on the shelf. At this point, previous issues with self-worth can often resurface. All of those who were autonomous from a young age are also susceptible to the risk of self-devaluation at the end of their working life. If their self-worth has primarily been shaped by their work identity over the past years or decades, they are certainly at risk of "pension death". Their will to carry on grows weak.

When faced with this scenario, some may elect to engage in voluntary positions not long after retiring as a substitute for work and consequently end up with almost more obligations than prior to leaving work. Others feel the urge to make up for everything that they missed out in life prior to now. This may be a long-neglected hobby. Or perhaps certain individuals may feel the need to discover the whole, wide world, to live like the rich and the beautiful, which only requires new clothes, a new car and new hobbies, and, ideally, a new partner.

Others prefer to revert to their student days by studying as a mature student in order to pursue old interests or to finally learn more about people, or, also, in order to become a little wiser. Certain individuals may reflect on their life and attempt writing a memoir. Others keep an eye to the future and volunteer in hospitals or hospices to care for dying people.

It is good not to constantly judge. Perhaps something needs to be caught up on first, which has been dangled in front of your face as something you have wanted to do your entire life, but something always got in the way (yourself?) before you got that chance to do it – like a carrot dangled in front of you that you never could quite attain, such as a desire to climb Kilimanjaro or to see the Great Wall of

China. Or perhaps to swim with dolphins...typically, this leads to satisfaction with the current situation and one's life thus far. Naturally, there is a psychotherapeutic need to process and understand the reasons behind this longing. In certain cases, if it does no harm to anyone, desires can be acted out, especially if for fun, which can be experienced as a reward and within the realm of possibility.

Many who retire in good health experience this period as a gateway into revived youth with a desire to conquer the world. They experience the freedom of time and a certain amount of financial freedom, which has unfortunately decreased significantly to become too little in recent decades for a demographic that is growing in size, meaning they need to invest their free time primarily in paid part-time jobs. Indeed, this is certainly the socio-political dark side to retirement.

3.3.11 Early Old Age

When a relative balance is struck between a new start and catching up, and simultaneously physical impairments require a calmer pace for a comfortable lifestyle, a retiree's sphere of life tends to become ever more restricted to those nearby. To family, caring for grandchildren, social activities, foundations, sponsorships, wards (also in animals), home and garden. Many retirees who participate in local cultural life with those of a similar age in events that promote intergenerational, traditional culture, may feel a sense of social integration over a long period of time despite their family situation.

This section explores the need for balance in life and a deeper, wiser perspective. Those dealt a bad hand in life can easily turn bitter. Psychotherapists are capable of counteracting this with resource-oriented perspectives and stimulating inner reconciliation.

As demonstrated by Antonovsky and his research into experiences of the Holocaust and the different approaches to the processing thereof, meaningful, coherent understanding and cohesive thinking can lead to heightened resilience.

3.3.12 Old Age

In old age, inner attitudes tend to diverge. Physical restrictions and an increasing dependency on caregivers experienced at the same time requires a certain level of sensitivity, tolerance for frustration and considerate appreciation. Those who have reached this age also start to become aware of the initial loss of mental capability. Is the individual capable of accepting this with dignity, or does it instead trigger fear, mistrust and aggression? Has sufficient trust been established in the individual's environment? Or perhaps justifiably not? This is often also hard for relatives to judge.

How much contact does the individual require and in what form? Conversations? Music? Gentle touch? Loving silence with each other, bringing issues to the fore, leaving behind objects with positive symbolism?

Does the individual want to talk about his life, perhaps his subjective highlights? Or would he prefer to find someone who also likes to talk about death? Does he find solace in religious notions or does he reject them, instead looking for a conversation on death and immortality in metaphors not concerned with religion? Is it perhaps most beneficial for him to be simply smiled at in peace?

3.4 Development, Preservation and the Interpretation of Wholeness: The Principle of Relatedness

3.4.1 The "Yes" in the Language of Waves

If two waves oscillating at the same frequency run parallel, i.e. in the same phase, their energetic fluctuations decrease, their amplitude increases, both in one direction and the opposite. This results in a high-energy synchronicity pattern as a merged unit of both waves.

The creation of the whole. In neurophysiology, as the phrase goes, "cells that fire together, wire together". The stimulation itself appears to trigger the materialisation process, especially for the growth of synapses in the central nervous station at the point of stimulation. Repetition (exercise) strengthens these synapses, whilst the absence of further joint stimulation weakens them.

With the firing of joint stimulation, the process combines the unconditional stimulus, which coincides in time and space, with the conditional stimulus to form a whole.

Depending on the presence and interests thereof, contact behaviour can lead to a crescendo or decrescendo of mutual resonance. The former provides inspiration, whilst the latter leads to fatigue.

Interpretation of the familiar. Psychotherapeutically, it is also important that the current reality is scanned by our brain in its search for pattern recognition according to similarities to previous experiences and interpreted in favour of familiar concepts (hysteresis principle). This force-economic principle concerns a precognitive projection onto the world. As such, the familiar is instead viewed as unfamiliar and a desired target is typically viewed as a less ideal real image, which leads to a correction of the ideal image. The greater the inner pressure to find a solution, the greater the willingness to accept rough matches. Conversely: The less the individual is emotionally dependent on the outcome, the more finely tuned the matching process can be, or the reality check can function. Accordingly, there is a certain flexible threshold in this regard. Distinctive synchronisation can act as a guide for finding similarities between the stored image and reality (Hypothesis). Constellations, whose striking light spots have been subject to entirely different concepts and images in the west and east are an example of extremely liberal, cognitive approximate interpretations. Indeed, this vast difference in interpretation gives us food for thought. Fortunately, this notion appears to be less arbitrary on the precognitive level of pattern recognition in our brains, which is based on Gestalt laws [26], to the extent that these have been recorded.

3.4.2 Cybernetic Feedback Systems

The art of control or feedback regulation (cybernetics) is primarily attributable to Norbert Wiener, who not only pursued this concept in technology, but also in living organisms and social systems. A key role in this regard is assumed by a thermostat, which is capable of comparing a current value to a target one.

Biological cybernetics is concerned with the research of physiological, primarily vital regulatory circuits of the inner milieus, the stabilisation thereof and adjustment to their target values. Furthermore, it is also interested in complex patterns of movement and the performance-related changes to the innervation patterns thereof in fish, birds and terrestrial animals (Erich von Holst, 1908–1962).

First-order cybernetics first emerged around 1945, followed by second-order cybernetics in approx. 1970 with the work of Heinz von Foerster. The observation of the observed gains its own intrinsic value that triggers recursion processes. As such, the observer is regarded as part of the system.

With regard to psychotherapy, there is a justified interest in the impact of cybernetic thinking in our field. Perhaps most clearly demonstrated in the systemic group of methods, which continued to be referred to as interpersonal and family therapy methods in the 1950s. These methods came from different sources and were subsequently merged. Proponents of humanistic methods then incorporated perspectives on the growth and development of relationships and organisations, and based their approaches on Gestalt psychology, with its understanding of the part and whole. Colleagues in the field of psychoanalysis contributed the multigenerational perspective and aspects of the collective unconscious and the strategic group of methods, which promote a focus on the system and the exploration of the understanding of the individual solely on the basis of his role in the system and the preservation thereof.

In the 1980s, the field of systemic therapy found a new fascination in the linguistic-constructive approaches (Luhmann's autopoiesis concept) to emphasise the intellectual-strategic aspects represented by the Palo Alto group in the 1950s and 1960s, whilst familial, social and emotional aspects were neglected. The latter were somewhat once again balanced with the emotional revolution in the 1990s. In this regard, the significance of the therapeutic relationship for a positive final outcome was also acknowledged and emphases and priorities were corrected. As demonstrated above, it is possible to identify periods of different priorities in the history of the development of systemic therapy.

The significance of various supraindividual perspectives due to affiliations to superordinate systems, or the influence of reference horizons shaped by emotions and their ideals (targets), are the key aspects behind the self-image, the processing of experiences and an individual's motivation for action. As a result of the identification thereof with a common whole, this whole receives its task to exist and the power to fulfil corresponding functions or represent values from its members. A healthy system fluctuates, provides feedback and readjusts its targets as required.

3.4.3 Recentring to Counter Disruptions: Psychotherapy Between Patterns of Chaos and Order

Kurt Goldstein (1878–1965) explored the theory behind the issue of losing and retaining balance at length. According to his findings, a mid-level of stimulation and relaxation is the physiological normal state, and the affected organism should exhibit a holistic readiness to react.

I would initially like to take a big, internal bow to all sensitive mothers who are able to intuitively establish what their baby is currently lacking, know how to remedy this and provide emotionally-regulating comfort. This is a stabilising, early experience that is increasingly internalised by the baby and turned into a resource.

The pedagogical and therapeutic approach to frustration depends on the dose, the age of the child, the preliminary consultation and rules, the resilience of the relationship and the instinctive feeling of the therapist or pedagogist.

Nevertheless, the following section sings the praises of imbalance. Dynamics and impulses for change only arise and develop through inconsistencies. As is essentially summarised by Heraclitus' much cited quote: "War is father of all, and king of all", despite the fact that this quote can only be heard against the background of his philosophy; then it loses all cynicism. For Heraclitus, imbalance is the tension between two poles, such as "becoming" and "decline", which is prudently controlled by a logos.

Kurt Goldstein (1878–1965, neurologist) was also primarily concerned with regulation and the re-establishment of balance within an organism after a disruption. This balance is characterised by an average level of stimulation. Goldstein believed in the dynamics of self-realisation, which can be found in each individual and urge for realisation. This need for development also creates necessary balance on the one hand, and the feeling of satisfaction when new, inner order is achieved on the other. Goldstein held the pre-Socratic thinker Parmenides, a contemporary of Heraclitus, in high regard. Parmenides focused on the enduring, all-encompassing being, the whole. Goldstein correspondingly also identified with the holistic thinking of Gestalt therapy and, following his emigration to the USA, was one of the founding fathers of humanistic psychology [27].

With his contact cycle, Fritz Perls ([28] assistant to K. Goldstein) determined the process behind how deficiencies on a physiological level can cause an organism such as a human to take various intermediate steps to compensate for this deficit in contact with the outside world and its resources in four stages. Contact needs have different emphases at each stage of development, such as the desire to find a partner, for social contact, for mental stimulation, for experiences of value and meaningfulness, etc. If the basic physiological needs have been satisfied, the need for social integration shall come to the fore (as per A. Maslow); followed by needs for mental and spiritual development, which seek orientation and consistent values.

Close observation of this interesting inner dynamic gives rise to the suggestion that perhaps Perl's cycle could be further split into small steps, and that an intermediate step should be integrated for pausing and perceiving the achieved balance. Otherwise, we programme a hamster-wheel pre-set. The contact cycles can be defined as spiral wheels, as what follows always builds on what can before.

3 A fully-developed contact process aims to select suitable and consistent material by default and draw a line against inconsistency. With regard to objectives, it should perhaps rather be referred to as the contact differentiation cycle (◘ Fig. 3.1, ► Sect. 3.4.2, growth cycle/contact cycle).

During periods of great deficiency or existential interdependency, contact with the unsuitable is accepted of out necessity. This often leads to inner dedifferentiation burdened with increasing disharmony. The unsuitable is able to enter. Internalised external material may irritate and become a source of conflicting interference.

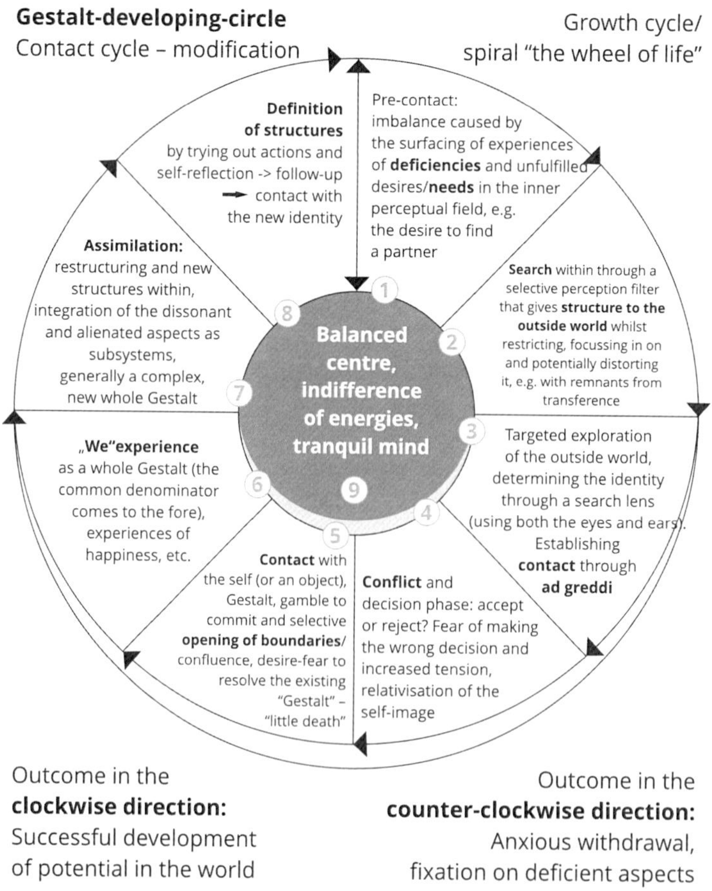

◘ **Fig. 3.1** Growth cycle

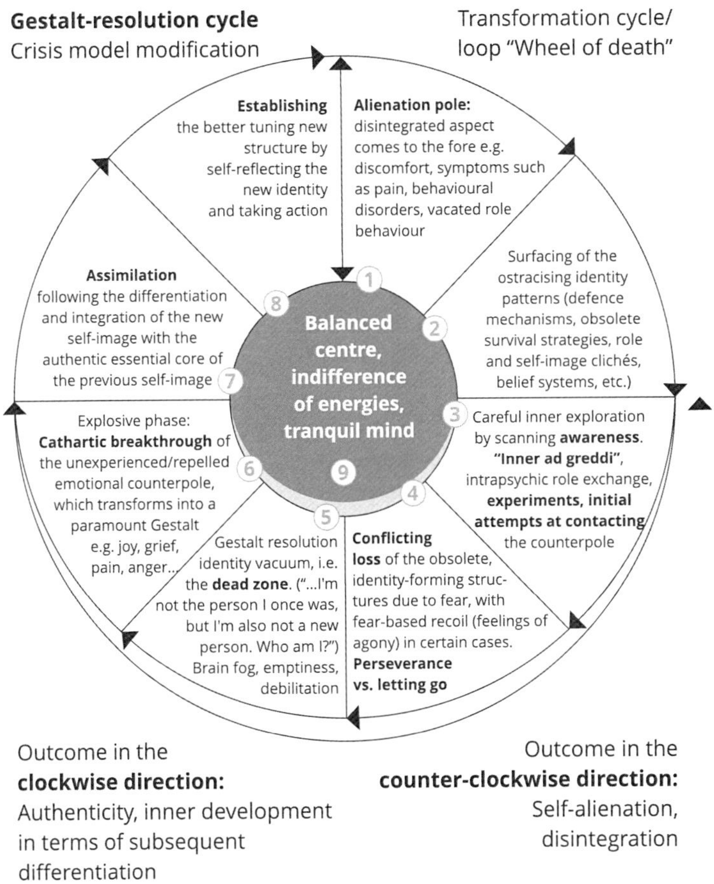

Fig. 3.2 Transformation cycle

Further development of Fritz Perls' cycle of crisis led to the establishment of a process similar to the previous cycle, but instead with a focus on inner inconsistent material, or rather a process based on symptoms, known as the transformation cycle. This cycle is divided in the same number of steps, i.e. 9, and defines the series of steps in traditional therapeutic treatment concerned with conflict solution processes. It provides direction for the sequence of steps and intermediate objectives. However, strictly speaking it is not a manual; it grants a certain level of freedom for creative adjustment to the individual and his process. The described method provides the opportunity for the individual to free or completely cleanse himself from unsuitable external material, helping the individual to treat himself with more sincerity and become more consistent. The terms transformation and development cycle are particularly fitting in this regard (see Fig. 3.2, ► Sect. 3.4.2).

3.4.4 Lock and Key Matches: The Material Counterpart to Polarity

On the purely material level, matches are preformations that support certain formations of the whole. These spatially matching constellations increase the chances of inversely formed structures finding one another. Throughout the school of physiology, particularly immunology, this is the leading principle of internal match dynamics. The materialised aspect of procreation also uses the lock and key principle to ensure the future of the species.

In chemistry, it can also be found in the attraction and matching of electrostatic patterns. In summary, electrical polarity results in superior integration potential, which inherently surpasses the purely material lock and key principle.

3.4.5 Oscillating Between Poles Versus the Neutralising Overruling of Polar Contradictions

Our body is configured to accept a certain amount of natural variation. Indeed, the rhythm of waking and sleeping is an adjustment to our environment, or rather, to the condition on our planet. The closer we live to either pole, the greater the adjustment to our sleep-wake cycle due to seasonal variation in the length of days. This rhythm, which is also capable of a certain range, is linked to other regulating activities such as the activity of the vegetative nervous system and the functional and organ systems that depend thereon.

Furthermore, our body signals losses of balance, such as thirst, hunger, uncomfortable changes in temperature and also a hunger for contact and touch, and later sexual desire and the desire for recognition, love and mental simulation, etc., through mental aberrations on the level of our basic needs – not just in babies. This enables us to register the deficiency and seek balance. We inherently oscillate on this level between deficit and satisfaction and use signals of a deficit as a way to help us navigate through daily life. How strongly do we feel exposed to the imperative of our needs, which we permit children to do? How capable are we of masterfully staving off desires if the situation appears unsuitable or a distressing situation emerges? A reasonable tolerance for frustration corresponds to a mature and adult personality.

Internal signals can also be dismissed for pathological reasons. Perhaps in the given situation, it was the lesser evil to suppress signals than satisfy them and this could now have turned into a no-brainer.

Psychologically, at the beginning of life, it is key that the baby sends clear signals of frustration and happiness into the environment to facilitate the understanding of caregivers. During this initial phase, polarised signals are vital for survival.

The more established early interactions are, the more nuanced they become and likely they are to expressively move from the poles to the centre.

Mothers experience the pendulum swing from picking up a screaming, hungry baby for breastfeeding, including its powerful suckling, to the relaxed, satisfied smile and easy return to sleep as a consistent and comforting act between two poles. Absence is replaced with abundance. Mothers who have internalised this process do not panic when the baby starts to cry, instead helping him to better regulate his feelings with comforting encouragement.

Certain processes are established in a way that entails they can only occur in chronological order and must be perceived in their primarily cyclic process as a whole. The menstrual cycle is an excellent physiological example thereof. It would not be logical to strive for the body to produce the same hormone levels in the same composition on all 28 days of the cycle, thus corresponding to the monthly averages. This would simply be dysfunctional. On a somewhat larger scale, we are used to the sequence of the annual cycle, with its periods of flowering, ripening, harvest and winter regeneration. Knowledge on recurrent processes enables us to take precautions, find solace, security and also anticipation.

These holistic patterns with cyclically recurring or polarised ups and downs have been reflected on by humanity for millennia and we have long attempted to use the experience for emotional stabilisation ("Everything has its time..., a time to be born, and a time to die" Ecclesiastes 3, Old Testament; Heraclitus, approx. 500 BC, Anatolia, Greece: Opposites as a necessity for life dynamics // Kybalion, Egypt, approx. 2500 BC: correspondingly: the swing of the pendulum to one side balances the other.

The notion that opposing poles of force that shape life are simultaneously united in life to form a whole is particularly widespread in the cultures of the Far East (Yin/Yang symbol, I Ching polarity, since approx. 200–300 BC) C.G. Jung's individuation as a process is based on the integration of opposing inner poles such as the anima and animus.

In Gestalt therapy, the hidden, latent pole of unconscious action is rooted out in order to be reintegrated. For each aspect that is accentuated by awareness, another is reversed. Duality refers to arbitrary differentiation, whereby both aspects are merged, cancel each other out and released.

It is also purported that the mutual cancellation of the natural poles, or rather their merging neutralisation, leads to an extremely stable and energetic state. The cancellation of opposites with concurrent high energy is referred to time and again as a characteristic of the experience of enlightenment.

In physics, this can be compared with the zero-point field where there is a balance between plus and minus between the charges and with entangled states, whereby a common system is formed (e.g. a quantum with spin +1 connects with another quantum with spin −1 as the result of a collision and forms a whole with an internally balanced, counter-rotating spiral motion.).

3.5 Fragmentation and Complexity Between Parts and the Whole

3.5.1 The "No" in the Language of Waves

3

When we project a wave that is in the same oscillation but delayed by half a phase onto a sine curve, the latter is cancelled out. The up and down deflections cancel each other out. Despite that fact that our basic understanding of waves, as demonstrated above, may be subject to significant gaps to the complexity of our cerebral charges, this notion nevertheless helps to further our understanding of inherent cancellation procedures.

When transferred to relationships, this cancellation is often observed when we do not want to resonate, as we are worried that we will be walked over and knocked down, or because we are not able to permit ourselves to resonate or have encounters, or perhaps because we strive for objectivity, neutrality and distance, for a number of good reasons. This cancellation process can be regarded as the intentional avoidance of resonant behaviour. Cancellation processes are found both externally and internally in visible and concealed forms.

The clear and voluntary form of "No" is just as crucial to a trusting coexistence as an honest "Yes". An absence of external resonance is often used as differentiation and a protective measure to safeguard the individual wavelength. It helps to set boundaries and handle confrontation. However, when it emerges in an excessive form and without a balanced empathic sister, it can quickly lead to social fragmentation, isolation and potentially dissocial tendencies and destructiveness.

A healthy "No" is one of the key components of internal development and cultivation. The age-appropriate and correct dose of frustration within pedagogical eros is a high art. This capability can be regarded as a modelling tool. A lack of variance in the inner "No" is accompanied by a lack of self-control, self-discipline, distancing, modifying self-observation and self-reflection abilities. With no inner "No", there is no inner freedom of choice and decision, no ability to accept responsibility. When experienced in excess and with a lack of balance and loving appreciation, it can lead to the division of an individual system into a dominant and subservient part, the former of which can pathologically compulsively terrorise the latter. On the other hand, addictive tendencies may become rife if the control capabilities of the inner "No" remain underdeveloped and insufficiently integrated.

The concealed "No" is predominately caused by fear. It often seeks hidden paths towards social withdrawal, intrapsychic tension and division tendencies and the manifestation of psychosomatic symptoms, and, in certain causes, towards delinquency.

As such, life-enhancing, healthy development requires an integrated, appreciative interplay of open, emotionally graded forms of "Yes" and "No", both internally and externally. With respect to waves, the ability to modulate the selected waves in both an enhancing and diminishing way is required.

3.5.2 The Refractive Reflection of the Whole in Its Parts

A whole is characterised by the allocation of its parts to itself in one form or the other, or the existence thereof in participation with the whole.

Temporal reference horizons. How do we define a temporal whole? One day? The entire period of development? An entire human life? The rise and fall of a thing? It is a question of semantics: When I let something begin and end, such as the interaction chain of justifications or finger pointing that gives significance to one detail or another. Is the basic structure of time linear or cyclic? Or a spiral, as a combination of the above? What do I perceive as the meaningful whole? Some view human development as the progressive increase in differentiation and higher development (Jean Gebser, Ken Wilber). This is also the stance adopted by the mainstream. However, does this understanding adequately prioritise the cornerstones of wholeness? According to another perspective on this topic, human development can only be grasped in great cycles that we overlook as insufficient (Vedic scripture, Mayan calendar). That being said, we are familiar with small cyclic fluctuations such as those between the striving to belong and autonomy during the course of life.

▪▪ Social and community spaces

For example, we believe that we are capable of differentiating between the average Englishman, Frenchman and Spaniard as their respective nations have cultivated particular preferences and lifestyles over an extended period of time, which are generally constantly re-internalised by the following generations. The common identity shapes its parts.

Both in a positive and a negative light, political parties invite the populace to identify with its stereotypical clichés that serve and strengthen its voter base. However, the guiding principles of these parties are susceptible to change over the course of time. A revolutionary fringe party may be required to adopt more centralist views to have the chance of winning a majority. Whereas an attractive party that enjoyed a high level of initial success may abandon its pledged values or perhaps violate these values, leading to a loss of the voter base who had always identified with the party prior to this point. The whole can also change its identity, flourish or shrink through the interaction with its parts.

There are also an array of other, more straightforward social groups and family clans that uphold particular ideals or attitudes towards life; those who value their membership in these groups orient their inner and/or outer attitudes in line with the group and suppress their own reluctant impulses in the case of a conflict in a crucial test. Isolation behaviours may cause the mainstream to shake their head: "The individual is clearly acting out of character" and has consequently been viewed as a "black sheep" and kicked out of the group... Indeed, fashion trends and the entire fashion industry rely on the desire of many to belong to the respective avant garde.

We also encounter the spirit of the whole in its considerate and cruel form in the world of work, be this in companies, schools, the authorities, a hospital, a clinic, etc., whereby it, alongside the work environment and atmosphere, requires and asks people to establish whether and to what extent they are willing to adjust to the entire system or instead adopt the position of an outsider. Outsiders can turn into leaders, counter alphas, and change the referential whole. However, certain outsiders may instead be subject to social exclusion or ruin.

Typically, an individual's place in the social referential structure is determined by their own self-image. As the popular saying goes: "In the land of the blind, the one-eyed man is king". In the land of the two-eyed men, the one-eyed man is no longer king. Some face a growing developmental struggle with regard to social roles. For others, social roles provide a supporting framework or hollow façade. Yet others view them as a prison. A change in a person's familiar social reference system can lead to an identity crisis, and likewise the opposite, the liberation to be oneself. Therapists tend to offer unfamiliar and/or beneficial perspectives in order to relativise deep-rooted viewpoints.

We have now explored a number of common examples from the larger social space.

▪▪ Psychotherapeutic groups

An entire book could be written on social spaces and the variations thereof experienced in group therapy. In which ways are social spaces used for healing experiences? How can the group leader ensure that the group forms a trusting, stabilising and encouraging matrix that can heal group members but nevertheless involves lively, creative potential and constructive, confrontational debates? Group therapy is a remarkable tool, which is just as fascinating as the attention to general life, which it reflects on a small-scale.

A dyad, i.e. a group of two people, forms a social space based on a special interest. The ideal notions of couples have emerged independently in various folk tales. They state that it is not an individual alone, but instead a dyad that constitutes that smallest, complete entity that is held together between two poles in loving attraction and appreciation. Certain languages also have coined their own terms for the dual entity, such as Sanskrit, which is surprisingly regarded by philologists as the most structurally developed language, despite its age!

Due to the increased complexity, the creation of a dyad dynamic verges on a higher life form, provided there is not inherently and innately synchronicity in numerous areas. That latter can cause the couple to sink into premature banality with a loss of attraction. As the saying goes: "Admire the exotic girl, but marry the one from your neighbouring village!" However, this does not guarantee success. Each party to the relationship is continually required to meet individual decisions on sufficient matches on various levels and provide adequate loving tolerance for differences in their partner. In many respects, couple's therapy is a tightrope walk, but it is certainly worth the effort.

The recording of interactions is addressed in its own dedicated section in ► Chap. 4 "Interspaces".

3.5.3 Intrapsychic and Social Space of Part-Whole Relationships

The referential whole refers to the intrapsychic dynamics of the current and potential experience of an individual. Countless variants on work on the part and whole can be found in psychotherapy. The author of this book obtained her knowledge on this topic from Gestalt therapy, where it has been applied since the 1940s, partly inspired by the passion for the theatre held by Jakob Moreno and Fritz Perls, who met several times, and also through recollections of lectures by Paul Federn attended by F. Perls in Vienna. Several variations have also cropped up over time, especially since the 1980s. These include Ego-State therapy, a number of variations on Inner Child work, work on the party, family therapy without family, therapy for the inner family, etc. All of these variations may be partially parallel inventions, partly modified plagiarism. The great advantage is that when setting the scenario, it is possible to select or change the current constellation in the family or another social space and the re-updated intrapsychic relationship constellation that has resonated with it back and forth. The subject is responsible for determining which scenario will be accepted on the basis of the constellation that mobilised strong emotions or the greater feeling of being affected with regard to the initial problem.

At any rate, the vast spread of this dialogical part-whole approach is certainly an indication of the great practical value and meaningful significance it can have for psychotherapy.

Some of these approaches work with relatively defined roles such as the inner validator or critic. Others, such as Gestalt therapy instead allow the roles to be reinvented, given a new emphasis and renamed on an individual basis, placing the subject in his own personal field of tension between his own dissonant parts that have shifted towards the outside. The traces of dissonance are a concentrate of his personal track record.

Note

The principal objective is to identify, personify, characterise and associatively explore inner partial aspects and to consult them according to their concerns, significance, time of emergence, impulses, desires, fears, needs, messages, loyalties, coalition efforts, bodily sensations and past and present relationships, etc.

According to the resonance principle, identification with key partial aspects can very quickly lead to an unresolved, formative core conflict. In an individual setting, empty chairs or symbolic locations may serve as backdrops for projection. Distance, posture and line of vision are highly relevant in this regard. In a group setting, forms of constellation work with representatives may be more helpful.

3

Note

The direct establishment of a relationship with the clarification of the relationship that promotes a dialogue between the individual parts is key in this regard. This can get rather lively in certain cases. In Gestalt therapy, the subject is expected or invited to assume or revert the roles for all of his parts. This change in identity and perspectives alone is capable of achieving great healing potential. The courage to speak directly, within the scope of a suitable scenario wherever possible, usually leads to an emotionally charged experience at the point where a process came to a standstill in reality due to trauma or conflict. The mere expression of preserved emotional content (in front of others) and the symbolic expression of heeded impulses is a relieving, healing and often identity-transforming event.

We shall begin by exploring a real scenario or a relationship constellation that has been symbolically condensed with regard to space that continues to burden the individual and cast a shadow. As therapists, we help the subject to view this scenario in a more favourable, genuine or open manner. How can we influence this? When we allow a conversation with the subject's father which he never would have dared enter into back in the day to take place in the mind of the subject, he will then gain the ability to consciously determine the distance at which the problematic person is allowed to be (he could be located outside the room), his posture when he appears, how he is required to stand or sit facing towards or away from the subject, etc., for instance. The same applies to the subject himself. Is he capable of imagining himself simultaneously as a child and the adult he is today? To help the child and put into words what he previously never dared to say? Would he prefer to summon a group of imaginary friends at his side to provide him with energetic tailwind? Or, if we are instead in a group setting, does he permit other group members who are able to put themselves in his situation to stand beside him to express their solidarity? As such, the stage has been set for the negative scenario to take a different course.

Tip

The inner child is incorporated in the scenario with love and appreciation.

It is possible to interpret certain body expressions and breathing to determine whether the event has become deeply engrained. In many cases, following a reproachable conversation, the subject may desire to embrace and reconcile with the problem person, who was previously experienced with ambivalence. This can also occur vicariously.

Note

The objective is ultimately to transform inner dissonances into appreciative (intra-psychic) coexistence.

Work with temporal regression or establishing contact with the inner child can consequently take place following an update of the backdrop to the conflict, which solely reflects the core conflict in a repetitive variation.

Note

It is important that the connection between the updated there and then is experienced in the here and now, emotionally adjusted with adequate expression, healed, and that the subject's own processing of experiences is differentiated from a newly obtained level of observation.

3.5.4 Psychosomatics as Levels of Reflection

Now we shall turn to the level of complexity in which the bio-psychosocial and spiritual condition of the individual being is additionally replicated and reflected on its psychosomatic-bodily regulatory level. Organisms depend on a complex interaction between physiological functional circuits and organs. This complex, cybernetic feedback regulation borders on miraculous. Indeed, it is certainly good that it is not responsible for coordinating human consciousness, as it would quickly become overwhelmed. More research needs to be done on the peculiarities of this notion. Regulatory tasks are primarily performed by the vegetative nervous system, the immunological system, the functional circuit for cell regeneration and genetic activation.

Subliminal regulation tasks benefit from a certain level of balance in the individual. According to the hypothesis, living tissues have a basic rhythm that is synchronised in an entire organism. Perhaps the most prominent thereof is the cardiac rhythm, which can be found throughout the entire body and flows through everything as a coordinating clock. Strong emotional decline can lead to regulation tasks deflecting and introduce pathological learning processes in the psychosomatic area. A loss of internal balance, especially in the case of trauma, can have a disruptive impact right down to the cellular level of the body and genetic activation. Somatic cells are likewise capable of communicating with and regulating one another. Degenerated cells have lost the ability to receive messages from other cells and feedback from their social environment. A mildly activating eustress and mindfully-meditative and, above all, caring treatment of the mind, body and soul can strengthen the heart's natural rhythm, counter any disturbances and re-establish the required balance.

In recent years, the polyvagal theory has shed some light on the graded reaction mode of the vegetative nervous system under fear and threat. The SES (social engagement system), the trusting and calming security found in caring family, friendship and like-minded groups, provides security. It strengthens solidarity through social interactions. If the protection from the group is not available for any reason and the individual feels in danger, he enters fight-or-flight mode, leading to the sounding of the alarm in the sympathetic part of the vegetative system to prepare the organism for maximum performance. If the endangered individual consequently no longer sees any chance of survival, the phylogenetically old branch of the vagus nerve mobilises tonic immobility or a rudimentary paralysis variation thereof. As such, the organism is functionally switched off; in the most severe cases, the energy field may temporarily detach from the body, a form of dissociation which is often compared to a near death experience (a merciful measure for victims of violence and torture).

In terms of psychotherapy, this entails that the chance of recovery is determined by the success of the reintegration of partial functions that have been driven apart. The trick in this regard is to reach out to these partial aspects through empathic identification, which revitalises and reanimates them, grants them a meaningful existence and enables them to once again become valued players in their rightful place in the superordinate whole. With regard to experience, this concerns the revival and reintegration of excluded experience material.

Conclusion

In summary, the outlined increase in the objective or subjective danger appears to go hand in hand with processes of isolation and disintegration. On the psychosomatic level, social isolation and disintegration processes are essentially one in the same. Conversely, it has been proven time and again that people with strong social connections tend to live longer and healthier lives.

3.6 Structural and/or Oscillating Field Either-Or vs. Both-And

3.6.1 The Bias of the Solely Cortical-Dominant Path: A Trip to Plato's Cave and Back

Our sensory organs only capture narrow, arrow-slit excerpts from the entire oscillation reality that surrounds us, e.g. our narrow, approx. 400–800 nm wide observation slit and relays these snippets to the cerebral cortex for processing. The stimulus thresholds of the synapses function according to the on-off principle, to be or not to be. This process simplifies information, removes nuances and creates sharp, indistinct clarity on the perceived. Our dominant cerebral cortex allocates these stimuli that have been filtered through the on-off sieve and puzzles them back together into a complete image from the excluded individual parts. The basal method adheres to the "Either-Or" principle.

Aristotle was a strong proponent of close observation and believed that A is not B, no third possibility is given between two distinguishable items. This is certainly true for the information obtained by our brain through sensory perception. Likewise, it also applies to classical physics with its conventional science with regard to counting, measuring, weighing and statistical calculation, which expands on this notion. Ever since the elementary particles of matter were declared an illusion and replaced by the smaller units of individual dancing threads or clouds of energy, classical physics has consciously come to terms with statistical approximations through the qualities and behaviours of its mysterious and unpredictable living smallest entities: Quantum.

If there is an extreme difference in the size between the perceived and the perceivable, such as gas molecules and humans, then the calculated probability is sufficient, or the vagueness is negligible. The information deficits pertaining to the individual parts cancel each other out in the large mass. However, if there are only minor or hardly any categorical differences in size, such as from human to human (as is the de facto state in psychotherapy), then the statistical tool based exclusively on on-off logic will fail ad hoc. It becomes almost trivial. This lack of difference in size can only be compensated for in the case of size-related questions on average behaviour (e.g. what age do people in this country or others reach nowadays?). If we assume the role of a scientist, it is necessary to reflect on the manner in which the information emerges, the sub-areas it can deliver truths to and which it cannot, or to which it can only provide rough approximations.

To me, it would appear that swathes of people, including scientists, are fascinated and captivated by the digital shadow play that takes place through the sensory-induced, binary processing logic of our brains and perceive this as the ultimate trust, just like the prisoners in Plato's cave. Findings on reality phenomena that appear to exceed this or take a different form seem to threaten "scientific" pseudo-security and are rejected with hostility. Plato outlined this in Book III of the Republic. His allegory is cited below in excerpt form (as per Hirschberger, Vol. 1, p. 103).

> Plato writes that we humans are comparable to prisoners in an underground cave who are tied to a bench from birth in such a way that they can never turn around and only ever face the wall opposite the entrance. A wall the height of a man runs across the cave behind them, towards the entrance. Behind this wall burns a fire. When people pass between the fire and the wall and carry pictures, statues, animal figures, equipment, etc., over the wall, the fire casts a shadow of these objects on the wall of the cave, through which the echo of the sounds produced by the passing people travels to the ears of the prisoners. As the prisoners have never witnessed anything other than the shadows and the echo, these are perceived as images of true reality. If they were able to turn around just once and see the objects of which they have only seen shadows for themselves in the light of the fire, and could hear the voices themselves instead of simply the echoes, they would marvel at this new reality. If they were able to look out of the cave into the sunlight of living people, animals and observe actual things for themselves as opposed to the objects projected in the cave, which were only images, they would most likely be completely dazzled by another entirely different form of reality. However, if they were to attempt to explain to the prisoners who remained in the cave that everything they see and hear is not the actual and true reality, they would most likely be met with complete denial and mocked for their beliefs. Anyone who attempts to release the prisoners from their cave and lead them towards the light of the true world may end up losing his life. …

■ Fig. 3.3 Plato's Cave

Unfortunately, beliefs or concepts that shape identities, as referred to by the Mayans as illusions, are often defended as if it were a matter of life of death. This occurs not just in religion, but also in science. The above similarly applies to many partial truths and approaches in psychotherapy. If they were instead grouped together and a good relationship was established between them, we would gain an abundance of knowledge and become more competent for our patients (■ Fig. 3.3).

3.6.2 Is Wave-Particle Duality Already Latently Reflected in Western Philosophy?

Questioning this combination may perhaps sound bold due to the fact that it couples ways of thinking from various eras and various schools of thought, namely the natural and social sciences. Indeed, prior to the emergence of quantum theory, nobody had directly addressed the wave-particle identity as the smallest unit in us and everything. Perhaps the closest to this ascertainment was Leibniz' (1646–1716) monad theory, in which the smallest substantial units were defined as "un être capable d'action", eternal, indivisible, spontaneously moving, original forces, "un mirroir vivant de l'univers", whereby the monads are referred to as a being (esse) from which action (agere) flows. (As per Hirschberger, Vol. 2, p. 139ff.)

Even if the creative-materialising pole is typically regarded as the most striking feature of human sensory awareness, it would appear that the central intellectual concepts of ancient philosophy also revolved around the potential for metamorphosis between the two modes of being.

Heraclitus (approx. 500 BC) observed the emergence of complementary aspects in opposite poles with fascination and watched them return to the being from which they came, which he perceived as most likely a form of energetic primordial fire. This constant flow, this materialising, matter-metamorphosing and dematerialising process is regulated by a distant, wise logos in recurrent circular motions.

Parmenides (approx. 500 BC), a contemporary of Heraclitus, instead focused his superior cognitive faculty on the pure force and energy of the indivisible, infinite, timeless and absolute whole of the being, which invites us to rest in its arms. According to Parmenides, the temporarily limited, sensually perceivable manifold units are subordinate apparentia, which are constantly subjected to metamorphic processes.

Plato (approx. 380 BC) Referring to Plato entails delving into his ideology and thus encountering his ideals of goodness, truth and beauty. Plato's viewed his notion of ideas as the stronger and more unadulterated reality than the world visible to us, which is only a flawed and shadow projection of the former. His "ideas" are not limited by space or time, they are constant, ideal structural plans of reality, deemed most comparable with mathematical associations. For example: A circle on a blackboard only closely reassembles the ideal circle, which does not include any elongation of the line, dents or unevenness. Ideas can be regarded as the ideal blueprints. According to Plato, we do not gain "knowledge" from sensory perception, but rather an opinion (doxa), which corresponds to common belief. Every basic concept of order (such as equality, diversity, unity, multiplicity) and each archetypal notion (such as the beautiful, the good, the just and the true) is available to each individual a priori (from the very beginning). The individual's soul is only required to become aware of and remember these concepts and notions through recognition; it already perceives them as pre-existent. It is perhaps key to mention here that sensory perception is unreliable; the world of the mind is in a constantly flow of metamorphism. Conversely, the source of trust continues to exist unchanged on the level of the pure mind, the level of the a priori archetypes. Plato regards the eternal being (potentiality) as superior to the constantly changing being (actuality and the process).

In order to bridge understanding, this can be linked to the physical zero-point field in science, which is the counterpart to the maximal energetic and ordered level of being (which corresponds in philosophical terms to Friedlaender's "point of indifference", Leibniz' "pre-established harmony" and Nicolas of Cusa's "unity of opposites", amongst other notions). This level of being may fall into the energetically-lower level of concrete being in a polar of duality-inducing manner.

An understanding of the different probabilities for the synchronisation phenomena of waves, including the harmonic series thereof, may be obtained for the primary emergence of preformed ideal forms from the all-encompassing. In order to illustrate his ideology, Plato almost always employed clear mathematical examples to facilitate the identification of relations in those with no prior knowledge. Indeed, Plato's in-depth approach to Pythagorean knowledge and the mathematical understanding of sound and waves is well known. Oscillation may be capable of creating harmony, and the opposite. Plato's world is based on the derivatives of harmony: the good, the true and the beautiful.

In "Timaeus", Plato envisions the creation of the world's soul, the eternal being, then, as a second principle, the becoming (actuality) and finally a third form of being, a mixed form that connects opposing principles of the eternal and indivisible on the one hand and the temporally restricted and divisible on the other. (The analogy to a quantum, which is only temporarily pseudo-corpuscularly entangled and predominately appears in a floating state between its wave nature and materialised form, is impressive.)

According to Plato's Timaeus, the creator uses forms and numbers to create space out of the chaos that existed beforehand. Five platonic solids can be created from the triangle, the simplest geometrical plane shape: Tetrahedrons, octahedrons, icosahedrons, dodecahedrons, hexahedrons (cubes) which achieve the highest level of beauty through their symmetry. They are regarded as the fundamental building blocks of the cosmos and their mathematical proportions form the basis of symmetry, order and beauty.

Werner Heisenberg [29] postulated that consensus would occur between Plato's assertions in Timaeus on elementary particles and quantum theory if the former is not viewed as a substance, but rather as a mathematical form.

Carl Friederich von Weizsäcker [30] admires Plato's uniform basic principle and the opportunity to achieve a deductive natural science through the unity of nature, which he viewed as clearly more consistent and desirable than the exclusively inductive science of the day.

In contrast to his teacher Plato, Aristotle (approx. 350 BC) demonstrated a keener interest in the sensory perceptible and the concretising, materialising becoming (actuality) thereof from an unlimited space of possibility (potentiality). The latter is not a nothing, but rather a thing that pre-exists. The first matter (gr.: prote hyle) is not yet a being, it shall only become one when a form is present (gr.: morphe). The soul is a materialising, formative, spiritual principle. Aristotle differentiates between the souls of plants, animals and humans, as they each demonstrate staggered spiritual properties. In humans, the anima vegetative and sensitiva are further complemented by the spirit, including self-consciousness, intuition and the rational cognitive faculty. According to Aristotle, the human as a physical being is formed substance, which is granted life and experiences of reality by the soul and acts as a host for the mind and spirit. Aristotle believes that the latter is imparted by an external divine being. As written in "De anima": "Becoming is inherently the realisation of a preceding possibility, a transformation of matter … The world did not simply occur in its material foundation, rather it is eternal and uncreated. … Its entire order can be attributed to a supreme God" (cited as per H. Beck [31], p. 8).

With his complementary pair of opposites, hyle and morphe (matter and form), Aristotle succeeded in accounting for both of the characteristic modes of being of a quantum. However, he was not quite able to grasp the notion of either of them as a whole, similarly to how he was also unable to fathom "actuality and potentiality". He contributed, amongst a number of other remarkable notions, captivating observations of nature, with which he transformed the primary line of sight and made the first steps towards an inductive worldview.

Thomas Aquinas (approx. 1250) combined the two poles of being (potentiality) and realisation as a being (actuality), which were separately established by Plato and Aristotle, into a fundamental principle of being. Everything that exists participates in both components and is permeated by both, even if in differing degrees. The unlimited being is regarded as God, which can be found in limited forms in everything, including people. "Great importance is placed on virtue or strength (lat. virtutes) on the path to perfection. … This encompasses wisdom, justice, courage and moderation. The more a person is able to enter into his possibility of being with the help of his virtues…, the closer he gets to God, who represents the being himself" (H. Beck, p. 12). The inferior soul abilities thus serve the superior, that is to say, the vegetative helps the sensory, which in turn helps the spiritual. The spiritual consciousness surpasses the physical and departs from it in death.

Heinrich Beck [32] expands on the above notion of being with the question of why being is based on self-expression, coming to the conclusion that the purpose of being is love and identity, and, furthermore, that without the necessary conditions of peace on the grounds of justice, humanity would cease to exist.

The introductory question of this section regarding the reflection of wave-particle duality can therefore be fundamentally affirmed if the somewhat different conceptual system of European philosophy is taken into account and translated with a different emphasis: In this case, "being" can be regarded as a spatio-temporally unlimited, universally connected oscillation identity that was referred to as "potentiality" in the antiquity and scholastic philosophy and viewed as a God or with divine qualities. The (creation) actuality that produces the "being" corresponds to the materialisation whereby parts from the network of the energetic basic substance are released and can be transformed into pseudo particles for a limited period. The all-encompassing network is permanently fragmented by the act of creation, which elevates something (ex-sistere/brings it into existence). Due to their oscillating identity, the parts nevertheless have the opportunity to stay in resonance with the "all-encompassing", or not to, as the case may be. They may concentrate entirely on their own creative construction, or the construction of the subjective experiential world. In principle, the separation process corresponds to the Heisenberg cut in experiments on the quantum level. However, it differs from the latter due to the fact that the creative process does not lead to the release of chaos, but instead a release preceded by a conscious and unconscious plan, model or an idea (that can be modified and adapted in a context), an outline that regulates the creative accentuation and formation process.

Mathematics can facilitate an understanding of Plato's vision of the level of innate ideas perceived as spiritual. The "ideea innatae" may represent the probability patterns of high complexity, which are subject to compromising adjustments during materialisation. In layman's terms, this is comparable to viewing the hexagonal basic pattern of the H_2O molecule as water as a design specification for a snowflake. Depending on the air's moisture levels and on the type and intensity of the drop in temperature, the crystallisation process varies, meaning that no snowflake is like the other at the end of the day. The mathematician Ian Stewart [33] dedicated a work to the beauty of nature, a beauty that he acknowledges as a result of the mathematical proportions, order and symmetries concealed within.

3.6.3 Tabula Rasa: The Blank Slate

Steven Pinker [34] entitled his updated examination of the positions in the environment hereditary disposition debate, which are not only scientifically, but also socially and socio-politically contested, The Blank Slate – The Modern Denial of Human Nature. His title is a nod to John Lock, but may also invoke an even older school of thought, namely the Aristotelian school laid out in "De Anima" (III, 4 – cited from [34], p. 626):

> » ... the nous is in a way potentially the objects of thought, although actually (entelecheia) it is nothing before it thinks (noein). This is perhaps best visualised as a blank slate on which nothing has been written.

Thomas Aquinas, a key figure in the medieval school of thought incorporated this notion of tabula rasa and encouraged the prevalence thereof. He celebrates the free creativity of the human intellect. This also includes the notion of almost limitless educability. Or is it perhaps better to refer to this as hope or an illusion? The notion of hope has been an inspirational force behind behavioural therapy for a number of generations, not to mention the sociopolitical projects and waves of spiritual revolution. Pinker struggles to reach a fair assessment of this potential for freedom in the case of regulations that nature asserts through our genetics.

This chapter concerns the free space within the human consciousness with regard to the neurophysiological properties of our brain. The astonishing construction of the brain provides us with both its possibilities and limitations. It goes without saying that with focused attention, we can erect buildings using logical formulas or create entire parallel universes, we can describe slabs in the most impressive of ways, which may, however, on closer inspection, turn out to be the walls of Plato's cave. As such, we remain prisoners in the cave as long as we solely identify with the tool of thinking (the mind), which drastically reduces our possibilities of perception and understanding as a result of its rules. However, this is not a call for the opposite.

In order to return to the point, the above may seem somewhat paradoxical: yes, we do possess the freedom to think logically, which appears incalculable to us, and thus the ability to act in both a constructive and destructive manner. Whether this leads to anything coherent and meaningful, and whether this will relate to the surrounding reality is amenable and not a given in any regard.

Since antiquity, this aspect also has constituted part of European philosophy; it was afforded great significance during the Enlightenment and currently determines the mainstream in the west. When taken as a whole, it concerns the partial truth of a fragment of reality.

Conclusion

The probability that our actions and thoughts are, to some extent, connected to the extended reality may be increased if we use our brains in a balanced way and not only as a blank slate (mind), but with regard to its potential for integrating interconnectedness ("mind"). However, this path requires just as intensive training as the former. With regard to the second path, at current we can be regarded as illiterate or perhaps even "special needs pupils".

3.6.4 The Rediscovery of Resonance as a Healing Factor with the Example of Heart Rate Variability

The significance of the therapeutic relationship as the top interpersonal, resonance-promoting dimension and healing factor has been discovered and proven in the field or psychotherapy and beyond. However, the brain itself has also become a focus of interest as a potential resonator [35].

The unbelievably powerful EEG synchronicity phenomena, as found amongst meditating monks, have also been discovered in other locations (Matthieu Ricard, 2005, Wolf Singer).

Neurocardiological research over the past one to three decades has led to findings on a self-regulating, neuronal vagal network surrounding the heart which, provided it is healthy, can change frequency depending on breathing. Decreasing heart variability is an adverse, prognostic indicator for the regulatory capacity of the vagally and sympathetically innervated heart. The cardiac self-regulating network is connected to the amygdala, the cortex and subcortical structures, which may directly affect the emotional qualities in the heart ([13], p. 52). The heart is also influenced by the cingulate gyrus, the orbitofrontal and insular cortex, which may cause upheaval.

3

Note

The coherence of the heart rate, i.e. its resonance, can be measured through heart rate variability (HRV). At a frequency of 0.1 Hz (every 10 seconds), the heart most clearly produces a harmonically ordered, sine wave-forming pattern. The state is brought on by positive emotions such as happiness, love and gratitude and diminished or destroyed by fear, anger, rage, etc.

It would seem that increasing the strength of physiological regulation is also a way in which the associated biological systems can be stabilised. Positive feelings help the system as a whole slip into a coherent and harmonious state.

Tip

Experiments have successfully employed self-regulation exercises to demonstrate that a breathing rhythm of once every 10 seconds leads to a deepening of HRV, an increase in coherence and an increase in EEG amplitude. When the vagal afferent nerve pathways fired more often than average, the pain pathways were blocked in the thalamus. HRV correlates to psychological resilience, particularly with regard to flexibility, self-regulation and the ability to socially adapt.

Parallel to the HRV modulated by the vagus, an increase in the ability to emotionally regulate, socially interact, be personally direct and the sense of coherence with regard to connectedness and commitment were also observed. Cognitive tasks are consequently easier to solve. There is a common phrase that advises us all on how to approach unsettling situations: "Take a deep breath first..."

Expanding on Porges' notion, it could thus be asserted that we, as humans, with our autonomic nervous system (ANS), are not limited to fight-or-flight and paralysis with regard to response functions, but rather that we can trigger and use the SES (social engagement system) through self-regulation (McCraty, p. 57). Vagus stimulation has also opened the door to a number of new treatment approaches for the depressed, epileptic and patients of pain in the clinical field.

Indeed, self-regulation strategies can be beneficial for those who are able to trigger the vagus as a kind of biofeedback method.

Tip

Research conducted by the HeartMath Institute has demonstrated that a breathing technique of six breaths per minute can lead to a coherent heart-brain rhythm with a number of subjectively beneficial experiences.

3.6.5 Traditional Oscillation-Based Healing Systems and Patterns of Order for Our Reality

In Southern India, the remnants of a previously highly differentiated culture surrounding sound systems, whereby the human body is used as a cavity resonator and understood as a form of musical instrument with its own fundamental oscillation, are preserved from the understanding of the oscillation of reality found in Hindi tradition ("Nada Brahma" – the universe is sound). Three octaves can be found between our heads and feet: the higher frequencies closer to our heads and the lower downwards towards our feet. The various organs are located at different intersections of the introduced oscillations and can be stimulated in the case of diminished function or dysfunction through special sound combinations, provided the practitioner is knowledgeable and carefully trained in the extensive range of the traditional scale systems (Ragas) with their half and quarter intervals. In this regard, the objective is to achieve adequately harmonic overall oscillation that matches the personal fundamental tone. Remnants of this culture can still be found today in popular medicine as far away as Mongolia. At the end of the last century, a version thereof simplified to the current western understanding and its much less differentiated hearing, "sonology" (see U. Pühn), was introduced in Europe by the Indian physicist (and Nada Brahma follower) Vemu Mukunda.

It has not been verified whether there was any correlation between the western, or the Mediterranean, and the ancient Indian culture, yet it has nevertheless been suspected.

Pythagoras (570–510 BC) is said to have extensively travelled the ancient world (Babylon, Egypt), even reaching India and compiling the knowledge of his contemporaries, particularly mathematical knowledge, which was also significantly influenced by China at the time.

According to him, each musical interval corresponds to a numerical ratio. The relation thereof was precisely measured and calculated by Pythagoreanism on the monochord, a resonating body with strings of equal length. For example, an octave corresponds to a ratio of 1:2, a fifth 2:3, a fourth 3:4, a major third 2:5, a major sixth 3:5.

Note

Each vibration generates harmonies that may or may not be audible. In other words, each vibration produces a mathematically harmonic sequence (of harmonies) that may potentially sound as a chord. This forms the basis of all tonal systems found throughout the world and is the result of the universal laws of vibration. The first six partial tones are a series of whole numbers (1, 2, 3, 4, 5, 6). They form the harmonic major chord that is familiar to our western ears. The partial tones 7, 11, 13, 17 are not integrated in this system. Some higher placed harmonics seem to be dissonantly floating, as if they are out of tune; musically, they appear to push for transformation and resolution, but nevertheless retain their natural place in the harmonic sequence. There is also an analogous series of subharmonics, which is inherently in a minor key.

Plato demonstrated a fascination with Pythagorean knowledge of ratios and the harmonies thereof, and created his five Platonic solids from regular triangles, squares and pentagons based on this geometric understanding. He regarded these solids as the fundamental basis of all elements; earth, fire, water and air, which were all firmly part of the ideas of the antiquity.

> In "Timaeus" (56 b-c), Plato writes: "Now one must conceive all these to be so small that none of them, when taken singly each in its several kind, is seen by us, but when many are collected together their masses are seen. And, moreover, as regards the numerical proportions which govern their masses and motions and their other qualities, we must conceive that God realised these everywhere with exactness, in so far as the nature of Necessity submitted voluntarily or under persuasion, and thus ordered all in harmonious proportion".

Johannes Kepler revisited the interesting theory presented by Plato, which combines vibrations and geometric figures, and expanded thereon with a ruler and compass. He completed the platonic series with the pentagon dodecahedron using the pentagon. Furthermore, his work proved that the 1:7 ratio could no longer be included in the aforementioned integer tonal system. Kepler attempted to apply his knowledge of harmonic proportions and vibrations to astronomical ratios. For example, he discovered that the ratios formed by two extreme points on elliptical orbits (perihelion and aphelion) "almost exclusively represent the major triad" (cited as per Schadel, p. 149).

3.7 Psychological Layers and Development Models for the Structure of the Personality

The understanding of the personal track record, ancestral development and ideas on how the world is structured overlaps with varying prioritisation in the individual models. In this regard, in a sense, in each detail, i.e. each individual, we encounter a reflection of the whole, even if this may often occur in an insignificant and/or not entirely comprehensible manner.

3.7.1 Onion-Skin Layer

A very vivid and popular way in which the onion-skin principle is symbolised is through the matryoshka doll, popularised in the Eurasian region. The doll reflects a varied experience of life and asserts that the innermost part of the person is, unfortunately, for good reason, concealed from the outside world. Blind trust, or blind confidence, does not appear to be called for in this world. Friendships and trust require time and a sufficient number of good experiences in order to grow. Only following this is the friend, who is no longer a complete stranger, invited to get closer to the innermost being through the peeling of another layer. The onion-skin principle is based on the necessity to protect a vulnerable core in an environment that cannot be entirely trusted.

In her work "The Interior Castle", the Spanish mystic Teresa of Ávila (1515–1582) evoked a similar notion of the soul, namely that of an "Interior Castle", a diamond in the shape of a castle that contains several mansions, which she interpreted as the personal journey of development to her own innermost part, where she hopes to unite with God in an even clearer, unadulterated way. Her work describes a spiritual path of purification and development. Nevertheless, it also provides a general guide to truthfulness and self-knowledge, a meaningful approach to become authentic, both as a person and psychotherapist, primarily for oneself, but also in order to be an encouraging and trustworthy model for patients.

Gestalt therapists are familiar with Fritz Perl's peeling the onion model, which is also referred to as the crisis model. The process begins with an external symptom, which indicates discrepancy between displays of expression and emotional messages, such as in the case of an empty behavioural role exhibited by an As-If personality and an underlying, palpable emotion that differs; it releases the repelled emotion from its latency, permits it to be revived and transformed in various stages during the role reversal and gives what was previously repelled the opportunity to achieve recognition and integration (A more detailed description of the above can be found in the section on treatment). This process aims to help the patient lose their identification with the estranged and emotionally fragmented, and to rework unresolved conflicts in favour of more coherent solutions and a more authentic mode of being.

► Example

The emptied, camouflaging As-If façade is a sincere, honest, responsible and, in general, paternally kind personality that is focused on scientific results, which can help local and professional politicians, for example, appear to serve the common good. A special facade of the As-if-personality aims to convince us all, that the individual acts happen in regard to common values. This kind of facade pretends to serve the common good and tries to manipulate the duped ones to vote for him to become their mighty leader. At this level, the hidden core is the desire to quash those who do not make him their leader, a fundamentally primitive reaction that is capable of activating a chain of negative reactions. On the biographical level, unprocessed experiences of social and familiar rejection and helplessness, which inevitably resulted in the overadaptation of the defeated behaviour at the time, resonate as sources of motivation. The injured, shamed and hidden part has not yet been cared for with understanding, healing and encouragement. Its stunted potential may include sensitivity, partnership and sincere solidarity. In this negative example, the unprocessed, repelled mental material appears to move towards the subjective centrepoint, like a "black sun", which the rest of the personality is required to organise itself around. The ability to reflect a superior reference horizon, as explored by Teresa of Avila, is lost as a result ◄.

3.7.2 Vertical Layering: Western Tradition

Greek philosophers of the antiquity set the course for the western understanding of the self and schools of thought, which shaped the western world view and concept and man as a result of the significance thereof throughout the middle ages right up to modern times in Europe. It goes without saying that this phenomenon is of major importance to the field of psychotherapy. However, the laying of the groundwork defined above is predetermined by our nature as a possibility and, as such, is not the arbitrary brainchild of a few pedantic philosophers; It is only through the concise formulation thereof that we become aware of these concepts and are able to assume a position on the matter at hand. As such, we have yet another reason to be grateful to our forefathers.

Accordingly, we shall now touch upon the works of Socrates (469–399 BC), Plato (approx. 428–348 BC) and Aristotle (384–322 BC) and explore their worldviews and concept of man.

Socrates, the teacher of Plato, was renowned for his critical search for the truth and focus on justice. Despite the fact that he was a proponent of democracy, he placed substantive decisions above majority votes. His portrayal of himself as a midwife (maieutics) was constantly directed at the search for truth on the inside, in order to help the inner knowledge to come into the light. This had a strong influence on his student Plato, who remained a staunch supporter of Socrates until the hour of his death.

Perhaps more than any other, Plato stands for a holistic, deductive world view, in which the most powerful unit of being comprises and supports the various layers of weaker ones. In the case of the latter, the gradually increasing fragmentation brings their multiplicity as substructures to the fore. It would appear that Plato obtained remarkable access to intuitive knowledge and possessed higher insight, as indicated in "Timaeus". In his later works, such as Parmenides, he examines the critical viewpoints of people who solely rely on their observations that depend on sensory perception and the logical processing thereof, on which his talented student Aristotle also relied. This inevitably results in an inductive worldview, that is to say, a worldview that can only be created from concrete, acknowledged partial aspects, or from bottom to top towards the generalisable. Aristotle observed the precision of nature with great enthusiasm and consequently became a pioneer for the later natural sciences.

The apparent incompatibility of the inductive and deductive worldviews led Bertrand Russell to describe Plato's "Parmenides" as the best collection of antinomies, whilst Carl Friedrich von Weizsäcker later employed quantum logic (as non-classical logic) as a resolution in this respect.

However, the findings of New Physics were not yet available to the great thinkers of the antiquity.

Suggestion

Let us imagine that the ancient philosophers, first Socrates, Plato and Aristotle were invited to a symposium on Mount Olympus, perhaps somewhat away from the main fray, with Einstein, Heisenberg and Carl Friederich von Weizsäcker in attendance as the representatives of quantum theory. The topic of discussion is "On Nature and Beyond" – or "Physics and Metaphysics".

In place a fierce confrontation, as in the following centuries, this symposium at Olympia could instead result in a complex and constructive synopsis of both worldviews. It is assumed that a higher, integrating insight could be gained at this exalted location and by virtue of the mutual inspiration amongst those present in such a group.

Which notions of the soul does Plato bring to the table, partly inspired by Socrates? To what extent does Aristotle adhere to Plato's teaching or instead find his own path? Indeed, this intellectual arc is worthy of its own book. Within the scope of this book, this can solely be implied. However, it remains significant for the subsequent history of European philosophy and psychotherapy, with its numerous approaches on worldviews and the concept of man.

It may at first appear that the notion that everything on earth has a soul, from humans, animals and plants to gods, demons and stars, and that the world's soul, which also animates the entire cosmos, and human souls, which constitute the self, are similar in character was evident and indisputable to Plato. The soul participates in both the sensory and spiritual, and bridges the two. As a self-moving element, it belongs to the principle of life and is indestructible, immortal and connects being and becoming. In death, it departs from the body. The soul is the discerning principle behind the notions of the good, the true and the beautiful, and for ideas (ideae inatae according to Plato) (Plato, Sophistes, 248 c-d). However, in order to perceive these ideas, one is required to turn away from the sensory; the soul is required to focus (meditate?) whilst avoiding the use of the sensory organs, which may present a source of disruption or errors in this regard (Plato, Theaetetus 184 c-d). The soul must first strive for a moral lifestyle, removed from injustice, intemperance, desires, etc. The final objective of the soul is purification, higher insight, the vision of God and to become God-like. This changes individuals, gifting them eudaenomia, harmony, intrinsic order, clarity and spiritual peace.

However, now we shall return to the lower depths. In his work "The Republic", Plato describes the stratification of the individual human being, as well as the entire society of a state, the superior polity. The simplest level in humans, that of desire, is based on sensation and perception (aesthesis). This concerns nourishment, rest, reproduction and sensual pleasure. Both humans and animals have this level in common. It is projected onto the lower abdomen. Desire also includes self-interest and greed for profit and gains. Level-headedness is required to tame these negatively accentuated characteristics. In a positive light, the common people are only required to provide food and reproduce to ensure the continued existence of the population. In return, they are granted a family life and property.

The political situation from which thinking originates is particularly evident in the second level in man (Athens is required time and again to defend itself against attacks and reassert its position): This level is dedicated to the courageous, combative, energetic victor. It advocates just order in society. At this level, aggressive instinct, anger, belligerence, ambition and the desire to deliberate, etc., are positively connotated as characteristics of a warrior. Courage is projected onto the chest. The warrior solely forms an opinion (doxa) by listening to debates and as a curious onlooker. The finest warriors should then be able to advance to the next superior group of state leaders through additional training.

The third and highest layer in man, similarly to in the state, corresponds to the wisdom-loving portion of reason in man, which is capable of regulation: the immortal spiritual soul. This part of the soul orients itself on knowledge and truth (noesis), and on the good and the beautiful. These qualities are projected onto the head. In the republic, this level corresponds to a form of philosopher king or a council of statesmanlike philosophers.

Note

Plato's renowned Chariot Allegory is used to describe the dynamics of the parts of the soul, whereby the charioteer (spiritual soul) needs to tame two horses, the noble white horse and the ignoble dark one. In his description, the charioteer fights and struggles vastly. The inner peace of the ultimate goal of eudimonie appears to be in sight, all be it far in the distance.

Whilst Plato may only split his descriptions of characteristics into three principal layers, it nevertheless becomes clear that his source for the regulation layer is not reliant on inflows from the sensory layer and the mental processing thereof, but rather on intuition and mental introspection, which he was evidently able to access through a contemplative attitude.

Critical Note

Plato's ideology reflects the view of a respected male Athenian of the period, who was constantly provoked by the contemporary democratic structures. Indeed, men of the time were tasked with defending society and engaging in politics, complete with all of the related politically-regulated power struggles, which, as with Socrates, could result in a cup of hemlock! Of course, slaves and women were not allowed to participate in these aspects of society. The role of women was largely ignored in the previously mentioned works. It would appear that "love" did not play a prominent role in this world. There was no indication of any particular appreciation of women.

However, at the same time, women played a key role in the secretive Eleusinian Mysteries. Any divulgence of the secret of these initiations was punishable by death. The cult in Eleusis was based on the mother-daughter relationship of Demeter and Persephone, dedicated to fertility and the becoming and passing of life. Eleusinian traditions also contributed to the myth of Orpheus and Eurydice. It would appear that Eleusis survived as one of the cultic remnants from the ancient, matriarchal, early Mediterranean cultures. This aspect of Ancient Greece is explored in more detail in ▶ Chap. 7 (which addresses the zeitgeist and ideal development). It seems that in Ancient Greece, two worlds with partially different values were able to develop to experience their heyday at the same time, perhaps even promoting each other in a mutually encouraging dynamic, or perhaps competing, which may have also spawned negative side effects on both accounts.

In a sense, Aristotle incorporated Plato's three parts in his work in the form of the anima vegetativa, the plant soul that is capable of nourishing itself, growth and reproduction; the anima sensitiva, the animal soul which also includes sense-perception, lower desire and mobility; and the animale rationale. It is solely the spiritual soul (Nous) that turns man into his characteristic self. Aristotle adopted a more holistic stance with regard to Plato's concept of soul. In most cases, in deviation from Plato, who appeared to solely identify with the spiritual soul, he did not differentiate between the lower parts of the soul and the spiritual soul (except for in death). Aristotle made the distinction between lower and higher aspirations. The lower consist of aspirations also found in animals, such as the desire for food, sexuality, bellicosity, vengefulness, feistiness, wrath, contempt, the boldness required for self-affirmation and a domineering nature. Higher aspirations instead correspond to desires the emerge through reason. As such, this requires employment of the decision-making freedom of the human will and control of the emotions that influence it.

According to Aristotle, the perception of the five senses (sight, hearing, smell, taste, touch) is united in common sense (sensus communis) and brought into consciousness as lower insight. "The spirit (logos) rises above this as the higher and actual human cognition, which is "understanding" as discursive thinking and judgement (dianoia), but "reason" (nous) as a look at concepts and principles. ... Nous (is) eternal, God-like, uncreated, immortal, undiffused, unpassionate, pure energy. However, if the spirit (mind) is described as tabula rasa with the signals from sensory perception and the concepts of pure thinking and observation, it is capable of suffering and also mortal" (Hirschberger, Vol. I, p. 212). Aristotle continued to adhere to Plato's approach with regard to the immortality of the spiritual soul and the mortality of the sensitive soul.

The subsequent multifaceted intertwining of schools of thought from the antiquity and Christianity played a decisive role in European intellectual history for the following two millennia. Despite dire blemishes throughout its history, Christianity is primarily based on altruism, humanity and the "amor deis", the love from and for God, culminating in the following citation from St. Augustine's

"Confessions": Inquietum est cor meum, donec requiescat in te (Thou hast made us for thyself, O Lord, and our heart is restless until it finds its rest in thee).

In summary, it would appear that the Plato's virtues of the warrior class, which remain idealised, were instead granted less significance by Aristotle, who also differentiated more strongly between the basic needs. Aristotle made a clear distinction between mental activity as a result of the processing of the five senses, which is perhaps useful for describing a tabula rasa, and, on the other hand, the cognitive abilities of the immortal nous, which facilitate higher insight and observation than the former. In contrast to the contemplative Socrates and the visionary Plato, Aristotle was more concerned with close external observation in nature. Overall, he shifted the focus of interest from Plato's notion of higher insight (mind) to the recognition of the closely observable perception of everyday life (mind) with its own strict and reliable logic. Christian values fill the relative absence of charity and higher love found in the notion of the human soul established in the antiquity.

We shall now attempt to obtain an overview of the general understanding of layers in the transition from one generation to the other with Socrates, Plato & Aristotle, proceeding from bottom to top.

Note

Individual soul in kinship with the world's soul (as per Plato)

- Basic needs: Nourishment, sexuality, bellicosity, anger/sense-perception
- Egocentric needs: Possessiveness, self-interest, striving for power, rational planning
- Higher needs: Courage, spiritually illuminated will, camaraderie
- a system of values, reason, the spiritual soul as a potentially responsible leader
- higher insight, with dimmed down sensory perception (according to Plato)
- Eudaemonia (peaceful ability to become more Godlike) (according to Plato)

3.7.3 Vertical Layering: Tradition in the Far East as the Understanding of an Energy Frequency Conductor in the Worldview and Concept of Man

In relation to the body, energy frequency gradations have been perceived as independent to one another on all continents at different points in time and in various traditions, predominately in a sensitive and tactile way, and less often visually with the human eye. This notion is prescribed to by Vedic advanced civilisation and derivations thereof can be found in Hinduism and Buddhism, and additionally in the Bon religion of Tibet, the Incas, the North American native American tribes, in Early Christianity, Sufism and Hesychasm (Greek Orthodox Tradition).

Predominately found in Japan and Russia, Kirlian photography is used to technically bring coronal discharges down an octave into the visual range of the eye, thus making them visible.

With regard to his frequency spectrum, the individual is solely regarded as a part and a reflection of the whole in all of these cultures.

The understanding of energy vortex systems spread through the body, which relate to specific functions, varies somewhat throughout the world, yet is surprisingly similarly differentiated for the most part. Most widespread is the ancient Indian seven-part system of the focal points, which is outlined here due to the fact that it can be assigned to other, scientific observations and contribute to a practical, somewhat rounded understanding of development, which is, of course, the aim of this chapter.

We intentionally do not delve into the deeper meaning of the individual chakras in this regard (instead we refer to special literature for this purpose), but instead focus on the bridges to current, western scientific knowledge, such as assignable neurophysiological functional levels, for example.

Tip

- Existential level: To be or not to be from a purely material point of view/concentration on the body, individual survival, self-preservation, the will to survive.
  - Muladhara chakra: Material nesting, putting down roots, building up matter according to plan, trust through a connection to the earth, strength;
  - Tonic immobility after exhausting the fight-or-flight mode (dor.nerv.x)
- The level of basic needs: Feelings, a zest for life through satisfaction of instinctual desires, preservation of the species, sexuality, parenthood, taking care of the brood, bonding, a concentration on needs;
  - Sacral chakra: Trust and security through belonging;
  - The social system provides adequate protection/differentiation between friend and foe
- The establishment of autonomy and power/control, intellect, plans, sets of rules, life-enhancing and unfavourable beliefs, self-orientation;
  - Solar plexus chakra: Intellectual power, will, the ability to persuade
  - Cortical top-down pathways; relativisation of the social system
- The personal ability to empathise and love, understand and comprehend, and to concentrate on others, the ability to devote and form couple relationships;
  - Anahata chakra: Empathy, personal love, social connectedness
  - SES: Social engagement system is fully developed (ventral vagus nerve)
- Creative self-expression/individual artistic creative abilities, mindfulness, inner observer, level-headedness, justice, beauty, good, a concentration on values
  - Vishudda chakra: the holistic soul. Expression, differentiated understanding
  - SES: differentiated, facial expressions, vocal comprehension

- Intuition/ receptive, open mindfulness/inner clocks/inner withdrawal as a prerequisite for interpreting information, a concentration on capturing information
  - Ajna/third eye chakra: Expanded perception, access to information;
  - Pineal gland promotes development (blocks sexuality),
- Experiences of higher cohesion/being free from contradictions and duality/active empowering compassion with respect to unconditional, higher love;
  - Sahasrāra chakra: The bridge to the cosmos, a concentration on connectedness
  - Potentially extensive, high amplitude EEG synchronisation pattern

The above short overview hopefully suffices as a mental bridge. Western science, which has almost entirely and exclusively relied on the development of cognitive processing, struggles to acknowledge other methods as of equal value and also, whilst in an entirely different manner, as effective.

3.7.4 Vertical Layering: A Quantum-Theoretical Mathematical Concept

The quantum theorist Burkhard Heim (1925–2001), who for a short period was member of Carl Friederich von Weizsäcker's cohort, presented an extended, unified quantum field theory based on his calculations as a depiction of reality in an attempt to unite quantum physics with the theory of relativity. Furthermore, he also experimented with artificially generated gravitation.

It is perhaps important to note that Heim lost both of his hands, his eyesight and the majority of is hearing at 19 years of age (1944) in a laboratory explosion during his time in the Wehrmacht. Nevertheless, with the help of his family, he went on to study Chemistry and Theoretical Physics in Göttingen with great success.

His legacy includes a 12-dimensional concept of the universe, which features a staggered transition between matter and spirit. Please refer to specialist literature to learn more about his calculations on this topic, which can only be comprehended by those with a background in quantum theory. The step pyramid or ladder is subdivided into four clusters of different sizes with the following emphases: Space-time, control, information and consciousness. Einstein's theory on the equivalence of matter and spirit remains applicable in this regard; however, everything is multidimensional, never belonging to simply one dimension. In death, the energy system solely departs from its product of space-time.

Dimensions 1–4 correspond to space-time (1 longitude, 2 latitude, 3 height, 4 time); they pertain to the world of the classical natural sciences, which are reduced to matter. They correspond to the daily world that we are able to sensually perceive.

Dimensions 5 and 6 are regarded as energetic fields of order and regulation. The formation of ordered patterns and structuring organisation as a formative force in relation to its organisational objective can be found in dimension 5. Dimension 6 provides the overriding sense and purpose in this regard.

Together, dimensions 7 and 8 form the universal information field. Information is assigned for an individual in dimension 7. Dimension 8 is responsible for collectively comprehensible information fields.

Heim was always wary of the need to provide more qualitative information on the final third of the dimensions, dimensions 9–12, the existence (according to Heim) of which was crucial for mathematical reasons in order to avoid feeding into speculation. The entire cluster that encompasses dimensions 9, 10, 11 and 12 corresponds to a higher, graded, holistic consciousness.

Exponents are of the habit of assigning hypothetical categories to the latter, such as: Dimension 9 for individual souls, dimension 10 for soul collectives, dimension 11 for divine will, intention, sense; dimension 12 for an all-encompassing (divine-energetic) presence of consciousness in all possible modes of being. These attributions are explicitly not backed by Heim's purely scientific calculations.

3.8 Basic Forms of Beeing: Tower of Variations of Energy / Pseudoparticle-Wave Tower as Reflections of Qualities of Consciousness

» additionally modified on the one hand
by security, attachment, life-enhancing love and integration
and, on the other side, by danger, fear and fragmented disintegration
as consequences thereof (please also refer to ▶ Sects. 3.3 and 3.5 in this regard),

taking into account the neurophysiological and psychological stages of development, aspects from attachment theory, intrapsychic constellations, psychodynamic defence mechanisms, learning theory, psychosomatics, interpersonal sociopsychological phenomena and humanistic questions regarding establishing values and meaning.

This chapter attempts to provide an overview of the previously mentioned underlying principles. First, we shall categorise the various modes of consciousness: (a) the laser-like powers of mindfulness that distinguish from the background temporarily create individuals with focused mindfulness and a more or less existential fear; (b) the mode of being and consciousness of the sea of waves appears to be free of fear, time, ego and limits.

The state of being of an individual quantum can be simultaneously described as the mode of the wave poised to connect on the one hand and the pseudo-corpuscular, spirally curled and folded-in energy thread reminiscent of DNA on the other and (c) a floating state that we encounter as a third mode between both poles. The phenomenon Quantum, beings and the cosmos can all be placed in a sequence of increasing structural complexity; they each display a mixture of pseudomatter, wave fields and floating states as flexible hybrids in various forms of order.

For example, we can take the image of a tower to symbolise existence in its entirety, with its more compact forms of being, such as pseudo-matter, at the bottom, and its interconnected, oscillating states of being at the top. As such, our tower possesses a horizontally layered structure. It should here be noted that the vertical connections, reflections, fields of resonance, blockages and growth impulses are just as significant for the course of therapy. The following compression level therefore represents bundles of qualities that are primarily ranked one above the other from dense matter and emotions to higher oscillating values, information and spirit. However, each level contains variations in development (partially delayed), which may overlap to a certain extent with the qualities of the higher and lower ranked qualities. In this regard, a chronological perspective is consequently abandoned in favour of a modal being. The forms project onto one another over time.

However, we retain the distinction between the variants attributable to life-enhancing security on the one hand and fear and danger on the other. They weave through all levels, but are addressed in most detail at the beginning of life as they are most directly observable at this point, and can be least compensated for at this age.

Allen Schore pioneered research into the crucial connection between suboptimal to disastrous attachment and abuse experienced by babies and an increased susceptibility to psychiatric illness later in life. As the right brain hemisphere, where early physical and attachment as well as vegetative and pituitary regulation experiences can be found stored with systems that process stress, demonstrates clear dominance with regard to development up to 2–3 years of age, it seems only logical that psychotherapy provides emotionally corrective experiences for those who have been neglected in their relationships or abused at an early age, and for patients with secondary regression as a result of trauma – primarily on non-verbal, body-centred levels – to facilitate the classification of the shift in identity only secondarily in terms of putting this into language and cognition. In this regard, it is perhaps useful to learn that the right or the subdominant parietal lobe registers spatial orientation, including any changes in perspectives; as such, spatial-symbolic work on the part and the whole, as well as work on the inner child within its spatial-symbolic world, as recognised in Gestalt therapy and constellation work developed from psychodrama, and all of the variations on changes in roles and locations included therein, is of increased significance and effectiveness in the treatment of early emotional stress.

It would perhaps not be conducive to attempt to obtain traces in a left-brain mode that were established in the right-brain mode. They either only partially convene, or do not at all. This provides us with a number of therapeutic opportunities. An inadequate method often demotivates the patient, costing him the opportunity to recover and quality of life.

If deep levels cannot be reached, despite that fact that they contain memories of shortcomings that significantly influence the current disorder, psychotherapy is reduced to a shallow courier of symptoms. This would be akin to shooting ourselves, as psychotherapists, in the foot with regard to the available potential of psychotherapy.

Whilst the left-brain approach alone (despite in-depth and earlier disruption) may initially appeal to some economists who are not knowledgeable in this field as the cost-effective route, the relapses and possible shift in symptoms make it precisely the opposite. Conversely, targeted and corrective experiences may be precisely the correct and appropriate method for treating disorders without major upsets. Admittedly, therapists must be comprehensively trained in order to be capable of recognising the difference.

Psychotherapy over a course of years is not necessarily required to establish contact with the deepest levels of damage and subsequently begin with healthy regeneration from these levels following consolidation. However, psychotherapists need to be capable of bringing their relationship into resonance with the various levels and coming into contact with the most hidden, damaged parts of the patient, also (additionally) through non-verbal contact where required. The therapist is first required to become familiar with this ability through his own personal experience before he is able to work with it in a sensitive and tailored manner. In this respect, the focus lies in being able to invite short-term regressions and control the process in such a way that at the end of therapy, the patient is once again released stronger and in full contact with her adult potential to help her cope with everyday life.

Note

Overview of the levels of consciousness and energetic condensation

1. Existential level of rootedness and materialising structuring
2. Level of basic needs dependent on care/interpreting inner proximal
   States of (im)balance, including their regulation of external help
3. Corticalisation with preferred focussed qualities of consciousness or actively differentiating, structuring, language-related and binary logic
   Actively creative shaping power
4. Cardiac synchronisation/empathy resonance/personal ability to love
   The interpretation of and relationship-shaping reaction to beings/altruism, humanity
5. Corticalisation through oscillating fields capable of integration/subjective selfhood
   Creative powers to create order, set values and shape structures through resonance fields
6. Conscious sensibility/present energetic receptivity/
   Interpretation of informed, energetic fields
7. The willingness and ability to synchronise with extensive reference horizons

   All-encompassing willingness and ability to love

Levels 1, 3, 5 and 7 have certain creative and formative properties at different frequency levels, whereas levels 2, 4 and 6 primarily facilitate perception and receptivity.

We shall now look at an in-depth description of the levels and a subsequent diagram.

3.8.1 Level of Existential Rootedness and Materialising Structuring

This level is primarily concerned with whether a new being indeed exists, or rather, whether a new being succeeds in holistically and thus also evidently materially consolidating. The process of embryonic development has been described in detail, yet the interplay of forces behind its development dynamics remains unclear.

(The mitogenetic (cell division stimulating) radiation of germinating cells based on a photon emission in the spectral range of 20 nm (nanometres) was discovered by the biologist and medical doctor who specialised in embryology, A. Gurwitsch, in 1923 and later verified by F. A. Popp's research group as fields of ultra-weak biophoton emissions for all living cells. A biophoton is a quantum of the ultra-weak light spectrum. A closer understanding of the interaction between these fields and the materialisation process requires further clarification.)

Subjective view on experiences (empathetically worded using an adult's vocabulary):

(a) Security mode

"I am seen, greeted, noticed, understood, welcomed and, above all, met with eye contact; I am verified as likeable and cared for, understood and influenced in part by my surroundings. Everything is in resonance. I can sense myself from within: I exist! On the one hand, I am my own whole with my own rhythms that appear to be considerately respected, but I am also anchored in the oscillation of the whole and a part thereof. I am an accentuation in the whole. My surroundings seem to aim to coordinate and appear to be on the same wavelength as me. I feel overall secure and supported. This gives me a sense of basic trust. I feel empowered in myself and life. My desire for function inspires me to try out new things that can be experienced or performed. The list is constantly growing. Me and my world are primarily the one in the same, only rarely does it dawn on me that there is clearly an outside world that has nothing to do with me".

Mental experience level: confidence, fearless attitude to life, joie de vivre, inspirational high soulfulness, incipient, targeted inquisitive behaviour (intentionality).

Structure-forming level: Overproduction of neurons, leaving only those that are used behind. Initial development of the inner/outer boundary as a result of stimulation pathways for initially different impulses from inside and those that signal aspects from the outside world using the far senses (Eccles, Damasio).

Physical-psychosomatic correspondence: Successful internal regulation. If internal regulation is not sufficient, help can be received from outside; healthy, undisturbed development.

Contact and attachment level: Finding joy in stimulating, sensitively dosed and regulating non-verbal and pre-personal contact which appears to be embedded in the synchronised, common wave pattern; joy in being accepted and establishing interactions; naive, unbroken trust. The foundations are being laid for a secure mode of attachment in the future.

(b) Uncertainty-triggering mode, which activates the fight-or-flight response

This mode results from recurring extended periods of being overlooked when suffering from hunger, thirst, lying in the wet, soreness, being exposed to an audible sensory overload, not being noticed, etc., in other words, a lack of care or even potentially neglect.

(b1) Fight mode: I am not granted consistently reliable, empathetic or affirmative attention, even when I scream in distress (postpartum) and reach a state of alarm, experience colic, fight back or potentially have a temper tantrum in the uterus. My alarm still manages to get help and improve the situation adequately, or at least often enough. This imparts a small feeling of success for this early learning: The impulsive alarm can consequently become a sure-fire success. Subsequent, more solidified survival strategies may revolve around the excessive willingness to attack. As the saying goes: The best defence is a good offence. In this respect, it can be difficult to decipher that this attack conceals a desire to be loved.

(b2) Flight mode: This mode typically arises when alarms do not lead to improvement. Instead, the fruitless alarm most likely results in further deterioration through disappointment and exhaustion. The next best compromise: To go silent and retreat from this inhospitable, overpowering or even hostile world.

This can save energy and disappointment. It feels like the lesser of two evils and facilitates survival on the back burner.

This may lead to subsequent survival strategies such as: "Dodging", "ducking away", "becoming invisible", "not having one's own opinion", etc. These patterns are perceived by the baby as helpful in coping with life, compared to portraying a clear stance in a subjectively hopeless situation, getting caught up in the moment and wasting energy on fighting.

Psychological level:

(b1) Breakthroughs in vitality as the equivalence to combat readiness due to the overcompensatory fear of life. Increased tension, restlessness, various forms of agitation, hypermotoric behaviour.

(b2) Flight mode: Anxiety, nervousness, increased general level of fear, tendencies to retreat and avoid situations.

Psychosomatic equivalent

(b1) Loss of balanced rhythms of activity and rest, of waking and sleeping. Increased activation readiness of the sympathicotonic vegetative system, such as the mobilisation of stress hormones, an increase in blood sugar levels, activation of blood lipid and blood sugar with a subsequent slope towards atherosclerosis. (At the same time, numerous experiments conducted on animals with early mother-child separation have verified the subsequent emergence of chronically increased stress patterns, which, in turn, result in increased morbidity and mortality.)

(b2) Early manifestations with regard to excessive defence reactions in the skin and immune system.

Contact and attachment level:

According to Mary Ainsworth, behaviours that are characterised by uncertainty in unfamiliar situations (forms of avoidance, ambivalence and disorganisation) appear to represent compromising hybrids between fight and flight. The form of avoidance appears to be the most developed of these compromises. An insecure form of avoidance, which may come across as calming, is developed during childhood, whereby we learn how to actively block feelings of pain/fear/grief/anger towards our mothers. It also constitutes a "No".

In the case of ambivalence, which often involves strenuous clamping reactions to the returned mother, the archaic clamp reflex comes to the fore in the event of otherwise confused emotional outbursts. According to René Spitz, this marked the final bastion in the course of anaclictic depression prior to death.

(Although Ainsworth obtained her findings from 1-year-old babies, it can be assumed that earlier precursive damage has occurred.)

(c) Paralysed mode at the same time as tonic immobility

"I am exhausted and give up entirely". Or: "Certain parts of me have become numb; they no longer exist to me. At most, there is a gap or hole" Schultz-Hencke, F. Perls). Resignation marks that transition from flight mode to a manner of powerlessness with functional paralysis. In a mild case, this will be accompanied by partial blockages, and potentially (perhaps at a later point) with a tendency to give up and suicidal thoughts.

Mental structural level: A failed approach to an inner/outer boundary, an insufficient start to the integration of individual functional centres and the failed movement of poled emotions towards a centre, perhaps as a result of the recurrent sensory overloads and lack of regulatory assistance No initial development of an internalised, self-regulating authority.

Psychological experience level (later impact in certain cases): acute, traumatic experience of being overwhelmed accompanied with a sense of powerlessness and potential dissociation; chronic traumatic experiences of powerlessness lead to self-insecurity, structural weakness, insufficient resilience, a dissociative readiness to collapse and latent suicidal tendencies;

As a subsequent remnant of this time, a mild form of the chronic feeling of being overwhelmed by the demands of life may continue with an underlying fear of life.

Psychosomatic-physical level: Significant disruptions to regulation, such as hypertrophic pyloric stenosis at birth, later: Susceptibility to vagavasal syncope (collapsing), potentially also a tendency for blockage phenomena in the neurotransmitter area, also possible: hormonal dysregulation, significantly increased vulnerability to infection and potential delays to development.

Social and contact level: Among the mother-child regulation disorders, babies have also been observed to experience certain disorders whereby the body stiffens and turns away in avoidance. Infants may run away and hide in paralysis, as if they no longer exist. They are able to distinguish between those for whom they can and would like to exist, and those for whom they would prefer not to exist, to answer to or to simply make themselves unreachable without establishing a relationship.

3.8.2 The Basic-Needs Level: The Interpretation of Inner, Proximal States of (Im)balance and the Regulation Thereof with the Help of External Care

A baby's inner state of balance and the registration and balancing of any deficiencies is key to the survival of the young being.

■ ■ Early phase

The baby is protected by caring love and inherently supported by the (oxytocin-regulated) parental care of the direct caregiver and also the extended personal network. Its own active components that start at the beginning of life but only fully develop at a later stage have been descried by Stephen Porges as the social engagement system. This system is inherently permanently functional through the myelinated ventral vagus branch and it helps us to develop loving interest and beneficial bridges of understanding towards our potential caregivers. Despite the fact that the SES cannot yet be employed with its full potential at this point, it nevertheless helps to ensure that caregivers display understanding and care for the baby's basic needs at this point in time. This primarily concerns basic physiological needs such as nourishment, sleep, hygiene, tactile stimulation, help with regulation, respecting natural rhythms, accepting encouragement and desires for interaction.

(a) Security mode: "My various bodily needs have been acknowledged and sensitively nurtured, including the need for attention and touch. Almost everything is allowed. Interactive repetitive games increase trust and security, especially between 9 and 12 months of age. The games are a little different with each person. Those surrounding me have become increasingly more personal and familiar in their individual characteristics. They help me to regulate my needs and also experience joy when I enjoy something and am well, and I can make them happy by giving them a radiant smile. They also help me to endure through my discomfort with solace if something has not gone my way, thus helping me once again attain balance. I learn what "We" want and don't want from my caregiver. Relationships become increasingly differentiated. Being on the same wavelength with my caregivers brings moments of happiness. They inspire me and make me curious in the world around me, which I find increasingly familiar. I feel safe. I learn to differentiate between the familiar and unfamiliar (at around 9 months of age) without approaching the latter with fear. I gain my initial awareness of myself when I recognise myself in the mirror. I learn to assign meaning to sounds and speak words. My initial experiences of self-efficacy allow me to feel confident in my own potential. If something doesn't go my way, I try to influence it in my mind. If this doesn't work, I may become angry. My desire to play with the aggressive forces that arise makes me excessively struggle and reach my limits, where I am thwarted, always with adequate care. Provided I oscillate in synchronised waves, I begin to gain an understanding of how my counterpart feels. I am able to interpret this to a certain extent, sense how I can influence them and how I can get back into resonance with them and reach a state of feeling secure again. My caregivers are lenient with me, even when I push them to their limits".

Psychological level: The development of confidence in one's own potential and an adequately good world.

Structure-forming level: (1) The development of the inner/outer boundary; (2) the merging of positive and negative experiences with respect to subjective self-perception, also regarding external representations; (3) the interlinking of emotionally opposite poles, such as happiness and unhappiness, with increasing fine tuning; (4) the development of an integrating, centring structure with a certain level of regulatory ability; (5) striving for experiences of self-efficacy; and (6) adequately good experiences of self-worth, which are interpreted as rewarding.

Physical-psychosomatic level: Undisturbed development, strong level of defence against infection.

Contact and attachment level: Secure attachments with average frustrations from outside. Aggressive reactions, which are typical at this age, are integrated and tolerated with care. Confident in establishing contact with strangers with no fixed outcome.

(b) Insecurity-inducing attachment mode

A loss of references and separation spur fear and insecurity. The alarm mode of the fight-or-flight reaction springs in far too early. The shared relationship field for the caregiver is lost far too quickly. Mistrust creeps in far too quickly, leading to the fragmentation of the social situation.

(b1) Fight mode

- Version 1) Fight mode directed externally: I want to get what I want if I need more of something. This has typically worked out fine in the past. In the case that no one shows any interest in me, loves me or cares for me, I'll fight to get what I want to grab your attention. Perhaps I may even take out my revenge on you for causing excessive frustration, even if I need to break something or fire in all directions before you finally notice me! Or I may elect to scream, so that someone will hear me and tend to me. I'd rather be hit that receive no care and attention at all! I may experience pleasure or fear in the face of an aggressive, dissocial lack of impulse, which is, in itself, an emergency signal.
- Version 2) Ambivalent fight mode directed inwards with external (over) adjustment I "fight" with opportunities related to my relationships. Time and again contact to the familiar wavelength is disrupted. I am rejected or even overlooked as if I don't or shouldn't exist. At the same time, the level of stress hormones rises. They activate fear and anger and together they lead to the courage to vent my feelings with aggressive impulses, or the want to do so. This results in tension and the risk of losing the caregiver entirely, which could be a little risky at this stage, considering that I am significantly reliant on my caregivers.

I have learned to interpret the wishes of my caregivers and to adapt to these with discipline, and even to actively block any feelings that I may have which could disrupt the desired harmony (attachment marked by insecurity and attachment). This primarily concerns aggression. I may fight to become someone's favourite, such as a daddy's boy, despite the fact that I am a girl. I learn to bring what is desired to the fore and to suppress anything that disrupts this in order to receive love and acknowledgement and experience a sense of belonging in an emotionally sparse world. However, this may not suffice. If this is the case, I retreat from the world and have suicidal thoughts.

If parentification is demanded by my parents, the situation can quickly become overbearing and cost me my freedom as a child. But, initially, the role shapes me, making me feel big and important. This can be tempting. I first lock the things I do not get to experience into a dungeon. They need to wait, even if the excluded parts consequently grow into inner demons. If these aspects become angry and rebel with psychosomatic manifestations, the opportunity may involuntarily arise for me to back out of the indirect obligation as a patient. I may receive a justified level of recognition for my altruism up to this point from my social surroundings.

At first, I resent not being seen. Following this, I fear expressing this and ultimately, I lose the will to see my own needs myself and erase them. This saves conflict and irritation. Compromises of this sort are partially a form of self-mutilation. However, the inner pressure is initially relieved by auto aggression.

3

(b2) Flight modes

- Version 1. Depressive flight mode into a substance-related replacement solution:

  I am yet to experience being acknowledged and loved, or haven't experienced this in a long time. I no longer believe that this is possible or will happen in the future. Therefore, I've stopped fighting. My previous efforts ended in nothing. It all resulted in misery. My self-confidence and mood have plummeted. I no longer seek happiness. I live without any real or, at least, fulfilling relationships. Now I instead care for myself with substances that provide me with a temporary feeling of happiness – always followed by a hangover and feelings of emptiness. My substances are my reliable friends who I can call on whenever I want. At least, this is how it feels to me, even if, in all matter of fact, I am unable to imagine my life without them.
- Version 2. Depressive flight mode into a dependency on clinging onto things.

  I am a nobody, I have failed. I look for someone to replace my entire world and social network (the SES) and cling to him strongly. He needs to replace my lack of security. I elevate him through my perspective, regarding him as an omnipotent being, which flatters him and allows me to participate in his ability to cope with life. I live through participation with him. This satisfies my shortcomings and compensates for my inabilities. (Deep down, I am frustrated with myself for this dependence, but I try to conceal this from myself as the destruction of my omnipotent being would cause me to lose my security.) Earlier experiences of this were unfortunately not reliable at all. Hopefully this will be different in the future. I won't let my safe anchor get away or allow myself to be rejected. Separation would induce extreme anxiety. If necessary, I'll allow myself to be humiliated. But I'll stay. I'll hold onto him, my covert prize, through guilt. He would never dare to let me go.

  (Insecure and ambivalent forms of attachment are often characterised by anxiety and dependence.)
- Version 3. Depressive flight mode with self-surrender

  I'm a loser. I can't get through life. Everyone else is more capable than me. I am a nobody. I have never received adequate validation. The little validation that I have received was an oversight from people who are not necessarily highly regarded. Anyone who acknowledges me is similarly worthless, something isn't quite right with them. I give up over and over again. I no longer find joy in anything and have no desires. It would be better if I didn't exist at all. I can't say how much longer I will have the power to lift myself up. Sometimes I simply cannot drum up the will to resist feelings of helplessness. Death would suit me.
- Psychological level: Various forms of depression through identification with being a failure

- Psychosomatic-physical: Vegetative dysregulation with reactions of stress through the avoidance of fights
- Contact and social behaviour: An avoidance of locking horns, withdrawal and voluntary submission in advance

(c) Numbing and blocking modes (with physiological side effects)

According to the findings of René Spitz, in the final phase, babies may experience apathy and, above all, resignative loss of the ability to satisfy their basic needs and a loss of the willingness to establish contact and an interest in the world. The clinging reflex, which may also lead to self-embrace in desperate cases, remains almost intact until the end, similarly to with Harlow's monkeys.

Individuals who narrowly escaped the above final stage and only narrowly survived neglect or trauma at a young age may retain a tendency to subsequently deflect their regulatory mechanisms at various levels.

Indeed, on the topic of traumatisation and the potential consequences thereof at any age, it is beneficial to look into epigenetic changes or trauma experienced by the individual's ancestors.

Numbing and blocking phenomena in later stages of development

- Version 1. Traumatic, acute, functional paralysis of the entire organism, initially without shifting the target values in the regulatory system: paralysed with fear and experiencing no feelings. Fear makes us "dumb" (blocks thoughts) and produces a white wall in our brains (exam nerves), etc. "I am standing next to myself; I am no longer myself. Everything has become unreal" (dissociation, depersonalisation, derealisation). Memory traces of traumatising events are primarily stored in the amygdala. As such, they are all but dissociated from other memories.
- Version 2. Partial blockage of the sympathetic vegetative system with phased, partially triggerable shifts in target values (hypothesis): e.g. Major depression
- "My mood has hit rock bottom. I am no longer to pull myself out of bed in the morning and nothing fires up for me anymore. I no longer have any drive or desires. Nothing resonates with me. I'm empty on the inside, as good as dead. I don't dream. I can't sleep at night. I have lost the rhythm between day and night".
- Version 3. Secondarily raised excitation thresholds/blocking in parts of the central nervous system at the neurotransmitter level (hypothesis): e.g. negative symptoms of schizophrenia. "Everything has become somewhat monotone and grey: people, feelings, thoughts and life in general. But this is an improvement on the previously confusing turbulence and unpredictability".

Can the positive symptoms of schizophrenia, including symptoms of delusion, also be regarded as a secondary consequence of a blockage, comparable to the flare-up of the secondary and tertiary excitation centres found in the myocardium when regular, cardiac excitation processes are blocked? This results in an unusual and disordered transmission of excitation and dysfunctional stimulation.

A loss of order for the regulated excitation processes in the brain may also explain the psychotic phenomena that occur during experiments on sensory deprivation (e.g. in Samadhi Tanks), whereby the average excitation input via the sensory channels is excluded and spontaneous nerve firing may occur.

Observations made in studies on young soldiers exhibiting the initial manifestation of schizophrenia in 1941/1942 by psychiatrist Klaus Conrad [7], who was a proponent of the tradition established by Karl Jaspers and the Heidelberg School also appear to support this. He found that patients who were pre-emptively observed with increasing anxiety and displayed the simultaneous onset of the initial symptoms of delusion lost the ability to shift their perspectives, i.e. the ability to assume the role of their counterpart or the revitalising bird's eye perspective in order to look at themselves from another perspective; they appeared to instead be imprisoned in their "Ego" and viewed themselves as the significant centre of the universe. Fear restricts the ability to experience oneself inwardly, that is to say, the ability to focus on oneself and others related to oneself with a kind of field quality. Parallel to quantum characteristics, it is also possible to refer to this as a lock-in state; a completely or exuberant, pseudo-corpuscularly bound mode of being, which provides no or insufficient opportunities for interconnectedness. Healthy interconnectedness corresponds to a floating state that permits the individual to be semi-located in oneself and simultaneously be the field that surrounds the target object. Consciously entering this field with one's consciousness and temporarily locating oneself there facilitates the ability to shift perspectives and achieve expanded social integration. This is also regarded as a sign of maturity on our development ladder.

Sufferers of catatonia become paralysed, as if they are no longer in their own body; yet they remain capable of adjusting their posture externally and bending into shape. It is only the inside that fails to move. What could be missing in this regard? It would appear to be the case that the brain works with the integration of quantum floating states, including the self-organising patterns and order thereof, and the chaotic states and turbulence, inside and beyond the materialised level. Could it be that the catatonic state is a form of energetic suspension? In this case does the water level of the free-floating quanta sink below the minimum amount required by our physical operating system in order to form and maintain an intra-cerebral interconnected field and to deliver targeted impulses to depolarisable, neural structures assigned on the basis of function from this field? (Hypothesis)

- **Supplementary biological aspects**

- Version 5. Tonic immobility exhibited in phylogenetically younger animal forms in order to escape from the somewhat shifting predator-prey system.
- Version 6. The hibernation of mammals found in regions close to the poles: gentle and orderly dimming down of the vital regulatory circuits and similarly gentle reawakening in a logical order sequence and dose tuned to the respective function.

Unique roles over time: Sexuality.

Sexual development is initially inhibited by the pineal gland. It is inherently only permitted to develop gradually with a delay, meaning that it is only first included amongst the other basic needs to a full extent at a later point in time. As such, we are granted an extended period dedicated to learning and development at a young age.

The young being experiences its first surge of hormones in the uterus, followed by another at the end of infancy or pre-school age, and the organism, including the brain, is then largely restructured as a result of hormones during puberty.

When we reach the age of (4)5–6(7), gender-specific differences can be observed in relationships. For example, we may demonstrate a noticeable interest in particular toys, even if no deliberate differences have been made to what is offered to us. Perhaps the most attention in this regard is paid on the attraction to boys or girls, provided this is not taboo in the respective environment. In the case of weak or prohibited responses, erotic interest wanes. In an ideal scenario, adults subtly value the onset of this erotic identity, but it does not tempt them to commit an assault. Does the child find a model in the adults whereby sexuality is integrated with love, or does he instead witness a split between these two components early on? This phase of childhood sex differentiation often concludes with an interest in "playing doctor" and subsequently disappears into latency for a while.

Puberty is a daunting challenge, both for the individuals going through it and those surrounding them. It marks the phase when children are required to break away from their parents, find their own path and potentially also new partners. Here, the topic of autonomy once again emerges, this time also in combination with the topic of self-worth, responsibly becoming an adult, confronting one's parents at eye level and the ability to help oneself. Saying farewell to the previous mode of being can cause old wounds from unprocessed losses to reopen. One key issue in this regard is the varying temporal gap between the start of initial biological puberty at the early age of approx. 10–12(−14) years old, and the delayed entry into social adulthood due to extensive education and studies.

3.8.3 Corticalisation with a Preferably Focused Conscious or Actively Differentiating, Structuring, Language-Related and Logical Binary, Creative Power

In the ideal scenario, the consciousness is used in the centring mode.

General information: Clarity requires the acknowledgement of and focus on differences. Aristotle was renowned for his close observations of nature. His focus on differences enabled him to adopt the following logical stance: A is not B, B is not A; tertium non datur (no third [possibility] is given). The emphasis is clearly on the definable in this regard. We trust the image that our mind creates. This behaviour is fit for everyday life. We do not question whether our mind and the neurological excitation processing with its on-off switch may deceive us. Are we, as humans, clearly definable? (Aristotelian logic does not apply to quantum theory in any case.)

Sigmund Freud determined that the secondary level of our clear conscious is a higher development that arose from the primary level, i.e. the unconscious. Conscious activity on the secondary level exhibits a temporal, local and causal structure: "If-then", "Before-after", and substantiated thinking. Reason was already held in high regard several generations prior to Freud in the wake of the Enlightenment.

The relationship to objective aspects was met with less appreciation by Martin Buber, who referred to it as the "I-It" level (not to be confused with Freud's "Id", the source of instinct). Whilst Buber recognised the necessity of objective references, he felt that they should not usurp the "I-Thou" pole in the interpersonal, which represents experienced relationships, from its pre-eminence, but instead be conducive to it.

Fritz Perls, who was deeply impressed by Buber's charisma, intuitively introduced the foreground and background dynamics to the common understanding of the creation of Buber's "It" dimension, and also contributed to the clarification and fragmentation of the world. The driving force behind the focus on the consciousness has long been a capability and remained a cultural asset throughout the history of humanity. Indeed, it is responsible for the creation myths of all cultures.

In ancient Chinese cultures, this determined, focused approach to energy (Chi) was assigned to the male form, the Yang. Several traditions of martial arts were developed and cultivated on the basis of this understanding.

The power of the energetically focused, mindfulness-based stream of consciousness touches on the foreground-background dynamic, which itself is the creative process, per se. Accentuations create imbalances, asymmetries and tension between polarities and dualities. This entails that the foreground encompasses everything that is accentuated by an energising and evaluative interest from its background and exists in contrast to it (Ex-sistere = stand out). This may potentially only be a purely subjective bias in the eye of the observer; it may be a creation of the mind with an outline, plan and denotation; it may create figures of the imagination and parallel universes that can expel reality into the background; it may revolve around a particular allocation of meaning that makes individual aspects stand out and suppresses others, thus creating a new relational structure within the otherwise familiar; or it may involve the accentuation of mindfulness in a social setting. It goes without saying that potentially all materially distinct, uniquely formed individual parts may benefit from a new attention that casts a spell over them. However, they can just as easily lose this capability once again when the attention disappears. Materially structured aspects lose their ability to enter the foreground when they fall apart.

Changing to the psycho-physiological level: The initial top-down corticalisation begins during the second to third year of life in right-handed individuals through the spurt of development that predominately occurs in the left hemisphere. Once a child begins school, the left brain hemisphere starts to gradually take control, and its potential increases until the individual reaches their mid-20s. It remains at this level for a decade, only discretely dropping, and then finally significantly decreases once the individual ends their career. This gentle decline is compensated for by practical knowledge over an extended period.

Now we return to the start of life. First to emerge is the delightful sense of increasingly gaining power over one's own body, initially through the sphincter muscles located in the bowels and bladder. Over time, motor skills are joined by verbal classification of things and the intellect, with its logical, predictable procedures. The ability to control with precision is experienced by the proactive "Ego" as power and strength. Accordingly, this desire to control can result in leadership qualities and potentially the will to power. The composition of the world according to one's own whims is experienced as validating and satisfactory. The way in which this initially neural process of exercising power is applied at a later point depends on the individual's system of values.

Differences in development due to security or danger:

(k) Security mode (early childhood): "I am increasingly able to control my body: I can walk independently and I can hold in urges to go to the toilet in order to keep myself clean, which makes my caregivers happy and helps me to receive praise. I feel powerful, capable, acknowledged and strong. I have experienced joy in discovering myself as an independent centre, in determining my own desires and conveying them the others in order to achieve something. I reach the limits set by my caregivers but do not exceed them, as I am simultaneously also my own counterpart, or at least can sense what was important to me and gave me security up until now. That is not the say that I don't get involved in a dispute from time to time in order to stretch my boundaries. However, when this occurs, I don't risk breaking down relationships".

"At the latest during my pre-school years, but also in the years following this period, I increasingly get the urge to measure myself against my new abilities and compete with others, particularly as a boy. Similarly to many other animals, we practise friendly scuffles and prepare for serious fights that may occur later in life. Controllable aggression leads to dominance. Sport is also important in this regard. When I have experienced security, then I can allow myself to play fairly and my opponent remains a friend (otherwise I may attempt to view him as an enemy or competitor and want to defeat him.)

In the worst-case scenario, I find myself in the position of the loser. The I look for opportunities to compensate for this inferiority in other ways."

"As an adult, I perhaps most enjoy my ability to plan, structure, control, order, to maintain a clear head and overview of the situation, to assume responsibility, to make announcements, to pay attention to differences and ranking orders, to have power and respect, to exercise power in the sense of "Law & Order", to remove any confusing matters from my line of vision, to budget time and money, to perform tasks on time, to fulfil my duties reliably, to channel and, if required, vocalise my feelings, to lead a normal life, to systematise thoughts and numbers and also potentially to pursue exact science according to reasonable criteria. I am inventive and creative when it comes to technical design according to a plan".

(l) Insecurity mode: In the case of insecure attachment experiences, there are several patterns that may be memorised; the relationship with the other person which is already on eggshells is always essentially lost.

3

(b1) "I assert myself with my will. I become a tyrant and strive to become the victor whenever this is achievable within my environment, which continues to applaud me despite my resistance. Everything needs to go how I want it to.

This may lead to the following pattern becoming fixed later in life: I expect new environments to give me the priority that I was used to during my childhood. It often suffices to adopt a confident attitude and assert a wave of leadership claims, as though this suits me – and then I usually succeed in achieving the distance I want from others at first. I don't generally shy away from the challenge of a competition. This subconsciously makes me feel hyped up, which has a psychosomatic impact".

(b2) "Earlier draconian restrictions from outside often made me feel like an impotent loser; I have come to terms with accepting the role of the defeated, inferior, dependent, conscripted, subordinate, servant and the recipient of orders, etc. within a certain context. Whilst not a particularly enjoyable role, it is nevertheless a familiar one that belongs to a superior set of rules that I adhere to. Maybe I'll be rewarded for this and exist on the praise received. I may only receive sympathy and recognition if I make sure to stick to the rules. I identify with them. Any impulses to go against the rules cause me to feel anxious. They emerge within me with a similar force as though I had experienced them from an external viewpoint and I am not in the position to fight against them. Strong waves of emotion that do not conform with the rules make me feel anxious. I am proud of my self-discipline. I am in control both inside and out and don't give up easily. My militant overdiscipline may even turn into a sure-fire success. Then I turn into an emotionally removed observer who has given up their leadership position. I only see how someone/something demands something from me and another part of me feels obliged to carry out their demands. This can become rather inconvenient and be come across as unreasonable to the outside world. The advantage to this behaviour is that my emotional side is suppressed and finds an outlet in my actions, to a certain extent".

(m) Experiences of powerlessness Severe external negative experiences of limitation permeate and the punishing (potentially sadist) introjects (internalisations) also convincingly or suggestively control the inner world. The individual primarily identifies with the underdog ("I surrender as the inferior loser. I don't stand a chance") and perhaps also with the executioner ("Punishment is necessary"). In this case, there is no (longer) identity with the inner, concealed counterparts, but instead: "I am guilty, I need to suffer and atone. That is fine. It is the only way that balance can be restored and the world can go back into harmony". In the case of psychotic decompensation, this process will likely lead to delusions of guilt and sin. "I am condemned and crucified".

The development of the individual function clusters overlaps leading to:

modifying integration or overlapping zones between corticalising power and the socially-oriented peer group phase:
- Enjoying competition and power struggles with the rules of fair play.
- "Theory of Mind": The individual is capable of not just emotionally but also intellectually grasping the perspectives of others. (Mentalisation / Fonagy)
- Ability to manipulate and the risk of being manipulated

Between awakening sexuality and the social-oriented peer group phase:
- Childlike-exploratory experiences of new-found gender attraction in the extended, social reference space stimulates and shapes the identity.
- Please also refer to ► Sect. 3.2.3 in this regard: Digression I Sexuality as the biological basis for physical love, which aspires to be integrated.

Between primary deficiencies, sexualisation of the childhood, egocentricity and social instincts:
- A combination of a persistent lack of existentially affectionate attention and the sexualisation of childhood with intellect-based empathy can lead to an individual making up for lost attention and interest through gender attraction on the social stage and through manipulation. The individual tends to fall back on this "hystrionic" pattern increasingly following puberty, which constitutes a suboptimal attempt at self-healing and is initially stabilised through rewarding experiences.

3.8.4 Cardiac Synchronisations/Empathy/A Readiness for Personal Affection: The Interpretation of and Reaction to Beings/Others

Key role: This compression level is defined as a suspended field in the concept of Buber's "I-Thou" relationship. It is a field that encompasses all others and modifies, similarly to the SES, which exists on all levels, but nevertheless appears to be most pronounced at the heart level (see also cardiac vagus or ► Sect. 3.2.9). The way in which concepts from entirely different fields are expanded on and verified by instigators who know nothing about the other is certainly impressive.

▪▪ Social space

The significant differentiation between relationship qualities on the one hand (I-Thou) and conveyed material (I-It) on the other is also observed in the large section of the primarily interpersonal space, to which it is also directly linked in this context with regard to interaction research, the communication sciences, couple, family, group and system therapies. The above also includes organisational consulting and large group work. A separate chapter is dedicated to this interpersonal space, which can be regarded as a bay window in the image of the developmental tower at the level that we can encounter (see ► Sect. 4.2.2)

Biological bases

The neurophysiological basis for interpersonal love appears to be provided by the social engagement system (SES), particularly through its cardiac vagus branches in combination with the hormonal oxytocin system. At first, this system seems to be responsible for taking care of the brood and ensuring continuation of the species, namely for serving basic needs. The SES starts to make an impact after birth. Previously, its signals were directed at the large familial group to ensure survival. Nowadays, we are required to communicate with one another in much larger contexts in order to continue surviving together. We require the principle of humanity beyond our families throughout our lives. It lays the foundations of our culture and the further development of each individual. Our inherently hardwired system of humanity enables us to develop differentiated contact behaviour. Differentiation occurs as a result of this behavioural ability to establish, dose and also block contact, in other words, to make responsible choices with regard to our inner balance.

In this section on human love, it appears that the concerns of the archaic level, which revolve around ensuring the continued existence of the species, affect the maximally spiritualised level. As such, this primarily integrative bridge connects two worlds.

For the purposes of this book, I shall look at the phylogenetically older part, namely the non-verbal understanding that conveys the experience of being in a relationship with someone and being in resonance (or perhaps not), separately to the ability to communicate linguistic content through the allocation of meaning to certain sounds.

In addition to the latter, we also require a growth spurt of the left hemisphere. This, in turn, facilitates the involvement of I-It communication.

Further detail on the heart's neural network, heart variability and heart-brain coherence can be found in ► Sect. 3.2.9. It is also beneficial to be aware that the heartbeat, with a calibration of 1 mV in the ECG, is ten times stronger than the heartbeat in the EEG; as such, the pulsation of the heart becomes the regulative pacemaker for all body cell activity and is relatively easy to detect as a radiating field outside the body.

Porges explains that mammals, especially humans, have gone through a development whereby they learn to communicate their emotional states and needs through facial expressions, eye movements, head posture, contact calls (prosody) and purposeful listening. This communication is of crucial importance for the safety of animals that live in the wild and humans alike. It is for this reason that higher mammals and humans have differentiated the phylogenetically young, medial vagus branch in its myelinated form.

However, appropriate, development-promoting stimulation, or rather, early experiences of being loved and cared for are required to develop this potential. Perhaps unsurprisingly, this potential appears to dwindle in young animals and children who grow up without love. In the social deprivation experiments conducted on Harlowe's monkeys, the deprived monkeys, if they eventually succeeded in becoming pregnant, exhibited either impaired or no behaviour of maternal care.

Relevance to psychotherapy: The topic of this section, love that provides a sense of familiarity and security has received surprisingly little attention in the field of psychotherapy, particularly in consideration of how many suffer from a lack thereof. Indeed, the constitutes the majority of our patients.

However, with this assertion, we do not seek to overlook the work of Sandor Ferenczi, Michael Balint, Erich Fromm, Alfred Adler's concept of community feeling, Jacqui Lee Schiff's transactional analysis treatment of psychosis, developments made by René Spitz, John Bowlby, Mary Ainsworth, and Mechthild Papousek, the fathers and mothers of attachment theories and research on observing babies, contributions from the family therapist Virginia Satir, the talk therapist Carl Rogers, Heinz Kohut's self-psychology and the creative psychoanalyst Gaetano Benedetti and Daniel Stern, to name but a few who have acknowledged the significance of love to psychotherapy, along with numerous others.

There has also been another common vindication in the field, namely the change in the attitude towards the therapeutic relationship over the past 70 years in all schools of psychotherapy, albeit to varying degrees and at different speeds. In my opinion, this change provides the clearest evidence that, above all, practising psychotherapists have grasped the key factor in the healing process as such. Here I would like to extend a nod towards Carl Rogers as the pioneer of "authenticity, congruence and empathy", whose therapeutic attitude – endorsed by Klaus Grawe – has been incorporated in behavioural therapy and likewise in humanistic methods as well as in depth psychology, where it has long been at home. His contemporary Heinz Kohut specified a warm and empathic relationship for his self-psychology, which was only hesitantly accepted in part by the mainstream, at least if my interpretation is correct. Indulging in a semi-authentic and relatively interactive therapist behaviour was initially attributed to structural weakness [36]. As pioneers of sympathy for one's fellow human beings, I personally met with Gaietano Benedetti and Stavros Mentzos. Gestalt therapy has been based on Martin Buber's notion of the I-Thou [37], which both challenges the therapist in his whole being and simultaneously supports him, since its establishment. A minor variant thereof can be found in schema therapy's limited reparenting, which can be traced the Jeffrey Young's attempt to manualise Gestalt therapy. -

The above list is by no means exhaustive.

In summary, the following can be stated: Despite the fact that interpersonal love has not become the focal point of theory for any school (perhaps due to the fact that it is regarded as too shameful for cliché scientists), it has always been practised by those who are psychotherapists at heart. As such, the neutral, scientifically objective observability advocated by Pavlov and Freud, as children of the Enlightenment, which provided the model for therapeutic relationships, has since been eroded and largely replaced by the personal strength of the relationship, often with a guilty conscience towards the respective "scientific community".

Now we shall return to the individual and his experience, initially to his formative years.

(a) Positive version following experiences of humanity in early childhood:

"I want to and feel confident enough to seek and try out empathic contact that provides me with a sense of security, which I have always received from my family and siblings as well as amongst groups of friends who are the same age as me. I enjoy roleplay. I am able to make new friends. I can also see things from the perspectives of others, including animals. I have experienced reliable, life-enhancing relationships from the very beginning, but have nevertheless felt noticed and validated in my own identity. I enjoy being with others. I don't feel threatened by them but instead often experience happiness. I have a strong sense of community and generally tend to understand what is important to my group and what isn't. In the majority of cases, I then adjust to this, but I may also risk having another opinion. I am hurt if I am unpopular, excluded or don't belong to a group. Therefore, I make sure that this only happens very rarely.

The adult version of humanity: "I feel more or less connected to the people around me and I appreciate them despite their shortcomings. Nobody is perfect, even me. This is how we stay on good terms. I live in harmony with them. We feel a sincere attachment and connection to each other, as if we are one big family. But everyone is different. We are very capable of integrating sexuality into our systems and ensuring that all other basic needs are satisfied and are permitted to become traits. I feel upset if one of us isn't doing well and I help him. I also accept help from others when the situation is reversed. This provides me with a sense of security, familiarity and confidence. I am a dedicated member of the group and assume responsibility for ensuring that everything remains constructive within the group and that it continues to exist. Purifying conflicts are necessary from time to time. We have since established a good system in this regard. It would not be good to be excluded. Isolation would be a cruel punishment. However, if this community was no longer sustained by humanity, I would leave it".

(b) Deficiency version "I have never felt loved, liked or right. I have no faith in people. My innermost self protects me. I won't reveal it to anyone. Resignations and pain have caused feelings of sadness, inferiority, anger, jealousy and bitterness to accumulate there in a dark corner, which sometimes turn into schadenfreude that I attempt to conceal. My lack of sufficient experience of being loved hurts; It makes me feel lonely and leaves me longing for these experiences.

(c) Destroyed faith in humanity: "Someone that I trusted unconditionally intentionally lied to me, ruining my life, everything I loved and achieved. I don't trust anyone anymore. I want to settle the score and make him suffer more than me. I want to see him tortured and in pain".

(d) Overlapping zones of level 2 (sexuality) when there is a deficiency on level 4: isolated sexuality

Overlapping zones of levels 2, 3, (4): Flirting, making a move, histrionic sex appeal

Overlapping zones of levels 2, (3), 4: All-embracing personal love

Overlapping zones of levels 2, 4, 5: Culturally transformed love

Overlapping zones of levels 2, 4, 5, 7: All-embracing spiritual love

3.8.5 Corticalisation with Oscillating Fields Capable of Integration/Subjective Selfhood: Creative Powers to Create Order, Set Values and Shape Structures Through Resonance Fields

We appear to experience the integrative self-organisation of our oscillating fields as subjective creative potential. This comprises moods, feelings, atmospheres, impressions, inner movements, impulses, non-verbal expression and other chains of associated images, etc., but not language. Language may be associated with this area, but this is not automatically the case. As a rule, language typically consists of fewer nuances than is required and often simplifies or distorts the message. The link between subjective experience and language usually paves the way for early, verbalised communication from a mother to her baby by responding to the baby's mental state. However, this canalisation is not obligatory. As such, certain individuals may find it extremely difficult to put their innermost feelings into words.

In artistic work, it is particularly evident if there is an experience that seeks a form of expression, be this musically, in colours and shapes, or poetically with the use of language.

Japanese poetry particularly cultivates the separation of mental awareness, or attitude, from the primary experience substrate through the following language-based forms as adequately as possible: the poem form of a waka (31 syllables) or a haiku (17 syllables) (see ▶ Chap. 7, [38]). This ability to differentiate may also be derived from the legacy of Vedic traditions.

The creative power of wave fields within inherent materialisation processes is a noteworthy topic. Particularly in the plant kingdom, whereby nature provides numerous examples for the clear creative power of force fields: Flower and leaf formations, cones and other seed formations, etc. The two modalities of being, namely the wave pattern and material structure, are clearly evident in the above.

The marginal resonance fields of the entire organism which are registered by and may even influence the whole, for instance in the case of ear acupuncture, foot reflex massages, are similarly impressive in the human body. The reflections of superordinate patterns appear to have an effect on the subordinate cell levels, be it in a stabilising or chaos-inducing respect. Our body places us in an overlapping mode of being of pseudo-matter, which is exposed to the constant flow or more or less ordered wave information. It is generally acknowledged that meditation can help the body to regulate and keep it healthy. The majority of people are able to sense their state or natural oscillation if they are at one with themselves in a positive sense, and become sick if they are unable to experience this for an extended period.

Whilst the unconscious cannot be located, it can be found most concentrated and able to resonate in areas where the conscious is not in focus. In the case of children, this is the entire brain, and the subdominant hemisphere in adults, which is usually the right hemisphere. The unconscious appears to be able to extrapolate and integrate the vast number of subconscious signals from the outside and inside world in a self-organising manner; a feat that would completely overwhelm the conscious. It provides the results of this process to the responsive "Ego", which fundamentally is free to follow its gut feeling or not.

At this point, it is worth returning to the fact that S. Freud, who successfully distinguished between the conscious and unconscious purely on the basis of observation, meaning that he was unable to find any spatio-temporal or causal structure in the unconscious in comparison, thus determining that it does not offer any logical functions. Instead, to Freud, the unconscious can be accessed in free association through the resonance principle. Freud's relatively inferior assessment of the unconscious compared to the conscious has not been adopted 1:1 in today's approaches, similarly to the assertion "Where id was, there ego shall be". The above only applies in exceptional cases and in the case of pathological processing of conflict and potentially also of trauma. As a rule, we can be grateful for everything the unconscious removes for us with regard to making decisions and coping with daily life.

The capabilities of the subdominant right hemisphere open up spaces of opportunity for imagination, compress chains of association and facilitate the communication of various emotional worlds with each other. This is most likely where artistic creativity is engaged in self-organising power struggles. It aims to achieve a harmonious expression of tension in addition to healing balances and hidden adjustments.

▪▪ Coherence

The observed synchronisation phenomena between the heart, breathing and the limbic system suggest that the generated pulsations and wave patterns take over the role of stabilising, biologically balancing order parameters in favourable circumstances. As the human organism – depending on the age thereof – consists of between 60–90% water, it can hypothetically be expected that various forms of water-geometric patterns, which may exhibit all forms from the chaotic to highly-synchronised are created (and decay) in the inner cavity systems (somewhat comparable to the EEG, which, as a vectorial sum likewise displays various wave formations, amplitudes, frequencies and degrees of order, and usually reflects a switching interaction between synchronisation and partial excitations).

Harmonic, inner wave images are also likely to occur in meditation and as a vehicle for the substantially verified, functionally stabilising effects of the popular mindfulness-based exercises of recent decades (within the context of John Kabat Zinn's mindfulness-based stress-reduction programme). The reverse path, namely achieving regulative tranquillity by concentrating on external, centred, harmonic versions of quasi-geometric patterns (the mandalas), has been and remains an equally reliable meditation technique for millennia. The above values are rarely able to find a permanent home in a chaotic brain or organism.

▪▪ Values

Values are commitments, beliefs and stances. Experience has proven that level-headed, balanced, responsible and empathic individuals typically gravitate towards values such as equality, forgiveness, an appreciation for the good, the true and the beautiful, unconditional love, etc., and seek fulfilment. It would appear that no one is entirely sure how values are represented in our brain. Our understanding tends to lean towards orderly patterns in which fixed and variable oscillations complement each other.

■■ Correction processes

It also remains unclear precisely how the organism manages to correct excitation patterns that represent a real event if they deviate from the ideal or target value. This process is referred to as an unconscious defence mechanism in psychodynamic terminology. An ideal is formed on the basis of an assessment; it is created and maintained as a "model" by the focused consciousness, and used for comparisons with reality. Any significant deviations from the ideal can lead to anxiety, pain, depression and/or even lead to a crisis of self-worth. During development, the notion of the ideal can also be overhauled, changed, refined and once again adopted. This can be compared to the fate of internalised, normative demands (Freud's "Superego"). The ideal is usually revised throughout the course of development, but may also become fixed in a pathological manner.

Notably, certain neurophysiological structures appear to prefer to take over unconscious correction over the course of some individual development. For example: Projection through the inner/outer boundary, or: Directing aggression against oneself instead of externally, or: Blocking the accompanying, affective responses. Perhaps the reason behind this phenomenon is the fact that there is a significant amount to be corrected in each case when the structures responsible for these abilities are in a vulnerable or particularly formative and adaptive phase. The emergence of structural levels for individual personalities can be visualised in this way. Previous challenges leave behind wide neural pathways or perhaps blockages. Furthermore, the corrective processes also suggest that a receptor for obtaining a balance between the actual and target value/ideal can be found in a certain area, potentially the upper limbic system, and that this receptor initiates a corrective process when signal anxiety (as coined by Freud) occurs as the result of a deviation. What could play out during this process?

Our body is resonating by our numerous organic cavities we are. There are cells and bigger structures, reflecting waves, empowering them by synchronisation or deleting them. Healthy cells are "in contact" with each other. A minimal change in the formation of the wall will change the pattern of the waves. You can tune to harmony or to chaos and you also can delete information. Maybe in this principle there may be soon found the key for the ability to repress. Let us have a look at an example at a lower level of complexity: on our voice. At the beginning of our lives, we all are able to learn unconsciously to innervate our pharynx with subtle changes in a very great variety. We succeed to imitate the vocalisation of the language spoken around of us. After a while, we nearly lose the ability to innervate in another way. The reward strengthens the innervation programme. Back to repressing: to get rid of an "ugly affect" is also rewarding.

We possess the following functional corrective structures: the inner/outer boundary, the integration fields of the emotional poles, the perception of our own needs, the inhibitory top-down neurons with regard to emotions and also those found in the sensory and motor areas. Learning to correct is only required if a failure to do so would result in anxiety, namely the fear of negative evaluation.

A prisoner's uprising? Following a successful corrective process, it still remains to be established how we should imagine the personal dungeon for the banished. The censoring receptor appears to no longer be able to register the banished. However, banished aspects may continue to resonate with similar aspects, growing into a monster and consequently transcending the boundaries in an untimely and disproportionately powerful manner in reaction to the appropriate trigger.

Unnecessary conflict? All that has been explored on the defence against undesired aspects in the above paragraphs concerns an inwardly directed battle against partial aspects, desires and impulses. Someone equipped with the power of focusing consciousness denies someone else his existence. This "someone" may be an internalised person, but it may also be a value held highly by oneself or another, or one's own competing impulse. The above notion requires further identification and clarification.

Mediating and changing instead of belittling? At this point, it is imperative to establish: Is only an "Either/Or" possible, or does everyone deserve a loving, appreciative look that facilitates modifying transformation and an integrating composition? Accordingly, inner executioners can transform into wise men, and prisoners can also turn into valuable "fellow citizens in the inner republic".

At its core, development primarily consists of an increased and reshaped cooperation between analytical and focusing abilities through the integrating potentials of the subdominant hemisphere (Minor association: "The eternal feminine draws us upward" Goethe, Faust II, final part).

▪▪ The collective unconscious

According to C.G. Jung's observations and concepts, our unconscious personal inner waves are reached and drawn-in by the universal waves of the collective unconscious, without the latter being required to transcend any particular barriers in this regard. Jung transformed the world of the unconscious into an enriching treasure trove that extends far beyond personal horizons and the horizons of the currently known world. He was the first psychotherapist to indicate the phenomenon of acausal connections, synchronicities based on the resonance principle. Jung demonstrated an interested and open-minded will to form a scientific understanding of this phenomenon, but he was ahead of his time in this regard. Today, this phenomenon appears to be somewhat less unnatural than it initially appeared to be to our colleagues a little over a century ago thanks to the advent of quantum theory.

▪▪ Polarity and integration

Jung made an intriguing contribution on the polar structure of the world and people. According to him, there are forces that strive to come into balance. He also contributed his archetype theory, inborn models or people, behaviours or personalities that play a role in influencing human behaviour, on the basis of the four brain functions: Thinking vs. feeling and sensation (i.e. sensory perception) vs. intuition. The latter activates the process of integration in each individual, referred to as "Individuation" by Jung. Intuition is fed by the unconscious and also by sensitive sensory perception.

In situations of need, shortcomings, anxiety and danger, many individuals switch to an emergency programme that is focused on survival, satisfying the basic needs and seeking a practical and smart solution. Humanistic values are often sacrificed one by one. In the majority of cases, petty theft is sanctioned first. (Brecht: "Food first, then morals".) Those who identify more strongly with their values resist this regression of values when in danger and the temptation to take an easier path. They may even risk their lives, in totalitarian power structures, for instance. These individuals merit great respect.

3.8.6 Conscious Sensibility/Present Energetic Receptivity/ Interpretation of Informed, Energetic Fields

■■ Intuition

Intuitive perception can be improved through mindful practice. Everyone is capable of developing their perceptive potential through meditation and empathy. (Taking care of a baby as a mother and/or father can represent natural intuition training for parents.)

We have previously explored the special ability of the pineal gland to perceive higher frequencies in ► Sect. 3.2.14.

The ability to interpret subtle and informative fields increases when the inner withdrawal of one's own mental activity is accompanied by high alertness. "Wu-wei", the state of unconflicting knowledge of the environment, was held in high regard in Ancient China. In Far Eastern philosophy, Yin energy is regarded as conscious receptiveness. Yin and Yang are not in conflict, but rather complement each other and refer to each other.

■■ Psychotherapy

This ability has always been trained in psychotherapy. However, unfortunately, our field also offers a number of negative examples in this regard. The psychotherapeutic mainstream has only recently acknowledged the beneficial form of the therapeutic relationship, as it has always been practised in humanistic methods, not least due to the impressive research findings [35].

Therapies whereby the therapist bases their approach around the high level of sensitivity at this level yet nevertheless permits adequate, stimulating activities tailored to the development of the parts of the patient's personality that come to the fore are the most successful in this regard.

As a result of its insensitive methodology, the interpretation of the therapist's relationship posed by psychoanalysis has fallen out of favour and is often experienced as counter-productive abuse. Truly sensitive therapists are capable of adequately establishing the material, the right time and the right dosage, and approaching these aspects with an economical attitude. Mindful ones among them also simultaneously inform patients that they may be mistaken and always ask to be corrected. This requires a certain level of deference to the individual counterpart, his subjective world of experience and perceptive abilities. In this respect, certain elements seem to have gone sub-optimally in the past. Things that have gone well are rarely mentioned. This applies to almost everything.

The notion of therapists as primarily young, inexperienced behavioural therapists with standardised, psychoeducational lists of questions and homework can often be met with comparative awkwardness and embarrassment by older patients, who have never been viewed as an individual or had their concerns understood by anyone. Young colleagues in our field seem to know no alternative to the quasi teacher-student relationship. In recent decades, opportunities have been wasted for many patients as a result of this.

It here needs to be explicitly stated that the transference process does not reflect the abilities of this level. (In the case of transference, we assume that the distorted perceptions communicated to us therapists by a patient in favour of a person who he values, with whom emotional harmony has not been achieved, may indicate a problem area that requires processing.)

Conversely, the healthy countertransference of the therapist should be used to sensitively interpret the patient and his unconscious. Indeed, patients also unconsciously interpret their therapist. As such, an honest, transparent attitude conveyed by the therapist is key for successful therapy, even if the notion of being semi-authentic is and should be maintained with regard to content. The harmonious relationship results in the emergence of a supportive, coherent interspace between the therapist and patient with various, parallel relationship levels featuring varying emotional tones. The strands of the relationship are increasingly modulated not only by the contextual knowledge of the world of the other, which deepens understanding of specific connections, but also by the sensitive co-experience, which is not usually verbalised but instead adopted in the overall behaviour in a controlled manner.

An analogoue field, such as the one found in couple contact, emerges in the "constellation work", based on the methodology of Jakob Levy Moreno's (1889–1974) psychodrama and his sociogramm from the 1930s. Of course this approach has much mor complexity. A whole group can be involved. In this situation, people without any detailed knowledge on the members of a family or organisation, or on partial aspects of an abstract problem, are placed together in spatial-symbolic fields and used as field antenna for the ideational realisation of emotions and impulses. The reactions, which are often accurate, are astonishing. The Gestalt psychologist Kurt Lewin (1890–1947) was the first to apply extensive theoretical reasoning for the field phenomenon. Similarly, the psychoanalyst (Jungian) and quantum physicist A. Mindell attempted to gain an understanding of the subtle fields. The cognitive phenomenon observed in family and constellation work remains not entirely understood and categorised by science.

Another area that has only partially been explored is the higher perceptive abilities of many species of animal [39] as well as the typically stronger intuitive perceptive capacity of many indigenous peoples, which suggests that civilised peoples have suffered losses in this respect.

3.8.7 The Ability to Synchronise with Extensive Reference Horizons

■■ General dissemination

Experiences that spontaneously come into contact with superordinate levels or only with parts of the individual's knowledge do not constitute part of everyday life in the current western world. Neither are they expected to by the mainstream. (This explicitly does not refer to substance-induced changes in consciousness.) We are only able to speculate with regard to the number of unreported cases.

Despite this negative response from the current mainstream and its zeitgeist due to a misunderstanding of the "Either-Or" approach, we aim to accept and collect indications and primarily evidence relevant to psychotherapy. Of course, this anthology remains incomplete.

Indigenous peoples continue to exhibit a different openness to a superordinate relationship. For example, A. Mindell [40] describes how Native Americans are brought up with the holistic understanding of nature, whereby Mother Earth is the channel and source of wisdom. The vision quest is their most important religious ceremony. During the rite of passage, young males entering adulthood seek support, leadership and knowledge from the higher power of Mother Earth, whereby they ritually fast, meditate and carefully pay attention to nature and their dreams and visions, discussing them with the Elders.

We have previously looked into the abilities of the ancient Vedic people to establish contact with higher spiritual levels and the knowledge found within, which were evidently collectively developed over multiple generations, earlier in this chapter. At present, the world in which we live and our zeitgeist is very far removed from the above suggestions, abilities and possibilities.

■■ Nevertheless

People continue to strive for inner orientation and occasionally experience spontaneous and unusual unity moments, most commonly through calm and relaxed moments in nature, but also after birth within the context of a deep experience of love, or also in the functions that accompany dying. These experiences are typically experienced as private and intimate, and safeguarded against dismissive eyes and ears. People require a certain level of trust in order to try to explain these unusual momentary experiences. Language, which is formed by our dominant brain hemisphere, appears insufficient for experiences that may have been permitted to enter the flow through the brain's wave potential. Again, it is hard to ascertain the estimated number in the population in this regard. In principle, they cannot be excluded as possible current experiences for each individual, particularly due to the fact that comparable experiences have been described throughout all cultures and periods, moments reminiscent of a gap in our space-time limitation. I am not aware of any statistical attempts to gather evidence of these experiences. I myself only know of a number of individual cases. It would appear that these experiences are subject to all possible nuances. It has been suggested that those who are well trained in the art of meditation and have practised it for a number of years are more likely to experience "Satori" i.e. awakening, than non-meditators. Nevertheless, meditation does not provide any guarantees that this kind of experience will occur, and conversely, non-meditators can also experience a form of awakening.

The changes in the brain that occur as a result of meditation have been specified in ▶ Sect. 3.2.12, the normal ranges of radiation in ▶ Sect. 3.2.13; a certain scope for deliberate intensification is referred to in ▶ Sects. 3.2.16 and 3.2.9.

As a collective experience, to the extent that it is possible to refer to an experience as such, a positively overwhelming experience of participation in an infinitely great being typically with only few details and perhaps with certain, intensive sensory experiences (light, colours, sounds, altered bodily perception), with the experience of the collapse of opposites, with the feeling of a sense of purpose and higher insight (which cannot necessarily be subsequently recalled or specified in detail) and a strong energetic, integrative and loving presence, often experienced as a God-like presence (however without a special form). In terms of mood, the individual usually experiences joyful and overpowering wonder and the certainty that they have just perceived something extraordinarily important. Any later attempts to interpret or compare the event are predominantly shaped by the cultural environment, the individual's reference horizons and the expressive abilities of his language skills. All of these experiences appear to have one thing in common: there are no fitting words to describe them and they cannot be compared to anything else.

When looking for classical models in this regard, it is not possible to avoid mentioning Meister Eckhardt here in Germany and a considerable number of other great mystics around the world. To list them all here would take us beyond the context of this book.

▪▪ Addendum

Connections to everyday life at present. It would appear that there are many nuances and intermediary forms of these experiences that we are unable to and should not yet view as awakening, but rather as pioneering, impressive visions, the content of which may appear to represent an influx of general knowledge. This phenomenon is also experienced in a purely scientific scope; Kekulé's dream vision of the benzene ring, whereby an understanding of the structure thereof had previously pervaded him, is a well-known example here. (However, perhaps this example of very limited, informative detailed knowledge is better attributed to the previous level.)

Tip

Specialist "vision quest" for psychotherapy-relevant active principles.

1. Wholeness as the origin of a polar world
2. Relationship or non-relationship – or: connecting or disconnecting
3. Re-establishment of balance as a principle of survival
4. The creation of a (subjective) world – constructivism
5. Revival work as a prerequisite to therapy
6. Relativising potential worlds
7. Chains of metamorphosis between death and life"

We shall now embark on a historical vision quest and recollect the descriptions from the founders of the school and their spiritual role models, provided they refer to ideas that they themselves found impressive and perhaps even turned into a concrete issue. I can name several examples for visions from the intermediate area related to structures and principles for psychotherapy in the field, especially with regard to Gestalt therapy. A logical approach is to look for related and complementary visions that may complement each other to form timeless clusters. It is perhaps important to note here that there are more visions that have not been included in my list as my overview is not exhaustive. As such, others may be added to the list.

1. Vision cluster: Wholeness as the origin of a polar world

In this case, there would primarily be neutralising wholeness (creative indifference/zero-point level), which may, on a secondary level, disaggregate into polar forces or duality in a creative act; these partial aspects may be reassembled to once again form a balanced whole by humans, primarily in themselves. As such, the constant objective is to remain in the correct centre. The return to wholeness is everyone's mission in life.

This notion was supported by the philosopher S. Friedlaender (1871–1946), the first spiritual teacher of Fritz Perls (1893–1970), who admired him for his visionary ideas, particularly as Perls struggled throughout his life to remain in his centre. Decades later, following a period of studying Zen, Perls also came across similar advice from Laozi, namely, to constantly stay between the two poles at the centre of one's being, in addition to another variation from Heraclitus, his favourite philosopher from the antiquity, who paid attention to the logos at the centre throughout the ups and downs of life.

The notion that we and our world are constructed from polar forces that we have balanced is also central to C.G. Jung's (1875–1961) teachings. The four functions of the brain: Thinking, feeling, sensation and intuition are equally in need of balance and integration as the various archetypes (preformed experiences). As far as I am aware, both of the above founders of the schools of psychotherapy did not have direct contact with each other.

2. Vision cluster: Relationship or non-relationship, or connecting or disconnecting

These functions can be regarded as the key modelling tools of reality. Relationships intrinsically create wholeness and fields. Conversely, turning down or breaking up relationships creates contact limitations, parts and fragments.

The epitome of personal relationships is an oscillating field, the I-Thou in the sense of Martin Buber's theory (1878–1965). This constitutes the common secular level. The English translation of Buber's work into I-Thou suggests that Buber, who was himself a mystical Hasidic Jew, believed there is a divine dimension that resonates in such a way that the greater whole is reflected in every individual, whether he is aware of this or not. This reflection may also become increasingly clear or clouded. It connects man to all other parts of the all-encompassing whole.

As such, this kind of relationship is highlighted by a flow of powerful unwavering and healing benevolence.

The supplementary dimension to the I-Thou is the non-empathic, oscillation-free, objective orientation towards a something, otherwise referred to as I-It by Buber. Focusing consciousness is required to interpret an objective relationship.

Martin Buber's differentiation, which he almost poetically described in 1923, appears to trace back to a vision. It became a continuous topic throughout his life. Gestalt therapy adopted Buber's I-Thou concept as a model for the therapeutic relationship (explored in more detail in ▶ Sect. 4.2.4), not least due to the fact that Fritz and Lore Perls attended one of his lectures in Frankfurt am Main in 1925 and were stunned by his credible, considerate charisma. This contrasted with the lege-artis abstinence of their teaching analysts at the time. (Verbal communication with Lore Perls.) This discrepancy was decisive in terms of experience. It paved the way for Gestalt therapy to subsequently turn its back on psychoanalysis.

Shortly following Buber's death, a time during which his work was once again the focus of attention worldwide, Watzlawick and his team adopted Buber's differentiation as types of relationship in their work "Human Communication" (1967/1979). Instead of I-Thou and I-It, they use the terms relationship and content level, and combined these with existent and non-existent. F. Schultz von Thun [41] similarly adopted this structure in a form complemented by self-expression and appeal in his four-side communication model.

3. Cluster: Re-establishment of balance as a principle of survival

Several approaches have independently arrived at similar healing processes. Kurt Goldstein's (1878–1965) visions, on the one hand, pertain to the righteous self-actualisation of an organism and, on the other, the notion of an average state of excitation of an organism if it is released from the state of balance through environmental stimuli, reacts to this in a holistic manner and strives to return to its average state of excitation. (In terms of philosophy, Goldstein was strong influenced by his admiration of the pre-Socratic Parmenides, whose viewpoint emphasises the wholeness of the being and integrates transformative phenomena; Goldstein also consequently studied Gestalt psychology and its application for the processing of excitation in the central nervous system.)

Kurt Goldstein's notion of the necessity to regain an energetic balance following a loss of balance can be viewed analogously to the treatment of the polar forces in Gestalt therapy, whereby focus is paid to balancing the forces and equilibrium. The conflict is first symbolically realised through body therapy, then it is subsequently verbalised. The same applies to blocking processes for energetic constructs (somewhat casually referred to by Fritz Perls as unfinished business), regardless of whether they are attributable to conflict or trauma. During therapy, the conflicting, resisting forces, the origins of which are associatively identified beforehand, confront each other in a tangible and dialogical manner. The situation of powerlessness that was previously paralysing is transformed into a lively conflict. The opponents are granted the opportunity to form a new third.

Freud's (1856–1939) neurosis, which is understood to refer to unresolved, unconscious conflict, also represents a coagulated imbalance caused by an intrapsychic (mini) traumatic collision between a powerful impulse (e.g. an archaic desire) and a more powerful counterimpulse from the evaluation level and denies the former the right to exist. The above also concerns existing and not existing even if only in the subjective world.

More recently, the trauma therapist Peter A. Levine (born in 1942), who initially transferred to the field of psychology and psychotherapy from biology, described an analogous vision of the healing process: In order to restore the natural, healing balance, the impulse that remains lodged in the trauma process must be allowed to abreact, at least symbolically, non-verbally and as specifically as possible. In this regard, emotionally-supported defensive action is often required to mitigate, relativise and even to outweigh the experience of helplessness. Indeed, this process is so fundamental that we can also find similar patterns of activity in animals.

Balance is a wave function. Before an etched-in trace can be changed, it first needs to be reactivated in terms of experience and thus become fluid. It is only during this re-experience that the corresponding psychotherapy can receive the opportunity to transform and heal the mentally straining memory trace in an emotionally corrective manner. Restoring balance in a psychological sense requires healing, relativisation and also the removal of the threatening gravitas of the experience of imbalance.

4. The creation of a (subjective) world

This process uses the modelling tool of separation through differentiation and allocating meaning. It is a constructive process. It is a function of focusing consciousness, which creates structures on the basis of definitions: Maps, images, opinions, beliefs, attitudes, objectives, judgements, biases, etc. As such, it is responsible for allocating values, definitions and names as well as creating languages. It creates differences and animates these through tension. It modulates, educates according to models, values or unvalues. It creates heavenly and hellish conditions, war or peace. It divides. It draws boundaries.

It is capable of blighting quantum experiments with its concentrated mindfulness energy, whereby it is capable of separating quantum from their wave bond to fold or bend into pseudo-matter (Heisenberg cut). Focusing consciousness corresponds to pseudo-matter space-time. It can use its intelligence to construct digital worlds according to the on-off principle. Piaget observed the formative chains of interaction between children and their environments as well as between insight and the individual that acquires it with great interest.

The focusing conscience is capable of setting rules, apportioning power and justice, and determining what exists and what does not. It can prioritise its own existence, be it privately or politically, over all others. It is capable of alluring an individual to omnipotence.

Whilst it may be an omnipotent tool, focusing consciousness is only one significant pole of the whole. It's counter and complementary pole, which is just as significant, is the conscious in the mode of the perceptive, integrating oscillating field. This field allows definitions to fade, levels differences and introduces the expanse without space and time to the oscillation. It appears as the eternal now.

Liveable values and a culture of empathic humanity as well as a wise use of power only arise in the combination of both poles, whereby all beings can live in dignity in their own unique manner with and alongside each other.

We now return to the imbalance of separation, differentiation and analysis.

What does this pole entail for psychotherapy? What benefits could it bring?

I am unable to say whether the fathers of cognitive therapy (Ellis, Beck, Meichenbaum) experienced their cognitive counter-concept to behaviouralism as a vision in the 1960s. In any case, they succeeded in recognising how stressful and life-impairing convictions in "false beliefs" could be, which they attempted to correct through reasonable counterarguments. However, they remained on the same level of cognition during this process. As such, it was unlikely or also not intended, that they would succeed in reaching and changing the roots of the false cognition in the deeper layers. It was only in the transfer to everyday life, provided corrective experiences set in, that these therapies were immersed in the holistic experience and succeeded in achieving disencumbering changes.

At the time, cognitive therapy was not only a counterpart to behaviouralism, but also to the psychoanalysis of the day, which, through its practice of turning its back on the past, primarily attempted to fix the patient in his pathology, rather than releasing him from it.

Learning, cognitive and behavioural therapy as a whole was established in the spirit of the Enlightenment. As such, it held science and rationality in high regard, and emotional and creative aspects correspondingly low, and surpassed itself in its approaches with regard to manualisation and definitions. However, it struggled to reach an understanding of the therapeutic relationship whereby the therapist is able to offer the patient scope for development and a shared field for encounters through his consciously cultivated tolerance for uncertainty. In the meantime, many encouraging positive signals have arisen in this direction.

Conversely, certain other schools of psychotherapy that originally emerged with a (r)evolutionary pioneering spirit, often quickly became fixed in dogmatic hypotheses that remove patients from their lives and consequently provide increasingly less help.

It is key to psychotherapy that each individual lives in his subjective construct, in his unique personal world, the associated links to which only he is capable of retrieving through his chains of association. In certain cases, universal and zeitgeist-determined traces may be interwoven in this approach, but they rarely lead to the personal problem area.

Focus section: Inner-outer relationship or: Self-actualisation vs. adapting to the environment

As previously mentioned, self-actualisation was a central credo and concern of Kurt Goldstein. In his day, it was still a pioneering notion to advocate that an individual has his own impulses and inclinations from the outset. Empiricism continued to prevail, with its conviction that humans are born a carte blanche and can subsequently be shaped and characterised by their environment at will.

At the same time, this also had to compete with old pedagogy, which worked with heteronomy and demands for obedience, and various reform and Gestalt pedagogy movements (Dauber), which aimed to support the free development of the young personality. The former was traditionally associated with the ability to impart more knowledge and discipline. Nevertheless, nowadays, in a comparison of the achievements of schools (in the PISA studies), the school systems based on humanistic and progressive educational approaches (from Scandinavia) appear to have done the best with their flat hierarchies.

What does this entail for psychotherapy? Is the focus in this regard on a transfer of knowledge within a teacher-student relationship? Or perhaps practised humanity, possibly in the meantime on catching up on shortcomings in parent-child care and self-help?

Various models are preferred and taught in the different schools. They concentrate on the understanding of the theoretical background and collective beliefs about what truly helps: Methods, adapting to the environment and orienting knowledge – or a life-enhancing relationship and new beginnings. It is not beneficial to split aspects that belong together into an either-or. People need both.

5. Revival work

Fritz and Lore Perls demonstrated a fascination with the notion of revival in the 1950s. We encounter the revitalising power of oscillations in this regard. How are we able to reintegrate those who have been scared to death, suppressed, suspected of terror, fled, wasted away, dissociated, aspects that have faded away with bitterness, or been delusively frozen, etc., back into a traceable and liveable existence? Indeed, is this achievable at all? It certainly is not as simple as a fast-track in passing. Instead, this requires an unwavering, resilient, reliable relationship and special amount of human empathy, solidarity and encouraging attention diverted towards the hurt being, who hardly exists any more behind his defensive barriers – as well as a vision of his potential. The trick in this regard is to initially invite the remaining personality into the common I-Thou field and provide it a safe place and right to exist there. Only following this should contact be established with it when the defence mechanisms start to appear and learn from these mechanisms the protection and will to survive at the time. In an equalised state, the symptom may become a source of power and self-acceptance.

Rectifying, reintegrating revival work incorporates influences from numerous generations and schools combined, including ancient Zen traditions. As is perhaps evident, I initially incorporate the suggestions of Gestalt therapy with its special therapeutic stance and the traditional style of shifting perspectives. However, it is also clear that it is also based on the psychoanalytic concept with neurosis; of course, the process also includes influences from body therapy, psychodrama and body therapy, which permit it to observe the non-verbal encounter level.

6. Relativising potential worlds

3

Using levels of possibility to help people who have experienced problematic starts in life in terms of relationships to put their formative years into perspective and encourage them to permit their compensatory wishful fantasies about what would have been required in order to be well received in the world as a compass for symbolic scenarios is often a delicate balancing act. Here, the shortcoming experienced in reality leads to the fullness of the symbolically staged realm of possibility (e.g. to experience a caring, supportive mother like other children). At the same time, this typically triggers sadness and disappointment with regard to what had remained hidden prior to this point. The above experience often constitutes a particularly relieving component for the patient, which often concludes with a certain sense of melancholy. Fullness and clear inner direction emerge from an aspect that was previously experienced as a shortcoming through the inner affirmation of what was missing. It creates a virtual experience of participation in the strived-for possibility, which subsequently can be incorporated into the individual's identity.

Irrespective of my background in depth psychology and Gestalt therapy, whereby these experiences were staged time and again, this process corresponds to a certain extent to a vision of A. Pesso, which he and his wife successfully practised in a therapy approach incorporating elements from role and body therapy in order to heal and trace difficult and deficient families of origin.

7. Chains of metamorphosis between death and life,

To exist and not exist?

Fritz Perls was particularly struck when he had the idea – or vision – that each individual, and of course the patient, is required to cross a (subjective) dead zone with each substantial change to his identity. This zone refers to the space-time in which he is no longer the familiar human, as he is required to gain a certain amount of distance from a certain aspect as it needs to be let go, be it internally or externally. As such, the patient feels insecure, in a mist. Who am I now? Am I still me without this part that seemed so important? At this point in time, the individual is not entirely capable of perceiving new, more coherent shape of the identity with a sense of security. Only then can the emptiness turn into fullness. During this time, it is beneficial if the relationship, which can tolerate the material insecurity and is therefore reliably there for the patient and helps him to understand this process, listen inwards and to greet and interact with the new aspect that wants to develop, is maintained by a perceptible therapeutic presence. It was at this point in Fritz's vision that the subjective desert starts to blossom with great joy and enthusiasms.

From this perspective, we may perceive the experience of each life and, above all, each therapy as a chain of "deaths" that we have caringly encouraged due to their ability to help us all let go, to return to ourselves and become more coherent as a result of the transformation. However, extending one's hand to "death" (or perhaps only bidding farewell) will most likely always remain a subjective adventure.

In this context, the metaphor of "death" does, of course, not refer to the materialised being that clings to its definitions and (over)identifies with them, as is common for humans due to our sense-centred and analogous construction that is unable to perceive oscillations. From the viewpoint of the energy flux, this process solely concerns games of metamorphosis with intertemporal materialisations of aspects that are shed like scales from time to time in order to continue participating in the general transformation.

In summary, the "Tower" section appears to reflect the basic interplay of the two modes of being of the consciousness in a multifaceted manner. On the one hand, in the lower area, this concerns the energetically focusing and structure-creating form and, on the other hand, in the upper part, the networking, oscillating form of energy and consciousness may possess the ability to interpret information.

If desired, this polarisation can be categorised as analogous to the ancient Yin/Yang teachings, without being based thereon. The principle of the two modes of being fundamentally corresponds both to the development and cooperation of the brain, and also to the smallest currently known unit, the quantum, which acts as a wave or pseudoparticle, but can primarily be found in a floating state, at least according to the findings of experiments conducted on this matter. As such, we are phenomenologically faced with a threefold manifestation. In terms of consciousness, we live precisely in the centre and on the limit, as transpires from the Heisenberg cut in quantum physics, in a place of energetic form metamorphosis. (The observer's biophoton laser beam causes the quantum wave networks to collapse due to its focussed attention and thus disrupts experiments on the quantum level.)

Our brain works integratively in wave mode and presumably stores information in pseudo-corpuscular mode. When we relive old memories, the oscillation is rekindled and connects to the current one to a form a new whole, which oscillates at the same time and therefore changes the existing content. This provides an opportunity for psychotherapy. It also conversely simultaneously represents the formation model for chains of association that are afflicted by neurosis. Everything that oscillates together is neuronally stored together. In the analogy of this section, the stored aspects correspond to being dead, the energy of the moment is the fire of life that occurs during metamorphosis and new beginnings (◘ Fig. 3.4).

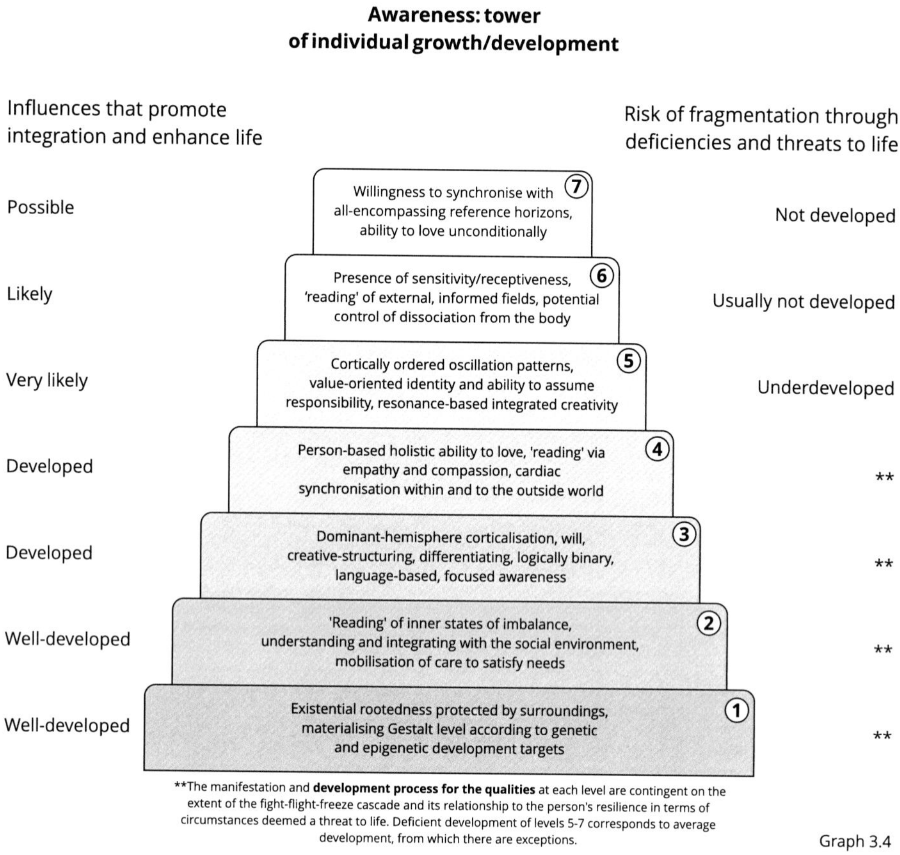

Fig. 3.4 Overview diagram of the tower of consciousness

3.9 A Small Theoretical Excursion into the Micro, Meso and Macrocosm

3.9.1 Energy Differences: Order Patterns, Part and Whole, Life and Death

According to our current understanding and string-like conceptual hypothesis [42], the smallest known unit, the quantum, constitutes a versatile, small energy unit, which can both structure and untwist itself as a strand to form a pseudo-corpuscular knot, similarly to a chromosome, and also naturally forms field-shape networks in its wave form. In between lie a number of various floating states with varying degrees of order. As such, this entity is multifaceted, or rather multiform. The familiar Either-Or categories suggested by our minds do not fit in this case.

In the cosmic vacuum, i.e. the zero-point field, the energy can be found in a virtually-neutralised, highly organised and connected form between negative and positive polarity. According to the calculations of the Nobel Prize winner Gordon Wheeler, the potential of this neutralising energy is enormous: there is enough vacuum energy inside a single light bulb to boil the world's oceans. This is simply beyond our imagination.

We are familiar with lower forms or energy and order from our everyday lives. The various forms of order of light help our perception the most: The photons of a conventional light bulb have an unconnected, chaotic relationship to one another. Their energetic range is low. However, in comparison, a laser beam is characterised by the fact that its photons link up with each other in fixed sequences and consequently form significantly more extensive and energetic structures. In plasma, ionised particles fuse through their conductivity in order to form enormous, holistically connected energetically charged plasma units, which may already be used for technical purposes to some extent, but which we otherwise encounter and marvel at in awe when they arise in an unpredictable form in the case of lightning, solar winds, the northern lights, the stars, etc.

Conversely, the notion of entropy, otherwise referred to as heat death, whereby all energy and information is presented in an equally distributed and unconnected form, to approach absolute zero. This would entail the death of all energetic proper motion.

The quantum physicist Erwin Schrödinger deliberated the principle of life and coined the term negative entropy (negentropy). According to this principle, life is the ability of an organism to develop and temporarily maintain its specific structures through energy (and thus counteract the entropy thereof).

F.A. Popp's findings (1984, 1992, 2000, 2002) on the ability of chromosomes, particularly that 90% thereof are not used for protein coding are certainly notable in this regard: the four opposite, essential amino acids in our cells from a kind of cavity that is capable of energising light quantum (analogous to the mirror principle for the technological production of laser beams). The physiological process is based on a biological reflection principle, whereby biophotons are temporarily lifted to a higher energy level. In the ultra-weak micro-range, structures such as myelinated neural substance, vessel walls and membranes. assume a mirror function (personal note). The fall of the biophotons back to a lower level releases energy that can be used by the organism.

Living beings in the mesocosm range are open systems that communicate with their environment by absorbing matter and also radiation energy from it and releasing a certain amount back in return. Living organisms are consequently capable of developing their own order structures, an energy-intensive process, and maintaining them for a period.

3.9.2 The Principle of Self-Similarity for Our Dual Identity

The brain is (a) a *structuring* and (b) *resonance-capable* organ. Due to cerebral plasticity, we are at liberty to promote more of one aspect or the other. In early childhood, neither brain hemisphere has been observed as the dominant one. We primarily react as "children of our time" and our social environment. In the west, by the time he starts school, a child exhibits much more interest in his structuring-intellectual side, or

3

rather in the functions of the brain hemisphere that is dominant at the time as a result thereof, than in the resonance-capable, artistically oriented hemisphere and, above all, the intuitive receptivity which is attributed to the subdominant brain hemisphere. Consciously child-centred pedagogues, in addition to certain progressive schools of pedagogy and humanistic pedagogy [43] attempt to counter this and provide creative breathing space in lessons and also to facilitate the self-development of the subsequent generation.

We remain unaware to what extent a millennia-old, epigenetic influence plays a role in the asymmetry mentioned above. Further research into the topic would no doubt be illuminating.

Far Eastern tradition distinguishes and cultivates the perception of the wave side of our identity through the perception of "Chi" the energy that flows through our bodies, its radiance around the body, its potential for harmonisation, its tone equivalence and even its focusing power in the martial arts.

Vedic advanced civilisation, which recorded a number of astoundingly precise findings, such as on the planetary system, the periodic system of elements, the age of the sun and other developments concerning the sun only relatively late in its existence, suggests that humans around three to five millennia ago engaged in meditation training when they were children as a matter of course. This can be compared to our basal cultural techniques of reading, writing and calculating. In Vedic tradition, this training may have even facilitated an axiomatic interpretation of information fields with a capability to resonate trained to receptiveness.

At present, the current form of our brain may primarily prevent this, or it would certainly be hard to achieve. We can observe it in an interpersonal rudimentary form in an empathic mother-child dyad, in anxiety-free, playful improvisation groups, in certain empathy-based therapeutic relationships, between familiar, mutually open-minded people, between lovers...

Exceptional talents to describe detailed knowledge that cannot be registered by our sensory organs have emerged from time to time in more recent history, and inevitably lead to perplexed disbelief. We have already referred to Plato in this regard.

Note

At this point it is interesting to refer to Leibniz (1646–1716). He describes a modell from the reality as if he was already in contact with the quantum physicists of today. He postulates the smallest substantial units of our reality, namend monads, as spontaneous units, indivisible, conscious, without expansivity, eternal "etre capable d'action" and "un mirroir vivant de l'univers" with gradual increments of their consciousness. All beeings, including the material, are regarded as a variant and phenomenon of spiritual activity. "Things are only phenomena, though well founded and well connected" (cited as per Hirschberger, Vol. II, p. 143).

The notion of a basic, structural sequence of similarity is a perplexing one. Are the complex structures mirrored in simple ones? And conversely: Can the basic structure of the smallest units be found in the largest?

It is perhaps worth considering here whether fractal structures (as per Benoit Mandelbrot, 1924–2010) can not only be found in geometry, but also in the entire realm of being, including the psyche. At any rate, the number of self-similarities that shape the development nature is bewildering.

The smallest fractal widely accepted by the current scientific community is a quantum, an energetic unit that can be found in us (thanks to Einstein) as an avid quick-change artist between a wave and pseudo-corpuscular identity. Hans-Peter Dürr, a student of Heisenberg, according asserts in his book "Es gibt keine Materie" (1999), that coiled or folded energy is perceived by our senses as solid, yet this is an illusion. When uncoiled, a pseudoparticle transforms into an electric force field for which there are no two spatio-temporal determinants at any one time. As such, the underlying assumptions of conventional geometry, mathematics and statistics are not satisfied; the logical structures established on the basis thereof are not applicable to the level of reality of quantum phenomena. The theoretical structure of quantum physics has been continually developed for over a century.

The brilliant, almost visionary Gestalt psychology Kurt Lewin defined the abstract noun "Gestalt", which constitutes a delineated unit for Gestalt psychologists, as a structure and field. This definition cut across all categories, which was regarded by many in the 1930s and 40s as overreaching, leading to opposition.

The sensory processing of stimuli that takes place in our brain prefers to place structure and field, in addition to matter and spirit, above all, in separate categories (regardless of what Einstein may have to say in this regard). In doing so, we rely on our own conceptualising thinking.

If we imagine a fractal that encompasses all categories similarly to a quantum in this regard, all cells, organs, particularly the human brain, the entire person, groups of people, races, nature and the earth would also be in field and structure at the same time. This does not appear to be an abnormal notion. Several experiences with field phenomena of the relationship level would consequently gain their natural allocation.

Indeed, we are familiar with notions that cross categories with regard to states of matter: Water as ice, snowflakes, liquid, vapour, mist and clouds are normal aspects of daily life.

What is a quantum's state of matter? Its structuring aspect is not only required to form rough matter, but it is also capable of creating fantasies, plans, concepts, rules, names and language. The creative energy of mindfulness is, in itself, a creative power for the development of the structure. It is responsible for daydreams, which we may find comical, provided the gap to reality seems too extensive and thwarts us. Furthermore, it is also responsible for creating constructs and plans for the material world to follow. The energy of mindfulness is a dividing force for the wave side of the being. It separates figures in the foreground from those in the back with razor-sharp precision. In quantum experiments, it is responsible for the Heisenberg cut, which enables waves first to collapse in order to bend into pseudoparticles. It is well suited to the structure of our sensory organs, which are capable of registering and separating the narrowest of excerpts from reality. The brain then constructs an overall picture using these puzzles to the best of its ability. We rarely acknowledge that it does so, and indeed, how it does so. We generally only become aware of this phenomenon when it is outsmarted by deceptive figures, as researched in Gestalt psychology. W. Metzger's work "Laws of Seeing" reveals and describes a number of the brain's construction rules. The brain focuses on the relational structure of what we observe and thus ensures that we are able to recognise things and that they remain constant, despite the constantly changing input of stimuli. Without doubt, this is a great achievement.

We are yet to detect any direct sensory organs or receptors that are capable of registering the wave side of reality. Nevertheless, we are able to perceive atmospheres and other aspects that are not easy to label. Language is used for distinguishable and separate things, not things that do not have a definition or differences.

The ancient Far Eastern concept of "wu-wei" is often translated as non-interference in order not to disrupt the existing interplay of forces. This attitude shows a willingness to respect the course of nature, which can be understood to a certain extent.

Conversely, the culture of receptive listening and feeling is not held in high regard in the western world. Attempts to be mindful and meditate often form a modest exception to the rule. Nevertheless, rudimentary forms thereof can still be found. For example, mothers are inherently required to practise receptive listening and feeling towards their baby in a sensitive manner. Indeed, Daniel Stern's plight for this behaviour towards babies was ceaseless. Variants of non-vocal, intuitive understanding have also been regularly observed amongst lovers, those who are dying and in animals. We also require these aspects to a certain extent in psychotherapy, as many are aware. Success demonstrably follows when they are integrated. Intuitive feelings occur outside of academic concepts, but are promoted by certain schools more than others.

3.9.3 Overview Diagram for the Background Dimensions for Field vs. Structure, with Reference to Einstein's Formula $E = mc^2$ (x-Axis) and the Whole vs. the Parts on the y-Axis

- **X-axis: Field and/or structure**

We shall take Einstein's $E = mc^2$ for the x-axis: The material possibility of existing for a quantum is at home on the left side, meaning space-time can be found on the far left. On the right side we have the other side of the coin of our quantum, its wave nature. This cannot be located, but always covers several realms of possibility at the same time and connects with other waves. An extensive area for the floating state can be found In the centre between both poles, a "Neither-Nor" and a "Both-

And". This is not a familiar notion to our western brains, which prefer clear relationships such as an "Either-Or".

As such, we get a three-way split, reminiscent to a certain extent of Popper's three worlds model. However, in contrast to his clearly separated partial worlds, our model, which is based on the analogy of the willingness of our quantum to transform, is somewhat more reminiscent of a watercolour, whereby separate colours merge with each other. A quantum is thus easily able to transform itself in one direction or another from its floating state.

We are aware of the natural radiation of radioactive substances in the decay of matter. We also know the energy release rate of fire and the natural emissions of biophotons (= quanta with ultra-weak light frequency according to A. Popp) in all living beings. According to U. Warnke, humans emit approx. 10^{27} quanta per second, as recorded with a type of Geiger counter or residual light amplifier.

Conversely, we also marvel at how plants are capable of converting the light quanta of solar energy into chemical bond energy. With ther chlorphyll they produce carbohydrates, packaging it in manageable portions. Indeed, it is a beautiful notion: To ingest the sun's rays, which have been coiled together according to a particular order, as a fruit and be capable of using the energy contained within when digesting them. Our skin is similarly able to utilise direct solar energy in moderation.

It is certainly impressive, even if only in distant analogy to the above, that mentally focused mindfulness energy can likewise lead to a consolidating formative process. Certain examples thereof include planning, conceptualising and artistic work. In this regard, it is initially not important whether the concepts remain daydreams or are actualised in the material world. The mere plan already possesses its own reality.

Almost every creation story across all cultures and times starts with something unshaped that is consequently differentiated from the rest, formed and linguistically defined by a higher, active intelligence according to its will. The opposite process, the fading of definitions, the loss of particular meaning and identity often slips out attention, remaining overlooked, which it lends itself to. Re-entering the blurry background is a quiet operation. This process is carefully observed, upheld and assisted perhaps most in the cultures of the Far East.

When considered in this light, it would appear that a continuous pattern of becoming and decaying can be observed on various levels. However, due to the limitations of our sensory organs, we are only capable of perceiving the part of the overall process that develops the material structure and enables it to decay again. Our sensory perception is not equipped to detect the compression of wave formations. We only register an increase in interest, curiosity, excitement or fascination and, in turn, the opposite phenomenon of disinterest and decreasing energy to a certain extent when we listen to others or participate in conversations.

Furthermore, we are only able to perceive differences. We lack the ability to perceive the existence of states that are free of form and differences. Instead, we incorrectly perceive them as empty. Perhaps if this were not the case, we would find it easier to understand becoming and decay as a permanent transformation and not one that leads to an end.

We shall now divert our attention with regard to this process towards psychotherapy. Who or what hinders the natural process of transformation? A fear of losing physical strength, beautiful, mental potentiality and ultimately a fear of death accompany almost all of us throughout life as a shadow and triggers impulses with the intention of stopping the wheel of change. The primary exception to this rule is found in almost all of those who have experienced near-death experiences. They often desire the next experience of being liberated from their body.

Impulses to fixate always arise when an energetic state cannot be balanced, when an internal plan did not reach fruition, for instance, perhaps with a negative sign (trauma was not repelled, an insult was not responded to, a life goal was not achieved, the chance to say goodbye was missed, etc.), or perhaps with negative or positive signs (a thanks for taking the blame, or finally wanting to be your true self, etc.). Flawed experiences lead to tension. As such, finally experiencing the unexperienced and shutting open doors can facilitate balance and allow life to move on in a carefree manner.

■ Y-axis: Whole vs. the parts

For our Y-axis, we shall take the dimensional bundle between the upper pole of "wholeness, consciousness, abundant energy, integration, connectedness, order" on the one hand, vs. the lower pole of the "isolated parts, fragmentation, non-consciousness, a lack of energy, disintegration, chaos" on the other. There are a range of options with regard to emphasis.

We shall first look at the lower end, physically speaking, the entropy level i.e. energetic heat death. All existing energy is balanced. The isolated parts, which we shall assume settled down in an intermediate form, have no relationship with each other and are not organised amongst each other, instead there is chaotic gridlock. There are no energy differences or energetic bases for the development of consciousness. Even in a figurative sense, life is not possible at this level. Energy differences are required to facilitate life.

If we were to place humans somewhere in the centre of a connecting line between the smallest and most simple and the largest and most complex, then we would analogously allocate a condition at the lower pole whereby all collaboration between the subsystems comes to a standstill, all contact is broken off, all relationships fade away, all experiences of emotional significance are plunged into apathy, all supportive systems dwindle away and the conscience sinks into dullness and darkness. This represents the pole of sickness and death.

Now we shall look at the upper pole of the Y-axis. In physics, this is referred to as the zero-point field. It consists of neutralised, connected power with the highest level of potentiality. (Nobel Prize winner Feynman is said to have calculated that there is enough vacuum energy inside a single light bulb to boil the world's oceans. cited as per Warnke.)

In philosophy, it corresponds to the level at which opposites cancel each other out, such as the coincidence of opposites from Nicholas of Cusa and Advaita, the level of non-duality mentioned in Hindu scripture. Laozi regarded the upper pole as the nameless source of the all-encompassing, from which the ten thousand things rise and return to their source after a certain amount of time. The psychotherapist C.G. Jung defined the lifelong task, or rather development objective, of each human as the integration of his inner, polar forces. S. Friedlaender (one of the spiritual fathers of Gestalt therapy) referred to this level of focus on development as "creative indifference", i.e. the state prior to any differentiation into polarities whilst nevertheless possessing the full creative power to do so. Certain individuals are said to have the ability to re-enter this inner superordinate state. This goal is highly reminiscent of C.G. Jung's in this regard.

This holistic level includes various forms of high, complex degrees of order. Matter-bound patterns of order such as crystals with complex structures can be found in the far left of our diagram.

On the far right we have the highly-organised, energetic wave formations; from subsystems alone we are aware that various, impressive synchronisation phenomena are possible at higher states of consciousness, either concentrated in the front in the EEG (M. Ricard, W. Singer),or synchronisations through the receding vagus fibres from the hearth to the brain, which are capable of creating holistic synchronisation phenomena without direct, neural connection with the field of the neighbouring neurons (HeartMath Institutes), or the re-entry circuits located above the limbic system observed by Tonini.

And emotionally? Love is the most binding element. In this regard we refer to all-encompassing, interpersonal love with its unwavering tolerance and benevolence (outlined in the song in Corinthians 13, 1–13); it is non-denominational. It also refers to the loving kindness of the Buddhists. Similarly to Buber's Hasidic I-Thou, which regards humans as particularly capably of achieving particular success in encounters with the I-Thou as a result of his field of resonance with the highest level.

Experiences of fear, violence, destruction, chaos, envy and resentment, etc., cause us to plummet in this diagram. Conversely, experiences of sympathy, life-enhancement, cohesive order, sincerity, goodness and love cause us to rise.

3.9.4 Entropy vs. Negentropy – What Is Life? Principle: Caring Love – A Centring Oscillation Pattern that Brings Order? When Is Psychotherapy Beneficial to Life?

Nature provides a number of beautiful examples of how harmonious oscillating fields construct the organisation of matter into beautiful patterns on the middle floating field, such as the illustration of cones in opposite rosettes and the symmetrically offset arrangement of the petals of flowers. It we regard this as a parable that the physical aspect of humans is also constantly shaped by the oscillating patterns of its wave side, whether in a positive or negative manner, this may go a certain way to explaining why certain people and faces appear to radiate a moving, joyful glow and others, despite external beauty, comes across as somewhat disturbed, unsettling or faded.

In ancient India, the concentration on the mandalas was used to increase the inner order of humans. In comparison, Gestalt psychotherapy also strives to attain coherence. Each centring leads to the creation of a new, harmonious pattern, which adapts to the basic oscillation. We hereby assume that the increase of the assimilated substance, namely all experiences that were organised and separated without stress, leads to the formation of a kind of self-oscillation from the centre, which becomes increasingly perceptible. This constitutes part of the development of the personality. A similar concern can be found in the quest of talk therapy to achieve congruence, or rather, to reduce incongruence. Ultimately, the conflict resolution of psychodynamics also aims to relieve tensions and harmonise the personality.

Coherence is reflected in both the social environment, in how the individual treats himself, and also in the undisturbed processes of the psychosomatic regulatory circuits.

The mindfulness wave advocated by the humanistic methods of the 1970s and 1980s that was also adopted in behavioural therapy facilitates the integrated, coherent development of the personality and the stabilisation thereof.

Negentropy – a condition for living structures. Above we explored the notion of entropy, which can be found in the lower pole. However, we shall now address its counterpart, negentropy or syntropy, which characterises life. It is not possible to ascribe a definitive definition to life nowadays. That being said, one thing remains certain: syntropy, or negative entropy, is a key aspect to life through its ability to use supplied energy to maintain its specific order as an open and oscillating system (which every living being is). (This corresponds to the understanding of the Nobel Prize winner Erwin Schrödinger over 50 years ago.)

What does this entail for psychotherapy by way of analogy? When is psychotherapy beneficial to life, and when, in unfortunate cases not? In this case, in the figurative sense, contrary to the downward slope of entropy (hereby this refers to contrary to an upsetting, destructively devaluing, functionalising and loveless environment), positive values and life-enhancing beliefs are consciously lived and maintained and may even lead to positive resonance in the environment through mental power for acceptance of others. The question in this regard is where does the energy and motivation to overcome the downward slope of entropy come from? And how can it be countered as a psychotherapist with a resilient attitude and inner work?

There may be a reward in this regard, which can be found on three levels, (1) it most clearly arises from the grateful resonance of the recipient, which has developed into a mentally-nourishing I-Thou field which required no words (2) furthermore, one's own intrapsychic field may oscillate between the trust and ideal ego, which may lead to the feedback of a sense of meaningful, harmonious efficacy and the dispersal thereof throughout the entire organism as a validating healing pattern; (3) provided, at least to some extent, a part-whole correspondence can be perceived between the individual and his mentally-spiritual holistic reference horizon, the latter is reflected, embraced and contained in the partial aspect, releasing it from its isolation. As such, this would correspond to Buber's I-Thou field. It is key in this regard that the therapist is well equipped with negentropic abilities and an abundance of provisions in order to avoid the risk of becoming dependent on or negatively influenced by a resonance or the absence thereof.

General psychotherapy should inherently radiate a high level of negentropy. In other words, it should be viewed as beneficial to life.

The schools of psychotherapy are rooted in different basic assumptions. In the early days of psychotherapy and long thereafter, the primary focus with regard to the key aspects of the field was placed on technique and methodology. Behavioural therapy textbooks tend to follow the historical path of their intervention attempts centred on will, cognition and new experiences or perhaps even an alphabetical list of previous approaches. As such, this corresponds to a basic position similar to entropy.

The psychodynamic groups of methods initially focused on partial findings that were overloaded with meaning and overly generalised, first and foremost Freud's libido theory. The mainstream initially viewed verbal expression as the sole acceptable method of intervention and communication. This narrowness based on the concept of psychoanalysis was only put into perspective step by step through the addition of further aspects and approaches, such as internalised object relationship experiences, the ability to empathise body-related, artistically creative and scene-acting methods, which were primarily cultivated in the humanistic approaches.

The most comprehensive understanding of the role of the psyche in its world was presented by C.G. Jung at the advent of the twentieth century with his notion of the collective unconscious, which incorporates the notion of connectedness and aims to achieve integration and individuation in order to facilitate the potential structural reflection of the whole in the individual.

The humanistic group of methods supplemented psychotherapy with its experience-oriented, targeted approach towards feelings and basic needs in the tension between meaningful values and personal responsibility. The systemic methods extended the psychotherapeutic perspective alongside the primarily individual-based approaches of psychoanalysis and behavioural therapy with regard to social relationship dimensions in group constellations and organisations. Each of the aforementioned groups of methods has made a key contribution to the joint development of the field; they have all given and taken knowledge – be this intentional or not.

Psychotherapy currently finds itself in an intermediate state between voluntary and predominately tradition differentiation on the one hand, and integrative tendencies backed by research and experience on the other. The latter primarily concentrates on the development of therapeutic relationships, which are allocated different statuses and styles through the various schools. They account for the largest share of variance with regard to positive therapy outcomes: As such, the appreciation thereof is indeed on the highest amount of attention. Conversely, the differences in techniques and methodologies employed within the schools appear to have lost significance in contrast to the early acceptance of their relevance to the final outcome and have thus become almost interchangeable. They have essentially transformed into the vehicle for the relationship level. The positive aspect to the current situation is that each therapist is granted a wider choice of expression and tools that he can assemble the range thereof at his own discretion, provided he employs them in a holistic and reliable manner (◼ Fig. 3.5).

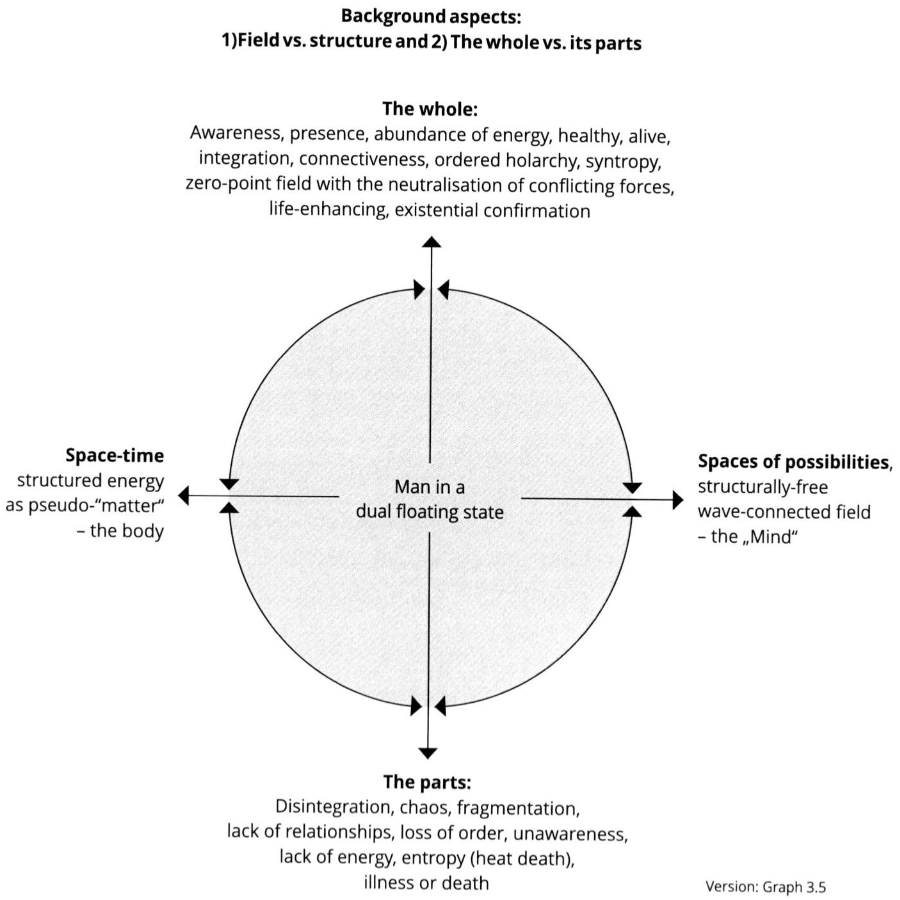

◼ **Fig. 3.5** Background dimensions

Conclusion

In an overriding sense, humans, similarly to essentially all other beings, exist with their psyche in a dual floating state, one the one hand (as per Einstein) as a hermaphrodite in a partially materialised state and on the other hand, in an energy field as two sides of the same coin. However, in the ideal scenario, an individual finds himself in a more holistic and securely socially integrated, energetic and life-enhancing state of being. Or, in the worst-case, in one that threatens to collapse into disintegrating, low-energy subunits, particularly under anxiety and psychological stress. Disorders are presented as acute fragmentations or as suboptimal attempts to repair these.

The cultivation of life-enhancing, empathic, authentic and harmonious relationships on the inside and out to caregivers, superordinate values and reference horizons, to the social environment and the body, including to intrapsychic needs, has a healing effect. In this regard, the establishment of relationships is not regarded as a levelling of all differences, but rather as the integration thereof.

If we are to observe the "part" of psychotherapy as a whole, then it requires significant energetic input with regard to the formation of the therapeutic relationship in order to truly release the healing potential thereof or to facilitate negentropy/syntropy; in themselves, the various psychotherapeutic techniques are largely meaningless and interchangeable as tools.

Literature

1. von Gontard A. Spiritualität von Kindern und Jugendlichen. Allgemeine und psychotherapeutische Aspekte. Stuttgart: Kohlhammer-Verlag; 2013.
2. Hart T. The secret spiritual world of children. Novato: New World Library; 2003.
3. Hay D, Nye R. The spirit of the child. London, Philadelphia: Jessica Kingsley Publications; 2006.
4. Roehlkepartain EC, Benson PL, King PE, Wagner L. Spiritual development in childhoodand adolescence moving to scientific mainstream. In: Roehlkepartain EC, King PE, editors. Sage: New York; 2006.
5. Tacey D. The spirituality revolution. The emergence of modern spirituality. Hove: Routledge; n.d.
6. Roth G, Strüber N. Wie das Gehirn die Seele macht. Stuttgart: Klett-Cotta; 2014.
7. Conrad K. (6. uned. edition from 1958): Die beginnende Schizophrenie. Stuttgart: Thieme-Verl.; 1992.
8. Heinl H, Heinl P. Körperschmerz – Seelenschmerz. Kösel-Verlag: Die Psychosomatik des Bewegungssystems. Ein Leitfaden; 2004.
9. Geuter U. Körperpsychotherapie. Grundriss einer Theorie für die klinische Praxis. Heidelberg: Springer; 2015.
10. Bateman WA, Fonagy P. Handbuch Mentalisieren. Gießen: Psychosozial-Verlag; 2015, engl. Orig. 2012.
11. Singer W. Der Beobachter im Gehirn. Suhrkamp, Frankfurt/M; 2002.
12. Metzger W. Gestaltpsychologie. Ausgewählte Werke aus den Jahren1950- 1982. In: Stadler M, Crabus H, editors. Gestaltpsychologie. FfM: W. Kramer; 1954/1986.
13. McCraty R, Shaffer F. Heart rate variability: new perspectives on physiological mechanisms, assessment of self-regulatory capacity, and health risk. Global Adv Health Med. 2015; 4(1):46–61.
14. Espinosa S. The therapeutic role of the practitioner's heart in classical Chinese medicine and modern medical science. A critical literature review. Eur J Orient Med. 2014;7(5):18–25.

3

15. Freud S. Das Ich und das Es. Gesammelte Werke, Bd. XIII, Fischer, Stuttgart.
16. Schore AN. Affektregulation und die Reorganisation des Selbst. Stuttgart: Klett-Cotta; 2007.
17. Capra F. Der kosmische Reigen. Barth-Verlag: Physik und östliche Mystik; 1977.
18. Mann F, Mann C. Es werde Licht. Die Einheit von Geist und Materie in der Quantenphysik. Fischer, Frankfurt/M; 2019.
19. Mann F, Mann C. Es werde Licht. Die Einheit von Geist und Materie in der Quantenphysik. Frankfurt/M: Fischer-Verlag; 2017.
20. Popp F-A. Biologie des Lichts: Grundlagen der ultraschwachen Zellstrahlung. Berlin: Parey; 1984.
21. Popp F-A, Chang JJ, Herzog A, Yan Z, Yan Y. Evidence of non classical (squeezed) light in biological systems. Phys Lett. 2002;A 293:98–102.
22. Popp F-A. Evolutionas the expansion of coherent states. In: Popp FA, Li KHGu Q (Eds.): Recent advances inbiophoton researchan its applications.World Scientific, Singapur. pp. 445–56; 1992.
23. Popp F-A. Leben als Sinnsuche. In: Dürr HP, Popp FA, Schommers W (eds.) Elemente des Lebens. Naturwissenschaftliche Zugänge - philosophische Positiuonen. Die Graue Edition. SFG Kusterdingen; 2000.
24. Hess W-R. Psychologie in biologischer Sicht. Thieme Stuttgart; 1962.
25. Porges SW. Die Polyvagal-Theorie. Neurophysiologische Grundlagen der Therapie. Emotion, Bindung, Kommunikation und ihre Entstehung. Paderborn: Junfermann; 2010.
26. Metzger W. Gesetze des Sehens. Kramer-Verl., Frankf./M; 1953.
27. Goldstein K. ist keine Literaturangabe, sondern Biographieangabe zur Gründung vom Humanistischen Manifest; 1962.
28. Perls Fritz. ist keine Literaturangabe, sondern Biographieangabe; 1925/26.
29. Heisenberg W. Der Teil und das Ganze. In: Heisenberg W (Ed.): Gesammelte Werke. Bd.3, Piper, München, Zürich; 1969/85.
30. von Weizsäcker CF. Platonische Naturwissenschaft im Laufe der Geschichte. In: C F v Weizsäcker: Der Garten der Menschlichkeit. Munich: Hanser; 1977.
31. Beck H. Akt und Potenz als philosophische Grundbegriffe. Freiburger Zeitschrift f. Philosophie und Theologie FZPhTh 66(2019):2; 2019.
32. Beck H. Dimensionen der Wirklichkeit. Argumente zur Ontologie und Metaphysik. Lang, Frankfurt/M; 2004.
33. Stewart I. The Beauty of Numbers in Nature. Ivy Press Brighton/ Dt. 2017: Die Schönheit der Schneeflocke. Mathematikin der Natur. Theiss, WBG, Darmstadt; 2001/2017.
34. Pinker S. Das unbeschriebene Blatt. Die moderne Leugnung der menschlichen Natur. Frankfurt /M: Fischer, -Verlag; 2017.
35. Wampold BE, Imel ZE, Flückiger C. Die Psychotherapie-Debatte. Was Psychotherapie wirksam macht. Bern: Hogrefe-Verlag; 2018.
36. Rudolf G Strukturbezogene Psychotherapie. Schattauer, Stuttgart; 2004, 2006.
37. Buber M Ich und Du. Verl. Lambert Schneider, Heidelberg; 1923.
38. Itzutsu T, Itzutsu T. The Theorie of Beauty in the classical Aesthetics of Japan. Martinus Nijnhoff Publ., The Hague; 1981/1988.
39. Sheldrake R. Die Wiederentdeckung der Spiritualität. 7 Praktiken im Fokus der Wissenschaft. Munich: O. W. Barth Verlag; 2018, ISBN 978-3-426-29288-4.
40. Mindell A. Traumkörper und Meditation. Olten: Walter-Verlag; 1992.
41. Schultz von Thun F. Miteinanderr reden. rororo, Reinbek b. Hamburg; 1981.
42. Greene B. Das elegante Universum. Superstrings, verborgene Dimensionen und die Suche nach der Weltformel. Berlin: Berliner Taschenbuch-Verlag; 2002.
43. Dauber H. Lernfelder der Zukunft. Klinkhardt, Bad Heilbrunn: Perspektiven Humanistischer Pädagogik; 1997.

Systems and Interspaces

Contents

L. Hartmann-Kottek, *General Psychotherapy*,
https://doi.org/10.1007/978-3-030-87466-7_4

4.1 Introduction and Survey

Ever since the works of Kurt Lewin (1890–1947), the innovative Gestalt psychologist, every existing thing, each "Gestalt", has been understood – independently yet simultaneously at the same time as Einstein's understanding of reality ($E = mc^2$) – as an equally alternative possibility of "field or structure", even if the fields can be neither satisfactorily nor uniformly defined at present. Kurt Lewin started to attempt to grasp the understanding of the dynamics of forces in social fields in the 1940s/50s.

I can recall the growing interest on the subject in couples and family therapy in the 1970s, which, at the time, was available in three different forms: a humanistic, that placed an emphasis on growth and development perspectives of couples and families (Maria Bosch Institute, Weinheim); a psychoanalytical approach, which enriched the psychotherapy concept through the multi-generational perspective and the collective unconscious thereof (E. Sperling, 1925–2007, A. Massing, G. Reich/Göttingen); and a systemic-oriented approach from H. Stierlin, which was based on the work of the Palo Alto group and the Milan approach. The latter was primarily concerned with strategic viewpoints and focused on the system as a whole. For example, strategic considerations included the separation of generations and strengthening the bond between subgroups of a similar age through tasks. Another approach was to start by completely "unsettling" the system through corresponding instructions and then observing its self-organisation with curiosity (M. Selvini-Palazzoli, 1916–1999). In this instance, the relationship characteristic reverts into the observer role, or otherwise settles on a multipartial role.

As such, in terms of psychotherapy, it is useful to first determine (1) whether a relationship exists at all and (2) if yes, using which means the relationship is expressed and (3) which substantive messages or information have come to the fore in this relationship, i.e. for which messages it functions as a carrier wave in the "interpersonal universe". (4) Discrepancies in signals: (a) the combination of positive content and the lack of a positive relationship or perhaps even a tangibly negative attitude is symbolic of a hidden threat (e.g. cold, flat politeness); (b) conversely, the combination of negative content in a tangibly positive relationship tends to signify humorous, tongue-in-cheek, potentially "chummy" and reliable acceptance, i.e. the patient could be considered a rough diamond.

As the relationship level can be accessed by the conscience to varying degrees, it is perhaps sensible to take an interest in whether problematic information, that is to say complexity, has been shifted a level higher or lower. Individual experiences of conflict can be both projected on social systems and acted out therein – they can also simultaneously find expression psychosomatically and trigger a loss of balance and potentially consequential damage in the corresponding regulation systems.

The other cornerstone for the encryption and deciphering of messages on the various levels is, of course, (1) the understanding of the contextual symbolisation, (2) the ability to "read" non-verbal secondary signals with empathy – and (3) the adequacy of the current neurophysiological and cerebrophysiological processing of stimuli. The latter may be entirely blocked by changes in thresholds or also moved to an over-sensitised area. Perhaps the most popular everyday example for the changing of thresholds is our ability to see: during the day, we could be convinced that the gleaming blue sky is entirely empty if we were unable to know any different, as at night, we can behold the same sky with the same dark-adapted eyes and it is instead be littered with countless stars.

4.2 Communication Levels in the Intra- and Interpersonal Realm: Relationship or Content?

The title of this section can perhaps best be construed as a picture puzzle or reversible figure: What is brought to the fore? The existence and quality of the analogous relationship or the issue of its information or content? Both components are not experienced as synonymous or exchangeable. As pedagogists are all too aware: if the relationship does not function, content cannot be reached. Research conducted within the field of psychotherapy has confirmed the precedence of the relationship; it is largely responsible for the success or failure of treatment.

The interpersonal ability that succeeds in integration, or rather, forming a whole, is more fundamental and consequently more powerful for the being than the information composed of fragmented parts which is brought to the fore in a supporting relationship.

4.2.1 Forms of Communication Within the Human Physiological Structural Unit

In this section, we shall avoid delving into detail and instead approach this extensive level in a cursory and exemplary manner. Indeed, an entire book could be penned on this subject and it exceeds the required scope for this book. Signals can be sent in a number of ways within the human organism: as messenger corpuscles, messenger substances, electrophysiological excitation patterns or as light signals.

Various neurotransmitters in the different neuronal functional units bridge the interspaces between the synapses at the junctions between two neurons. These neurotransmitters can be enriched or inhibited by medication and form the basis for psychiatric psychopharmacology, and also for the medical treatment of vegetative dysfunctions.

Hormones bridge the typically large distances between their original location and the location of their target organs in the body. They migrate via the circulatory and lymphatic systems. A special chemical surface helps the hormones and target organs locate each other. The release of hormones is, amongst other functions, a part of the reaction cascade that is released at the suitable point in time during the development of the organism. A certain stimulatory or inhibitive influence from the organism's environment can often assist in this process.

The immune system functions according to the specific lock and key principle and helps the body to defend itself against contaminants and harmful microbes. Certain overactive forms with "decryption errors" that recognise the body's own protein as foreign and attack it emerge.

Our haematological system supports the defence of the organism with its anti-inflammatory white blood cell fraction, it helps wounds to heal quickly through the production of platelets and adjusts the number of red blood cells which carry oxygen for the organism's oxygen requirements, particularly when the organism moves to altitudes with a low oxygen content. We can also find overshooting regulation processes that take on pathological dimensions in the field of haematosis.

The vegetative or autonomic nerve system works in a highly complex manner with a combination of quick-acting transmitting agents at the neuronal synapses (adrenaline and acetylcholine) and with the slower-acting hormones. Stress and threats mobilise the sympathetic part of the autonomic system, facilitating quick fight or flight reactions. This temporarily immobilises the digestive tract, mobilises the cardiovascular system and increases circulation to the locomotor organs. After a slight delay, this is then followed by elevated blood sugar levels as a source of energy and increase platelet agglutination in the case of an injury.

Once the threat has passed, the ventral myelinised vagus branch, i.e. the social engagement system (as per Porges), gives the all-clear and reverts the organism back to digestion, sleep, relaxation and regeneration.

4.2.2 Resonance Levels in a Development-Psychological Context

Existential level A shared feeling of beeing affected by a force majeure can happen, combined with the threat of which takes hold of everyone involved. This mobilises their survival mechanisms, which can lead to the emergence of a close sense of community. For example: Those who share the same fate, be it war, exile, terror, natural disasters, employees of bankrupt companies etc. are often united by their common fear. Perhaps the most coherent phenomenon on this level is the panic-like mass reaction (Le Bon). Conversely, the most de-coherent phenomenon in the intrapsychic field is the paralyzing tonic immobility experienced by the individual, and the ruthless murder of another in the interpersonal field. Prior to both of these events, empathy and the ability to resonate need to be switched off.

In the best-case, life-enhancing scenario, the processing of an existential threat may lead to empathy-based, interpersonal solidarity and the elimination of the notion of friend and foe.

In this instance, the primordial interconnectedness is revived or fortified as the basic foundation – for some individuals who were not exposed to an adequate number of coherent attachment experiences at a young age, this experience of solidarity can sometimes result in the rediscovery of their existence.

In the case of babies, it is key that their initial natural helplessness is catered for by sensitive caregivers, therefore welcoming the baby into its existence. Babies that are deprived of emotional attachment face death, even if they are provided with adequate physical care.

Basic needs level The shared satisfaction of basic needs such as eating and drinking can be counted amongst the rewarding conditions that simultaneously satisfy the need to belong. Indeed, there is a clear thirst for interaction in humans, as actively demanded by babies. Successful chains of interaction evoke feelings of happiness. The notion of sleeping next to another individual requires a sense of basic trust and puts this to the test. In this regard, it is key to distinguish between whether the primary motivation for proximity (on the part of the parent) is based on parent–infant care, or rather sexual in nature.

One of the pioneers in the differentiation between early childhood desires for affection and closeness on the one hand, and the later, eroticised desires on the other, was Sandor Ferenczi (1873–1933), who attempted to develop on the work of Freud, which sadly did not suffice in this regard. Indeed, Freud was not subject to any lack of attention during his childhood.

Sexual development is inherently delayed and controlled by hormones in gradual phases. The unconscious is aware of the reaction to sexual signals, the transmission of the body's own signals and the corresponding attraction far earlier than the conscious, which influences gender-specific behaviour and interests. The same also applies if the sexual-erotic components are excluded from or repelled by the conscious awareness. First and foremost, Sigmund Freud's (1856–1939) psychoanalysis has made a lasting contribution to our understanding of this field.

Parent–child care as a basic need, or rather as "care of the brood behaviour", is, at least during the early phases of parenthood, controlled by hormones (oxytocin) and therefore safeguarded to a certain extent. It is also triggered in humans by the inherent characteristics of "cuteness". Parent–child care initially serves to preserve the species, but can also grow or transcend into the ability to develop all-embracing, unconditional love.

4.2.2.1 Power and Control Level

Polarising experiences of power and powerlessness occur in interpersonal relationships. The former is experienced through an increase in self-worth, centring on oneself, leadership potential, control and autonomy, whereas the latter is instead experienced in connection with a decrease or complete loss of self-worth and dependency. Empathy and compassion subsequently largely pale into insignificance behind the subjectively experienced factual constraints, rules and rational justifications. Fellow man is reduced to the necessary functions. The signals and resonances of this level can adopt a harsh, strict, upsetting to downright degrading tone.

Alfred Adler (1870–1937) addresses the topic of inferiority and overcoming these feelings, regardless of whether the inferiority is experienced due to organic, social or other reasons. According to him, the successful path initially goes through a stage of egocentricity and overcompensation, ultimately achieving the capability to integrate in a loving and unifying manner in a community.

4.2.2.2 Early-Childhood Intermediate Integration on the Preschool Child Level

Empathy Empathy can be viewed as the force and ability that keeps the fascination of powerful ego-centredness in check, or rather in balance. Interactional empathy was already present in the personal "tuning phase". In the meantime, the autonomy process has started, creating a certain degree of freedom of choice and permitting empathy – or not in certain cases. Empathy, a coherence phenomenon, resonates with the emotional disposition of the counterpart and spontaneously invites him to reverse roles, led by emotion. Altruistic love can develop to a certain extent on this basis.

Cortical development: The growing centring of attention is promoted through positive resonance in terms of education (school readiness criteria).

At this age, a (left) cortical growth spurt leads to "mentalisation", as mentioned in the Theory of Mind [1]: A child starts to imagine the viewpoints of others towards the current, real situation, provided that he lacks certain information about reality. This change in perspective is an intellectual act and requires certain development of the dominant brain hemisphere.

Hormonally stimulated, gendered identity perception and the interest in differences between genders requires resonance: am I treated with positivity and do I receive validation from my key caregivers as a boy or girl? – (without encroaching?) Does this new quality inspire me to draw attention to myself (justifiable, primarily due to the fact that prior to now, I received a lack of adequate, phase-specific validation and experiences of positive self-esteem)? It would appear that at this point, the opportunity arises to balance earlier deficits. This new potential tempts the child to try out things through luring and frustration, which may lead to the development of a stimulating flirting component. Compensations for earlier deficit situations are often found behind the extreme forms of the behaviour outlined above.

Discovery of the imagination: In playgroups – or also in lone individuals – the imagination incites creative improvisations in parallel worlds and "As-If" scenarios. These scenarios can be used to test reality to a certain extent, and in other cases for the abreaction of stressful experiences or to compensate for deficits. For example, in imaginative play, a missing, ideal family and the longed-for positive resonance therewith can be created, helping to overcome and bypass a deficiency. In the case of a fixation on this illusive escape mechanism, pathological lying, whereby the grasp on reality is blocked due to conflict, is only one small step away. Delusional misjudgement and the illusion itself would mark a glaring pathological increase.

Developing fine motor skills: The individual's own and, above all, other's responses to the increasing dexterity of motor and fine motor skills spurs the individual, helps him to invent competitive games, to be appreciated within his peer group and also constitutes an established criterion for school-readiness.

Extended social reference system: The familiar reference group and their judgements are increasingly extended to include and sometimes replaced by the interest in the peer group. To some extent, the quest for resonance amongst those of the same age mobilises old formulas for success. These include the (over)adjustment of the child in order to achieve a sense of belonging, or achieving a high ranking in the social reference group through an independent sense of entitlement as a result of competitive behaviour, power struggles or the precursors to ranking sequence clashes. The above behaviour constitutes an attempt by the child to determine his status as an equal person amongst his peers.

4.2.2.3 Integration and the Ability to Love

In its currently developed form, the level of the ability to love and integrate constitutes a series of peaks of maturation throughout life. It can be found in different variants depending on the respective life phase. Each variant has its own designation in different languages. The German language is dismayingly somewhat simple in this regard, making use of loan words to differentiate between different variants (sexual, flirtatious, erotic, passionate, capable of love, charitable, caring-altruistic, philanthropic...). The biologically secured, caring parent–child love has already been mentioned several times. It forms the basis for our ability to develop attachments and a mature ability to love. Even though the adult ability to love is also biologically secured, it nevertheless relies on its integration in a mature personality to achieve full development. At the same time, it triggers integration processes in the external field.

Erich Fromm ([2], p. 28/29) writes: The "aspiration to achieve interpersonal unity is the strongest desire for humans. It constitutes man's most fundamental fervour..."...however, "do we refer to the love given by a mature person in response to the struggle to survive, or are we addressing the immature forms of love that can be referred to as symbiotic unity? ...Or should the word 'love' be reserved for the special form of unity that has been regarded as the highest virtue of man by all of the major humanistic religions and philosophical systems spanning the past four thousand years of the history of the western and eastern world?" – (p. 33) Love is, "first and foremost, giving and not receiving". Here, giving refers to giving in the interpersonal realm. It triggers responses and overcomes separation, as love produces love. It features the following fundamental elements in all of its variants: "Care, a sense of responsibility, respect for others and awareness". (p. 36). In this case, awareness does not refer to knowledge. Awareness of oneself and others arises in the act of the loving commitment to the other you; a devoted attitude can emotionally develop towards an intimate group and potentially also expand to include all other humans or even all other beings. In this sense, the emphasis is also placed on the giving of invigorating compassion. Mystics from all backgrounds expand their readiness to commit to the whole or its representation, provided this appears to them as fundamentally capable of being encountered.

4.2.3 Contact and Interrupting Contact as a Fine-Adjustment Tool of the Relationship Level Both in Inner–Outer Relationships and in the Intrapsychic Internal Area Along the Mental Development Ladder

Buber's outline provided the blueprint for the contact-centred concept of Gestalt therapy and related blueprints, a concept that does not solely aim at the fundamental duality of communication structures but rather places an emphasis on the fine regulation between both complementary poles. In this regard, the focus is on both the current establishment of contact and the habitual mode of establishing contact, interrupting contact and modulating contact externally and inwardly – both as an ability but also as a defensive mechanism and symptom. Preferred establishment or interruption patterns may be fixed in each of the phases within the contact cycle. For instance, during a certain phase, an individual may have experienced that it is beneficial to not acknowledge signals from within in order to prevent conflicts with the outside world. Others may experience that actively striving towards what is desired is punished harshly, so they may instead prefer to block spontaneous impulses to seize and grasp things. Yet another individual may be so scared of making the wrong decision that he avoids clear commitments or rejections. This can lead to the emergence of another problematic influence (potentially through introjection), whereby each contact cycle features a devaluation of what has happened shortly before it ends, meaning that no joy or satisfaction is derived from the contact, regardless of how well everything went.

In principle, we all have the ability and liberty to block the usual sequence at any stage or to remain on one stage for a longer period. Depending on the situation and constellation, this can be a very useful ability. However, it can turn pathological if establishing or interrupting contact turn into automatisms, meaning that the freedom to decide whether or not to develop, enter into or leave contact is lost.

Fine adjustments In order to cope with life, beyond fundamental freedom of choice, at each stage we also require the opportunity to achieve a finely adjusted mixture of acceptance of a relationship and the ability to reject it. As such, with regard to content, communication is another dimension, which is nevertheless shaped by the modulation of the relationship.

Statements from others shaped by discrepancy. In cases where the non-verbal evaluation contrasts with the one expressed in terms of content, the non-verbal one is consequently understood to be the real and most truthful version. For instance, if someone wishes another person "have a nice day!" with a cold facial expression and harsh tone, in general, the content is typically construed as meaning the opposite. An adequate relationship is required for the content, otherwise the contextual message may be skewed into a contrary message. When we discuss "good girls" and "respectable society", and there's a clear indication of quotation marks in the speech, we automatically think of the opposite, in the latter case of organisations such as the mafia. The reversible figure phenomenon can be found in all areas whereby the designation of the actually intended pole appears too risky or fear-laden, leading to the contextual use of the opposite pole. For example: In French, stepmothers are referred to as "la belle mère", arguably a perfect instance of collective appeasement... In certain regions that are strongly influenced by religion, the "devil" must not be mentioned by its name, instead by the expression "the Evil One". Many punchlines used in the political cabaret, especially in dictatorships, owe their exonerative guaranteed laughs to the same principle.

4.2.3.1 Contact Cycle

The contact cycle facilitates the vital establishment of external contact in order to remove defects, to encourage growth and provide development-promoting stimulation.

Franz Perls' contact cycle ([3], p. 193–221) is divided into (1) Fore-contact (awareness of an internal deficiency or a needs-stimulating external stimulus, (2) contacting (desire, scanning the environment for a suitable object to satisfy one's needs, reaching out), (3) full contact (fascination with the chosen object, unity, (4) post-contact, featuring an evaluation and satisfied fading of the event. In this cycle, something else that draws the energy of attention to itself is constantly in the foreground for the self. The entity in the foreground, or rather the excerpt of reality, is referred to as the "Gestalt". In other words, the shortcoming on the needs level motivates the individual aspect to contact the corresponding excerpt of its environment in order to establish new balance within the organism. Above all, this simple system takes care of the basic needs on the basal level to remove material deficits such as eating, fluids and clothing as well as tactile and other stimuli. For example: I go for a drive along a street in the evening with an almost empty fuel tank, right before the petrol stations close. At this point, I pay attention to hardly anything aside from the signs for petrol stations, despite the fact that there is plenty to marvel at out the window on my left- and right-hand side. This specific restricted perception changes abruptly once I've successfully filled up my tank.

Extended contact cycle It is important to differentiate between, duplicate and elevate the phases of the contact cycle into nine phases in line with the progression along the "developmental ladder" in terms of function. The final phase deals with the pause between two cycles and the experience of being "central" or once again "in balance". The duplications take into account the primary complexity that can be carved out as a result. Furthermore, there is an increasing focus on illustrating interpersonal relationship needs [4]. Despite its principal schematisation, the step sequences in this cycle are grasped according to content, pace, quality, goal-orientation etc. as an individualistic event for which schematic manualisation would not be adequate.

Note

Growth cycle as a formula for the satisfaction of needs (◘ Fig. 3.1, see ▶ Chap. 3)

1. Fore-contact. The perception of imbalance, physical discomfort, flaws, desires, tension, atmospheric disturbances etc.
2. Recalling a search image, scanning the environment with selective perception, with the risk of projective distortion and also the expectation of transference.
3. Targeted, speculative and assessing scanning (ad-greddi contact) of the outside world
4. Conflict phase marked by a fear of making the wrong decision: select or reject?
5. Full contact with the "Thou", commitment to risk with the desire or fear of confluence or even solely of the selective opening of boundaries, "la petite mort"
6. A "We"-experience as a common being, potentially strengthening the experience of happiness
7. Post-contact: Evaluation and assimilation: Integration of the foreign and dissonant parts of the rest into the extended "We", restructuring of the self-image
8. Fixation of the new structure through self-reflection and trial-and-error in everyday life and social environments
9. Completion: Return to the level of indifference; stage of peace and balance

4.2.3.2 Additional Remarks to the Steps of the Growth-Cycle

Growth-cycle as a formula for the satisfaction of needs (◘ Fig. 3.1, ▶ Sect. 3.4.3). The phases 1–9 are corresponding with the steps of the Growth-cycle.

On 1) conscious outward denial of contact or interruption of contact. This concerns a vital skill that requires development and cultivation with fine tuning. It begins in the preliminary phase by not allowing oneself to be inundated by tempting external stimuli offered by advertising, which are capable of mobilising latent needs. Other temptations and manipulations also need to be recognised as such in good time and the individual needs to learn how to dim the internal responses to these stimuli, insofar as this seems more coherent.

On 1) conscious contacting in internal relationships that promotes life and development. Due to a range of different attitudes, internal stimuli may also be perceived to a more or lesser extent as well as over or undervalued. For example, a young woman who wants to lose a few pounds is glad to feel hungry. In her mind, the unnecessary fat cells are being destroyed and she will look younger any day now. She happily sips on her glass of water. Another example is a young mother falls asleep exhausted, despite the fact that there is a good amount of street noise outside, above all from the pub directly opposite the house. When her baby begins to quietly whimper, the noise, which is objectively much quieter than the pub, wakes her up immediately. As such, it would appear that we are capable of programming which stimuli are important and which are not to a certain extent.

Unconscious, anxious introspection, regardless of the motivational reasons behind it, can generate discomfort or at least distort reality and lead to hypochondriacal symptoms that are experienced by the affected individual as a real threat. For example, slight swaying can be anxiously interpreted as an imminent risk of falling over and responded to and strengthened with unsettling anxiety.

On 1) alexithymia. The opposite can occur when certain feelings and sensitivities are not only addressed within a family but rather tabooed instead of defined as something embarrassing or too intimate. Individuals who are exposed to the above situation tend to develop the opposite of introspection, a form of inner blindness and speechlessness. If internal perception is halted when the individual is still a child, it will likely remain rudimentary, i.e. underdeveloped, later in life.

On 1) interrupting contact that promotes life and development in internal relationships.

The suppression of pain perception and emotional impulses can be found in several approaches to education exercised around the world. “An Indian knows no pain” [literal English translation of the German phrase “ein Indianer kennt keinen Schmerz”, meaning “One mustn’t be oversensitive to pain”]. “Boys don’t cry”. “Work till you drop”. Being cool is synonymous with showing no emotions. This has been further perpetuated by culture and zeitgeist. Certain parenting styles promote the ban of certain feelings, e.g. “You don’t know how lucky you are, you have no reason to cry. Be happy that dad is gone, you don’t need to and shouldn’t miss him. He was a crook!”

On 1) patients suffering from endogenous depressions experience absolute emptiness when they look inwards. In this phase, they also don’t dream. When they wake up in the morning, their sympathetic nervous system fails to start up, meaning that they usually struggle to leave bed in the morning and have a particular lack of initiative (morning low).

On 2) when scanning memories, mental blanks of a physiological, traumatic or conflict-related nature may emerge.

The opposite may also occur, whereby an unprocessed memory disproportionately comes to the fore and dominates the individual’s perception, or rather abnormally changes the search filter. For instance, walking through their childhood home during an inheritance dispute in comparison to when they experienced a family party in the same house. Memories are not objective. They are new creations based on old traces and new assessments.

A harmless form of this allows lovers to view the world through rose-tinted glasses. They are permitted to generalise and project their being-in-love onto the world. As the saying goes, “love is blind”. – Those who unhappily fall into unrequited love often find themselves under the pressure of the hopes and expectations of their desires, which can easily lead to the over-interpretation of neutral events.

If an individual succumbs to a delirium of interpretation, he may “see” hints of a person of significance to him or validation of his hypothesis of the situation everywhere, either with positive or negative connotations. – This process is not dissimilar to the popular fairy tale opening line: “....In the olden times, when wishing still helped...” – Unfortunately, in the case of delirium of interpretation, the individual is afflicted by a very serious disorder, similar to psychosis. Nevertheless, this possibility reminds us of the creative potential within that is capable of obscuring or stimulating the perception of the life we have lived thus far.

On 2) The paralysing effects of fear are widely known. "Fear makes us dumb" and produces a white wall in our brains etc. This represents the form of paralysis encountered at this level. – This functional obstruction is different to the lack of access to memories due to the organically conditioned structural loss thereof.

On 2) Forms of pseudo-dialogues whereby one conversation partner showers the other with details from his life thus far and his current situation in a bias manner and without being prompted, without taking into account whether the other conversation partner has signalled receptivity to this information, can be regarded as a fight response. The majority of these instances are characterised by rushed, fast-paced diction. In this scenario, the inundating party uses the energy of his, often only politely engaging, listener for his own self-staging.

On 2) The (sub)manic flight of ideas that overruns stop signals and overwhelms the boundaries of others in an associatively loose manner can be viewed in a purely phenomenological sense as a clear, mixed form of fight or flight mode.

On 3) Positive "ad-greddi", i.e. approaching a boundary, inquisitiveness and spontaneous interest in discovery. The individual begins to wonder: is there anything that is suitable for my unsatisfied need? In which environment could I find it? Do I need to enter unknown territory and make a risky gamble?

On 3) Carefully delving out of protection and exploring the world when desire outweighs fear. This can lead to a gratifying rebound effect and trigger a positive chain reaction. It can cause a valuable emotionally corrective experience to occur, which is often observed in well-supervised, behavioural therapeutic exploration.

On 3) Combative and destructive interaction with the world; conquering strengthens the ego, but the individual is nevertheless left unhappy when no reference objects/people remain.

On 3) Mental withdrawal when faced with too many possibilities that are all worth considering and the subject is unable to decide between them. In this regard, there is a lack of criteria for differentiation. This can cause the subject to feel uncertain and at a loss.

On 3) Social withdrawal, escapism following discouraging events or encounters – up to paralysis in unexpected or shocking situations

On 3) Paranoid denial fantasised about beforehand and projected blocks the individual's own initiative to arrive at the counterpart's pre-imagined denial in advance.

On 4) Positive obstruction facilitates a slow, conscious and happy process towards the objective or encounter: showing stamina in the case of differentiated deliberation

On 4) Spontaneous, (almost childlike) rash decisions based on the individual's trust in his own gut feeling/intuition

On 4) Reckless decisions based on feelings of non-commitment, of a permanent readiness to correct; for example, making a decision without really getting involved against a background of the individual's own, insecure, i.e. non-coherent, attachment experiences.

On 4) Avoiding decisions; each commitment seems to be dangerously restrictive. It would be better to look for another solution than to seriously consider the objective without wanting to obtain it. It's better to live without restraints than in a self-inflicted, inner prison.

On 4) The inability to make a decision due to the fear of making the wrong decision, which the Superego would resent, can often erect barriers. Making the wrong decision would be inexcusable for a somewhat obsessively structured individual. In order to avoid getting truly involved, the individual rejects his options and instead leaves everything as it was.

On 5) Positive commitment to risk with a transformative identity; full presence in the joy of satisfying needs and happiness; arriving at the desired place requires affirmation of oneself and others, despite that knowledge that this feeling of happiness will not last forever but fade again at some point, and despite the knowledge that perfection does not exist, neither in oneself nor others – (even though the other individual may appear to be almost perfect at this point in time.)

On 5) If the individual has too much fear of the wave of satisfaction fading away, he instead prefers to not allow himself to experience the high points at first and is not present in the decisive moment. If the individual experiences too much fear of life's ups and downs and prefers to only accept things that will definitely occur, moments of happiness may be construed as deceitful.

On 5) Do I engage in a rewarding or satisfactory encounter with the sense that I am not entitled to it? Are only others entitled to it? I am too young, dumb, inferior, insignificant etc., which makes me feel like a fraud.

On 5) In the opposite case: the satisfaction was sub-optimal, even if it was perhaps in line with the situation; the encounter with a counterpart brought adequate unison and joy, but not full harmony: can I still view the experience as positive or does it lose its value as there are comparatively more optimal situations?

On 5) Do I have the feeling that the great other experience that happened to me was so powerful and great – and that I am so weak and faceless, that I am unable to offer anything adequate or counter this. Despite the underlying friendly tones of this encounter, I need to fight against it in order to stop myself from being destroyed. Or do I instead run away at the last minute in order to lose myself in the encounter, to avoid dissolving into this great experience and practically ceasing to exist. This stance is often palpable in structurally weak individuals who are close to psychosis (Mentzos described the ambivalence between desire and fear experienced in this situation in "Konflikt und Psychose").

On 6) The feeling of being satisfied and quenched sets in on the basic needs level, a situation of increased frugality which facilitates the emergence of happiness and relaxation. An empowering "We" experience occurs on the level of interpersonal encounters. We are able to understand one another, to get something right and discover what can be achieved in this common space of interaction. Unprecedented chains of interaction may emerge that are only possible within this collective "We". Each "We" is its own being. This new addition provides exhilarating new territory. The abilities and qualities of both individuals are scanned to determine whether they can contribute anything towards common ground. Similarities are nourished and propagated through increased attention. In certain cases, previously unfamiliar aspects from the other individual's world are consequently experienced as compatible and accepted or even invited into the "We".

On 6) In the case that the individual has previously experienced that relationships typically end in a disappointing manner, the acceptance and searching for potential similarities may only take place in a half-hearted manner and reduced responses will be experienced, indicating the pre-programming of a vicious circle.

On 7) After a certain period of time, aspects that have been relatively excluded from the "We" will pipe up once again or register a need for a space to live. On the basic needs level, this pertains to the other needs that were required to take a back seat for a while. On the partnership level, this instead refers to the qualities, aspects, preferences that are either no longer or barely able to resonate with the counterpart and have therefore been cast to the side in order to not disrupt the process of aligning with one another. Indeed, it can be regarded as a great feat to not entirely banish these aspects due to the fear that they may endanger the happiness of the "We" experience, but instead to grant them their own adequate living space. This process tends to succeed best when both partners mutually acknowledge that there are certain areas that are not best experienced together but should nevertheless remain, albeit in a modified form. The perception of a certain degree of parallelism benefits the partnership. Generally speaking, the formation of a "We" entails the rearrangement and restructuring of both identities, which is required at this phase.

On 7) In the event that assimilation is only expected from one person – due to ascribed roles or other notions of inequality that may appear to justify this scenario – the person who is required to adjust may feel overstrained and subsequently leave the "We" through a crisis reaction. (This constellation is often observed amongst couples with a narcissistic and depressive structure). Structurally paralysed and fixed combinations that originally appear to be complementary partners put a halt to the development of the couple and each individual due to the inflexibility of the relationship. This often leads to painful yet ultimately liberating crises. However, this does not entail that the couple needs to separation, but rather that the individuals need to break away from their previous partnership approach. (Jürg Willi's concept of collusion made significant contributions to this topic).

On 8) This requires trial and error. How can I experience myself in everyday life with the new knowledge or quality? Will I still be able to recognise myself? Do I want to identify myself to others with it and stand up for it, even if I am unsure whether others are aware of this new aspect (it wasn't without reason that I have kept it hidden until now)? For example, my aggressive nature. This trial and error also help the individual to find the appropriate degree of expression for the newly integrated aspects.

On 8) If I were to skip this concretisation phase, could I perhaps change my ideal and live under the delusion that I've already taken a big step and become more truthful; however, in real life, I'm most likely to revert to my old ways in a knee-jerk reaction. Concrete experiences open up and reinforce approaches. Therapies that are based on experiences, especially when carried out in groups, often make use of small consolidating exercises whereby the group members assume the role of the trainee following individual work that has led to a new stage of awareness. For example: Someone has found out that, despite his previous opinion, it's ok to say no from time to time and that this will not lead to his social downfall; the group members then immediately start to demand unreasonable things from the subject: "Give me your watch" – "Let's go on a date tonight without the others!" – "Kiss me on the hand" – "Kneel before me" – "Tell us all how stupid you are!" etc. Of course, there is a certain element of playfulness to the above and these questions can often lead to laughter, but it nevertheless can be a serious and effective experience despite the exaggeration. The individual learns that indignant refusal can be justified and requires a clear negative response, particularly as it all takes place with conscious eye contact and conscious contact with the individual's inner child, who was previously unable to dare to respond negatively towards his problem person. Once primarily the blocked, individual trace has been worked out and fittingly resolved, the individual will no longer struggle to assign analogous invitations and to emotionally use them to consolidate the newly permitted response and approach. Furthermore, we have also gained a wealth of experience from the arsenal of exercises such as self-confidence training employed in the field of behavioural therapy, as well as from the pedagogical field.

On 9) A moment of reflection is required at the end of the cycle to allow the individual to become aware of what has just happened. When successful, this will reinforce the process. Following this, the individual may experience joy and gratitude. If the response fails to meet up to what the individual had expected, or is disappointing, a small amount of reflection can help the individual to consciously become open to alternative options.

On 9) If conscious reflection does not take place, there is always the risk of becoming lost in unconscious and unsatisfactory actions and wearing oneself out like on an exhausting hamster wheel.

4.2.4 Buber's "I-Thou" Concept (1923) – Differentiation Between the Relationship and Objective Levels: Part of a Universal Principle?

The pathway for Buber's philosophy of dialogue was pioneered by the contemporary philosophers of dialogue in German speaking countries. Buber named the following thinkers as his forefathers and contemporaries in the field: Friedrich Heinrich Jacobi, Ludwig Feuerbach, Hermann Cohen, Franz Rosenzweig, Ferdinand Ebner. Gabriel Marcel was also a proponent of a similar approach. (cited in [5], p. 43)

The "I-Thou" relationship level is set apart from the objective level, which is referred to as "I-It". "I-Thou" refers to Buber's extended, spiritual background. In his widely circulated, almost lyrical, narrow yet nevertheless substantial booklet "I and Thou" (1923), Martin Buber, a philosopher of religion with a Hasidic background, described two fundamental qualities of the human attitude towards the world, others and oneself.

Note

The "I-Thou" approach, which Buber understood to be a conceptual unit of its own that can be found in the willingness to be aware of others in their existence in an affirmative and acknowledging manner and to supportingly and encouragingly acknowledge them. When reversed, this helps the individual to feel validated in his own being through the confirmation of his counterpart, which is of vital importance. This approach extends beyond purely being able to perceive empathy. It includes a certain degree of self-encouragement and the ability to develop the individual's own potential.

The fact that Buber's English translation for his work is "I and Thou" refers to his spiritual background. Since the days of Shakespeare, the use of "Thou" has been solely reserved for divine addresses in English. According to Buber, the divine God is ever-present in each individual person, regardless of in which version or whether the individual is capable of perceiving it or not. This perspective imparts particular dignity to the individual. According to Buber, this creates a double resonance, (1) the secular, interpersonal and (2) simultaneously, the spiritual through the common, superordinate reference point of the all-encompassing. By definition, psychotherapy is limited to the secular approach. However, it can provide people from a spiritual background with connectivity without causing them to experience any dissonance.

Buber's complementary pole can be found in the objective "I-It". It is important to note here that the latter explicitly does not refer to Freud's "Id" but rather to the rational perception of the world of objects. Emphasis is placed on the objective, emotionless observation process, which is a key function for regulating life, even if in an entirely different way. The "I-It" is responsible for separation and dissociation. It removes any resonances, or rather potential common ground. It enables the mind to be observed, designated, categorised, labelled, diagnosed from a safe distance and facilitates the development of its full potential to conceptualise.

Tip

The classification into a relationship and objective level particularly stimulated psychotherapeutic understanding. Even without a verified source, this classification also found resonance amongst the Palo Alto group of systematists, particularly Watzlawick, and can also be found in continuing systemic traditions (Schultz v. Thun).

Within the field of Gestalt therapy, the Buber concept has since been adopted as a requirement for the therapeutic relationship. For the Avant Garde school of Gestalt therapy, i.e. relational Gestalt therapy (Rich Hyncer, Lynn Jacobs, Gary Yontef et al.), Buber's existential relationship approach acts as the sole therapeutic agent and as a general effective component.

Rich Hyncer's doctoral thesis was supervised by Maurice Friedman (1921–2012), Buber's biographer in the English-speaking region.

In the context of the history of science, the turn of the twentieth century marked a period of particular synchronicity on different levels of knowledge.

4

Note

Martin Buber was a contemporary of the first, innovative generation of quantum physicists. He may not have been involved in the field, yet his approach, which may correspond to a vision, describes a similar, alternative basic structure for interspaces amongst humans as the structure that emerged in the field of quantum physics: the energetic basic substance oscillates, thereby connecting to form a greater whole – or it collapses and forms a distinct structure, or rather 'receives information'. On an interpersonal level, a relationship either forms or it does not.

In principle, the specified poles should be considered value-neutral. Ending a relationship enables freedom, intensifying a relationship imparts a sense of security and confirmation of its current existence. As such, this perspective is particularly required in psychotherapy and is viewed as vital and therefore superior. The underlying energy behaves in a fundamentally transformative manner. We are a part of it.

4.2.5 Empathy and Compassion as the Basis for Therapeutic Relationships

Many view empathy as one of the key factors behind healthy development and healing. Empathy forms a bridge of resonance between people. Empathetic understanding usually enables individuals to pick up on their counterpart's state of mind, to resonate with it and "read" it in an understanding manner.

Somewhat astonishingly, empathy in its direct form did not play a role in early psychoanalysis or cognitive behavioural therapy, namely the forms of psychotherapy that strive to achieve or uphold a "scientific" identity. The use of countertransference in psychoanalysis can be regarded as an indirect form of empathy. In this regard, the focus is primarily placed on the therapist's information and not on establishing a form of empathetic relationship with the patient.

When Kohut placed empathy in the fore in the 1980s, he undoubtedly faced fears of being barred from the psychoanalytical association.

Carl Rogers (1902–1987) was the first to recognise the significance of empathy for psychotherapy and tailored his education with a focus on this topic. According to Rogers empathy encompasses sensitive understanding and a non-judgemental interest in a person. In addition, he was also a proponent of unconditionally positive attention, i.e. appreciation, and congruence with regard to the therapist's authenticity and genuineness within the therapeutic approach. The aim of empathy is to view the patient through his own eyes. For Rogers, the focus and aim of his therapy was on the patient as a subject and not his issues.

Rogers belonged to the group of founding fathers of the humanistic psychology movement that entered the public sphere in 1962. Since then, this notion of empathy has generally become common knowledge along with the alternative humanistic approaches. Greenberg's "Emotion Focused Therapy (EFT)" is officially recognised as a combination of client-centred therapy and Gestalt therapy.

With regard to the therapeutic relationship, original Gestalt therapy is based on the "I-Thou" concept, as previously discussed, and broadly corresponds to Matthieu Ricard's notion of compassion. However, empathy is used to connect with the inner child. In the case of empathy, the hidden pain of the patient can be traced and initially shared in an identificatory manner. The brain's pain centres also resonate when empathy is experienced in MRT examinations; which does not occur when the emphasis is placed on compassion. A compassionate relationship consists of empowering, positive centring on the patient. We can thank Matthieu Ricard [6] for the precise distinction between empathy and compassion. We are now aware of the need for both empathy and compassion in the right place and in the correct amount in psychotherapy.

The role of empathy was first introduced to psychoanalysis by Heinz Kohut (1913–1981) and his self-psychology in 1957. He discovered the significance of empathy for the healthy development of individuals and gave it centre stage. According to his work, empathy can be defined as vicarious introspection, which is not entirely synonymous with Rogers' notion.

The mentalisation concept developed by Fonagy and his cohort is not dissimilar to empathy, but nevertheless remains primarily bound to the corresponding stage of mental development proposed in the "Theory of Mind".

An empathetic approach (as per Rogers) was recommended by Klaus Grawe to his colleagues in the field of behavioural therapy as of the 1990s, thus leading to an emotional turning point in BT due to Grawe's conviction that this combination would increase the effect sizes of BT. The author can still recall conversations on the future optimisation of psychotherapy with K. Grawe between 1989 and 1991. At that point in time, he clearly stated his individual sources. Unfortunately, he failed to continue doing so in later years. The full incorporation of Grawe's approach into the field would require systematic, additional training and credible models of self-awareness, or self-reflection, to facilitate learning by doing. This may have occurred in selective cases, however, in the meantime, Grawe's recommendations have somewhat faded into the background. A generally friendly stance amongst therapists does not quite fulfil Rogers' notion of empathy, but nevertheless marks an improvement on the cold, pseudo-scientific poker-face behaviour derived from psychological testing, where it was most likely used to ensure equal treatment.

4.3 The "Self" as a Field in Dynamic Equilibrium, Its Challenge of Integration, Temporary Structures and the Dual Role of Its "Inner Observer"

4.3.1 The "Self" Is a Field in Dynamic Equilibrium That Is Capable of Consciousness

At this point, it is conducive to mention Kurt Lewin, one of the pioneers of the "field" in the school of psychology, whose works influenced Fritz Perls in the 1920s.

(Perls was made aware of Lewin's work by his colleagues at the time, K. Goldstein and S. Foulkes (alias Fuchs).)

The self performs its functions through resonances as an interconnected network in action. It communicates with the currently received sensory perceptions, basic needs that have not yet been satisfied, updated stored memory traces, fantasies, reflections of the reference groups at that point in time, mobilised and familiar patterns for processing experiences and evaluation priorities. As such, it also updates chronified material that characterises the recurring accents of a personality. Deep-rooted processing patterns influence events with biographical narratives of success or failure.

The self also includes (desirable and undesirable) structural elements, similarly to how water often contains floating particles. Internalisations (introjects, guiding principles) are also considered part of the self, as both an opportunity and risk. Additionally, the self also encompasses traces of experiences of conscious and unconscious influential and learning activity, age-appropriate identification, utilisation and the removal of outdated foreign elements. The notion of the former and refined self pertains to increased clarity.

In the most widespread model, namely Freud's structural model (S. Freud, 1923), Freud states that all internalised rules are fused together to form the potentially punishing "Superego"; all ideals form the potentially tempting and rewarding "Ego-ideal"; and instinctual drives form the enticing Freudian "Id". The "Ego" takes on the role of the organiser. When the Ego is functioning as it should, it is able to keep a realistic grasp on reality. It is required to assess which of the registered needs can be satisfied under the given circumstances and what needs to or can be put off and postponed. An increasing tolerance for frustration is regarded as a sign of maturity. Due to the impact of conflicting reference persons and experiences, entities may take on a split personality. For instance, the Superego may exhibit both an absolute imperative to be honest alongside a pseudo-legitimate right to deceive. The subsequent generation of object relations theorists were primarily focused on the internalisation of relationship experiences. Kohut's self psychology established the significance of empathy, which was – at a later point – reiterated by Fonagy.

The staggered cerebral reward and punishment system, which features phylogenetically old and recent cerebral physiological carrier correlates, plays a comparably important role for the development of the personality, from a learning history perspective. It provides navigation systems for survival under conditions of dependency.

The magnitude of the self varies with regard to breadth and depth. C.G. Jung asserts that the awareness of the self – depending on its cultivated receptivity – is capable of accessing the entire experiential knowledge of humankind to a certain degree.

4.3.2 The Centre and Its Resonance Versus Its Rebounding from Structures

Although we can assume that at the beginning of life, we are capable of roughly polarised affectivity in terms of happiness and unhappiness, the influence of a caregiver who is capable of achieving regulation can lead to experiences of centricity and affective undertones. The initially unstructured interspaces between the poles are finely scaled through traces of experiences, which are stored together with a mindfully responsive "Thou". The quiet happiness of satisfaction enables the experience of being centred to shine, to become aspirational and be experienced as the focus of the individual's identity.

A counterpart that takes the impulses of the new aspects seriously and encourages them in a responsive manner is required in order to form a stable "centre" to subsequently form a subtle awareness of an "inner voice" or "inner sound".

This may also potentially be referred to as "natural oscillation", an "inner melody" identical to the self. Another notion could be that of an authentic source with protective capabilities that needs to be secured at a young age. If this succeeds, it is surrounded by coherent experiences through assimilation and develops into a secure inner foundation that has its own identity and imparts a sense of being at one with oneself. "Being at one with oneself" and simultaneously resonating with a greater whole, such as in the peace of nature, can evoke feelings of increased belongingness and even security with happiness in many people, even if it would appear that this reference value cannot be defined in everyday language as it exceeds the categories thereof. However, experience can also be found on either side of the categories that have been recorded in language.

Provided all goes well with development, dissonances between internal desires and external requirements within the scope of a loving relationship can be solved and internalised through compromises that are found to be acceptable by both parties. In certain cases, development is not so successful: the inner source may get lost in internal dialogue as a result of the requirement to over-adapt and anxiety, which causes all antenna to extend outwards.

From the moment of birth, babies learn to differentiate between good and bad experiences on their own accord and to assign different designations to these experiences through their reward and punishment system. These assessments lead to the development of an inner observer, who is initially responsible for determining whether experiences the baby comes across can be regarded as good and beneficial, or a bad and perhaps even adverse. At first, this system equips us all with a survival mechanism for uncertain times in threatening environments. Unfortunately, due to its solely outward orientation, it comes at a price: the risk of developing an over-dependency on external factors.

Sub-optimal development processes may predominately be subjected to chaotic upheaval and rarely or never reach a balance.

Valuable findings on conflict as a disposition, which often leads to "breaking points" for coping during development, have been compiled in research undertaken in developmental theory, baby observation and mother–infant research, both with and without references to the breadth of literature from the psychodynamic approach. As such, conflicts between previous structures used to cope with everyday life and newly emerging needs, abilities and opportunities are pre-programmed. The challenge would appear to be integration. The former order patterns are overwhelmed by the emergence of new impulses, which causes them to enter a state of crisis and potentially collapse. They often trigger internal and external ambiguities and fight battles whilst retreating in order to save anything that appears to be good and intact, or conversely, they may be forced to adopt the new position in order to survive in the case that the former situation appears sinister. A new order pattern subsequently emerges from the chaos with a slightly higher degree of complexity and receives the opportunity to establish itself self-organisationally. The new patterns may be less stable in crises – or perhaps the opposite. Either way, this characteristic will have an impact in stressful situations. Anxiety is sparked in volatile situations of upheaval, which may also irritate or overwhelm the caregiver if the stability in the main relationship that used to provide security is noticeably called into question. In response to these situations, our fight or flight responses are inherently mobilised. Each point of upheaval tends to feature behavioural patterns that can perhaps best be referred to as the behaviour of a victorious warrior or a fleeing loser. The latter experiences fear. Accordingly, fear and failure can be found at every stage of development. Similarly to coping mechanisms. –

We shall now return to the development ladder of breaking points:

The newborn battens down the hatches against an unfriendly interpersonal environment and protects his inner milieu; he finds adequate security in his own fortress, which will perhaps even grow into an interesting, mental or creative private world over time. Another newborn likewise attempts to batten down the hatches but is instead overwhelmed by the external influences which prevent the development of his inner order. He feels powerless. The last resort is surrender and self-abandonment.

The discovery of personal interaction can be utilised in an aggressive and challenging way, which is more likely to happen when games replace genuine loving attention. Of course, it also depends on the counterpart, whether he permits himself to be excessively dominated, indirectly invites this and does not understand how to set boundaries in a loving manner. As the focus is on satisfying basic needs in this regard, the demanding attitude at this level can facilitate addictive tendencies. Dependent beings experience the significance of the relationship that they rely on at this phase as alarmingly strong and are prepared to make over-adjustments to their own detriment in times of crisis.

The initial phase of autonomy marks the ideal opportunity to aggressively and defiantly turn against the whole world. Enticing experiences of power can have a significant impact, especially if they draw special attention to the individual and the consequences thereof are largely unproblematic. However, in another case, defiance from parents can lead to anxiety-laden quarantine measures that can cause the already fragile sense of security to collapse entirely, causing early childhood phenomenon of anxiety to permanently dominate.

Pre-school and school starting year are a time at which aggressive dominant and competitive behaviour can be found in a group context, where it is used to compensate for other shortcomings. Imaginative ranting also serves this purpose of emerging victorious, be it in the form of physical confrontations, in the battle for attention or to receive validation in newly discovered gender roles. Egocentricity is generally in balance with empathy that has developed within the social system, which regulates it. This may counteract a distant destruction which nevertheless unfortunately occurs. There are many "losers" at this phase, which subjectively remain on the run for their entire lives in certain cases. In some instances, they find a breakpoint or even turning point for their willingness to run through a relationship or a skill that they develop, e.g. in sport, and succeed in overcoming their aggressive nature and returning to the balance of a secured identity.

Traces of confrontation, in the case that it was a loss for one side or the other (in order to stick with the image of confrontation), shape the identity of the personality and are highly likely to lead to analogous processing of experiences in later, analogous constellations. Traces of competitive points of upheaval can be found in all approaches that work with structure-changing psychotherapy and address personality disorders.

It would appear that individuals who, based on experience, tend to enter into flight mode and consequently are likely to experience their own powerlessness, are susceptible to quickly experiencing a psychosomatic overreaction due to their emotions. This is always the case when "warriors" are required to block their fight response, for whatever reason.

Tip

The aforementioned inner melody can usually only be heard (once again) if the former battleground is perceived from the distance of the meta levels, perhaps a deeper meaning will be discovered and a new type of inner peace can be found.

Therapy aims to recognise old, deep-rooted (survival) fight strategies, to acknowledge the individual's concerns, to question polarisation between the concept of friend and foe, to strip away bias identification and in the process to discover new, previously hidden, empathetic changes in perspectives and try out alternative approaches.

Conclusion

Development is a hurdle in the sense that new experiences constantly need to be integrated and once again plunge the recently acquired balance into chaos. Perhaps development, provided it is successful, can be viewed as a sequence of phases of chaos and balance, whereby each phase obtains its fascinating characteristics through its own colourful harmony.

4.3.3 The Critic, Who May Transform into a Passive Observer

The critic is simultaneously the potential creator of the conceptualised subjective world. Experience and survival are initially dependent on what is granted a positive or negative connotation. In this regard, the critic ensures survival through this subject-oriented viewpoint. The external circumstances as well as people are skewed by this assessment in their inner depiction. Everything that occurs in the external field is assessed by the critic and his reward system with regard to whether it is beneficial or detrimental. The events are then correspondingly conferred a meaning and stored. Detrimental behaviour can cost lives on an archaic level. Aspects that neither support one nor the other are not assigned any significance, remain faint and tend to be overlooked. This leads to a constriction of perception in terms of the economisation of energy.

An evolutionarily analogous yet subtler variant with regard to the question of survival is the issue of how to optimally cope in dependency on caregiver(s) and in the encountered social environment. Certain experiences receive a positive response from the environment and come across as beneficial, whereas others may be regarded as detrimental. This creates basic structures for evaluation on the basis of encountered social constellations, such as assessing the desired and beneficial behaviour. Internally, this can lead to the installation of a form of "target" structure. A failure to achieve the target may mobilise fear in the same way as existential threats. The concrete behaviour can be referred to as the actual value (if we allow ourselves to take a glance at cybernetic control engineering). Those familiar with psychodynamics will most likely recall Freud's signal anxiety and his censorship at this point, notions that appear to be fully compatible with the above. Only the moral accentuation may be somewhat relativised in this regard.

A newborn may be confused when the assessment criteria change; either when two caregivers advocate different principles and styles to the same baby or the primary caregiver is inconsistent and switches from one extreme to the other. As a result, the child does not build up an internal sense of security in a reliable and consistent world.

Changes to reference systems. Certain upheaval may occur throughout the duration of an individual's life that challenges, relativises or changes the previous reference system: familiar people may leave the reference group, others may join; the individual's peer group may change, potentially only the leader or teacher, the individual's professional sphere of influence may change, the desire to be accepted by his in-laws may arise, he may wish to change the assessment criteria, a trip abroad may open his eyes to new perspectives, perhaps even culture shock etc. Confronting strangers can, of course, also lead to a stronger appreciation and closer grip on everything from the previous inner circle. A special emphasis is placed on this decision between various options.

Consistency in relationships and values. Under emotionally secure and reasonably consistent conditions, the newborn is granted with the relative "luxury" of no longer being required to fear for consistency but instead able to explore and expand his horizons with inquisitiveness. These kinds of excursions are not advantageous to survival, but instead constitute age-appropriate, unbiased insight.

Non-judgemental observation. In an anxiety-free relationship field (at a later development stage), under safe circumstances, the inner observer is able to liberate itself from almost automatic judgement. This provokes a certain level of authority over the environment. It is able to register its own impulses that were previously understated and depart from the exclusive dictation of the social environment. The affected individual is able to consciously engage in the separation between the observation on the one hand and the judgement on the other. He is able to intercept the latter. Corrections to embellish reality in favour of the target, otherwise known as repelling, may be omitted.

Important

Non-judgemental observation in psychotherapy. This approach typically first needs to be learned. When successful, the therapist must be able to lead the way as a credible model, opening an objective field of perception. As the popular paradox states, "true perception changes the picture". Perception in this sense refers to non-judgemental vision. If I look inwards with this view, I am no longer required to fight or defend myself as there is no condemner and no condemned. It becomes easier to (once again) recognise particular, predominately hidden sources of motivation behind individual actions and attitudes.

As perhaps goes without saying, the potential of inner judgement belongs to the fixed structure of a mature self. Inner judgement should only be consciously applied when necessary and similarly dismissed or softened where appropriate. Furthermore, inner rules need to be constantly updated and adjusted in line with the increasing level of maturity. In this case, the following also applies:

There is a time for everything. These structures grow with us, are distinct and may start to fade over time to a certain extent due to loving understanding. An impartial view and the corrective perspective thereof are required time and again for the inner dialogue of the work that takes place to make changes following distinctions.

The inner critic is a creator – but can also take a back seat. It creates and modulates the inner world by assigning significance, analogous to how the external observer uses its attentive energy to shape the outer field. Foreground figures are created through the particular interest when we look inwards and outwards, transporting the remaining details into the foggy background. The notion of subjective truth is based on all aspects of experiences that have been illuminated and let through the evaluation filter on the basis of their personal significance. To this extent, we each only register a completely personal excerpt from a range of possible ways in which the same situation can be accentuated.

When we reminisce and recount old stories, our current creative ability to transform narratives noticeably comes into play, for a variety of motivational reasons. This may be received with ambivalence by the environment, especially when the focus is on objective processes. Humans are subjects. They are capable of forgiving themselves and others, reconciliating and emotionally letting go. Objectively registering machines do not possess these abilities. However, humans are also prone to missing out on their opportunities. At the end of life, if details become blurred, fade and become less important, this does not necessary signify loss.

4.3.4 The Self's Dependence on Context

Each being is integrated into and influenced by a greater whole, something that has been long acknowledged. This principle is equally valid in biology, physiology, sociology, ecology, etc. An impressive example can be found in the growth and distinction of a cell, which is also controlled by signals from the neighbouring cell cluster. In the case of malignant degenerate and proliferating cells, the receptor for controlling signals from the local environment no longer functions.

The human "self" is similarly shaped by its dialogue with its environment, which provides it with the opportunity to adjust well to its surroundings and integrate. Albrecht Boeckh [7] broached the subject of the dialogical character of the self between its core and its social environment with particular interest and questioned whether the self possesses a unique core at all or whether the substance of the self is not derived from these various environmental references.

4.4 Communication Structures

4.4.1 "Pragmatics of Human Communication"

This is the title of the renowned book from Paul Watzlawick (1921–2007), J. Beavin and D. Jackson which was first published in Germany in 1969 and in America in 1967. In addition to his famous quote "one cannot not communicate" – because even when one communicates in a non-verbal manner (digital), he is nevertheless sending non-verbal (analogue) messages – Paul Watzlawick emphasised the particular significance of the difference between the relationship and the objective level. The former is the most decisive of the two with regard to communication, which harks back to Buber's "I-Thou" concept. However, surprisingly, I was unable to find any reference to Buber, despite the fact that "Pragmatics of Human Communication" was only published 2 years after Buber's death, namely shortly after Buber's works once again gained worldwide media attention. Both men were predominately long-term natives of Vienna.

Paul Watzlawick, who was extraordinarily linguistically talented and not only educated in philosophy and psychotherapy, but also in philology, addressed systemic perspectives and radical constructivism, which he knew precisely how to make a mockery of in a humorous manner to public appeal (e.g. "The Pursuit of Unhappiness").

4.4.2 The Communication Square

The spirit and objectives of the 1970s, especially those advocated by the humanistic-systemic groups of methods including group dynamics, thematic interactions and also individual psychology, can be found in this concept, which aims to promote cooperative relationships.

The communication square, which is also known as the four-sides model, by Friedemann Schulz von Thun draws on Paul Watzlawick's concept of relationship and objective levels, and further adds the aspect of self-revelation as an I-statement as well as the option to appeal to the you.

In terms of internal communication, this provides four quality characteristics into which messages can be categorised. – Of course, there is no guarantee against projections and false interpretations. This makes the calibration between the sender and receiver exciting and informative. – Those familiar with the field will be able to find the original Gestalt approach in the inner team model and its interactions, especially considering the author also attended corresponding seminars on the subject.

4.4.3 The Indian Tetralemma (Insa Sparrer, Matthias Varga von Kibed)

This approach is an adoption of an ancient Indian logic school of thought (Nagarjuna) to the possibilities of systemic structural constellations to solve conflicting problem solutions. – Tetralemma provides four possibilities for any logical proposition:

(A) affirmation, (B) negation, (C) both, (D) neither

There are several variations for (C), for instance, compromise, a supra-summative connection, an iteration, a transfer of resources, the elimination of pseudo-contradictions or allowing the power of the elected to influence the unelected etc.

A fifth possibility may depart from the existing category system in terms of reflexive pattern interruption and approach the process from a different perspective. This is often met with the question: what precisely is the tetralamma?

The tetralamma is regarded as a logical-epistemological tool that is used to open new perspectives when it comes to making decisions. It is constructed with symbolic spaces, meaning that each possibility is assigned a space which is entered by the subject. When doing so, he pays attention to his mental state, thoughts and psychosomatic condition.

(To this extent, this approach can be intellectually linked to Moreno's sociogram, Gendlin's focusing and, provided the frame of reference is changed during the step towards the fifth possibility, to the tradition of Kurt Lewin's research on reference systems from Gestalt psychology.)

4.4.4 Range of Communication in Transactional Analysis

Inspired by the Freudian structural model (1923) with its three parts, the Freudian "Id" as a pool of needs, the "Ego" as the organiser and the "Superego" as the normative condensate, Eric Berne's transactional analysis (TA) split the interpersonal interaction space into three "ego states": (1) a needs-controlled or spontaneous, healthy and curious child identity, (2) the boundary-setting or encouraging parent role and (3) an adult identity capable of taking on responsibility. The patterns of interaction outcomes are particularly interesting in this regard.

This TA concept incorporates the entire range of development stages and invites observation of at which level interactions primarily take place, when and why the level is changed, if and whether appropriate responses are provided, which patterns are used as a defence mechanism or as a means of persuasion, or which also impart a certain amount of authenticity to the interaction.

4.5 Adaptable Breathing Spaces Between Rhythmical Connections

Despite the fact that there has always been an interest and observations of interaction and also for the leap from one rhythmical numerical pattern to another in physiology and biology, anti-stress research and in rhythm theory, this focus of inward and/or outward coordination has remained somewhat limited in the field of psychotherapy. It would appear that, in physiological terms, on the one hand, there are preferred rhythms and connections yet on the other hand, a certain level of flexibility with regard to adaptation can be a survival benefit.

This merges observations from all areas. One external clock is the earth's orbit and the difference between day and night, which relies thereon. In addition to the role it plays on our sleeping-wake rhythm and sympathetic activity, daylight impacts numerous parameters of our metabolism. Certain specialised fields of occupational medicine are already delving into the issue of the harm caused by the elimination of light-dark time rhythms as the result of constantly changing shifts or flying across several time zones, for example.

Empirically, the subtle control of female hormonal cycles and the appetence behaviour that is likewise regulated by this has been attributed to the moon. Hunger, anxiety and stress have the capacity to destroy this connection.

The biologist Erich von Holst addressed the natural coordination phenomenon of movement when swimming or flying, and movement with various gaits (trot, gallop etc.) in the 1950s. Fixed connections within the body between functional sub-fields create an ordered, internal activation process with secured relations. They can be regarded as integral, holistic sequential patterns. Points of change are characterised by chaos. These points are required to facilitate new self-organisation or a change to a relational pattern that is more suitable at present. When faced with an increase or decrease in performance requirements, the body constantly changes its ratios between heart rate and breathing rate. Let us imagine a set of interlocking gears, "toothed" analogously to the harmonic series as the hypothetical options for both of the systems that will be connected. As we are aware: everything oscillates (nada brahma) and everything that oscillates has a harmonic series. This opens the door to various possibilities to parallelise waves and thus synchronise their potential. Some of these patterns may be more physiologically and emotionally linked with anxious excitation, or with standby tension, with refreshing alertness, and others with relaxed tranquillity or peaceful bliss etc.

Around the globe, lullabies are sung in 6/8 time, a natural ratio of breathing and resting heart rate. Across the world, triple time evokes upbeat dance steps. Confrontational-opposing rhythms that inspire joy in resistance can also be found all around the world etc. Dance and movement therapy is particularly capable of tapping into rhythmic and movement coordination for diagnostics, expression and therapeutic means.

Anxious stress can loosen or even destroy coordination within a system; meditation and introspection provide an opportunity for coordination to reorganise and develop. A common oscillation pattern inherently supports the integration of a living system, be it an organ, a functional subsystem, an individual, a couple or a social organism; keeping it in one of the possible balances. Each connection is generally provided with various opportunities to stably align with its joining elements to a greater extent through rhythms – and also for the elements to mutually miss each other and become broken.

4.6 From the Previous Illusion of Interspaces as a Void to the Main Location of Encounters – Observations from Quantum Theory

This entire chapter initially follows the naive, human perception, which distinguishes between solid, material substance and the interspaces in between. This notion was also followed by developments in the field of psychology and psychotherapy, which have gradually addressed these interspaces in their practical research over time, as has already been described in the prior sections. However, these experts stumbled across phenomena from time to time that did not fit into their scientific view of the world, such as intuitive insights gained from the representative role in constellation work, and instead elected to taboo the necessity for them to do so. At the time, this represented, and continues to epitomise a stance that conforms with the mainstream.

A small sideways glance at the potential raw material of this phenomenon. Radiation emitted by people (with the exception of radiation from the heart and brain) remains a neglected topic by science. Residual light amplifiers are capable of measuring the fact that, unbeknownst to us, we constantly emit radiation, and even the levels we emit. There could indeed be an intense encounter taking place in our proximal and large-scale interspaces! Analogously, the impact of the interested presence of the therapist and/or patient in the psychotherapeutic individual setting is clearly palpable, along with the opposite thereof.

"New phenomenology" (Hermann Schmitz, Friedhelm Matthies [8]) provides an unconventional and original interim category for the introduction of "half-things" for emotions, for affective concern and, above all, for atmospheres. These half-things are also referred to as temporary partners. They can be understood as bridge supports between material existence and the spiritual realm. It is both astonishing and notable that phenomenology, a descendant of philosophy, has unintentionally touched on the school of thought of quantum theory, when the latter inquires into in-between existence.

We shall now inquisitively explore the micro-world within the context of the notion of complexity. Initially, researchers used models to introduce their understanding of particles and assumed that electrons orbited around the nucleus like planets around the sun. Upon closer observation, the definable matter appeared to increasingly break down and transform into wave formations.

Note

In 1950, Schrödinger wrote that the two notions of waves and particles need to be merged. They never existed as separate entities, they are two sides of the same phenomenon. Indeed, both conceptual notions of waves and particulars were not really suitable, but instead somewhat confusing. The correct concept remains beyond our grasp.

Our imagination is also stretched by Heisenberg's uncertainty relation, namely that we are not capable of simultaneously ascertaining the location and velocity of a quantum at the same time and must instead seek satisfaction in probabilities.

Schrödinger poses the further challenge of understanding "particles" solely as pure forms, which are continuously produced and destroyed, and may emerge as waves – both within "matter", such as free electrons within metals, and in interspaces.

Note

As such, the interspaces become living spaces of encounter and transformation for the energetic basic substance in its mode of existence, that we are likely unable of adequately recording with our current coordinate system. We focus on the poles of manifestations, on waves and particles, on mind and matter, but we struggle with regard to the interspace substrate, despite the fact that more or less everything exists as an intermediate substrate, we ourselves most definitely, and despite the fact that as a result, the interspace provides us intermediate beings with our established "home" as the location of our "centre".

In the analogy to the quantum, this corresponds to its intermediate form, i.e. its cloudlike floating-existing between interconnectable and structurally fixed manifestations of different levels of order. Our mind is incapable of providing its own concise category system for this form of existence.

Bibliography

1. Förstl H, editor. Theory of Mind. Neurobiologie und Psychologie sozialen Verhaltens. Heidelberg: Springer; 2007.
2. Fromm E. Die Kunst des Liebens. Frankfurt/Berlin: Ullstein; 1956.
3. Perls FS, Hefferline RF, Goodman P. Gestalttherapie. Wiederbelebung des Selbst (Bd.I) und: Lebensfreude und Persönlichkeitsentfaltung (Bd. 2).Klett-Cotta, Stuttgart. (Engl. original 1951: Gestalt Therapy, Exitement and Growth in the Human Personality. New York: The Julian Press; 1979/1951.
4. Hartmann-Kottek L.: Gestalttherapie. Ein Lehrbuch. Heidelberg: Springer; 2004/2008/2012.
5. Wegscheider H. Dialog und Intersubjektivitätin der Gestalttheraoie. Vom jüdischen Hintergrundzur methodenübergreifenden Bedeutung. In: Hartmann-Kottek, L (Ed.):Gestalttherapie - Faszination und Wirksamkeit. Eine Bestandsaufnahme. Psychosozial-Verl., Gießen; 2014.

6. Ricard M. Allumfassende Nächstenliebe. Altruismus – die Antwort auf die Herausforderungen unserer Zeit. (Franz. Original title: Plaidoyer pour l'altruisme – La Force de la bienviellance.) Edition Blumenau, Hamburg; 2017/2013.
7. Boeckh A. Die dialogische Struktur des Selbst. Perspektiven einer relationalen und emotionsorientierten Gestalttherapie. Gießen: Psychosozial-Verlag; 2019.
8. Matthies F. Leiborientierte Gestalttherapie und Psychopathologie. Der Beitrag der Neuen Phänomenologie zur Gestalttherapie. Bergisch-Gladbach: EHP; 2021.

An Understanding of Health and Illness

Contents

L. Hartmann-Kottek, *General Psychotherapy*,
https://doi.org/10.1007/978-3-030-87466-7_5

5.1 Attempts at Classification

5.1.1 Health and Illness WHO Definitions – Ideal and Reality

The World Health Organization's understanding of health is broad and holistic. It corresponds to the ideal notion of an individual and associatively contextual balance. As this also incorporates a mental dimension (in whichever way this is defined), it can be closely linked to the objectives of this book on "General Psychotherapy".

> "Health is a state of complete physical, mental and social well-being and not merely the absence of disease or infirmity" (quote from the Internet, keyword WHO/definition of health).

The healthcare systems in individual countries have come to an agreement on, or been required to unite on, modest goals, primarily due to financial reasons. In many countries, mental illnesses are not refinanced by the healthcare systems or insurance providers, or only covered to a certain extent. As a result, socially deprived swathes of the population are unable to access psychotherapeutic help.

Situation in Germany: In Germany, the costs for psychodynamic treatment have been covered by healthcare providers since 1967, with behavioural therapy included as of 1987. Lobbyists from both of these well-established schools have managed to successfully exclude other psychotherapy approaches, such as humanistic and initially also systemic approaches, despite their superior scientifically proven effectiveness, from statutory patient care in Germany since 1998 by law. Since then, representatives from the established approaches have been conferred the ability to rule on requests for recognition from the excluded approaches on the basis of arbitrarily high and conceptually disproportionate screening criteria. This panel has never fulfilled its legally required neutrality. This situation in Germany does not provide a shining example for other countries, despite the fact that attempts have been made to generalise it by representatives from the established approaches.

5.1.2 The International ICD and DMS Classification Systems

Psychiatric WHO experts began to call for the development of transnational diagnosis criteria and the establishment of corresponding conferences in the 1960s. These efforts led to the creation of the International Classification of Diseases (ICD). The 1970s saw clear coordination between the various schools of psychiatry and their national idiosyncrasies. The APA (American Psychiatric Association) developed an operational classification system known as the Diagnostic and Statistical Manual of Mental Disorders (DSM). In 1980, a German translation of ICD 9 (compared with DSM III) was published by R. Degkwitz and his team. At present, we find ourselves in the transition phase between ICD 10 and ICD 11, and DSM IV and DSM V.

A clear advantage of shared, international systems of denotation is the interdisciplinary and transnational scientifically oriented communication and comparability. Categories that are laden with hypothesis such as neuroses and psychoses were elimination in favour of pure descriptions. Attempts were made to determine severity on the basis of the additive aspects of symptoms. The transition from quantitative measurements to qualitative diagnosis remained and often continues to remain unsatisfactory. All too often an insignificant accompanying symptom leads to a categorical leap and receives disproportionate importance. It is widely agreed that these diagnostic measurement systems are workable compromises that are unavoidable at present yet nevertheless conflict with reality. The current categorisation system is not based on an inner, meaningful structural order. Instead, it resembles an arbitrary, additive list based on external similarities. There is a need for continuous development and improvement in this regard.

Amongst other aspects, there appears not only to be a lack of a symptomatic, statistical description of details in the specification of phenomenon, but also a lack of recognition for discrepancy and tension patterns. In addition, observational sequences require somewhat deeper and processual summarisation, e.g. observations on relationship structures, processing frustration, tolerating insecurity as well as the discrepancies between conscious signals and unconscious non-verbal expressions etc. This also encompasses the dynamic notion that superficial behaviour is contrasted by a qualitatively conflicting, complementary opposite, provided that there are signals of tension instead of well-balanced equilibrium. It is possible to learn and teach awareness of stress patterns. The phenomenological diagnostics of Gestalt therapy are based on this awareness training from Gestalt psychology.

The concept of progression patterns is not entirely new: mood swings throughout the day are a common psychiatric differential diagnosis. Morning lows typically characterise endogenous depression. This observation takes the normal rhythmic course of events behind the activation or obstruction of the sympathetic nervous system into account.

Most people are also familiar with the psychodynamic observation of the complementary counterpart, whereby their counterpart brings out and shows its size, strength, integrity, friendliness, altruism or modesty etc. when placed under tangible internal stress.

It is also important to determine at which levels we encounter semi-fixed or even solidified patterns of distortion and whether this is the level at which they emerged or one to which they were shifted. The topic of patterns is explored in more detail in the following chapter.

It is conceivable that diagnostic procedures may acquire precision and validity through holistic pattern recognition and progression criteria. Meaningful diagnostic entities are not discovered through the addition of individual symptoms, but rather through the recognition of biologically embedded dynamics, or rather of the blocked dynamics for the processing of experiences. An interesting interplay occurs between the neurophysiological "hardware" and the "software", which has partially or entirely leaked into the unconscious.

Unconscious patterns for solving conflicts are fixed, learned ways in which we are able to cope with unsuccessful situations that outlast the situation, both in terms of sub-optimal educational objectives, delegations, socialisations as well as unconscious, automated patterns for solving conflict. Behaviour and experience are often only categorised as borderline pathologic, depending on the adjustment to the updated reference system. The absence of the full presence of all options that could potentially have led to a more suitable resolving behaviour is a common observation. To a certain extent, the automatic fallback on resolution and defence mechanisms that appear to have worked before in order to avoid conflict can be regarded as "leaving the field".

Conflict resolution is one of the key tasks of psychotherapy. As a starting point and definition, a symptom can be understood as the part of a fixed, energetic pattern of discord that is close to the surface. At its core, it contains a compressed, unresolved issue. Further classification requires the diagnostic "reading" of a pattern of discord through inquisitive interest in the associative background of the patient. A symptom can also be regarded as a solidified pattern obtained from sub-optimal crisis management. Internally, it corresponds to an energy-binding, dissonant aspect that is semi-isolated from its context. The reason behind the disorder may lie in trauma, a shortcoming or a conflict – or perhaps a combination thereof.

Therapy concentrates on the context and the emergence of the symptom. Externally, the symptomatic manifestations may appear to be of a similar nature or feel related to each other, despite the fact that they originate from different internal dynamics and consequently require different encounter approaches (e.g. anxiety has many faces: anxiety may emerge from the faithful absorption of anxiety symptoms according to the model;– and: anxiety following trauma; fear of loss;– or also fear of punishment after an admission of guilt;– or: fear in the face of impending structural collapse and a loss of identity; – and, especially: fear of not being good enough .... etc.).

The "symptom" itself – or also the entire pattern of discord – is generally perceived by the affected individual as a new, often disconcerting experience and behaviour which is out of the ordinary and disruptive. It is experienced as removed from the individual's usual self and may change his self-image, motivating him to seek therapy, amongst other help.

Disorders resulting from trauma manifest in a wide range of manners and vary depending on the individual's age, the frequency of traumatising events, the relationship constellation between the victim and the abuser and also potentially cushioning aspects of resilience. Indeed, the inclusion of specific trauma diagnoses in the international classification systems over the past 30/40 years despite initial resistance marks a welcome development.

Traditionally, healthcare deals with diagnoses that are centred around the individual. The notion that the primary cause behind a disorder may lie in the individual's relationships with his partner of family or an "unhealthy" environment has long been acknowledged in the field of psychotherapy. At long last, this contextual reference is finally being incorporated into the DMS-V revision.

Conclusion

The respective internal dynamics, or rather the development of symptoms, are more significant for insight and therapy for than the external manifestations.

5.1.3 Alternative Suggestion for a Future International Classification System

> (On the hardware basis of biological-neurophysiological complementary poles, including their various stages of development – as well as the software from individual learning processes and patterns for processing experiences)

Nature has made several provisions for the survival of the species. On the one hand, we observe complex patterns of experience and behaviour that facilitate brood care as well as secure protection and attachments in higher-level phylogeny. On the other hand, the same beings are equipped with survival mechanisms that are biologically embedded to activate fight or flight responses depending on the situation.

The inherent patterns of experience or behaviour to avert danger exhibit all transitions from healthy and rational reactions to pathological variations, which may become solidified in unfavourable conditions. All reactions need to be assessed in light of the respective context. They may also reflect certain cultural, age, maturity, gender and zeitgeist influences.

Since Porges has started to elaborate on the neurophysiological principles for polyvagal theory in recent decades, contributing to the classification of vegetative regulation mechanisms, it seems conducive to view the onset of illness in its entirety with its various levels of symptoms (the social behaviour level, the intrapsychic dynamics, the psychosomatic (dys)regulation) from this perspective. These deeply rooted objectives to ensure survival of the species and the individual being may provide a natural basis for pathology without automatically reducing it to solely a biological level. However, the potential to transform to higher levels of maturity, which is of particular importance for human development at present, remains preserved.

The resonant commitment to connect, which induces empathy, integration and bonding, can be viewed as a basis of life that provides a sense of security.

Irrespective of his genetic and epigenetic traits, a healthy individual has typically gained adequate experiences of sensitive and positive attention, coherent relationships and regulation throughout his life.

Conversely: Those who do not experience a sufficient number of emotionally stabilising events when young may react with increased vulnerability to the same stressful events as an adult and risk losing their balance and ability to integrate. The social empathy field connects, strengthens and "nourishes" individuals. It establishes basic trust in the world, it allows individuals to enter into interpersonal communications with curiosity and interest and it imparts a sense of joy in unity within communities. It may consist of a single person or also a group, large family, clan, kibbutz etc. The caring, brood-care dynamic of this caring field is biologically embedded from the outset of its existence and guarantees survival for the living being during its initial years on the planet.

At a relatively young age, i.e. amongst children, an understanding of reciprocity for giving and taking which is based on empathy and on the satisfaction of basic needs begins to develop. Later in life, this understanding expands to additional areas, takes on a value and responsibility-oriented manner and matures into a general ability to love, or at least is capable of developing into this ability. The interconnected field can develop beyond the level of personal love into a general interpersonal love. This encompasses adequate self-love to facilitate inward and outward tolerance, and thus inner peace.

Interconnected fields have the tendency to oscillate, or rather synchronise, on the same wavelength. Indeed, people commonly refer to "being on the same wavelength", or not. Anxiety can initially bring people together in a regressive manner. If this does not appear to be helpful, fear instead mobilises fight or flight responses, which consequently isolates the individual.

Deficiencies. Many people are required to get by without any enthusiastic attention, encouragement or supportive concern for their personal development. As perhaps goes without saying, in this regard, the duration of the deficiency has an impact on the outcome. Shortcomings for a short period as a result of a loss or break-up can lead to shock and deep insecurity; however, they can also draw the individual's attention to the value of supportive relationships or whether the shortcoming instead represents a permanent state that may threaten to shape the individual's identity. If so, his identity will most likely reflect the insecurity and fragility of the experienced state and his lack of trust in the world. In turn, the future may subsequently appear more frightening than promising. People who experience this may feel insecure and unconsciously cause themselves to be undervalued and indeed to undervalue themselves.

Compromising. Creative transitional solutions can be found even in extended periods of a lack of social empathy. Adults often retroactively recall the fantasies they used to keep themselves emotionally in check during difficult periods during their development. They may have imagined friends or twin figures who easily understand, appreciate, trust, save, strengthen as well as respond with solidarity to the individual. These figments of imagination may potentially also retaliate and provide the required emotional balance. – Fantasies of belonging to another family with a different atmosphere, namely a loving, honest, trusting, happy and fun one, usually the opposite of the experienced reality, are a common occurrence. – Needs for compensation not only lead to fantasies, but also establish elective affinities in addition to the idolisation and idealisation of real people, where there is a risk of disillusionment at an inopportune stage. The primary driving force is the desire for positive involvement in a relationship. Even if this may not provide an adequate long-term solution, these forms of compensatory reactions help to bridge deficits and prevent the shortcoming from becoming fixed.

There are also pathologic versions whereby the individual unrealistically identifies with the sugar-coated and overstated fantasies. The affected individual may struggle to recognise that his pseudological illusionary worlds are daydreams, to depart from these and swap them for the comparatively unattractive real world, which instead provides solid ground.

Inherent conflict dynamics. During the phase of growing autonomy, initially during the phase of defiance, self-referential impulses are regularly at odds with impulses from the relationship dynamic, which may cause the latter to be overlooked from time to time. This can test the tolerance for frustration of stressed parents in addition to their capacity to remain in control and love. In this regard, at the end of infancy, provided everything has gone well up until this point, the child will usually achieve a reasonably integrated state of balance. Adolescent detachment can once again place the conflict in autonomy back into the limelight with somewhat more fundamental importance.

When there is a lack of empathy in the adolescent's environment and he is placed in stressful situations, or rather his alarm system is constantly on alert, the ability to live in a way that is open to relationships gives way to an increased focus on the self. Being closed off and not trusting anyone may then appear to be more logical to the individual. Early experiences can have a particularly long-term impact on the growing personality.

Attachment research has highlighted the significance and even vital nature of reliable, coherent attachment experiences that provide security, especially at a young age, with regard to establishing inner stability and later being able to develop secure attachments. Deeply rooted experiential dispositions are behind this phenomenon. Attempts to separate mothers from their infants at an early age in experiments with animals have also induced existential stress that permanently changes the physiological system and leads to premature mortality when compared with the comparative population.

Deficiencies are damaging. The mere lack of positive, personally experienced attachment and attention may represent a factor that appears entirely unspectacular from the outside yet nevertheless causes significant stress and uncertainty. The individual usually develops low self-esteem. Children who are emotionally deprived are often more susceptible to accepting abusive relationships from adults or being lured by dissocial peer groups as a result of their lack of attention. They are also more prone to ending up in parentification relationships at the risk of self-exploitation.

Detachment and integration: two intertwined, partial development goals. During the subsequent phases of life, during adolescence or young adulthood at the latest, concrete detachment is the order of the day. In principle, the development of each individual switches several times between various desires to belong and the need for autonomy. Maturing requires the experience of both poles, the socially networked and the independent, self-referential, secluded, and socially fragmented.

Patterns of threat and stress management as stimulating principles in the fight for survival

Our survival is ensured by our alarm system (ARAS = ascending reticular activating system), which triggers fear in response to threatening stimuli and provides a physiologically and hormonally supported cascade of fight or flight responses in order to save us with the corresponding activity. We should be thankful when it works as intended. When the alarm system is activated but the event is registered as positive, this triggers curiosity and fascination.

The stimulus thresholds of the alarm system may shift towards the under-sensitive or over-sensitive range. This may lead to the anomaly whereby early trauma causes a short cut pathway for anxiety-inducing stimuli to find its way across the nuclei of the amygdala. Consequently, the excitations are then registered as symptoms of shock in the periphery before the cortex is granted the opportunity to recognise them as insignificant and to dim the excitation.

The cortex usually evaluates and modulates the incoming sensory perception on the basis of its experience and assessment of the situation. It indicates a potential threat and then gives the all-clear signal once again. Even in the case of a persistent threat, habituation will usually occur after a certain period of time.

However, stress stimuli can simultaneously be so unpredictable and intense that activation of the alarm system becomes permanently fixed and the all-clear signal no longer functions. In turn, this stress can lead to the onset of psychosomatic overreactions.

We shall now explore the different forms of expression and collect our thoughts in clusters. In dangerous situations anxious stress arises, which can transform into a bold attack (ad-greddi) in fight mode and, in flight mode, into a fearful retreat in order to become as invisible as possible. The waves of excitation mobilised by the alarm may lead to balance in the affected regulation system through the strong resonances.

Clustered collection of thoughts:

- Fixed in fight mode towards the outside world (interpersonal and social world)
  Dissociality, loss of control over impulses, propensity to violence, vandalism, rampages, hyperactivity, activism without any goal or purpose, logorrhoea, aggressiveness, dogmatism, excessive competitive behaviour, narcissistic arrogance with the need to put down others, agent provocateur (instigator), sadism; active and passive rejectionist attitude with regard to communication; in childhood and adolescence: Denied satisfactions of basic needs by the caregiver, e.g. inadequate soothing from picking up and speaking to a screaming baby, food intake (eating disorders during infancy, anorexia nervosa, bulimia), sexuality: Hyperactivity or loss of appetite; lack of interest in learning;
  Sublimated and ritualised combat: Fair play in sports, martial arts, overcoming frustration through performance – Professional critic.
- Inward fixation on the intrapsychic field (psychic level) in fight mode
  (It first needs to be determined who is fighting against whom with which motivations and justification, "who" is wielding power from which position, who defines the right to exist, i.e. who defines who or what is allowed to exist). Obsessive thoughts, compulsive acts, obsessive-compulsive disorders, all forms of symptoms of compulsion, experiences of self-esteem, personal capabilities and accomplishments, the individual's right to exist;
  In the case of pressure to outwardly over-adapt: eradication of internal needs that are not in line with the relevant environment. In the case of pressure to adapt to norms, ideals, a false self, sacrificing one's own values and goals. Confrontation with the discrepancy, provided there is one, with subsequent adjustment of the individual's self-perception or concealing and eliminating inconvenient aspects (defence); which aspects are permitted to exist?
  Hidden struggle in the case of an internal change of position: Underdog power struggles, inner protest and unruly behaviour: Addictive behaviour, passive resistance;

Sublimation level: requiring a principled, value-oriented life
Physical level: Physical secondary reactions: misinnervations in the interplay of neuronal complementary systems and psychosomatic reflection error: Symptoms of sympathetic overexcitation in the cardiovascular system, respiratory tract, gastrointestinal system, genitourinary system, immune system, musculoskeletal system etc.
Physical accompanying or rather outlet mechanisms may turn into a sure-fire success.

- Fixation related to the outside world in flight mode (interpersonal/social environment)
  5 Attitude of a loser, experiences of weakness and a sense of inferiority - The threat appears to come from the outside world
  - Undirected, free-floating fears often unconsciously fixate on specific items and transfer to the outside world. This results in the formation of phobias of certain situations or items (phobias of spiders, snakes or dogs etc.), which may enable us to believe that they can be controlled or avoided to a certain extent. Affected individuals may find relief in this notion. Agoraphobia and social phobia can typically be somewhat moderated by the influence of a companion. However, this approach is prone to resulting in dependency. A fear of desire is often regarded as the reason behind agoraphobia. "Being alone in vastness" may trigger former fears of loss, or, on the other hand, the exposed and highly visible position may cause desires characterised by shame to surface. In the case of social phobia, projections on the imagined crowd are of central importance to the affected individual.

    Those who mobilise the flight response tend to employ a protective smile or use childlike characteristics (voice, posture) as gestures of submission (e.g. tilting their head to the side and exposing their neck, similarly to dogs) and instinctive protective behaviour to prevent aggression, which can be equated to surrendering by waving a white flag.
- Inward fixation (on the intrapsychic field/ psychic level) in flight mode
  - Calming rituals to bind anxiety; in the case of issues of guilt: mood-dampening self-punishment rituals, acts of reparation, making amends (with an internally recognised, judicial entity) with subjective loss of value, potentially also characterised by ostracism, impoverishment, sin or potentially paranoid fears of devaluation.
  - Attempt to flee into the crowd/masses ("everyone would do the same"), fleeing into enforced conformity with the denouncement of any individual responsibility, follower identity seeking alleviation (in the case of a latent or suppressed sense of guilt);
  - Attempt to flee into the insignificance of a certain event and the lack of interest therein, which makes it easy to forget. Anything that is not emotionally significant loses its intensity and outline, eventually becoming forgotten. Distraction and relativisation can play a role in this regard. The latter is only successful if the reference framework can truly be changed.
  - Unconscious protective trauma or conflict-related amnesia prior to flooding from unprocessed emotional contents

  - Amnesia as a result of post-hypnosis or suggestive instructions
  - Fleeing into deceptive mimicry if the motivation to survive is paramount: routine paralysis or becoming subjectively invisible through adjustments despite full presence in order to survive.

    Physical level: Psychosomatic level. Anxious fixation of initially undirected fears on the body's response, usually with a sympathicotonic nature, e.g. heart phobia, erythrophobia (fear of blushing) and other symptoms that may appear to fabricate and increase symptoms with hypochondriacal, worried observation and locally bind free-floating anxieties. The localisation is experienced as relief by the individual, as it limits the phobias and seems to satisfy the need for causality.
- Tonic immobility as premortal surrender behaviour

  Tonic immobility as the final premortal chance of survival, or rather mimicry: becoming invisible to enemies in order to divert their aggression elsewhere.

  Psychosomatic: Activity of the phylogenetically old, non-myelinated dorsal vagus branch. Mentally, this leads to a blackout as paralysing freezing. Physically, it leads to the body losing control over its ability to regulate, potentially fainting as vagovasal syncope with paralysis of vasoconstriction, a loss of blood pressure, and over-innervation of the pylorus (pylorospasm with vomiting). Partial functional obstructions are observed in the symptoms of traumatic amnesia, which may persist for months or years, in post-traumatic stupor, which blocks motor, sensory and mental functions, and, very markedly, in the catatonic behaviour of schizophrenia.
- Healing fight or flight responses – reversion with victorious coping.

  Spontaneous self-healing occurs in all situations whereby tough and challenging situations and tasks mobilise so much energy and so many tasks, that they may end in success despite temporary anxiety and perhaps also interim defeats. A major part of pedagogy consists of offering age-appropriate barriers and simultaneously providing the required abilities and encouragement to overcome these barriers. This encouraging spirit also applies to the emboldening impact of self-confidence training in behavioural therapy.

  Perhaps the most important principle in psychotherapy is the experience of achieving the objectives of a process that was started yet blocked and unable to reach its goals for whatever reason and consequently enabling the individual to conclude, finalise and let go of this issue.

5.1.4 Summary of the Suggestion for New Classification

The dynamics between health and illness run parallel to the following dimensions in all living beings: (1) Integration vs. disintegration and (2) voluntariness vs. involuntariness. This applies both in an intrapsychic and interpersonal sense. Voluntary integration – and also voluntary isolation are both conducive to life. On the other hand, involuntary disintegration and integration are hostile to life. Illness is often attributable to an involuntary loss of relationships or integration – or also involuntary dependency, attachment, entanglement etc.

Psychosomatic. Higher life forms are structured on the basis of numerous, progressive regulation systems. The responses of these systems are characterised by specific averages and targets, and unexceptional deviation from these. It is only when this margin is exceeded that critical issues emerge. The body's regulation can be disrupted at different levels, causing it to deflect and become fixed at a level that induces stress.

Certain subsystems are separable within the entire personality to a certain extent. Subsystems that are reached through the fight, flight or freeze mode and their regulation work can be destabilised or event partially switched off.

5

Social relationship structure: Despite the fact that each personality is embedded in its social reference groups and relationships, for which the above principle also applies, and despite the fact that we know how much conflict potential can arise for the individuals at this level, we traditionally start with the individual understanding of illness and initially regard the personality as the referential whole. It would perhaps be beneficial to widen our horizons beyond this perspective, which is limited to the individual. The revision of the DSM has already made progress in this regard.

Consciousness. If our consciousness is overwhelmed as a result of trauma, the centre of self-efficacy may become blocked by a mechanism resembling paralysis. Those affected by stupor are unable to react arbitrarily or respond to any stimuli. Memories may be numbed or disabled for an extended period due to traumatic amnesia. This leads to dissociation. – If the personal-initiative centre is voluntarily surrendered during hypnosis, it becomes clear that arbitrary ideas also have the creative ability to feign their own reality. The brain creates its own worlds and is capable of erasing received excitation patterns on command. It dissociates and ostracises them. This is of interest to us, as we can use the phenomenon of unconscious self-hypnosis to achieve natural access to conversion symptoms. Manipulation through hypnosis can be used as a coping mechanism for life, but can also constitute abuse.

Post-traumatic fugue may be considered a symptom of a flight response on this level, or perhaps also aimless wandering. Both anxious activity and the behaviour when faced with a lack of attention could be regarded as equivalent to fight and flight responses.

The intolerant, hostile struggles of the normative islands (superego), internally legitimised for this purpose, against all different impulses and information within the semiconscious intrapsychic dynamics are generally recognised as anancasm. The underlying fear-driven counterpart often varies in terms of psychosomatics and also actionism (cleaning frenzies, perfectionism). In certain cases, the receding part of the identity collapses in terms of strength, but nevertheless fears being blamed for not fulfilling its duties. The general dissociation of feelings provides the principal line of fragmentation, including empathy directed inwards and outwards.

We are familiar with the notion that emotions are filtered on the basis of whether they are compatible with our self-image or not in terms of the target ideal (referred to by Freud as a censor) from psychodynamic hypothesis. Discrepancies trigger signal anxiety, a term coined by Freud, which is namely the fear that the self-image is no longer stable and may be destroyed. The struggle in this regard is to maintain a self-image that conforms to the target. During this adjustment process, certain elements are swept under the table or only retained in the periphery. The consciousness is offered the sugar-coated version of events. As such, constant fragmentation occurs. The unconscious is aware of this. The conscious would rather not know, provided it wants to be something other than what it is.

Tip

In psychotherapy, all key emotions, including those that appear to be negative, are scanned for their covertly hidden opposite pole. Hate often conceals love, coldness conceals suppressed emotion, apparent disinterest conceals secret interest and appreciation; as though it could be dangerous to be truthful and show one's own feelings. This can be linked to a certain extent with talking in opposites and is also a variation on the flight response and denial.

The vegetative nervous system, featuring the complementary poles of the sympathetic and parasympathetic systems, organic areas of innervation and the hormonal cascades to cope with stress, is the primary driver behind fight-or-flight and freezing responses, which are also always accompanied by fear. An aggressive fight response may lead to the onset of a bold desire, fear or winning spirit, which provides a reward in the form of increased life intensity. This is a desired attribute amongst athletes in extreme sports. Apart from this, a shame-ridden feeling of being a loser that leads to a loss in self-value can often be observed in addition to a trace of mortal fear in the fear induced by flight responses.

Practically the entire range of psychosomatic symptoms can be regarded as symptoms of over-regulation and dysregulation in this case. The line of fragmentation can be found where the misinnervation of the organs has unconsciously developed its own momentum; for instance: a middle-age woman seemingly abruptly experiences a rapid skipped heartbeat and the alarm is sounded. (Almost entirely unbeknownst to her, someone has just walked past the window who resembles her former lover who she continues to pine for.)

One way to escalatingly exacerbate symptoms is to observe them with careful assignment of meaning. – The release of aggression in vascular contractions combined with spikes in blood pressure is common. In this case, vascular innervation also occurs entirely dissociated to provide relief for the conscience.

Calming thoughts that stimulate a sense of security in combination with physical relaxation techniques that are usually capable of tapping into the unconscious can help to counter this.

We shall now take a look at the basal mental subarea, which is responsible for regulating mood and drive. These two aspects partially overlap in the sense that subdued mood and drive, and the respective elevated forms, more often emerge together than separately. Nevertheless, they may also be observed separately in abnormal manifestations. Psychiatric disorders are all united in their polarity and polarisation, namely the deviation from the centre, which appears to be poorly formed. If a patient is at either one of the poles, he may appear to be entirely cut off from the other in terms of experience, despite the fact that he is familiar with it from his former life. Its attitude to life, activity level and way of experiencing appear to be completely inaccessible to him.

Finally, we shall now briefly address the notably inhomogeneous, broad field of schizophrenia, previously also referred to as the splitting of mental functions. Schizophrenia encompasses the basic notion of illness as fragmentation, or a loss of integration. In this instance, this may also additionally refer to a loss of integration and functional order. Sensory deprivation experiments are capable of triggering psychotic reactions in healthy individuals. The brain requires ordered de- and repolarisation from the sensory periphery to the centre (similarly to the heart muscle which responds to an obstruction in the conduction system in the myocardium with peripheral spontaneous excitation discharges.)

Note

If a patient who presents the initial manifestations of psychosis receives a companion at their side for long-term social psychiatric aftercare who provides a constant, affirmative, reliable relationship, the patient is virtually certain to never again react in a psychotic manner. (Ciompi, Soteria project Ordered, social fields provide considerable healing potential through their resonances, something which has been undervalued time and again in recent decades [1, 2].

5.2 Diagnostic Tools – Beyond the Two-Sided Understanding of Illness

5.2.1 Biographic Anamnesis

Standardised questionnaires Almost every psychosomatic clinic and many psychiatric clinics have created their own anamnestic questionnaires that collect information on the situation at home, in the patient's private life, at work and on illness-related development, predominately on the basis of experiences. In order to be more patient-friendly, these questionnaires are filled out together in advance to facilitate further inquiry. As can perhaps be surmised, this approach is contingent on the therapeutic relationship. The inquiry can quickly devolve into quick therapeutic sequences. Due to the usual lack of time at present, patients are typically required to fill out the questionnaire alone. It is then discussed with the therapist at the next session. This constitutes the most common compromise currently in practise.

Representation of the current situation as a diagnostic lens. Michael Balint [3], the master of diagnostics and founder of the Balint groups, which focus on the understanding and resolution of entanglements in therapeutic relationships, recommended that therapists permit new patients the initial 5 minutes of therapy as a "free space" and to take note of how and with which concepts the patient's fill it and describe their situation. To a certain extent, this recital mirrors the patient's personality, life circumstances, ability to cope, resources and any chronic or current deficits. Of course, this assessment is increasingly successful for those with experience than beginners, whose inner reference system is only now starting the evolve.

Otto Kernberg's (*1928) recommendation for a semi-structured interview is based on a similar experience. He recommends permitting the patient a 10 minute "free space" to project their inner structure. This is followed by a structured component for anamnesis. During the 10 minutes, the therapist is required to observe how the patient appears to feel in this free space. Does he concentrate on the point? Does he get carried away in the chaos or associatively get lost in trivial matters? Does he switch between poles? Does he portray himself as particularly autonomous, important and prepared to pass judgement? Or does he remain mousy behind an imaginary wall and shun direct contact? Does he excessively blame himself for depicting something incorrectly in his opinion? Does he meticulously express a high level of detail in a monotonous voice? Does he display an increased craving for recognition or representation and indicate a desire for admiration? What feelings does he trigger in the therapist? All of these observations are based on proven factors behind the development-orientated emergence of conflict and on experiences from dealing with structurally weak individuals.

Anamnesis in a cross-section. Jacob Levy Moreno, founder of psychodrama (1889–1974) established a particularly helpful diagnostics technique for anamnesis in a cross-section, otherwise known as the social atom. This technique is based on his observations from relationship and group dynamics. The social atom illustrates the relationship constellation of a particular stage of life as drawn by the patient or illustrated with figures (small items, coins, etc.) in the form of a collage. Significant deceased individuals may also be included in the active relationship cosmos.

Subsequent additional criteria: In addition to the spatial symbolism, it is also important to pay attention to the size of the selected figures (circles, symbols) drawn by the patient, and their distance from one another, their colour symbolism (if the patient is offered coloured pens – the interpretation of colour by the patient), ambiguities (in the case of double strokes or connecting lines) as well as the desired dynamics, i.e. who would the patient prefer to be closer or further away? If the connecting lines were to be headlines or key statements on the respective relationships, what would they say? Are there official and taboo (yet no less true) statements for certain relationships? What would potentially happen if one of these taboos was broken? It is useful to ascertain the relationship constellation in this way for (a) earlier infancy, (b) for the triggering situation in the current crisis, (c) for the present day and potentially (d) to produce the desired constellation. The sociogram of the start of life, namely its relationship constellation is the key to understanding to initial formation of the personality.

In a group context, a sociogram may also be created with human representatives, potentially as sculptures including posture and gestures. If the representatives take on their roles, it would perhaps be conducive to ask them about their experiences and the impulses triggered by these experiences. This approach, namely

Moreno's sociogram, led to the development of widely used constellation work, which has since become one of the key methods of systemic therapy.

Brem-Gräser's Family in Animals free drawing method [4] has remained a tried and tested method used with children for over 50 years supplemented by statements on relationships between the selected symbolic figures as well as desires, dreams and also forbidden fantasies. Even though the specific animal selected for each person may often appear to be meaningful at first, the patient nevertheless needs to be asked each time: which of this animal's qualities fits this person particularly well? What does this animal represent, if it stands for XY? Individual association is key in this regard, not the typically standardised interpretation or input from the therapist. A cow does not need to represent stupidity, as is colloquially the case, but rather could instead represent reliable care and familial security.

It may be helpful to merge the social atom and the Family in Animals symbolisation technique. A further qualitative step in the exploration of the patient's world has been provided by dialogical part-whole research, as primarily developed by Gestalt therapy since the 1940s: in the case of both adults and children, a depiction of the currently experienced situation as a collage (using a range of materials such as tissue paper, stones, cones, shells, symbolic figures made from wood and potentially also coins etc.) or as a sand painting can provide great insight, in particular if the patient breathes life into the partial aspects sequentially, allows them to speak, reverse roles and also explains their associations and story. In Gestalt therapy, the patient often animates all roles himself, providing a representation of his psychodynamics without any external additions. Empty chairs are often used as symbolic placeholders for his subpersonalities.

Sandplay therapy can be regarded as a related method, whereby the current situation is symbolically expressed as a concrete manifestation in the sand (provided in a rectangle, table-like sandpit), complemented by symbolic figures to represent people, animals and plants, etc. This concept finds it roots in the analytical tradition pioneered by C. G. Jung, and was established by Dora Kalff [5]. It was initially solely used in child therapy but was later adopted into adult therapy. In diagnostic terms, it begins as a non-verbal, projective test that provides the perfect introduction to therapy.

In a group context, a so-called group-dynamic family sculpture of the participants can offer striking insights. The patient is the sculptor (or director). The gestures, position and posture of his living marble figures are intended to express the key characteristics of the relationship dynamics amongst his reference group. The task can also be modified if required into the (a) official, (b) secret and (c) desired relationship structures. Which impulse has the patient prevented from travelling from a to b and from there to c thus far? How could this occur experimentally? This alone would be worthy of its own therapy session. The sculptor is able to experimentally change the various aspects of his work of art and pay attention to the change in his feelings and impulses in the process of doing so.

Anamneses in longitudinal research. Experience-oriented anamnesis in longitudinal research [6]. Charlotte Bühler's life event research was met with a range of responses in humanistic psychology and its psychotherapy approaches, of which she was a proponent, and was widely incorporated in behavioural therapy. These methods of representation are suitable for the ups and downs of life, whether in the style of a timeline or life span perspective, possibly taking into account the additional aspects of health and resilience, hindrances, fears and coping, work situation, friends, family and partnerships, sense of purpose, happiness and satisfaction, etc.

Life can also be painted as a whole with freely selected accents or symbolised as a collage. This offers a bird's eye perspective on the patient's journey through life, perhaps as a river of life. It may weave through the various landscapes that represent the previous attitude to life. What does it look like at the source? Perhaps there are inlets. Or perhaps not, but instead the river trickles over rocky ground that only travels further underground. Perhaps there are waterfalls that crash or exhibit magnificent power or lucid beauty? Do any rocks get carried along with the flow? Are there any cliffs to traverse? What does it look like at the banks? Are there inviting areas to play and relax light-heartedly? Perhaps also secret mystical islands? Or does it constrict and become too strenuous to squeeze between the rocks? etc. What is the character of the river now? Does it follow its protected path in its riverbed or does it encounter turbulence, potentially in the form of drying up or excessive rainfall, which it is barely able to endure? Does it keep in mind that it's flowing towards the ocean? Does this thought worry or perhaps calm it? Where would it like to pass by en route, if possible? (This approach is suitable in both an individual and group setting.)

In addition to painting, the journey or river through live can also be wandered or danced through using fitting rhythms. However, it is useful to first reflect and draw a sketch that separates life into various stages of different attitudes to life. Once transposed, the path then passes through the group space. It may stop at certain points to reflect and potentially take care of something that had previously been missing with an imaginary person. A form of waking trance often occurs when the patient engages in this exercise, enabling him to temporarily interact with an earlier phase of life and his former identity. If it becomes palpable that a supportive and stabilising support person is required, the therapist or another group member, in the case that they are not all present at the same time, can provide support.

Who am I? Would I like to be friends with myself? No-one is able to provide an exhaustive answer for these questions. This requires a self-image based on the situational context (e.g. is this question asked during an admissions interview or within a group that has existed for a long time in which there is a high level of trust?). The friendship perspective provides relief: the majority of people find it easier to say something positive about their best friend than themselves. This additional variation for anamnesis is well suited to individuals with at least an average level of introspection. The subject is asked to stand up from her chair, to stand next to it confidently and imagine that she is her own best friend and standing by her side. "How did you end up becoming friends with this person? What do you particularly admire about her?" ..."Good. That makes sense. Does she also have other characteristics that you admire less? Perhaps walk around her once and take a look at her – symbolically – from her dark side. What do you find? What does this trigger in you? Have you ever discussed this or is it a taboo topic between the both of you?

A trip to the carnival. Imagine that you could go to the carnival with your friend and wanted to rediscover this old habit. In this scenario, the carnival in question is a special carnival procession whereby everyone who turns up "dressed up" as every side of himself that dwells inside will most likely change for the good. The question remains as to whether every concealed side remains dormant and has never become outwardly visible, or whether it needed to be locked away and disciplined, and has perhaps even adopted the form of a shamed, sad character or a threatening, wrathful monster, or something different entirely. What would that trigger within you and would you continue to tag along if your friend was to bring out this or every hidden side? Would anything change between the two of you?

Further suggestions for an expanded anamnesis can be found in the section on projective tests, which assist both in diagnostics and also, when required, with the introduction to therapy.

5.2.2 Diagnostic Assistive Techniques for Reviewing Multi-person Constellations

Following a verbal description of the matter – or also as the introduction to a discussion between the couple – non-verbal anamnesis of the current relationship structure in the couple's situation can be very illuminating.

It can be particularly helpful in couples therapy to freshly put the problematic pattern of interaction behind the couple's decision to seek therapy into concrete terms in a non-verbal form, something which usually occurs involuntarily as both parties are usually only able to perceive and describe the pattern from one point of view. There are a range of options available in this regard.

Couple's interaction illustration: a table with two chairs facing each other is prepared with a large drawing sheet and colourful crayons. The couple sit down. The therapist suggests that the drawing sheet symbolises their shared space and the next 5 minutes are their time together as they currently experience it with each other. Taking different colours is good in order to better differentiate between the drawings of each individual person. What follows is highly interesting. Who goes first? With or without consensus? Who encourages or potentially challenges the other? What is the first impulse? A thick dividing line? A mark in the space directly in front of the other person, which may be interpreted as transgressive? A subtle, reclusive circle in a corner. Something may also be drawn in the middle of the sheet, which opens the door to painting together. How close are the strokes? What do they express? Aggressiveness? Provocation? Cautiousness? Affection? A pleading invite? Fear of being hurt? etc. It could be that an encounter does not take place at any point. After 5 minutes, each participant is asked to provide a statement on their experience and their self-interpretation. If certain elements of this drawing could speak, what would they say? Has this underlying message been understood? At which points yes, and at which points no? How did misunderstandings occur? Are they common? Are there any similarities to everyday events? To what extent? What has been expressed in a way that means it cannot be understood? Or also, which messages do not receive a response from your partner for certain reasons (anymore)? What needs to be expressed first?

When was the best stage of your relationship? What was different? Take a step back in time and relive this stage over the next 2 1/2 minutes – now draw together on a new page. What has changed? How do you feel?

Do you still feel up for a third experiment? If you both were to draw how you would feel in your heart if you were allowed to live and behave as you truly want over the next 2 1/2 minutes on a new sheet, would we be able to take a look and see what is communicated and emerges? We shall now explore the side that is yet to find its proper place in the relationship. …. What do your strokes at the particularly pronounced points indicate? What do they say about themselves on their own behalf and what do they say about others? What does this entail for everyday life? What needs to change for you to be true to yourself and simultaneously part of this or perhaps another relationship of a different nature? What is preventing this from happening? This needs to be taken seriously. However, the impulse that was indicated in the previous experience needs to be taken just as seriously. What would happen if both of the aspects that appear to be relatively incompatible on each side were to engage in a dialogue? This clarification would require four chairs. The aspects that work well in the relationship and those that struggle and are at odds with the relationship. You alternate between your two sides. We shall see who grasps who or not and in which direction you will all change... etc. (this is followed by conflict-resolution chair work with double occupation of the seats, whereby the understated aspects are granted an equal chance by both individuals from the outset.)

"It seems very probable that you will both separate, but not necessary as people, rather from the previous form of your relationship." ...

A variation of this exercise can be carried out in the same setting – with the exception that only one (!) pen is provided. It can often be illuminating to observe the interaction in this case.

In both variations, a concrete change of position following the initial surge of accusations with imaginary role reversal can be a very enlightening intervention to enable the individuals to experience putting on their counterpart's shoes and seeing what feelings their behaviour triggers.

Provided it has been made clear that projections have been made onto the partner in this case, the therapist will place the projection chairs diagonally in front of him/her in order to sit the respective transference person from his/her life story in this chair and release his/her partner from this projection. (Transference separation: "What is the same between these two people? But what is clearly different?") "How does your partner change when he/she is allowed to be completely her/himself for your eyes as well? Does this lead to any new impulses?

Alternative introduction: find and occupy a seat. Another approach provides the couple with the option of taking a look around the room at the start of the session (provided the room is large enough for this experiment) and each finding a space where they feel sufficiently comfortable and adopting a position that feels most fitting at the time. Then the therapist shall ask each partner for their opinion and a link to their everyday relationship. The partners often stand far away from each other. Insults and accusations from both sides pile up in the space in between them, which are given a concrete form as "fragments" or overturned chairs. This sort of distancing with designations can already provide relief. At this point, therapy objectives may be written down on accompanying notes. The prognosis is sufficiently good if the notion of taking a break from the figurative fragments, leaving them there in the therapist's office and taking a ride into the unknown, as if they are both meeting each other again for the first time, causes the sun to rise in each of their faces. Of course, the fragments need to be processed using every trick in the book (see ► Sect. 5.6 on treatment) to enable a less stressful outlook for the future to emerge. The idea of taking a break from the fragmentation can be given priority as a ritualised rewarding retreat from everyday life to increase motivation and positive expectations in cooperation with the couple.

Sound variations: If musical instruments, i.e. sound generators, are provided and this level of expression is feasible for the couple (who aren't necessarily professional musicians), this medium can provide a great opportunity to express themselves and reflect. Initially, both individuals need to select one or, even better, several instruments that appeal to the individuals as aspects of their identity or generate resonance. Percussion instruments (drums) and small stringed instruments (psaltery) or warm, low-frequency wind instruments (ocarina) are often selected together in order to enable a range of appropriate responses to both the melodic self-portrayal and the interaction. Following this, the task is largely similar to the painting task: "...The next 5 minutes are solely dedicated to you and your relationship. Imagine that you are able and want to use this time to express your feelings towards yourself, your partner and this relationship. You are both invited to do this. ...." "Could you try to summarise your experience of the first 5 minutes into words, if possible? Did anything surprise or touch you, or affect you in an entirely different manner? Do you still need a closing section for clarification?" This non-verbal level requires additional music therapy training, which is not uncommon within the scope of Gestalt therapy training.

Family. Version 1: Each present family member is given the opportunity to create the family as a sculpture, depicting (a) how they experience the current, real relationships and (b) how they want them to be.

In order to prevent any influence from other family members prior to the sculpting, each member receives a notepad beforehand and records their point of view using circles and direction arrows. We shall now take a family with parents and two school-age children as our example. The therapist asks each family members how they feel, what their viewpoint is, whether any impetus to change has arisen etc. at the end of the sculpture session. Once the therapist has spoken to all 4 members, they are then each asked where (in which sculpture) they felt most at ease? What do I still not understand? What led to resistance in me? What clarification is still required in this regard? If we were able to look at ourselves from above: who has it the best and who has is worst? What have we all failed to try, despite the fact that it would also be possible? How will everything most likely look in a year's time? Is that what we want? Or do we not want that?

Family. Version 2: The problem child (patient) creates a sculpture of his family. He swaps roles entirely with every member of his family in turn. A slow pace is required, as empathy for the other family members takes time. What does the individual experience in each case? This can be refuted in order to allow projections to be revoked. Missing aspects need to be completed. Empathy is practised. What had previously not been experienced or not experienced enough? What appears to be better or worse in the other role, in comparison to your own role? What have you now learned to see differently? What does this mean for everyday life?

Family. Version 3: Quiz: Everyone writes a sentence on a notepad (possibly with a pen hidden by their hand in order to ensure anonymity) to the effect of: what I have never been brave enough to say to my entire family or to xy: ….". The slips of paper are then collected covertly or folded, revealed in turn, read out and each family member is invited to guess who wrote this message and who it is directed at. What is the background behind the message? What needs to be clarified? Do both members need support? What are the facts of the matter? What remains unresolved? Is it possible to pin this down? Could both opposing members please swap seats and roles? Does anything change? What can be said from the new perspective? Now please return to your previous seats. Has anything changed regardless? Can it be put into words? Or is there perhaps a better gesture for this?

Group situation mirrored by family dynamics. The group as a family. For the clarification of functions, emotional contributions, lateral transferences in the group's relationship network. "If we were a family, who would be best suited to which of the conventional roles? A few cards can be distributed at this point. Everyone receives the same pile. Each person receives two jokers on which they can write their own idea, which could be an aunt from America, an uncle who lives under a bridge, a modern-day robin hood, the girl next door or the mysterious stranger. The description is of no particular importance. Write the name of your selected group member on the cards and, where possible, also the key quality that led to your decision. (The following standard cards are prepared for everyone: Father, mother, grandmother or grandfather, siblings (brother or sister), best friend, distant cousins, pet (which?). The cards are first filled out then handed out. This can be done one at a time or for each role. For example, what are the choices for the father role? Does anyone have any comments? Questions on the qualities? Or perhaps astonishment if someone is not selected contrary to their expectations? Does it fit their self-image or not? Has anything hurtful emerged that requires further clarification?

Group situation mirrored by Freud's structural model. Cards similarly need to be distributed in this case: (1) Ego-ideal as the illuminating objective, (2) Super-ego as the setter of standards, (3) the Ego as the man of action and organiser, (4) the Id in its hysterical/representable/wanton form, (5) the Id as obsessive/meticulous/truthful/reliable, (6) the Id as subdued, unhurried, tranquil, (7) the Id as somewhat distrustful, peculiar, original and artistic, (8) censor/defence specialist, (9) joker. Following the reflection, lots of open discussion is required on the resonant aspects. A free selection of roles with a trial run and the adoption of the role of one of the previously particularly latent inner sides may be required in a second sequence. This can lead to free improvisation within the group.

Group situation: we as a town. Topic: The town with two faces. There is a set of prepared slips of paper for everyone. This time, there are more slips than participants. Each slip corresponds to a building. At the start, each member selects a building for themself with which they most identify. What lead to the appeal? Does this belong to my everyday life or initially only as part of the dream world? What type of community would this result in? Each group member is the assigned a building by another member. This results in a reflection of the self-perceptions and views from others. What is most striking about the discrepancy between the two viewpoints? Do I view myself completely differently to how others view me? What could my part be in this? Do the self-perceptions correspond more to reality or an ideal? Or are there also provocations and paradoxes? If yes, what are the motivations behind this?

The following buildings are offered: (1) a representative town hall with an assembly room and lecture theatre, (2) an old abbey with a concert and cultural programme, (3) a courthouse with an arbitration board and social office as well as a prison at the back, (4) a library with a reading room in the philosophy tower, (5) a hotel and inn with good traditional food and its own winery, (6) a house on the market place with a jewellery and clothes store, (7) a traditional butcher's with its own slaughtering facilities, (8) an inn with a dance club, (9) a milk and good shop with a coffee corner and small play area in front of the shop, (10) a nursery that sells flowers, fruit and vegetables in front of the town gate, (11) an art craft workshop that works with wood and clay at the town wall, (12) a half-timbered house with bicycle repairs and rentals, (13) a veterinary clinic and kennels on the city wall, (14) factory buildings for the town's cleaning department with a recycling yard for hobbyists, (15) a single impressive, modern factory building (laser technology) in the industrial outskirts, (16) a sport and leisure centre with sport facilities outside the town, (17) a summer house with a sheep pen and outhouse in the field, (18) a former royal summer residence with a pond and park for festive occasions on the romantic riverbanks in the outskirts of town, (19) a well house with a bench and chapel at the edge of the nearby forest, (20) miscellaneous...

The same building analogies can also be assigned several times in the group. It may be the case that two or three candidates contest entry into the group town hall or the summer residence, or that several group members are already sat on the bench.

Group situation as a shoe sociogram (according to Moreno): Everyone takes one shoe off and places it in the centre (symbolising their presence), changing its position until it is in contact with the shoes of those who the respective group member feels particularly at ease around or would like to be close to. Debriefing.

(B) Group dynamics sculpture Following the same principle as the self-organising group sculpture. At first, everyone in the group walks around in confusion, then gradually they all come together, enabling the key associations to be solidified, either through contact with hands, feet or backs, or eye contact. This self-organising arrangement of people may change over time. Needs for distance and control may also arise on the side-lines. Everyone should be granted to voice their opinion on their experience during the exercise and on their final position in their debriefing session. Unwanted approaches, potentially subjective transgressions as well as undesired distancing or subjective rejections can be delicate matters. It is often beneficial for the group's leader to point out before the exercise that these moments require sensitivity, yet should nevertheless also be treated with a certain level of cautious inquisitiveness, as they may reveal a lot about the person's own judgements.

5.2.3 Projective Tests

Projective tests can be found in individualised diagnostics. They are based on the psychodynamic school of thought and rely on the ability of what is currently unconscious to be projected with ease onto insignificant perceptual matter. Projections need to be returned to the world from which they were shifted. Of course, it is also beneficial to determine the motivation behind this shift, which can prove to be illuminating.

However, projective tests have been increasingly less "in" over recent decades in the mainstream. These tests primarily provide qualitative assessments, with quantitative statements only possible in a limited number of cases. The fear of misinterpretation is notable. However, certain tests such as Family in Animals and Sandplay have already been mentioned. They can be beneficially integrated in anamnesis, which takes place on an individual basis, and used in a Socratic approach with their self-interpretation. As such, the fear of misinterpretation can be widely mitigated.

The scenotest (G. von Staabs) has become an indispensable tool in child therapy since 1938. In this therapy, the child is provided with the material to depict their life situation, including their fears and aspirations, in the form of doll house equipment.

Projective tests can be roughly split into several categories according to the media they use:

- Verbal tests: e.g. the patient is asked to finish the sentence. – or a story is started that breaks off at an ambiguous point and the subject is required to continue imagining how it continues. – in a group setting, stories may be developed by several people in both big and small groups.

  Picture material. The picture interpretation technique of the thematic apperception test (TAT) developed by Murray in the 1940s is renowned in its field. It involves ambiguous photos or drawings. The subject is asked to describe the situation or scene, including all of the feelings and impulses, hopes or feats that are triggered in him. Does the interpretation of the picture provide similarities to coping or not coping with the current issue? Is there any indication of motivations that remain concealed? – The children's apperception test (CAT) provides a version for children.

  Even more well known is Hermann Rorschach's inkblot test [7]. This test consists of 10 cards on which figures can be seen by the subject. The subject is then asked whether they are able to see anything in these blots and if yes, what did they see first? Following this, the subject's answers are analysed on the basis of various viewpoints, e.g. according to the suitable shape, responses to colour as well as shadows, the originality or commonness of the interpretation, movement or freezing, overall impression or focusing on details, spontaneous switches to the opposite viewpoint, whereby the space between the figure in the foreground and the blot disappears into the background. Those trained in analysis may be able to make relevant statements on the structural development of the tested individual, their current stressors and reserves.

- Drawing tests. Requests to draw a tree, man or house have long been used predominately in the field of developmental and school readiness tests. Age-appropriate viewpoints, standards and signs of maturity are involved in these tests.
  The Wartegg drawing test [8] based on the instruction to complete drawing simply elements, e.g. dots, lines, circles, angles, is well established in this regard.
- Motor skills and balance coordination are also taken into account in school readiness tests.
  Posture and body language provide mirrors of our minds at all ages. In psychotherapy, the body can be read like an open book. Conversely, the "mind" can also be influenced by the body. Body therapies and improvisational theatre specialise in this field. At this point, we have already departed from the section on projective tests. Bodily expressions are generally quickly and easily unconsciously understood by others and also promptly answered. In humanistic approaches to therapy, non-verbal understanding and pantomime-like displays are used in many aspects. This field also uses deceptive manoeuvres to test the subject's precise powers of observation and sense of dissonance.

5

5.2.4 Psychometric Tests

This classification comprises (a) performance tests (to measure attention and concentration, such as the d2 test), including intelligence, development and school tests, (b) personality tests including testing the structure of personality and (c) clinical tests that focus on psychopathology.

"At its core, classical test theory consists of a set of assumptions on the measurement errors of test values, which are not set in theory or empirically verifiable"..."Literature on the subject remains divided to this day on how the assumptions should be stringently ordered" ([9], p. 463). In other words, a person is assumed to have an intrinsic, constant test value (e.g. intelligence) as well as a situationally changing, in many respects independent, measurement error.

High reliability is a prerequisite for test accuracy in addition to the guarantee that the test truly measures every characteristic that can be measured (validity). This is certainly logical. However, are isolated characteristics always independent entities or are we instead observing artefacts?

Psychometric tests assume that these characteristics represent individual characteristics that can be measured constantly and independently. The variability whereby individual characteristics may develop depending on the currently experienced relationships, or perhaps not, (whether in relation to a partner, friend, teacher or person of authority, including as a mirror of the individual's inner treatment of themself), appears to not have been researched thus far, but would nevertheless not only be important for psychotherapy but also, of course, for education. This variability may correlate to the extent of the reactive attachment disorder as well as the agitation of the social system that provides the subject with security. Only once we have a reliable test instrument at our disposal for the latter, would it be logical to assess distinguishing characteristics, e.g. assessments of mental capability, in combination therewith.

Norm-based diagnostics with standardised questionnaires can teach us a lot if we want to pose questions regarding the tested population. However, we can learn much less when this is scaled down to the level of the individual. He will most likely learn that he is close to the top of the Gaussian curve with his test results. If he is far from the median line, that does not indicate anything about how he subjectively allocates this, whether he feels good or bad about it, whether it makes him proud or ashamed, whether he is sad or happy about it, what led to this, whether this is a permanent state or exception etc.

Although immense work is required to provide norm-based, psychological measuring tools and they are extremely informative for research in general and for insights for large numbers, the results they provide unfortunately only provide us with limited knowledge for individualised psychotherapy. Many psychotherapists consequently largely avoid the use of test diagnostics.

5.3 The Two-Sided Understanding of Health and Illness

5.3.1 The Holistic Approach to Maturing and Personal Development

For this section, we shall initially once again refer to the WHO definition of health:

> Health is a state of complete physical, mental and social well-being and not merely the absence of disease or infirmity.

This broad, individual definition includes the regulation of intrapsychic and interpersonal relationships as well as those that are directed toward the spiritual. It takes our interconnectedness with our social environment and various spiritual worlds into account. The WHO definition of health extends beyond the currently still very moderate definition of illness provided by the German healthcare system, which is solely based on the individual in terms of dimension.

At present, in the field of psychotherapy there is a non-conducive rivalry between (1) a disorder-specific and (2) a more holistic understanding of disease processes. This polarisation has traditionally been observed throughout the history of philosophy and sparked several wars of scientific opinions, which remain unsolved. Both of these rival poles, the (1) variety and (2) the reason behind the unity, can also be found reflected in the politics of the field of psychotherapy as though they are incompatible. These ideological wars unnecessarily tie up resources that could otherwise be used to benefit research and help patients. Wars narrow the sight of the individual parties. Behavioural therapy primarily supports the disorder-related pathogenic perspective and fights against approaches such as psychoanalysis that are led by hypothesis and, above all, the holistic approach of the humanistic methods, particularly Carl Roger's talk therapy with its empathetic approach and also Gestalt therapy, which is inspired by Martin Buber's I-Thou concept in terms of relations. However, BT (at the suggestion of Klaus Grawe) incorporated the Rogerian approach, at least in a light form, as an additional recommendation during the emotional turning point of the field. Despite this, the appreciation of the trinity of empathy, authenticity and consistency/coherence appears to no longer be found in the professional politics amongst the mainstream and representatives from the field of behavioural therapy.

In this book, the dimensions incorporated by the WHO also belong to the concept of health and illness. We have already mentioned and discussed the healing function of Porges' social engagement system at several points. This system has an impact from the very beginning of a human life. It is biologically secured amongst our caregivers as a result of the baby schema [10] activating care instincts and increasing the production of oxytocin. Without this inherent factor, babies would have no chance of survival. It increasingly enables the baby to form personal connections and learn to integrate with joy. This ability to form relationships ensures the survival of us as individuals and our species. In this relationship, we learn how to regulate frustration and regain balance time and again, provided everything functions sufficiently well. The ability to obtain trust and help, or rather to accept it, also forms part of this process of learning regulation. A sufficient level of basic trust is also required in this regard. This process of learning how to regulate and handle frustration also helps us learn how to integrate. Generally speaking, babies experience which difficult feelings and situations are a part of their lives and yet that everything still turns out ok in the end. This is what we hope to achieve. Our patients may have experienced a number of situations that did not end well, that did not establish basic trust, that did not provide a reliable experience of attachment and the survival strategy for which became engraved in precautionary isolation, etc.

Animal experiments: The consequences of a lack of experiences of loving brood care have been particularly strikingly exemplified in the renowned experiments conducted on rhesus monkeys by the primatologist Harry Harlow [11]. Monkeys that grew up without a mother and were required to obtain milk from a replacement wire-mother at least cuddled their fake, replacement soft cloth-mother from time to time. They exhibited above average levels of anxiety and attempted to stay in contact with the cloth-mother even during exploration. Later in life, they displayed disturbed behaviour towards their own young, provided they were capable of mating or produced any. The monkeys that were forced to solely grow up with a wire replacement mother were quick to freeze in the case of minimally frightening stimuli and hold themselves tightly.

Without going into further detail, we here refer to the countless stress tests carried out on animals through the separation of mothers from their infants. A life-long vulnerability to stress can often set in primarily amongst small mammals, whereby increased stress hormones have been recorded in infants as the result of postpartum experiments on separation from their mothers.

These are just two extreme examples that highlight the immense significance of experiencing a coherent, supporting relationship for mental and psychosomatic development. A lack thereof is perhaps the most important factor in psychopathology and also for the healing practice of psychotherapy.

Any form of psychotherapy that neglects to take this factor into account can neither be considered universally applicable nor scientific.

Genetic and epigenetic stresses form a specialist area that would exceed this framework, yet they also belong to the potential stresses required for the development of sustainable, coherent interactive relationships with caregivers.

Standard conditions that are conducive to human development are described in more detail in the section on developmental psychology.

Unconditional love fosters strengthening, compatible and synchronised interpersonal oscillating fields. Similarly to the I and Thou components in Martin Buber's approach, we understand unconditional love as an oscillating field with the typical twofold hypothesis of the conscious: the active, life-enhancing component can be compared to strengthening encouragement that wants to help the individual develop their potential, the receptive component helps with understanding or rather empathetic introspection and "reading" the individual's needs.

Now we shall return to the ideal presentation of health: when everything oscillates, we are able to imagine a healthy being as a powerful, complex yet nevertheless primarily synchronised wave pattern. Acoustically (symbolically speaking) a prevailing, clear undertone can be detected. Everything aligns in tempo and key. The basic chord may be rich in harmony and incorporated additional melodies (information), it may risk and endure dissonances and may venture into other keys. The process always reverts to the basic chord in order to find peace there. This completes the acoustic comparative image.

The notion of synchronised common existence entails that the human being forms an oscillating system with many interlaced subwholes and is integrated in broader social and spiritual connections to the whole. Ken Wilber would use the term holarchy in this regard. Each separable, complete unit is both a part a greater one and also provides the reference framework for smaller units within its realm. For example, the integrated smaller units (subsystems) in the human organism would be: organic functional units such as the immune system, endocrinological system, cardiovascular system etc., organs such as the liver, spleen, pancreas as well as cells and cell organelles. The smallest energetic unit, the quantum, and the biggest unit, the all-encompassing, which exceeds our knowledge and imagination, constitute exceptions to the above dual perspective.

Note

Health encompasses the concept of coordination and mutual relatedness. In terms of tempo, the heart acts as the pacemaker that can be found in all cells and also detected beyond the physical boundary of skin in the interpersonal realm. The hypothesis is that interlaced systems function more or less in sync, therefore strengthening each other and maintaining their conciseness.

By implication, increased chaos can weaken the entire system. In the case of synchronised system hierarchies, it can be expected that the pattern is fundamentally reproduced and mirrored in less complex functional levels, even if in a somewhat dimmer, vaguer manner.

In summary, the notion of healthy development comprises the natural development of existing potential with the well-developed ability to form relationships and integrate both internally and externally, an at least average tolerance for frustration, an anxiety-free (therefore tolerant) judicial entity that can also be suspended and an ability to find one's own inherent order pattern through centring, which may strengthen one's own structures in a synchronisation process.

Inner vitality and balance can be maintained for an extended period, even under stressful circumstances, just as the internal dialogue continues even in the face of temporarily dissonant aspects. The latter are not denatured and removed as a result of this.

Conclusion

The conditions for starting and developing adequate regulation are experiences of loving, coherent relationships, satisfaction of basic needs and the setting of age-appropriate protective boundaries. This initially relies on adequate and empathetic care.

5.3.2 The Differential Disorder-Oriented Approach Leads to the Understanding of Symptoms

5.3.2.1 Pathogenic Fragmentation as a Result of Exposure to Fear and Danger

General dissemination: If we were to summarise the many different frustrations and traumatisations faced throughout life, those who have not or hardly experienced stress as part of their development tend to form an exception to the rule in comparison to the majority of those who have been traumatised or impaired to a greater or lesser extent. Experiencing a certain amount of anxiety is entirely normal and helps us to cope with life. However, it is key to distinguish between:

Isolation: Developmental psychologists and pedagogists argue that normal, age-appropriate frustration is important, realistic and promotes development. According to them, it helps us to cope with life and makes us suitable for integration into a particular culture. It helps us to internalise culture-specific taboos, realistic and healing boundaries and limitations. In this regard, it is crucial that developmental frustration is experienced against the backdrop of a fundamentally accepting attitude. One of the key goals of education is for young people to learn to test their boundaries and become disciplined. In order to do so, they are required to identify with those that limit them and continue their role intrapsychically. There is often a fine line between this internalisation and identifying with the aggressor, which we view as a defence mechanism that requires treatment.

Pathogenic stress: When we address trauma and deficits in this section, we do not refer to insightful, balanced, beneficial frustration, but rather a frustration that partially or, in certain cases, significantly exceeds the individual's stress boundaries.

In the case of disintegration triggered by an internal correction process, fear was initially previously triggered by a discrepancy between the internal current situation and the target. This can also initiate an unconscious conflict-related corrective defence process. It would appear that we possess a central point of registration that is close to the conscious mind for the sugar-coated version of a conflict constellation – potentially in the thalamus, i.e. the limbic system. However, engrams of the uncorrected, unconscious versions of the same event, which may be triggered occasionally with delays or recollected through hypnotherapy and brought back to the surface of consciousness, are stored in the periphery. This discrepant process whereby parts of an event that contradict how the individual thinks things should go are "corrected" to how things should be lies at the heart of conflict-based psychotherapy. The deletion and correction process employed as a defence mechanism is yet to be clarified in neurophysiological terms. In this regard we continue to rely on hypotheses. This is the domain of psychoanalysis. Of particular interest is the fact that stage-specific defence mechanisms form throughout our development, for example turning against oneself at depressive stages or at stages during which the interpersonal oscillating fields provide security as an internal home, or, as another example: the lack of emotions at the top-down stage. This enables us to speculate on whether or not the new capability to modulate excitation patterns develops with the functional use of a newly formed anatomical structure.

As a hypothesis, we could imagine a locally corrective counterpart for the subarachnoid space which is padded with glia cells as comparable to the modulation of voice production in the air-filled nasopharynx due to unconscious muscular microchanges. It may be capable of minimally changing existing oscillation patterns towards a target. In the example of the corrective counterpart, in the case of a baby, this would be the voice signals of the dialect from the environment into which the baby was born.

Neurobiological basis: In the past two to three decades, Stephan W. Porges has mapped out the biological prerequisites for coping with trauma and compared these with the work of modem trauma psychotherapists such as Bessel van der Kolk and Peter A. Levine, as well as with findings from attachment research. Initially he looked at the peripheral course of the branches of the vagus nerve, the tenth cranial nerve, but over time, significant recurrent branches to multiple parts of the brain became apparent.

A lack of loving affection and adequate satisfaction of needs at the start of life can initially lead to a broadly fragmented hardware in babies. These shortcomings mobilise both stress outlet axes, first the sympathicotonic axis of the fight or flight response with its unsettling activity and potentially also later the axis of archaic, parasympathicotonic, paralysing freezing. Both switch off the social system of the ventral vagus in part or entirely and allow the already traumatised individual to be further transported into the visual circle of his social fragmentation. In his work, R. Spitz described emotionally neglected babies who no longer wanted to reach out for contact after a certain stage and soon died. Besides the lack of contact, they were otherwise adequately cared for and fed. He referred to their state as anaclictic depression. This term indicates that the final bastion before death is the tendency to want to cling.

5

Note

We shall now take another systemic look at polyvagal theory:

(a) The level of the archaic freezing mechanism (activation of the myelin-free, phylogenetically old dorsal vagus nerve). Primary symptom: Dissociation of the mind and body in critical situations in addition to depersonalisation and derealisation. In order to protect against the trauma of birth and the potential consequences thereof, DMT (dimethyltryptamine) is produced by the new-born's pineal gland. It has become apparent that this also occurs when we die. Victims of torture and abuse as well as those who have gone through near-death experiences report dissociation phenomenon.

    Amongst other symptoms, shock is characterised by an overload in the hippocampus, rendering the individual unable to record memories as usual. Instead, the amygdala takes over the function of recording memories, however, only registers fragmented aspects.

    Animals that hibernate (mammals) have "reanimated" or retained this phylogenetically old ability to drive down their metabolism to extremely low values and only first reawaken from hibernation as a result of the pattern of sunlight increasing on a daily basis (possible in an extremely high frequency range) as a survival mechanism, in cooperation with the pineal gland. The notion that seasonal affective disorder may be a rudimentary remnant from hibernation has long been speculated.

(b) Fight-or-flight response level. (Activation of the sympathetic part of the autonomic nervous system as well as the hormonal stress axis into alarm mode). Primary symptom: Hypervigilance. Resisting highly dangerous situations with the readiness to engage a fight or flight response. Fittingly, this leads to fundamental mistrust and a subsequent polarisation between friend and foe. The collapse of social unity leads to social isolation or fragmentation.

(c) Activation of the social engagement system and the binding forces as a survival mechanism. The SES is controlled by the vagal nerve, phylogenetically young, gives danger the all-clear and conveys a sense of trust in familiar faces, voices, gesture, favours proximity and evokes a sense of security. It assists with regeneration, digestion and reproduction.

From another perspective: the ventral, myelinated vagus branch, which is fully operational as of birth in every (competent) infant [12] in terms of socially interactive and consequently integrative integration into the group that provides protection, may be hampered by stress or other unfavourable circumstances. It serves the purpose of promoting social integration and attachment security. In dangerous situations, the individual experiences fear (in addition to a need for clinging and dependency) that the protective community that provides security, acting as a social "uterus", may be lost.

In order to maintain this protection, all impulses that may be disruptive are suppressed and outwardly adjusted. Those who fail to adapt lose the protection of the group and become a fragment of the group. However, not all groups are equally supportive or tolerant, as is similarly the case in political environments that exhibit a vast range, from tolerant democracies on the one hand to totalitarian regimes on the other. Interestingly, the etymological root of sin comes from "to segregate". A segregated person lives dangerously on the surface whereas over-adapted people live dangerously within, as everything they suppress inside becomes an inner, disintegrated fragment.

In this regard it is perhaps necessary to continually weigh up which risk of fragmentation is more dangerous. Darting into isolation may provide the opportunity to experience oneself as more coherent and strengthened. In this individualised state, is becomes easier to give new responses and form alliances than when part of a close-knit group.

Additional external risks: The protective SES can also be used as a manipulatively placatory deceptive manoeuvre despite the fact that danger is imminent. In the popular fairy-tale "The Wolf and the Seven Little Goats" the wolf eats chalk in order to deceive the children with his high, feminine, innocent and trustworthy voice. False kindness that conceals aggression, or overprotective behaviour aimed at overshadowing anything to the contrary is observed on a daily basis in psychotherapy on the topic of conflict.

Our primary dimension runs between voluntary or involuntary fragmentation on the one hand and coherent integration on the other. The ability to integrate typically grows with increasing maturity and sophistication.

Somewhat interestingly and astonishing is the fact that psychoanalysis and behavioural therapy originally failed to acknowledge the biologically embedded significance of relationships that provide a sense of security for patients, namely those in need.

It remains difficult to assert that this notion has been fully incorporated by the mainstream branches of these groups of methods, despite the fact the importance of the therapeutic relationship with its healing and integrating potential has been acknowledged by the avant-garde branches.

5.3.3 Synthesis of the Two-sided Pathogenesis

Important

(a) A lack of empathy and compassion for the existential centre as the common cause with consequences on development with regard to structures, the ability to integrate, resilience, the ability to love and form bonds as well as the likelihood of genetic and epigenetic manifestations.

(b) External stress caused by (sub)trauma, conflicts or due to internalised stress dynamics from unresolved constellations, including the aforementioned likelihood of genetic or epigenetic manifestations, each with subsequent disintegration, including their variety of disorders.
(c) Connection: if the individual is capable of empathy, compassion, secure attachment experiences and building social skills, they typically have broad resistance to the stresses mentioned in b), including any genetic and epigenetic vulnerabilities. In the reverse situation, the risk of becoming flooded with fragmentation and the onset of symptoms during the processing of experiences when faced with relatively minimal stress increases.

Note

Both threats and life-promoting impacts can come from both outside and within, such as through conflict-generating internalisations, fixed wrong conclusions, pathogenic cognitions or needs.

On (a) Rogers regards a lack of empathetic understanding, above all in the early stages of development, as the primary damaging and development-hindering factor in the development of the personality. Significant contributions towards a greater understanding of this dimension that are consistent with Roger's stance have been made in recent decades by psychodynamic development research such as attachment research and the mentalisation theory. In Gestalt therapy, the relationship factor constitutes the principal axis of the healing process for development and recovery. The power of a compassionate field extends beyond even that of an empathic one and conveys a message that provides encouragement for life and the individual themself. Both are important and have their own indications. Connections are experienced on all sensory channels. Connections are inherently nourishing and conducive to life. Ashley Montagu [13] successfully proved that 10 minutes of caressing touch per hour can significantly decrease mortality amongst newborn babies. Accordingly, mammals instructively caringly lick their new-borns.

In this regard, Porges' trauma research contributes to these findings with the security and resilience provided by the social engagement system (SES), which is biologically and neurophysiologically regulated by the ventral vagus system. This system is developed starting at birth, initially providing caring protection and later forming the foundations of our personalities, ultimately culminating in our suprapersonal ability to love.

Even in the case of genetic and epigenetic vulnerability, the aforementioned general sensitive empathy provided by caregivers is the factor most likely to protect the individual from developing manifestations of illness. With an average disposition to schizophrenia, this factor is now estimated to be 50%. This encourages the provision of maximum positive emotional input. It's highly probably that the effective component in this regard is freedom from anxiety that permits to organism to establish and stabilise its regulating order patterns.

In conclusion: it can be said that a human being (and every high mammal) requires a sufficient level of this affirmative, life-enhancing influence on the relationship level for healthy and resilient development and maturation. This personal care can be conveyed through different sensory channels with visual, acoustic or tactile encouragement.

This finding also further corroborates the findings of contextual research [14], according to which the vast majority of variance in positive outcomes in psychotherapy is the result of a combination of the therapeutic relationship (30%) against the backdrop of the correspondingly convincing personality of the therapist (20%).

On (b) The disorder-specific perspective also provides a wealth of hidden knowledge that we can and should use. According to the common assessment and doctrine of this approach, the primary concern is what needs to be relearned: symptoms need to be combated, abolished, eliminated, reversed, forgotten or overlearned etc. Leaving aside the fact that nothing in nature can truly be demolished, this represents a waste of energy and information. Symptoms are disintegrated forms of energy that transport the information of their genesis as well as the potential strength of their original being, which can be used for their retransformation. When processed, they can form a valuable building block in their original system. The realisation that the individual in question wants to consciously and decisively stand on the side of humanity and not be drawn into any spirals of violence as a result of certain ideologies (anymore) sometimes only occurs at a later point. This would represent a significantly different and more respective identity than continuing to live in trauma-related, partial amnesia.

We do not solely encounter external threats as traumatisation. We learn to live and to survive in difficult times. Our reward and evaluation system helps us to establish targets that incorporate our basic needs in addition to the social values of the group to which we feel we belong, guiding principles, our own survival strategies, special experiences as well as how to survive in the particular environment in which we find ourselves and values that we have obtained as a result of our empathy with others and by identifying with our respective reference framework. This complex evaluation system grows as we do. Its target values rely on the situation at hand for certain aspects or may be less dependent on context for others. This system, which carries our identity, appears to unconsciously help us make decisions and act as a compass. It alerts us to fear and existential danger when its thresholds are exceeded. Freud referred to this as signal anxiety. It attempts to erase or partition off any part of an experience or behaviour that deviates from the accepted picture of our identity. We remain unsure on the precise neurophysiological mechanism behind this process. The notion of a censor is used to denote a defence mechanism in psychoanalysis.

Note

This part of reality is denied the right to exist. This can be regarded as an internal battle and referred to as a traumatic existential threat from the perspective of the respective part of subsystem that should not exist.

Traditionally, this process applies to instinctive desires. As such, they shall be taken as an example below.

► Example

This longing, driven by instinct, is split off from the experience in the case of a rigidly set value and encounters a random element, for instance, the melody that was playing at the same time in the next room, and experienced with this new element together as a new energetic unit (similarly to how Pavlov's hungry dog was trained to react to the sound of a bell with expectant salivary secretion after being fed to the sound of a bell.) Now the melody may transform into an abruptly triggering stimulus of forbidden, love-stricken longing. ◄

In the case of inner dissonances, conscious or unconscious, the vegetative nervous system becomes excited, similarly to a fight or flight response, which leads to norm-deviating swings in our internal lie detector, for instance. Internally, this corresponds to primary relief from conflict when attention no longer needs to be pinned down on the forbidden content but can instead naturally be elevated to one of the aspects of the alarmed vegetative system that comes to the fore, e.g. the increased heartbeat or panting, and this dynamic can be increased as a result of this attention. Furthermore, in panic disorders, the reaction from the anxious environment often intervenes as a rewarding component. It is entirely natural that palpable components of fear and desire are intertwined in these kinds of anxiety and panic disorders. Patients require an appreciative, honest and strong therapeutic relationship to openly address the usually shame-ridden dynamic and open the way for alternative solutions.

However, it may be the case that an impassable wall has been erected against draconian forbidden desires or impulses, preventing any response to corresponding requests. Harald Schultz-Hencke addressed the notion of an aggressive gap in his writings on the "Inhibited Man". This likely corresponds to internal paralysis, or a freezing reflex. Internal, unconscious conflicts can rampage with comparable violence and a lack of empathy for our counterparts, comparable to traumatisation from external factors. As such, we are here dealing with internalised auto-aggressive, non-empathetic perpetrators, or rather, with particular motivations with partially lost inner contact.

Internal, unresolvable conflict prompts exclusions. Disintegration forms symptoms. This is the disorder-specific level.

Tip

The great opportunity provided by psychotherapy lies in its ability to decode and reinterpret symptoms (with help from the patient and his cooperation) in terms on their concerns, symbolism and meaning. This matter is addressed in more detail in the section on treatment.

A strong, trusting therapeutic relationship as described above is crucial for this retransformation and reintegration work. Disorder-related work without it will not succeed in achieving its objectives.

On (c) A two-pronged approach is the overall aim, whereby (1) personality development is primarily encouraged by specific direct contact and (2) where the symptom has been "read", i.e. understood in terms of its concerns and hidden meaning, for which the avoided intermediate stages can be subsequently made up for and the symptom may even potentially be greeted as an enhancement when integrated into the overall personality.

Or, from another angle: Those who have only experienced minimal encouragement and acceptance within themselves are more prone to external pressure and internalising constellations that tolerate or are even capable of causing imbalance. Emotionally aggrieved individuals are more susceptible to self-harming. Conversely, experiencing being palpably loved can act as a life-long protective mechanism.

5.3.4 Destructive Fight, Flight and Freezing Within

Examples for the onset of anxiety, obsessive-compulsive disorders and depression

Note

From external to internal threats: at a less sparse level, the sequence of fragmentation continues with phenomena of disintegrating exclusions adopted from traditional psychodynamics.

As a result of the fear of needing to correct its self-image to the negative and being disparaged for this by the real sociophysiological environment, the internalised reference system or from its own, relatively autonomous moral entity (as a variation of Freud's Super-ego or Lawrence Kohlberg's developed moral entity), the mind usually manages to "correct" mental functions (including their mental substance) in terms of internal targets or ideals, or in other words, to make any discrepancies from reality invisible to the inner observer. Freud referred to signal anxiety in this regard.

Note

It is often useful to approach and treat the theory of conflict and neurosis from the perspective of fragmentation as a result of internal flight, fight and freezing.

If we take a moment to imagine the organism of the soul, mind and body as a whole, the immeasurably manifold interlaced structure of the subsystems of various categories may become clearer to us, at least to a certain degree. The whole appears to be connected through both light signal patterns, neuronal impulses and humoral messages. In analogy to the external formation of states, all variants on real democracies and totalitarian regimes can be envisaged in terms of internal leadership. The notion that non-conforming subsystems or impulses can be forced to flee or excluded in the latter seems self-evident. For example, in Victorian times, sexual fantasies and the desires of respectable women were forcibly repressed in order to not bring dishonourable disrepute to their external reference system and be forced to accept social devaluation. Another example: In the Third Reich, boys were not permitted to cry or exhibit any significant emotions if they wanted to at least somewhat correspond with the collective, heroic ideal of masculinity. As perhaps goes without saying, open or even indirect guidelines from direct reference and leading figures from an individual's personal environment can have a strong impact.

Internal concealment and shifts correspond to external escape. Defence mechanisms used at an early age remain in the unconscious as familiar loopholes and back doors and can be easily retrieved for later instances of distress. According to psychodynamics, these former defence mechanisms lead to neural destructuring.

Developmental sequences can form among the defence mechanisms. In order to defend against subjectively threatening content and emotions, the mechanism that is most employed is the one that was in its vulnerable phase with regard to its differentiation, learning and adaptability in the corresponding development stage.

The fight or flight system features a number of progressively graduated obstructing variants.

(a) Depression due to a fear of loss or separation.

During the initial high point of the social engagement system in early infancy, all impulses that may pose a serious risk are blocked when there is a lack of security. Above all, aggression must be blocked if it cannot be taken out on substitute objects, and either redirected to the individual himself or, in the best-case scenario, it finds a psychosomatic outlet. This depressive behaviour that pushes away aggression aims to protect the significant counterpart and the existential relationship with them from the individual's own aggression. First and foremost, it protects the connective field between two people or the connection to a group due to the fear of loss or separation at the price of internal dissonance, which is repressed, displaced or somatised.

Multiple variations of the described problem may emerge later in couples therapy.

The compromise for the partially blocked fight or flight dimension may take the form of any kind of unceasing agitation, which may be primarily motor accentuated or present itself as the inability to engage with someone or something for an extended period. The patient appears to be actively on the run, seemingly searching for something. In this case, the increased impetus is acted out against the background of sympathicotony, but stopped from heading in a hostile direction. Together with the classic flight response pattern, this indicates a relationship disorder that promotes social isolation but nevertheless prevents adequate stabilising contact within. According to the hypothesis, this behaviour mirrors an earlier lack of existential, coherent and affirmative attachment experiences.

Another mechanism behind the onset of depression is explained in combination with auto-aggressive tendencies at the end of the next section.

(b) The development thrust of the top-down neurons with their increased autonomy and control can relatively easily curb and modulate the affective bottom-up inflow. It can also excessively obstruct and crushingly fight against it. Early on in his work, Harald Schulz-Hencke conjured an image of the "Inhibited Man" with various gaps in drive, a notion which is congruous with these observations.

The impetus of infant tantrums that can bring adults to the edge of their ability to withstand stress when not directed outwards can disrupt their loving relationship bonds and enable the rebellious side of unloved teenagers in distress, who seek to dethrone their elders or are symbolically after their parent's blood, to strike inwards from outside.

Processes whereby norms are internalised come into play at a somewhat later stage of development; they can cause chasms and barriers to form between people due to their willingness to put people down, intolerant assertiveness, harsh judgement etc. What is the situation on the internal battlefield? Who do the battle lines separate and which battle cries or convictions are used the justify them? Aggression is often cloaked in righteousness in this regard. The unkinder and more ruthless the unfulfilled demands of the outside world were experienced, the more strictly they are vicariously passed on to internal subrepresentatives. This process marks the starting position for the development of a variety of obsessive-compulsive structures and symptoms. Accordingly, the near-conscious centre of the identity assumes the role of a warrior, identifying as the perpetrator in a certain sense. The warrior aspect experiences self-efficacy, which provides a certain level of satisfaction. In the case that attention is given to the subjugated aspect, the experience is subsequently combined with its doubt and resignation.

When the centre of the identity shifts to the role of escaping and being a loser, i.e. the role of a victim, this leads to a depressed mood and lack of drive. In this case, therapy can convey the notion that only truly strong personalities are able to suppress themselves relentlessly and it is perhaps worth revisiting who this fighting power was originally aimed at. Changing perspectives between victim and perpetrator is helpful against the backdrop of the subsequent clarification of the relationship with the detrimental counterpart. If ambivalence is shown towards this person, the life-enhancing, positive aspects of the relationship are consciously protected in therapy, where they are temporarily symbolically relocated to their own empty chair as a protective space before the pent-up resentment towards the remaining aspects of the person, who is projected onto another chair, is expressed in the subsequent confrontation (usually inner child work). Otherwise, the ambivalent aspect may block the expression to the other. At the end of this process, both sides of the coin should once again be recognised together and the relationship appreciated as a whole.

(c) If the escaped victim has been inwardly hunted to death, it may, if it had adopted the perpetrator's perspective, leave behind the belief that he is a bad, worthless, untalented, incompetent, unattractive person who is not entitled to anything positive. Emotionally he feels almost numb, in a dark gloomy mood with a lack of drive. No resonance, or hardly any resonance, is felt in terms of interpersonal relationships. This state should not be reached in the first place.

Every psychiatrist poses questions to depressive patients about their morning low, namely whether they are able to get out of bed in the morning in order to determine whether the mental depression has already been somatised in the vegetative nervous system and blocked the daily activation of the sympathetic system, which is responsible for fight or flight responses. With the support of psychiatric drugs, the re-animation subsequently receives the opportunity to therapeutically respond to the smallest relational impulses towards the world and others and develop a positive spiral of resonance that allows the individual to once again increasingly experience self-efficacy and self-esteem. A therapist can successfully help in this regard if he is able to apply rules in the smallest of steps and, if he is capable of simultaneously cocooning the patient in a charitable, encouraging I-Thou field with appreciation for his existence.

(d) Conversion disorder typically first develops at the late stage of a child's pre-school years, a time at which the mind is also developing, and later we encounter it again in adolescent and adult versions. The parallel worlds of fantasies allow the mind to creatively depart from reality. They can also help us to tolerate and survive unbearable situations.

Neurophysiological correlation: In the case of histrionic paralysis or sensory loss, mature inhibitory cell populations appear to be capable of mobilising for extended periods. Psychosomatic symptoms of failure at this level are largely characterised by ideation, which usually deviates from the neurophysiological innervation units, e.g. a glove pattern of deafness.

Even if defence mechanisms inherently belong to the realm of psychoanalysis, they can also be understood as inner learning processes that are rewarded internally – and outwardly in social fields – by the system as a whole with a reduction in conflict tension and usually with a secured social standing for the patient. However, the defence process also constitutes a self-sabotaging, functional self-amputation of the mind, or rather, an intrapsychic mini-fragmentation undertaken for the sake of better alignment with social or individual values. To this extent, it should only be regarded as a sub-optimal, unconscious attempt to resolve conflict.

5.3.5 Mental Crime Scene: Partial Ban on Existence and Micro Emotional Numbing, the Onset of Symptoms from Traces of Unresolved Remnants from Tonic Immobility

Is there evidence of a stunted, chronified loss of balance in terms of microtraumatic impairment? This perhaps sounds like something one would come across in a mystery novel. Which it is, to a certain extent. In this case, we are the perpetrators and victims in addition to being the director of the tragedy. What are the perpetrators hoping to achieve?

Let's take this opportunity to establish at which levels we can find semi or completely fixed mental "material" following mental damage.

Overview

(a) (aa) Psychological experience, processing and behaviour,
(ab) Permanent internal and external psychological attitudes (personality disorders).
(ac) Lack of intrapsychic regulation (lack of comfort and risk of addiction).
(ad) Lack of intrapsychic structure

(b) (bb) Psychosomatic functional misregulation,
(bb) Psychosomatically fixed misadjustment of a regulation system.
(bc) Psychosomatically caused organ failure

(c) (ca) Social withdrawal in particular trigger situations.
(cb) General social withdrawal (fixed omega position, submission).
(cc) Socially over-compensating outsider modes (e.g. dissociality, rampages)

On (aa) Existence and non-existence at the mental level.

The focus in this regard is the recognition of dissonance patterns, the most striking parts of which have formed a conspicuous charged relationship with regard to the situation and/or the updated issue. The tension is the result of one aspect counterattacking another and potentially denying it the right to exist. Perhaps the simplest model in this instance is a conflict that arises due to the protagonist wanting to have or do something which is judged to be unacceptable by his internalised authority, e.g. aggressively defending himself against assaults. Initially, the protagonist does not convey a clear opinion of his own. As such, his wish is banished into his personal dungeon, where it leads the muted existence of a corpse. There is a chance that it will fade away entirely in the dungeon under the repressive ban. The aggressive gap in the behaviour of the person as a whole is noticed by everyone.

However, if the person as a whole is able to maintain minimal contact with the banished element, it could reveal itself as a secret accomplice of the leader in the dungeon, develop into a dangerous monster, become increasingly vengeful and adopt the common fight mode to fight introjection from the authoritative figure in order to force the expulsion thereof. A renewed, sympathetic attempt at contact could provide the monster with the opportunity to transform into a useful, supportive protective being.

It is important to clarify the motives behind the primary act of suppression in each case. Has it been adopted externally or delegated? Is it based on adopted prejudices or the patient's own experiences and convictions? Is it correct as a measure with regard to dose? Is it consistent with the system as a whole? Why has no Both-And solution been found thus far?

On (ab) In this case, a previously obtained survival strategy is rigidly generalised to a universal recipe for life, overhyped as a life-saver and worn by the patient as a protective shield. It may take the form whereby it seems beneficial to the patient to immediately insist on things without deviation, namely to be correct; or perhaps the form whereby it makes no sense to trust anyone except for animals at most; or that it seems clear that no person in the world voluntarily takes interest in others, wants to listen and certainly not grant them the limelight. As such, the patient feels required to take a stand and make a clear statement to the people, provided they haven't all already run away.

Reaching the shadowy remaining person hidden behind the overhyped shield, showing him respect and understanding for his barriers, and above all sparking curiosity and desire for his unexperienced sides is no easy task. The person is then able to return to interpersonal life once he is able to drop his shield from his range of behaviours, except for in extreme situations, and if the remaining person takes the chance to become a living protagonist. Prior to this point, a resilient, therapeutic relationship is required to give this patient solid, trustworthy "ground" under his feet and allow a continuous encounter to take place in her heart. Liberation from an internal, almost lifelong, prison of a personality disorder as witnessed above verges on miraculous. But it can occur.

On ac) We learn how to control our emotions with the help of empathetic caregivers when we are young. In the majority of cases, this is facilitated by our mothers, but others can also take on this role: fathers, grandmothers, nannies, neighbours etc. The goal is to incorporate, slight modify and calmingly restore childlike emotions and lovingly encouraging the child through communication. Some babies are not granted these experiences and opportunities to internalise emotional restraint. When learning to control emotions, a semi-independent observer is created within, who increasingly permits a certain level of inner dissociation and freedom from impulses and needs that arise. This facilitates the ability to remain in balance somewhat longer later in life, even when need fulfilment needs to be postponed for a while. This intrapsychic possibility to dissociate helps to cultivate and regulate the personality. People who are susceptible to addiction work hard to develop this ability to channel thoughts. We also possess exaggerated regulatory programs that train themselves to become insensitive to needs and strive to erase them. Sufferers of anorexia learn to ignore feelings of hunger. Others are intent on no longer having any sexual desires in order to avoid getting into trouble again, and certain people may want to feel nothing at all. There is a difference between consciously downregulating certain needs and slipping them into a deep sleep for good reasons and attempting to eliminate natural needs, usually unconsciously and out of fear. What does the basic need that has been denied the right to exist transform into? Sublimation only succeeds once in a while. Instead, we most often encounter substitutive gratifications, often in the form of perversions. A certain section of the clergy has provided a somewhat controversial example of this for centuries.

On (ad) The structurally weak part. This part is already established in the primarily insufficiently integrated, or also the (trauma-related) secondarily disintegrated structural part of the personality. A normal development goal at around 3 years of age consists of the combination of (1) consolidation between the internal and external boundaries and (2) the incorporation of the positive and negative aspects of experiences of both the child's mother (caregiver) and the child himself into the overall picture. Both of these development steps should be capable of withstanding an average level of frustration.

In stressful or suspicious situations, integration only occurs on a sub-par basis, if at all. Security, trust and a form of "ad-greddi" or purpose, namely a curiosity and desire to approach a given task, are all required to build integration. In order to determine the classification possibilities of an integrating field and grade it according to the respective situation, a decentralised observer is required above this field who is capable of introspection and motivated to check which intensity and emotional temperament appears to be suitable at present. These are all unconscious processes that rely on the green light of security to run as intended. Old, unprocessed experiences that oscillate in resonance may enter into the process and trigger bias. The latter is often experienced more markedly in structurally weak individuals than those who are capable of adjusting their emotions.

When the individual lacks an integrating centre developed from love and security, there are multiple opportunities for disintegration to seep in. The polarisation of the axis of self-esteem is regarded as the most developed and widespread form of structural weakness. Experts have agreed on the term narcissism to refer to this. The identity fluctuates between an inflated sense of importance and insignificance. We also encounter the poles of closeness and distance, love and hate, activity and passivity, self-care and self-destruction, manic happiness and over-excitedness or deep-depression to some extent etc. All of these aspects present simultaneously in the case of a borderline personality, which can be regarded as the archetype of a structurally weak, disintegrated being.

They can only be outwitted to a certain extent by multiple personalities, whereby several mental personalities that typically have no similarities and no interest in interacting with each other appear to reside in the same body. These personalities often have different intentions and come to the fore at different times in order to allow contact.

In this cursory overview, we have not yet touched upon the disintegrating consequences of traumatisation. When the usual process of recording memories in the hippocampus is overwhelmed by emotional flooding, the core area of the amygdala helps out. However, it only records fragments and is unable to subsequently provide any connections.

Initially, the aspect of the fragmentation dimension that becomes increasingly compact over time can be linked to common psychopathology. From there, the ability to dissociate leads to the subtle breaking off of contact and ultimately to fine-adjustment in contact behaviour.

At this pole, the fragmentation dimension turns into a useful, carefully separating tool, which is capable of being subservient to a higher order and serves coherence.

On (bb) Psychosomatic functional misregulation.

Misregulations caused by exaggerated innervation in the vegetative nervous system and also due to the interaction of muscular agonists and antagonists are commonly observed.

The latter can often result in painful tension in the musculoskeletal system, shoulder-arm syndrome, sacroiliac pain and also herniated discs.

The former mobilises acute alarm responses such as hypertensive crises, skipped heartbeats, fleeting erythema, hypotyalism, globus sensation, temporary tinnitus, cold sweats, imperative bowel movements etc. These acute alarm responses are generally recognised as psychosomatic and, despite the fact that they are uncomfortable to experience in the moment, understood as psychosomatic secondary reactions. As such, they are not classified as life threatening.

It is simplest to determine which relationship constellation triggered the activation of the fight response but did not permit it to be put into effect through association at this stage. Which impulse blocked the activated fight response and displaced it to the psychosomatic level and with which tone is it heard within? "You don't stand a chance; it is stronger than you."? "If you fight back, you will only make things worse."? "If you protest, you'll lose the only person who still supports you to some extent."? "Disobedience is dangerous."? "Protest is a sin."? "Keeping your mouth shut will help you keep your job."? "As long as you live under my roof, you'll do as I say."? etc. This opens the door to a dialogical scenario for relationship clarification, potentially as a variation on inner-child work. In an ideal scenario, the blocked counter impulse can still be located and somehow expressed, as integratively as possible. The use of solely verbal cues is often not sufficient, but can achieve results, provided the response truly comes from within.

As can perhaps be expected, chronified patterns as described in the following sections are not as easy to address causatively. They tend to require reconciliation with their bleak struggle for survival in order to reduce inner tension.

On (bb) Psychosomatically fixed misadjustment of a regulation system. The chronic misalignment of a regulation system, e.g. blood pressure levels, is problematic for the individual's health. This leads to the thickening of the blood vessel walls, which may in turn suffer from a lack of oxygen supply. When adjusting to chronified alarm situations, activating systems may also reset their target values for blood sugar and fat levels and no longer be capable of sinking to a healthy resting value. All hormone releases can also be blocked or overactivated and the activity of the immune system may be brought to a state of acute or chronic deflection.

On (bc) Psychosomatically caused organ failure. Perhaps the most well-known example of this is the long-term damage from the aforementioned chronified alarm situations inflicted on the blood vessels that start to form arteriosclerotic plaque, which can lead to inadequate circulation and ultimately infarcts and embolisms.

These chains of damage are also observed in other organ systems.

On (c) (ca) Social withdrawal in particular trigger situations. Here, we primarily encounter frustration from a lost power struggle, from lost competitions that are of great significance in the animal kingdom. The primary fight ends in retreat, which is usually bitter and shameful. What can be done? In psychotherapy, we are able to change the reference system, put it into perspective, question it and suggest confrontation in a new framework with healing potential. Often, we are required to reverse fixed rules that have been rendered absolute. How can defeat be integrated whilst allowing the individual to nevertheless continue to exist with dignity in the same society?

In certain cases, it can also be a matter of the patient reintegrating aspects from his own devaluing issues projected onto rivals and social groups from which he feels rejected.

On (cb) General social withdrawal (fixed omega position, submission).

In this case, it is conducive to determine the biographic, potentially epigenetic (fates of migrants or refugees) or traumatic reason behind the onset of this fixation on submission, the margin or desire for change regarded as possible by the patient in the case that he is able to overcome the fantasised dividing line and achieve emotional resonance, and to what extent the former identity is able to permit or incorporate social acceptance under different rules.

On (cc) Socially over-compensating outsider modes (e.g. dissociality, rampages) In the case of rampages, the perpetrator's profile usually refers to narcissistic wounds that had led to his expulsion from the reference group. In other contexts, it may be the case of the rebellion of aggressively pent-up experiences of rejection by minorities.

5.3.6 Freezing of Defence Mechanisms: Personality Disorders or Over-Identification with Defence or Survival Strategies

Note

Over-identification with solidified, excessive survival strategies and protective measures can lead to disorders that were previously referred to as character disorders and now as personality disorders.

In this case, former protective measures have turned into an issue themselves as a result of their consolidation and overloading with meaning. The individual no longer has the sufficient flexibility to look at the situation, the current problem person or himself from another perspective or to correct the primary, schematically fixed understanding of the situation. In the case of different opinions, the original threat is unconsciously triggered and deep-rooted defence mechanisms are remobilised en bloc to an excessive extent like a routine for sure-fire success. This freezing resembles mental tonic immobility. The excluded part of the personality is analogously buried in the background. There is no longer any subtle oscillation between the foreground and background of the mental structure as usual. It is hard to reach any other parts of the self apart from the specific, overemphasised formation. Yet this is precisely the point of entry into the system that enables other perspectives to resonate.

It is worth trying to reach the identity of the (typically) child who has previously encountered the particular situation that led to the onset of the disorder but remained frozen in his emotional response. It is most likely that emotions from this time will be encountered that have since been suppressed. They need to be brought back to life with positive connotations. Then the overpowering and overvalued defence mechanism needs to be differentiated into a useful and helpful part that fits the individual's personality and into an overstated part that is controlled by fear and others, which needs to be left behind.

▶ Example

Anja Müller, 36 years old, excerpt from the fifth session. D: Obsessive-compulsive personality disorder.

Therapist "I'm currently trying to imagine you as a small girl, as you were in kindergarten. At this point, it wasn't yet the case that everything had to be as meticulously in order as your mother wanted. You were still able to make spontaneous decisions, elect to play with something other than what someone else selected for you. If I remember correctly?"

Patient – "Yes, exactly"

Therapist – "You sounded a little pensive then, as though there were other feelings involved. Can you transport me back into the world of the young girl a little to help me understand what she was really feeling? If I have understood correctly, it seems that outwardly she wanted to please everyone. But perhaps things were a little different on the inside?"

Patient – "I can't really remember any more, but sometimes I had the urge to throw all of my toys into the corner. Once I even broke one of my doll's arms by doing this. That scared me and made me feel bad. My mother even called me a bad child".

Therapist – "I'm assuming the girl had usually experienced something bad before she vented her anger?"

Patient – "Not always. Well, something was always wrong. Not always, but often. I don't know what was going on back then. One time, I really wanted to wear my new Sunday dress. It had a wide, bright blue organza skirt that bobbed up and down when I walked. It looked like a cloud. I wanted to pretend I was a cloud. It was my dearest wish. But it wasn't a Sunday, so my mother said no and shut the doors to my wardrobe. I lost my temper and started screaming and stamping towards her. She responded to me saying that bad children don't deserve such pretty dresses. She said she wanted to return it to the store. "Nelly, your cousin, behaves far better than you and is much kinder than you. Maybe I'll give it to her". This made me scream even more desperately at the idea of something that I couldn't let happen. My mother took the dress, left the room, locked the door from the outside and only unlocked it again in the evening. I was completely exhausted and fed up with everything, especially my mother. I did everything she asked listlessly: Eating dinner, brushing my teeth, going to bed. At night I was visited by powerful spirits who took things from me and judged me. Sometimes I screamed in my dream, which waked me. My mother told me to stop screaming so loudly as it was too noisy and may wake the neighbours. I've never forgotten about that cloud dress. I couldn't really say whether its linked to the doll at all. It fits emotionally.

Therapist – "It feels very hard to put myself in the girl's shoes. I could burst into tears right now. Was no one there to comfort her and stand by her?" (holds a pillow in her arms like a small child and strokes and rocks it lovingly, which first makes the patient look at her longingly).

Patient in a stern voice – "Big kids don't cry! They are level-headed".

Therapist (spontaneously taking ov er the role of the girl) – "But I'm still small".

Patient – "No, your brother is still small. He's still in nappies. Anja is already five. She needs to grow up".

Therapist – "I think I just heard to voice of your mother from back then. But what does Anja have to say? What does she want?"

Patient – "Is she allowed to want anything?" (looks into the distance with somewhat glassy eyes) "wanting something yourself is like emerging from a dark corner and discovering the green grass on the other side...but is that necessarily good? Probably not..." The patient once again averts her gaze to the pillow in the therapist's arm.

Therapist – "If we both wanted to take care of the little girl, it would be much more likely to succeed than back then when no one was truly on her side. Symbolically, it would be like both of us rocking this pillow, as long as the little girl is ok with that. Try to carefully listen to what she wants to catch up on. Being loved is one of these things. Pretending to be a cloud is too. Running in the green grass another. Perhaps she wants to dismiss someone from then or now. Something dear to her doesn't need to be broken just because it's nearby. I thought it was great that she wanted to fight back. It even made me happy. This side is also necessary in life and valuable.

Patient (looks somewhat sceptical at first, then lights up and takes a deep breath.) – "Introspection? How does that work, like in this case? Going into green fields? Dancing like a cloud?"

Therapist – "I just had an idea: maybe you could show the young Anja where the green fields and dancing clouds are in your current life during next week's session. You could also show her when "no" is the right thing to say and that it doesn't necessarily have to break up a relationship, but rather can lead to clarity and help a relationship gain ground and stability".

Next week, Anja turned up to her session wearing a long, breezy light-blue silk scarf and a happy smile. ◄

The key aspect in this process is the positive acceptance of basic needs that are concealed by the overemphasised concept of the personality disorder – in this case, existential and loving affirmation of the person, her spontaneity including aggressiveness and self-determination.

5

> **Note**
>
> In addition to the patient's overall personality, the therapeutic alliance particularly applies to the suppressed part that has fallen by the wayside.
>
> In the case of narcissistic personality, this refers to the devalued, unloved child ego; in the case of paranoid personality disorder, the child ego that is smothered in constant and secure love; in schizoid personality disorder, the child-ego that can be reached and responded to with sensitive resonance; in dissocial personality disorder, the child ego that acts out with loaded demands as a substitute for its frustration about its unsatisfied desire to be loved, which requires a strong con for this path but nevertheless requires a pro from the same person for its innermost longing for acceptance. Emotionally unstable and borderline personality disorders require assistance in developing their centre between the pole of emotions, the balancing act of closeness/distance and the axis of self-esteem, to identify with this and subsequently receive affirmation as a result. Anxious-evasive personality disorder requires a long-term therapeutic experience that provides stability and security as well as the processing of frustrations to combat the insufficiently coherent initial care provided by the individual's parents in combination with a self-parenting concept. In the case of dependent personality disorder, the child ego has buried tendencies to be independent, including aggression, that it would be conducive to re-experience and integrate to a certain extent under layers of over-adjustment. Histrionic personality disorder is characterised by excessive attention-seeking, which is often the cause of envy: in reality, a child that received inadequate attention or has been disowned is concealed behind the often amusingly dazzling and entertaining layer, for whom the applause from the current pantomime act fails to offer a true replacement for his long-pervading inner void, but instead causes the emergence of painful inner tension.

This section shall conclude with another example that looks at mixed narcissistic and paranoid personality disorder with a lack of impulse control.

Jan, 28 years old, has experienced an aggressive lack of impulse control since going through a difficult stage in life. Small frustrations can be enough to set him off. Sometimes they appear to crop up almost out of the blue; sometimes those around him are surprised by how quickly Jan (as an adult) is able to go about his daily life with calm composure and even radiate engaging friendliness despite the unfathomable intensity of his outburst, whilst those around him still need time to recover from the tsunami of negative emotions and get their emotions back in check. The result that Jan clearly revels in is the large amount of attention that he receives in these situations and commanding respect as someone who is capable of making himself heard and does not tolerate any answering back. As a result, he becomes the boss of the situation. It appears to be emotionally rewarding for him. He barely registers the fact that private, friendly relationships that would otherwise occur on equal ground are withering away or no longer being established in the first place. Sometimes he wonders why none of his friendships last in comparison to how most other people experience friendships and he feels alone and isolated. He has previously been fired from jobs due to certain work relationships and atmospheres or voluntarily quit due to atmospheric disturbances. He is proud of the fact that it's better not to mess with him but instead steer clear of him. It appears that his is incapable of properly integrating. It is inevitable that he feels hostility between himself and the world for the most part. He is not willing to acknowledge the part he has played in this development. He is not capable of spontaneously switching roles and looking at himself from how the world sees him. Everyone else is wrong and has something against him. Paranoia creeps in. It's best for him to be independent and in charge. The high turnover rate amongst his employees fails to prompt him to consider whether there may be something about him that could be worth changing as it clearly affects people. He does not regard the latter as an adequate reason to change, similarly to when his wife abandons him and he loses the support of his family. Crisis can no longer be averted. Jan struggles to accept the downsides of his survival concept, a strategy that appears to offer three significant advantages: a release of aggression, attention and a narcissistic boost in self-esteem. Everything he needed as a child. He did not receive enough emotional security. He did not have a loving mother who was constantly there to help him control his emotions. His mother was preoccupied with her constant marital disputes and career. The young Jan found a substitute in his grandparents, who lived in the same household and carried a peaceful atmosphere wherever they went. They provided him with a sense of kindness, simplicity, humanity, honesty and sensitivity. Out of his need for this, Jan is able to establish that he was full of anger at the time and didn't receive enough attention from his mother, which was what he wanted, in addition to her appreciation. This also applied to his father to a slightly lesser extent. Jan later expresses this to each of his parents individually inside. He allows himself to cry for the first time. During the course of his life, Jan has managed to keep his needy, miserable and softer sides well hidden from himself and the world and showed utter contempt for them. He has since tentatively learned how to show and receive loving appreciation, to experience life on equal terms without being required to sink into a corner or appear larger than life. He has discovered a form of loving relationship whereby spontaneous role reversal is a natural prerequisite for reaching an understanding. Initially, the healing relationship axis for therapy is an appreciative alliance with the frustrated infant ego, which is in need of love and understanding. This is joined by the individual's understanding of his choice to hide behind the mask of an intimidating god of thunder in his life thus far and the encouragement to become himself and establish relationships as his true self, without his mask. Parts of himself that were previously subject to hostility instead turn into treasure troves. The happiness that sensitivity can bring to relationships is truly impressive!

The reconstruction of the person, particularly in the case of personality disorders, needs to start in the protective space of a healing therapeutic relationship, where the course of the pathology was set.

Frames of reference: Interlocking of two principles on the understanding of health and disease pathogenesis:

1. The holistic principle of the development and growth-promoting, empathetic and also age-appropriate limiting care approach in terms of unconditional love, which establishes structural centring and emotional fine-adjustment as well as

resilience, the ability to experience forming nuanced, coherent relationships and to develop a holistic ability to love.
2. A lack thereof manifests itself in relatively elusive developmental delays, insecurities and diffuse fears.
3. The disorder-specific principle is based on risk avoidance, trauma, outlined deficiencies or conflicts. The time at which the damage occurred within the scope of development is significant in this regard and sculpts the strategy for defence and survival.

Digression: we possess phylogenetically and biologically embedded templates to avert danger both externally and internally; a certain aspect is also denied the right to exist in the case of conflicts. *Carried genetically by humans.

5.4 The Diagnostic, Therapeutic Reversal of Symptoms

5.4.1 General Observations

The diagnostic process for the reversal of pathological patterns on a trial basis always contains certain therapeutic aspects.

Note

The symptom is the most direct path to the location of the event. Principle: the ego-dystonic, emotionally paralysed symptoms are associatively traced back to the previously unresolved focal point of the living relationship by the patient. The therapist's emotional support enables the patient to endure this and discover new ways to solve the underlying issue. An intrapsychic change in perspective helps the therapist to assist the patient in discovering the hidden meaning in the banished impulse, which wants to be integrated and can be used as a resource!

The patient first needs to be familiarised with the reversal process by the therapist. Success relies on both of them and their relationship. The patient's stuttering and pauses can be used by the therapist to obtain deeper insight and convey appreciation. Resistance provides a healing force. The patient should always be allowed to progress at his own pace and defend himself. Who is he defending himself against within? (If required, it should be ensured that negative transference is supplanted early enough in order to be able to help the patient!)

5.4.2 Specific Observations – Causes, Timeline, Development Profile and Coping Strategies

It is important to distinguish between the following in the reversal process: is the patient suffering from

- A former, recurring or subsequent lack of empathy?

- Damage as a result of trauma: current damage or has the event become chronic? Potentially with compensatory fantasies and survival mechanisms?
- Damage as a result of conflict or a history of mislearning: predominantly current or chronic conflict-related compromises, possibly with advantages from making a compromise? Possibly with morbid gain?
- Damage as a result of an overtly protective and defensive attitude: solidified defence?
- Structural weaknesses as the cause of damage. Lack of integration and order for processing experiences and regulation?

5.4.3 Practical Examples: Dialogues Between Opposing Parts

Note

An event diagnosed as a dissonance pattern typically features two opposing parts and a contentually defined impulse that creates distance.

Example for the diagnosis of anxiety:

The symptom is repetitive nightmares in the case of an anxiety disorder.

Consensus: the dreamer is reflected in all details of the dream. (Subject level).

Version (a) dream: ...A scary monster is after me...

Version (b) dream: ...A jury has found me guilty...

Version (c) dream: ...I'm falling into a bottomless abyss...

On version (a): Ego-syntonic aspect: I'm being chased, I'm the victim. Ego-dystonic aspect: I'm doing the chasing, I'm the perpetrator. As the chaser, I feel wild rage, strength, entitlement and even joy at threatening others and seeing their weakness. Dissociation aspect: You are not allowed to be (openly) aggressive and (crudely) mean. Reality check (with an accepting therapeutic relationship!): I am both aspects. Solution: learning how to express a genuine "yes" and "no", albeit with fine-adjustment in relationships.

On version (b): Ego-syntonic aspect: I'm flawed and inadequate. Ego-dystonic aspect: I have the power to put down and judge others, which I even enjoy. Dissociation aspect: You do not have the right to put yourself above others and be arrogant and overbearing. Reality check: I judge and put down others secretly to feel better for a while. Solution: (1) Loving and supportive recognition of the small steps taken by the patient that succeed; (2) the patient's performance only needs to be good enough, which also applies to others. (3) Discovering the nurturing resonance of friendly acceptance on equal terms.

On version (c): Ego-syntonic aspect: I need ground and stability – both externally and internally. Ego-dystonic aspect: I am the emptiness, lack (of acceptance) and the void. I am the awareness of what is missing – and therefore also simultaneously the indication of what would have been needed for good development (via the opposite pole within and its fantasy desires): loving, reliable attention. Dissociation aspect: I no longer want to have anything more to do with the previous times of need. They make me weak. Reality check: Banishing shortcomings from my consciousness will only make me weaker. Solution: The shortcoming taught me to value what I lacked. I elect to live and even cultivate what I would have needed with regard to myself and other people and enjoy the resulting resonance. In the meantime, I have managed to give and love, despite the fact that I didn't receive it back.

Example for the diagnosis of obsessive-compulsive disorders:

The symptom of obsessive thoughts

► Example

5

"I get scared at the sight of sharp objects such as knives or scissors as I'm worried that I could do something bad to my young daughter or perhaps even want to do something bad to her. When the window is open, I think about throwing her out of it. It completely overwhelms me. I can't handle it anymore".

Ego-syntonic aspect: Shock at the possibility of committing an atrocity – accompanied by the mobilisation of loving, protective impulses and preventative (self-)condemnation of the dark scenario played out by her unconscious. Ego-dystonic crime scene as an associated daydream: Someone kills someone else, which doesn't only feel gruesome, but also satisfying. Identification with the perpetrator's incentives: "I can't stand it anymore, it's too much. I'm not playing along anymore. I also want live from time to time!" In the role of the murdered person: "I didn't do anything to you! You're crazy! Everything had been resolved". A formerly suppressed aspect rebels against the continued existence of the previous, decades-long interaction.

This crime-scene scenario corresponds to the previously neglected rebellion within an over-accentuated, obsessional basic structure, the problematic nature of which is currently resonating through the burden of intensive infant care.

Solution: Dialogue between the victim (=set of rules) and the rebellious perpetrator (own basic needs) leading to the establishment of a new way of living: specifically, it provides the patient with self-determined sanctuaries of freedom and regular relief when caring for the baby. The rebellious killer within transforms into the patient's life-enhancing advocate, helping her to find her way out of her inner tension and back to her ability to love. Following this process, the symptom disappears. ◄

Example for the diagnosis of depression.

Symptom: lack of motivation, lack of self-confidence and low self-esteem.

Version (a) with over-adaptation and self-devaluation tendencies.

Version (b) with high moral and performance requirements placed on oneself.

Version (c) with particularly fragile self-confidence.

On (a) Ego-syntonic aspect: Taking pleasure in harmony, participating in the greatness of others and in their positive resonance/gratitude. Ego-dystonic aspect: Putting self-will and one's own aspirations of greatness into practice. At best, defined as self-sacrifice, self-denial and great pain. Dissociation motive: Avoidance of aggressive disputes, competition and conflicts related to dissonant, internal desires that could destroy harmony.

On (b) Ego-syntonic aspect: being honest and willing to help, reliably useful and constructive for the reference group. Ego-dystonic aspect: Self-indulgence, addictive tendencies, crossing the line. Dissociation motive: Avoiding any violations of unwritten rules of the reference group in order to avoid being kicked out of the community as a result.

On (c) Ego-syntonic aspect: "I'm handicapped". The patient does not believe that they entered the world as a precious, amiable, special and positive being that is regarded as such in this world and capable of developing into an able being. Everything positive inherently appears to be ego-dystonic and undeserved. Dissociation motive: Negativity is familiar, the patient is not equipped to handle the other pole, even if it triggers longing. It feels foreign at first.

The diagnostic process for the reversal of the pathological pattern on a trial basis always contains certain therapeutic aspects.

Gestalt therapy provides perhaps the simplest and most direct way to reverse a symptom into the original field of tension by providing the patient with guidance on identification techniques after the establishment of a stable working relationship. Dependable teamwork between the therapist and patient is also required in this regard. The symptom is the most direct path to the location of the event. Principle: the ego-dystonic, emotionally paralysed symptoms are associatively traced back to the previously unresolved focal point of the living relationship by the patient. The therapist's emotional support enables the patient to endure this and discover new ways to solve the underlying issue. An intrapsychic change in perspective helps the therapist to assist the patient in discovering the hidden meaning in the banished impulse, which wants to be integrated and can be used as a resource.

The patient first needs to be familiarised with the reversal process by the therapist. Success relies on both of them and their relationship. The patient's stuttering and pauses can be used by the therapist to obtain deeper insight and convey appreciation. Resistance provides a healing force. The patient should always be allowed to progress at his own pace and defend himself. Who is he defending himself against within? (it should be ensured that negative transference is supplanted in order to be able to help the patient!)

It is important to distinguish between the following in the reversal process: is the patient suffering from

(a) Predominately current events, or has the event become chronified? Potentially with chains of repetition.
(b) Does the patient primarily identify with a defensive formation? Or is there still an element with an authentic core, even if it has been damaged?
(c) Is there not a sufficiently enduring inner structure (anymore), but instead only changing regulation by rotating impulses? (Therapy outline: Relearning a readjusting attitude in terms of self-parenting, partially including observational learning in therapy with the setting of boundaries and basic acceptance. Support with experiential learning in terms of corrective experiences in groups with special activities, in everyday life or related to the behaviours within the therapeutic relationship (pay attention to micro-changes and use them to reflect!)

A diagnosed pattern typically features two opposing parts and a contentually defined impulse that creates distance.

5.4.4 Special Diagnostics: Recognising Patterns of Dissonance as Excluded Gestalt

Around a century ago, the perceptive ability of humans was the subject of much interest in research conducted by Gestalt psychologists at numerous universities around the globe. The special holistic ability of the human brain to recognise patterns that are not only found in the visual realm but also in all sensory channels remains a fascinating concept for neurophysiologists today such as Eric Kandel, Gerhard Roth et al., who enjoy using vivid examples for this complex structural form of cerebral pattern recognition within the scope of high-level information processing, which is typically controlled by the right-hand side of the brain, in their lectures.

A pattern (a Gestalt, to use the technical terminology from Gestalt psychology) is an individual entity that clearly stands out from its background as a whole and possesses its own special, internal relational structure that is not related to the structure of the background field This form of relationship structure is transposable, similarly to how a melody can be transposed into another key.

The growth of malignant tumour cells is often used to exemplify how the human body can no longer be reverse controlled by its environment through pathologic changes to its receptors. As a result, the tumour cell no longer acts in harmony with the organism as a whole, falls out of its oscillation patterns and consequently separates and excludes itself from the body as its own entity. Patterns of dissonance are usually perceived by the organism's environment as an irritating or disruptive foreign body with a relatively independent existence in relation to the organism as a whole. If we transpose this concept to the immune system, any entities identified as foreign bodies are fought and eliminated.

The term "dissonant patterns" used here should be interpreted as comparable. It denotes a psychological or neurophysiological unit of information or excitation that cannot (any longer) be integrated into the mental state and oscillating pattern of its environment as a result of conflicting forces.

Functional processes across a number of levels can cause symptoms of illness to stand out as disorders and consequently as not normal, non-compliant, not age appropriate or inadequate in terms of personality. The applied reference system plays a crucial role for this assessment. Behaviour that may be regarded as acceptable amongst peers may be perceived as rough aggression in another social setting and is often intended as such. Conversely, familiar and intimate greetings and behaviour towards business or political partners can often come across entirely out of place.

► Example I: Instinctive, Tic-like Micro-movements

A young woman throws her head to the right and looks upwards, briefly closing her eyes, whilst simultaneously shrugging her right shoulder. The process appears to indicate abrupt aversion. This symptom often occurs in conversations on the topic of self-value, self-discipline, independence, denied needs, self-encouragement, bitter pride and loneliness.

Generally speaking: both the patient and therapist can benefit from repeating the non-verbal subtle statement several times and experimenting with it, e.g. by slowing it down or experiencing it with exaggerated expression and listening to which impulses, feelings and scenarios it evokes internally. When which principle or person arises as an inner counterpart does it appear to "click"?

We shall now return to the example taken in long form from the minutes of a group session as an excerpt from individual work in a group setting: The micro-movement symptom "I don't need anything" is a consolidation of an initially helpful survival mechanism. During childhood, the forced autonomy aspect excludes and kills the aspects of needs and desires to receive care and belong. It is key that the therapist does not take away anything stabilising, even if it has previously been a crutch or if the therapist is unable to simultaneously foster the development or unblocking of the corresponding potentials. The great need and longing to finally receive attention at long last emerges from "I don't need anything". Until the deepest, most dissatisfied basic need has been reached, therapy cannot be successfully concluded.

Example of individual work from a psychotherapeutic group in a weekend block setting. The group has already been meeting for over half a year. They have just discussed what has happened in the meantime and how they are all doing. Marianne is now in the spotlight.

Participant Udo – "Marianne, what's wrong? You said earlier that you had a headache. Is it bad? Can we help at all?"
Marianne (throwing her head back, smiling a little tensely, says in a reserved, brittle voice): "Oh no, everything is fine, I don't need anything" (concern in the group).

Group member – "Don't be silly! You can tell a mile off that something is wrong!"

Therapist – "Look within yourself, Marianne, and decide whether you want to take some time and space for yourself or not. The group would be happy to give you this if you need it, as you can see. If you would like to say no, establish what this no represents. Apart from that you know the rules of the game: Your decision will be respected".

Marianne – "I'm a little confused right now because I don't know what I want".

Therapist – "Perhaps it would be best to start by looking at this confusion. How about that?"

Marianne – "Ok"

Therapist – "Which feelings or impulses are you able to comprehend at the moment? What is at the fore right now?"

M – "Actually the no" (she says throwing her head back with almost annoyed pride.) "It's now kind of like: I don't need anything, it's just to make the others feel better!" (her brittle, now almost sad voice is once again apparent).

Therapist – "Is that all ok with you? "Needing nothing" seems to be important to you. You've already said it once".

M – "Exactly" (now in a more interested, lively voice) "Needing nothing is the key. It always makes you feel good – ok maybe not always, but usually. It leads to independence" (her voice now sounds firmer and more confident).

Therapist – "Would you like to look at this in more detail?"

M – "Maybe, now you mention it" (dwelling on it a little).

Therapist (asking with a smile) – "Are you sure? Have a look at your decision. Only a second ago you felt completely differently. (Marianne takes a deep breath and nods) I think this may have a very serious and meaningful message behind it (Marianne glances up with interest)

M – "Ok, how do we begin?"

Therapist – "We have actually already being. There is a being that embodies this "I don't need anything" attitude. Take over its role and see what occurs to you. Pay attention to which posture and voice it likes to use to communicate. If a counterpart suddenly appears in your mind's eye, that's also fine!"

(Marianne stands up, looks around and appears to find the situation not so bad): "Look guys, I don't need anything from you, which I've already made clear. I don't need any extra time, interest or attention (she stumbles, gulps, takes a deep breath and continues). No, I don't need anything from you (and looks slightly glazed into the distance)".

Therapist – "Is there a counterpart?"

M – "Yes, my father. He's sleeping on the table and spilled beer all over it in the process. My mother, who is annoyed anyway get upset about this and my small disabled sister, who makes strange noises in situations like this".

Therapist – "What does that trigger inside you?"

M – "Nothing but they want to get away. There's nothing to be gained here. But also: can I abandon them? Above all my mother. She would also like to leave. But what would the others do then? My mother needs me. I'm her "sunshine", as she always likes to tell me when she wants something (Marianne stumbles again and looks down)".

Therapist (relatively softly) – "There appears to be another Marianne deep down... (Marianne node) Can you allow her to join us and give her a voice? Perhaps she would first feel most comfortable talking from behind the protection of your back? Why not try to establish where her place could be?"

Marianne retreats back into her own background and listens inwards. She acts shy like a little girl with a lowered gaze. "I don't have much to say about this. My parents aren't normal like others. And if my mother leaves, everything will collapse and become worse. (Although her voice sounds feeble, her right foot makes a clear stomp).

Therapist – "I think your right foot has something important to say".

Now Marianne stomps strongly and starts to cry: "Why can't I get my own way for once? Not just you! You are awful parents! I want things too from time to time!" (She sobs and collapses into a small heap on her chair. Her hands hide her face.)

Therapist – "That must have been hard for you. Take all the time you need for yourself. If you would like and be ok with it, you could perhaps reach out to those in the group who are able to understand the feeling of being left alone and overwhelmed and learn what will help now in this situation from them".

Several members of the group have been sitting for a while looking as if they want to show Marianne solidarity. Someone grabs a blanket and places it over Marianne's shoulders after using eye contact to make sure that it would be a good idea. Two women carefully place their arms around her back, causing Marianne's childlike sobbing to once again intensify. An older group member slips his fatherly paw into her hand and gently presses it. Her hand initially responds with counter-pressure, followed by a hesitant touch of the back of his hand.

Marianne – "I wish I had a father like you. A father who cares about me and yet never needs anything from me and wants to be taken care of. And the mothers holding me are truly lovely. They can continue doing so until this evening as far as I'm concerned!" (Marianne starts to look a little more relaxed)

Group member – "What else do you like?

Marianne responds without taking a moment to contemplate – "Being rocked! I haven't had that in ages!" and is almost shocked by her own absurd idea.

Group member – "Why not!", mutters an otherwise reserved man, grabbing a blanket and waving over three other members, and before Marianne is able to deal with her qualms, she finds herself sitting on a swinging blanket in childlike delight with most of the group members rocking along in solidarity. For a couple of minutes, the atmosphere is just as high-spirited as at a children's playground.

Therapist – "It's great how resourceful you all are. Marianne, how are you doing?"

M – "I'm doing great. Completely taken aback by how everything has gone".

Therapist – "What did you do to allow things to take this turn? What did you do differently to normal?"

Marianne – "Me?"

Therapist – "Yes, you!"

Marianne's face transforms into a puzzled expression – "But everyone else came up with the ideas".

Therapist – "That's one side of the coin".

Someone from the group – "Marianne, you let us know something about your desires, you revealed yourself and reacted to us. You were spontaneous and not so led by your head and trying to act grown-up. It immediately made you seem more natural".

Marianne – "That's lovely of you to say, but I have mixed feelings and am worried about returning into my old rut when I'm no longer surrounded by the group".

Therapist – "Could I may a little suggestion to finish up? A small experiment, perhaps even an adventure?"

M – "Sure, what do you suggest?"

Therapist – "Remember how we started half an hour ago with your "I don't need anything" side facing your former family and then, with lowered eyes, the girl who got a raw deal came carefully out of the shadows. Can you remember whether they both wanted anything from each other? I can't. Now it has become clear – as everyone here has witnessed – that something else is hidden inside you: joie de vivre. Could you imagine both of them meeting and getting to know each other here? You could be both of them in turn (Marianne nods.) See which spatial distance, which is also symbolic, seems appropriate as a starting position for this conversation between them. Do they speak to each other back-to-back, from the side or face to face? What is the right amount of distance between them? Which voice, which posture? Start with the side that spontaneously comes to mind".

M – "I want to be the lively girl that I just discovered in the group, or perhaps have rediscovered after a long time. I'm facing the "no-needs" woman. There's a relatively big distance between us. That's just how it is. It's not exactly pleasant. We look each other up and down from the corners of our eyes. Hey, you over there. I don't enjoy you being here. You look cold, dismissive and unkind. I know that you're committed to a lot of things and that you take on a lot of responsibility, but you look unhappy, poisoned by bitterness and arrogance. You act superior, but you're not. If I give it to you, I worry that you'll tower over me".

Therapist – "So you would like to fight her? An either-or scenario? Or could you imagine taking on the other side from time to time? You seem to be hesitant? Listen to what says no first within. That's important".

M – "I don't want to lose contact with this newly discovered, lively source. That's one aspect. I don't want to continue existing in the phoney shroud of the "no-needs" woman. Now that I've put it into words, I can give it a go. (Moves slowly) I'm now facing the "no-needs" side. My feelings are confusing. Your existence in this way is new to me. I am not used to acknowledging you. I don't like how you talk down to me. You don't know how much strength it took for me to psych myself up. I could have just succumbed to resignation. You should be grateful that I kept the family together. It was demanding for me!" – Therapist: "Time to switch roles".

Girl Marianne – "Don't play the martyr! You've profited well from your prominent role. You put everyone else in their place and ensured that they didn't manage to leave it. However, I can agree that it wasn't fun. But did you ever wonder how I was doing in my shadowy existence? It wasn't just simply not fun: I was often desperate or, worse still, I didn't sense my own existence at all for extended periods. I was as good as dead. You deprived me of life. I put the final remnants of my vitality into your headaches and irritability. That simultaneously constituted my final hope and revenge. But what will you do with me if I won't allow myself to be suppressed any longer?" – Therapist: "Time to switch roles".

No-needs Marianne – "Suppression – do I want to do that? That's what you're implying. You're reaching out to me, which no one has done before. Afterwards I'll be able to find myself. However, I don't want to be there for your needs. I'm completely against that, as you can probably understand". Therapist: "Time to switch"

Girl Marianne – "Yes and no. I want you to see me in my true colours and don't want to be confused with your parents. I want to ask you to make this effort".

Spontaneous role switch. No-needs Marianne – "You are really unique. You've vibrant, like a breath of fresh air. Can I offer you anything that may be useful to you or perhaps even make you happy?" Spontaneous role switch.

Girl Marianne smiles "Your friendship" (imaginary hug)This is followed by lively feedback from the group on Marianne's impressive transformation with subsequent responses from those with similar experiences (sharing) and a tangible atmosphere that Marianne was able to sense of being understood by the group and clearly belonging to it, which was less pronounced previously. ◄

► Example II: Stuttering

Lisa, 7 years old, stutters when reading aloud at school and has a fear of failure. When she was barely 3 years old, her beloved nanny left to go and get married. Lisa refused to speak for several weeks, hiding in her play corner with a teddy bear. If she spoke at all, she only spoke in a baby-like manner. The previous loving and supportive relationship with her nanny abruptly died for her. Otherwise, she had tantrums, banging against doors, spitting at the wall, chewing through pencils and destroying toys. When she was left alone, Lisa would rock back and forth throughout the day (hospitalism) and wake up from nightmares at night. She only started to speak again once her next nanny had settled in and become familiar with her. Lisa usually found her mother to be cold, overbearing and intimidating. During a trial therapy session, Lisa found a small crocodile in the sand pit, which scared her at first, but she then curiously engaged with it. When asked what the small crocodile would like to do the most, Lisa responded that it would like to chase away its mother and nag her until she asks the crocodile for mercy. Lisa then acted this scenario out with the crocodile toys. When doing so, Lisa sang a melody that she had heard her new nanny singing to herself. She explained that she regards her new nanny as a new, secret mum. She has already kissed her on the cheek and told her that she loves her. Lisa's voice becomes more fluent once more, causing her stuttering to end.

In hindsight, Lisa attributes her earlier stuttering to her mother: I hate you, oppose you and want to be able to hide from you out of fear.

This small example illustrates the interplay of the empathy field on the one hand and the attempt to cope with frustration through fight and flight responses on the other. The latter is also reflected in the inverse innervation pattern of the speech muscles. ◄

► Example III: Erythrophobia

5

Johann (16 years old) turns bright red in the face in response to every trivial event. He is often teased at school as a result. Every time he is required to take a test at school, he now thinks that everyone looks at him as though they think he will cheat. Now he has no option but to learn his stuff. It started to turn into a real issue around 2 years ago. His neighbour Grete (15 years old) was sunbathing in her garden wearing minimal clothing. He was tempted into throwing a small stone over the garden fence onto her stomach and then ducking under the hedge. It was only meant to be a joke. However, Grete interpreted it as a hostile action, spotted him, told her parents and made him come over. When questioned by them he didn't know what to say. He just stood there like a drowned rat. With the wisdom of hindsight, he was able to interpret the "triggering situation" quite well. Throwing the stone represented a plea for attention and to be taken seriously. Ducking behind the hedge was to avoid the risk of any potential rejection: it you don't want to take me seriously, then it wasn't me. When Johann found himself standing red-faced in front of his neighbour's parents, he didn't know how to explain what he did. He felt as though he had been caught red handed, hyped up and at a dead end. It wasn't an act of aggression, as the others suspected, but it also wasn't a declaration of love. Not in this sort of situation. Johann retained the feeling of not being able to defend himself even when falsely accused and that he is the subject of mockery when it comes to matters of the heart. As such, he would be better off hiding these emotions than allowing them to be expressed. Johann interprets his blushing as "the desire to be seen is risky and puts me on alert". ◄

► Example IV: Hypochondria

Gregor was the fifth of six children. He received little attention growing up. His mother was sick for a long time following his birth. His elder siblings cared for him, albeit poorly, but found it a strain.

He was only given hand-me-downs from his siblings to wear. After being handed down so many times, they were already in pretty bad condition. His low sense of self-confidence mirrored his situation. His next eldest brother got in an accident as a child and consequently received more care than the others for a long time, as was necessary. Gregor was jealous of him for this. He also wanted to be cared for from time to time, even though he realised that his brother would suffer from permanent damage as his bone fracture healed in a displaced position due to carelessness. Gregor wasn't entirely sure whether his resentment was also partly to blame for this as he failed to help his brother enough when he shouldn't have put any weight on his leg. In this respect, he viewed it as fair if something were to go wrong for him too. Gregor then started to introspect. When Gregor takes the time to take a closer look at his introspection, he determines the following: "I've always been somehow not quite as good as everyone else and a miserable guy. I'm sure something bad will also take hold within me soon enough.

I want to be on guard, perhaps I'll manage to catch it early enough, otherwise it's my turn. That would be too soon though. I don't want that!" The issue here is old certainty about the negative, destabilising identity, which is usually concentrated on an individual physical place or function, a process that subjectively removes the need for causality.

Treatment is initially based on discovering resources for attention, allowing experiences of self-esteem to be cultivated in a sufficiently beneficial area and consciously including the inner child in this process, enabling his previous experiences to mix with new evaluations. The power of the previous reference system to define is put into perspective if the patient, in this case Gregor, is officially permitted to imagine a dream family that satisfies his needs. The Gregor in this dream family scenario is a friendly twin brother of the other Gregor, who invites him to switch roles. Surprisingly, both Gregors benefit from the broadening of their horizons. This appreciation causes them to merge into one another... ◄

► Example IV: Childhood Theft as the Equivalent of Aggression and Somatisation as the Equivalent of Self-accusation

A family was living in cramped conditions and the father was furloughed for operational reasons. At the same time, his recently widowed, increasingly demanding mother was at the early stages of dementia and required care. The family decided to let her live with them and take care of her in order to lighten the financial burden. The 8-year-old son Erich was required to give his room to his grandmother and make do with his corner in the kitchen area, which could only be separated by a screen when required. He would listen in on his parents talking in the evening, often featuring complaints about his ungrateful grandmother, primarily from his mother. Erich adjusts his behaviour, only secretly begrudging the fact that he was made to leave his beloved childhood room as a result of his parents. Forthright revenge was not an option due to the unwritten rules upheld by the Christian family. Erich secretly spied on his grandmother's terrain when she was asleep. When his grandmother died a year later and his father's siblings came over to see whether any of the family jewels from better times that they all knew about were left to be claimed. To their surprise, nothing could be found in her room. Erich's mother was suspect number one due to the fact that she was the primary carer. She found herself in the hot seat and assured them all of her innocence, to no avail, even from her husband. Behind the screen, Erich endured even more suffering. He suffered from nightmares, insomnia and stomach cramps. One of the nights when his mother found him crying and came to comfort him, he confessed to her that he had thrown his grandmother's jewellery box into the old well behind the house.

Erich's hidden aggression against his grandmothers manifested externally in stealing and throwing her possessions into the well; once his mother, the person he had originally intended to retaliate against, was accused of the wrongdoing, he somatised both his fear of punishment for the action, if it were to be revealed, which he initially did not have the courage to do, and fear due to his self-condemnation because he put his mother into a desperate situation through no fault of her own. ◄

► Example VI: Forced Compensation for Humiliation

Alex, 46 years old, finds it important to prove that he has made something of himself. This is expressed through his appearance (clothes, jewellery, choice of car, choice of wife, choice of holiday destinations and hobbies), as well as in his behaviour (business practices, private relationships). He wants to avert people's attention from his simple background, how insecure he is about that fact that he only achieved a basic school education, that he had never even heard of things that others take for granted, and that he was often unsure of how to act properly in certain circles. In the past, he had experienced his fair share of humiliating experiences to seemingly confirm that he was not one of those people who seem to have it easy from day one as they come from better backgrounds. On the other hand, he wants to make sure that no one is aware of his old jealously, instead diverting their attention towards admiring him and envying him and his future prospects. However, with this subtle intention, Alex is unconsciously boycotting precisely what he wants to achieve the most, a sense of belonging to the very groups whose behaviour he mimics and constantly tries to outdo. Depending on their structural maturity, someone like Alex is likely to be diagnosed as a narcissist or perhaps only as someone with a histrionic personality disorder. The wound that he conceals from himself is the lack of appreciation, love, existential recognition and the joyful incorporation of his new personality into his social, protective group. This may mirror the life situation of the reference person or his small group, which unconsciously conveys this to the individual. In this regard, being excluded represents humiliation, the consequence of a symbolic fight with a loser who has internalised this outcome. It is worth revisiting this "fight" and reliving it from both sides using role reversal, to end it from the perspective of the meta-level and to allow both sides to exist and be appreciated. This would consequently provide a healthy focal point for a new identity. Other common specific examples beyond (motherless) orphans at a general social disadvantage include migrant workers, immigrants, asylum seekers, politically persecuted persons, war, economic and climate refugees. ◀

5.4.5 Diagnostic Levels of Classification According to Functional Areas

Generally speaking, healthcare systems provide us with the following functional levels for the now commonplace methods of diagnosis based on symptoms and disorders, which often merge together or affect all of the cited levels. In these descriptions, the regulatory disorders, over-regulations, dysfunctions etc. do not have an effect on the description of the diagnosis. The development of an understanding of diagnoses with the inclusion of neurobiological functional presentations remains to be achieved and likely constitutes a rewarding task. At present, making do with the current systems of classification is both required and binding.

(Diagnoses based on ICD-10):

Predominately organic level: Voluntary motor function, vegetative nervous system, functional systems in the cardiovascular, respiratory, gastrointestinal, genitourinary and immune system.

Examples of symptoms: Stereotypical movement disorders, tics, stuttering, cluttering, psychogenic aphonia, nail-biting, hospitalism amongst children, hyperkinetic disorders, trichotillomania, enuresis, premature ejaculation, dyspareunia, vaginismus, sleeping disorders, psychalgia, cardiac neurosis, hyperventilation, indigestion, pylorospasm, gastric neurosis, irritable colon, psychogenic pollakisuria, undifferentiated psychosomatic disorder, dissociative movement and sensory loss, erythrophobia.

Predominantly psychological level: basic mental organisation (borderline psychosis), development of cognitive structures (lack of structure is relevant to diagnosis), patterns of emotional regulation, conflict resolution and lack of integration. Examples of symptoms: Anxiety, panic disorder, phobias, agoraphobia, depression, pathological mourning sorrow, anancasm, emotional instability, lack of self-confidence, asthenia, anxiety dreams, night terrors, sleeping disorders, anorexia nervosa, bulimia nervosa, hypochondria, multiple personality disorder, dissociative stupor, dissociative fugue, dissociative amnesia, post-traumatic stress disorder, structurally deficient personality disorders.

Predominantly social level: primary or secondary social exclusion phenomena and also insufficient dissociation phenomena; potential lack of switching between these poles.

Examples of symptoms: Dissocial behaviour (cleptomania, pyromania, gambling addiction, impulse control disorders with delight in destruction and aggressiveness, adaptive disorder with conduct disorder), social phobia, lack of self-confidence with over-adapted subservience, narcissism with a tendency to devalue, deficient emotional fine-adjustment in contact behaviour in the case of structural weakness, for example, in the case of borderline structures.

5.5 Identification and Recovery Techniques: Reversal of Aspects That Have Become Solidified Through Flight Responses into Constructive "'ad greddi" and Secure Relationships

The trick in this regard is to also inquisitively address unsightly, rejected, devalued etc. aspects and enquire about their original function and significance. This process incorporates both a diagnostic aspect and a therapeutic one.

In therapy, it is useful to initially acknowledge excluded, psychologically "dead" matter that has been dissociated as a substitute as such, revive it, discover its dissatisfied energetic focus and value and to once again reintegrate it to a certain extent in a suitable form in a suitable location. The dissonance pattern that contrasts against its other background of experience has a symbolic power and an internal relational structure that it is beneficial to discover and revive as the key to resolution. This process amounts to the regenerative disentanglement of energetically entangled strands. Methodically speaking, it concerns tracing associative chains of meaning in the patient's unconscious to the respective partial aspects. This often requires detective prowess. In this case, the dissonance pattern is not only observable from outside, but also, above all, has its own subjective assertion. Learning processes, emotional experiences including their resonance as well as personal evaluation dynamics encounter one another in it. In order to ensure the conflict-related dissonance patterns are successfully resolved, the patient needs to feel in safe hands in a secure, anxiety- and judgement free therapeutic relationship.

Note

The therapist asks specifically: What is particularly important to you in the situation you have just described? Or perhaps: What is rarely or never successful, despite the fact that you try hard? It seems as though there is a counter force. Could you perhaps transform yourself into this force? What does it want to achieve? What are its convictions? Has it already appeared to you at some point in your life? Is there a person who defends it in a particular way?.

Or also: Sorry for interrupting you, but you had just started a sentence and broke it off. How was it meant to finish? The blocking impulse just became stronger. If it could comment on this subject in this situation, what would it have said at the time it stopped you?

or: You just made a particular gesture with your hand, which I've noticed you doing a few times before. Could you repeat it once again slower, quicker and also once in an exaggerated manner? How does this make you feel? When you let your hand continue to act out this gesture, which is what it wants to do the most, what does this lead to? What does this entail for your emotions?.

Tip

Pay attention to everything that conflicts with the overall trend of the situation.

Usually the patient presents an issue, concern or self-representation. Are there any non-verbal signals that contradict the verbal message (e.g. a cheerful tone despite the fact that the topic of conversation is death)? Are there any non-verbal signals that contradict each other? (e.g. smiling and a tense foot movement). Are there any flaws or sidestepping in the narrative flow, indicating that something needs to be avoided every now and again? Are there monotone sequences despite the fact that something emotionally significant is being described? When does the patient break eye contact and when does he attempt to regain it? Can this occurrence be assigned to the natural thematic flow or does it lead to curiosity about its repetition and possible meaning? The therapist could suggestively ask at this point: Have you noticed that the amount of eye contact between us has varied throughout this conversation? Sometimes there is a lot of open contact and at other times there is barely any contact at all. Have you noticed anything in this regard? How do you interpret this? Perhaps it is somehow connected: Looking away and this issue, and then in relation to this person! It may be useful to reconstruct this scenario in this room. Is that ok with you? Perfect, that's what we'll do. You're the director! There's a problem person, an ego-syntonic entity and a dissociation impulse that is looking away. You can allocate the space. Is anything missing? Is there a motive behind why this contact is taking place in spite of the dissociation impulse? You're nodding? What concern do you think it could be? Your facial expressions already speak volumes. This could turn out to be a lively and beneficial discussion. …

Subtle dissonance is resolved for personality traits following the same principle as in dialogical part-whole research. The sole difference is that the conflicting impulses and forces initially appear to be somewhat vague and first need to be associatively assigned to significant aspects from the patient's life story and current life circumstance. Only then can they be gradually recognised as mostly fundamental concepts that dominate the life of the patient.

This subjective aspect of diagnosis leads directly into the therapeutic process. It requires the therapist to exhibit a high level of empathy and also a Socratic attitude, in this scenario this entails showing encouraging interest in the patient's exploration process and revoking their own ideas as interpretations or constructions. At the same time, this also requires nuanced, diagnostic background knowledge for the allocation of the patient's various aspects.

When successful, the transformation from numbness back into living often corresponds to a crisis-like rebellion of the banished aspects that increasingly penetrate into consciousness. The potentially purifying storm offers the chance for a new, healthier beginning. Sometimes the rebellion on the quest for internal coherence can result in a loss of contact (possibly only temporary) in the outside world. Development can be painful.

Certain starting situations can also be tainted with sadness, whereby the loss of former happiness, vitality and courage to face life can resemble the extreme shock.

For example: Marga, 33 years old, teacher, suffers from attacks of self-uncertainty or blackouts, which she views as abnormal and humiliating. When around friends, she comes across as confident, eloquent, resourceful, quick-witted and funny. She is able to sustain her appropriate level of authority as a teacher in most of her classes. However, in classes where teenagers are in a power struggle and like to pose attacking questions, she sometimes finds herself turning numb, as if she faces a white wall inside and doesn't know how to respond, just like when she was a child and later in her exams. This allows the adolescents to triumph, sparking the cue for them to break into a spiral of attacking behaviour, causing the lesson to go off course. As a young child, Marga was very shy and only spoke to people she knew very well, mainly with her grandparents and their friends, but rarely with her own parents. They radiated an emotionally cold atmosphere, fought often and consequently failed to provide their children with reliable support. Marga hardly ever spoke with her brother (3 years older), who let off steam with outbursts to receive some attention from his parents, which he would otherwise not get. These loud rampaging outbursts frightened Marga. She didn't know where to escape to in the small apartment and often huddled in the corner of the room as if turned into a pillar of salt, pretending to herself that she wasn't there. This lack of any reaction was the most likely way to cause her brother's behavioural pattern to fizzle out, at least it would not further ignite it. Later, when at school, she would freeze and become silent when she had to go to the front of the class or stand in front of authority. However, in groups of friends, she unfroze and was even capable of taking the initiative. Around the time she left school, she got over her fear of doing presentations to the extent that she felt capable of becoming a teacher. Now the old shadows are re-emerging. Throughout therapy, the therapeutic relationship, which tied in with her relationship with her grandparents, was extremely important. The fantasy of growing up solely with her grandparents instead of her parents resulted in a stable, security-bound alternative identity that accompanied the original identity like a good, sisterly guardian angel. As this pair together, she felt empowered enough to send her boisterous brother signals to stop, to scream at him and hold him with adult strength. The role reversal helped her to determine that her brother's rage was an expression of shortcomings and despair, not of strength. This realisation changed a lot. He wanted to be taken seriously and to experience considerate encounters. So did her students. This led to the turnaround. She didn't need to escape or fight. She felt interest in and, to a certain extent, an affectionate, considerate understanding for their needs.

The process, which daringly leads to an open, sometimes vehement and sometimes only explicit discussion with the imaginary problem person and to changes in the relevant relationship(s), or at least has the power to lead to this, can be welcomed as a phase of transition following one of the modes of numbness, even if anxiety levels increase intermittently. It requires sufficiently consolidated self-awareness and self-confidence, which may have been established with the help of a resource-oriented approach in therapy. Often, the acceptance experienced within the therapeutic relationship and its momentum is enough for the patient to escape flight mode and start resisting, fighting, clarifying and regaining security. The focus in this regard is always on self-esteem, equal terms and dignity. Following this, security takes over the situation. Although we are referring to conflict here, the primary aim is always intrapsychic clarification. The patient's new viewpoint alone is usually able to transform their charisma in such a way that the quality of the interpersonal relationship with the real problem person also changes.

However, the process of transformation can also grind to a half, become chronified, cause psychosomatic aggravation and pose a risk to the entire system. (See below) We need to use our antennae to stay alert for out-of-breath rushed and driven behaviour. We need to attempt to address it. A chronified flight pattern is conveyed through a rushed manner of speaking, fidgety eye contact, absentmindedness in correspondence and body language etc. "If I imagine being as tirelessly on the move as you are, I think I'd quickly be worn-out. You must have a lot of energy to be able to manage that for so long now. I wonder what would happen if you were to stop, sense your strength, turn around, take at least three deep and slow breaths and face what you've been running from up until now. I'm would stand right next to you and also have your back. It would be interesting to see who or what appeared in front of your inner eye and whether they or it would not change if you were to look them or it directly in the eye from your centre and strength.

5.6 Inherent Fields of Tension in the Development Process

General observations. Fields of tension pose challenges for the ability to integrate as predetermined breaking points. In periods during which adequate support is provided by the social engagement system or if the internalisation thereof has led to the development of sufficient ability to regulate, they can induce a growth spurt.

If this is not the case, an unresolved deadlock of conflicting impulses remains stored, which can resonate again when the opportunity arises.

Our development process consists of multiple stages and phases that overlap to a certain extent. (See also ▶ Chap. 3) The location of the most intensive growth changes over time. In each case, this is the location where the most energy, proliferation, functional flexibility, malleability and vulnerability can be found. To this extent, the inner energetic centre also changes in phases. (A good illustration of this is the medieval custom of German emperors to move around their realm between different imperial strongholds in order to carry out government affairs from various locations throughout their kingdom. This opened up their perspectives.) Where could the very first "centre of government" be for a developing human being? In this regard, it is certainly impressive that the embryonic heart starts to beat a mere 3 weeks after fertilisation. The stages of development follow each other at a breath-taking pace.

In the perinatal period, our brain develops an incredible surplus of neurons, of which only the neurons that are aroused by environmental stimuli and fire together are retained. This can be regarded as a generous measure from nature to enable us to adapt to the world into which we are born. We shall now take a look at the further process in more in-depth sections:

5.6.1 Field of Tension A: The Organism and the World

Provided everything is going well, we develop a certain curiosity about the world around us and turn our attention towards it in order to discover it (intentionality). If the environment behaves in a way that is not conducive to a baby, the baby organism inhibits its readiness to respond to external stimuli and, in the best case, turns a little more towards internal sources of stimuli. Later in life, this can become a source of artistic impulses. However, this cannot be guaranteed. Later contact often contains a remnant of this withdrawal from the world, which is mistrusted for the most part and may be experienced as a "glass wall" phenomenon or general anthropophobia by outsiders. In psychodynamic tradition, this is referred to as a schizoid structure.

5.6.2 Field of Tension B: Toddler and Caregiver

Toddlers enjoy sharing a stimulating field with their personal counterpart (usually their mother or caregiver). Personal contact leads to the creation of shared oscillating fields (the parallelism of which can even be demonstrated in an EEG) and to interactive games that the baby helps to shape on the action level. As of birth, a baby is capable of activating the social engagement system in his environment and tuning into his counterparts. If the counterpart only sub-optimally fulfils the toddler's expectations, the toddler will initially try to maintain the relationship by adjusting his approach and suppressing all deviating impulses that may endanger the community. This typically affects his own aggression and initial impulses of independence. Identification and energetic attention are located in the interpersonal, stimulating oscillating field, on which the baby is dependent. This dependency can come into conflict with his basic needs and could disturb them. In this case, a learning process whereby the baby's own needs receive diminishing attention could set in.

Pathology: This adverse course of events is traditionally referred to as depressive structural development and an autonomy vs. dependency conflict in psychodynamics.

5.6.3 Field of Tension C: Free Will vs. External Demands in Addition to (as a Secondary Process), Will vs. Inner Needs

The first neocortical top-down growth spurt has commenced, causing the centralising experience of autonomous ego-involvement to be called into play, (1) as free will vs. involuntary demands from the outside world and also as adversaries of the adaptable, internalised SES (social engagement system) and (2) as a potential counterpart to the level of the patient's own emotionally driven basic needs. The latter constitutes a consequence and compromise from the experience of tetchily confronting boundaries in the outside world and only being able to integrate through inner compromises to adapt to the outside world. Compromises whereby the desire for rewards takes on a material aspect (money, possessions in addition to a career, status and power) to substitute for withdrawal of love that has been experienced in the meantime in the social environment. This field of tension is the reason behind many conflict variants.

Pathology: In psychodynamics, these tendencies are traditionally regarded as the development of obsessive-compulsive or anankastic structures (previously referred to as the development of the anal character by Freud) if the battlelines are directed inwards. If the focal point of tension remains between free will and involuntary demands, we then enter the problem area of dissociality.

5.6.4 Overlapping Field of Tension Combined with C and SES Joy in Competition (Analogy: Freud's Urethral Stage)

In this case, we initially encounter functional desire and joy in combative defiant stances. This initially constitutes an expression of security and excess power. In the animal kingdom, an analogous joy can often be observed in rough-and-tumble play amongst adolescents.

This field of tension has found a widely diverse and almost ritualistic outlet in sport, where it is met with social recognition, primarily if the incorporated aggression can be controlled through fair play and the rules of reciprocal social appreciation are not compromised. Rule breaking is ritually sanctioned. Competition aims to achieve an evaluative ranking order, whereby the individual is judged according to the respective criteria and reduced to this specific ability. As a result, the individual's multiplicity and multidimensionality can significantly fade away. Provided competitions remain solely playful they can be stimulating and motivating. However, when they are used to compensate for a lack of self-esteem, they can be overemphasised and turn into a source of annihilation anxiety. Individuals who are loved for their performance and proficiency instead of for themselves are susceptible to falling down this rabbit hole.

Sensitivity towards this topic has emerged in education, whereby the conclusion has been reached that – at least in the early school years – the fear of failure resulting from constant grading can lead to more harm than good and destroy the much-needed anxiety-free room for development.

Although largely unspoken, competition shapes the careers of the majority of adults in a manner of speaking. This may occur along the various career larders, in relation to external status symbols or beauty contests etc. In some instances, healing midlife crises put a stop to this external competition mode and refers to other levels and values.

There is also a collective rivalry within commercial enterprises, political associations, fields of psychotherapy, religious groups and nations, which sometimes improves quality, but, unfortunately, usually is mutually antagonising to the point of hostile destruction. Wars are often provoked out of greed.

Reciprocal social responsibility found in humanity is the antidote to a competitive mode that leads to destruction.

5.6.5 Overlapping Field of Tension Combined with B (Delayed Sexuality) and SES

Newly discovered attraction to a certain gender is tested on the social stage, a process which may override earlier shortcomings with regard to self-esteem. Signifiers of attention-seeking behaviours can often be observed. In the animal kingdom, these behaviours lead to the development of an animal's initial attractive and inviting mating behaviours, particular movement sequences, colours, changes in calls etc.

Do I feel a positive sense of belonging as a boy/girl that assures me that this now increasingly palpable state is completely ok? Increased erotic attraction between a mother and son as well as a father and daughter can typically be observed in the core family. (pronounced cross-entanglements are often referred to as the Oedipus or Electra complex in psychoanalysis). Adults are required to confirm their children's gender orientation in a friendly way whilst clearly observing limits. In the best-case scenario, children are able to sense that their parents are mutually attracted to each other and that each generation includes its own level of special intimacy.

The ability to emphasise, which has developed in the meantime, opens the door to manipulation (analogy: Freud's histrionic stage). As a result, the child discovers the popular drama triangle: if I adopt the role of the victim and correspondingly complain, I will trigger rescuer impulses in others who will then concentrate their aggression on those I want to harm without me being required to get my hands dirty.

Solid, genuinely sensitive and empathetic relationships within the child's core family and its environment, whereby the child is required to switch roles when he oversteps the line and determine whether he actually wants to harm the other person or not, form the antidote to the manipulation of the ability to empathise.

5.6.6 Overlapping Field of Tension Combined with B (Delayed Sexuality and D: Personal, Erotic Love (Analogy: Freud's Genital Stage))

This stage marks a direct continuation of the previous one with the objective of enabling the pubescent and adolescent to combine, fine-tune and learn how to integrate their emerging sexual-erotic passion with the emotional capability to love. The second component helps to compensatively modulate and secure the mentally stronger ups and downs of the first component.

If there is a lack of experience of supportive long-term relationships, the individual is at risk of becoming incapable of forming attachments and fragmentation between sexually gratifying brief encounters on the one hand and attempts to form emotional bonds on the other.

Tragically, children who have previously experienced sexual abuse are sometimes no longer able to distinguish between sexuality and normal contact behaviour, leading them to unconsciously open the door to future assault, despite the fact that they seek emotional and friendly contact.

5.6.7 Overlapping Field of Tension D: Interpersonal Love vs. Self-Love (Egotism)

Field D is at the core of the SES. It consequently takes on a special status, meaning that it may potentially operate at all times during all growth spurts from birth to death. Experiencing reliable, trustworthy and supportive attachment at the start of life is regarded as a resource – or deficit. It shapes subsequent behaviour in relationships.

Emotionally neglected children, or those who grew up in an emotionally scant environment first require a sufficiently convincing, emotionally corrective experience. Unfortunately, this rarely occurs. In this state of distress, the child's reliance on their own survival is the only thing that matters. As a result, the child's mistrust of the outside world and deep-rooted belief that he isn't worthy of love prevails.

In cases of parentification, giving and taking switch far too early and can become fixed in pathological altruism. Unfortunately, this field of being compassionately loved may also be primarily underdeveloped, insufficiently internalised or destroyed as a result of trauma. As such, it lacks the ability to counterbalance the interplay of forces, above all in the conflict with powerful egocentricity (C) and with the consequently easily inundated demands of basic needs (B).

Interpersonal love is the matrix on which every culture rests.

5.6.8 Cerebral Integrative Field of Tension E

Individual, centring integration of holistic wave identities experienced as the self, vs. the analytical, language-based, pseudo material functional mode that can be allocated to field C (activity of the dominant hemisphere). In terms of neurophysiology, this field can be found in the area in which the highest limbic functions, which are connected to pre-orbital fields, are active. According to the hypothesis, this is the location where the emotional and value fields integrate with the experience of the "self" or "being centred", which is also communicated to the SES environment, allowing inner-outer integration to take place. All artistic activity that is simultaneously a form of self-expression is assigned to this formative and pattern-generating level of integration. It is irrelevant which medium is used for artistic expression in this regard.

The current zeitgeist generally tends to downplay the potential of this sub-area of analogue experiences and their artistic integration in relation to the "digitally processing" dominant cortex. In other eras, the reverse applied (see ► Chap. 7).

In line with Jungian psychology, the field of tension consists of thinking and sober perception (referred to by Jung as sensing) on the one hand, and feeling and intuition on the other. The latter bridges into the next field of tension.

5.6.9 Field of Tension F: Receptive Mode vs. Independent Structuring Mode

This field is concerned with the individual scaling back on their own egoistic tendencies and becoming receptive to other balances of power and external information. One of these withdrawal ideals can be found in the wuwei attitude, in which the forces of nature must not be disturbed.

We also need this attitude on a daily basis in the early mother-infant relationship. Daniel Stern referred to it as sensitivity, which is also required to handle the topic of death. Arnold Mindell made strenuous efforts to subtly communicate with the energy field of coma patients.

As has already been mentioned at certain points: the nowadays widespread constellation work on social relationship structures with representatives is, not always but nevertheless clearly, coincidentally able to pick up on valuable knowledge, impulses, emotions etc., as if there is an energetic field that can be brought into the here-and-now through concentration and "read" by receptive beings.

Last but not least, many of our colleagues are convinced that they are capable of reaching their patient's unconscious field and consistently coming into contact with aspects of their patients that enable them to increasingly trust their countertransference.

5.6.10 Field of Tension G: Ability and Willingness to Participate in Higher, All-Encompassing Associations or, Failing That, Blindness Towards Extended Reference Frameworks

This level is characterised by the widespread notion that has been present for over 5000 years that the whole is mirrored in the minutia (even if this is usually in a sub-optimal quality). (Broken down to the simplest level of the human body, this means analogously that the whole is reflected on a small scale in the sub-whole, perhaps in the ears, eyes, hands and feet; and at a far more basic level, the whole is represented in each cell.) Conversely, which higher associations do humans represent and reflect? Does our composition, or rather the interaction between wave aspects and pseudo-matter indicate potential cosmic relationships? This constitutes an objectifying question. This approach is regarded as possible for those who have received a conventionally scientific education.

The subjective experience of connecting with an all-encompassing, energetic framework has been described rather similarly throughout time in all cultures by extraordinary people. In more recent times, the bridging vision of Martin Buber's I-Thou set a forward-looking double impulse for psychotherapy, one for secular connectedness and one for resonance with the all-embracing.

5.7 Inherent Resources at Various Stages of Development

Sources of security and trust within oneself and in the world in general help to avert fear and consequently (re-)integration. They help individuals to get (back) in touch with their own potential, which has previously been partially or entirely lost due to crises, trauma or due to mental overload from processing experiences.

If we incorporate these aspects into therapy, they are primarily beneficial when they slot into the therapeutic relationship as a whole and above all into the specific situation. Do we, as therapists, convey a genuine positive attitude towards patients at this point in time and do our non-verbal signals correspond to the fact that we want to support and encourage them and give them strength and dignity? As is always the case, authenticity is key here.

Resources can be found on all levels. Not everyone is capable of accessing every level equally due to their background experience. Not every level is suitable for each problem area in the same vein. On the other hands, we can also rely on the fact that every person unconsciously picks out elements that affect them particularly through personal resonance.

However: The range of capabilities is somewhat differently accentuated for structurally weak people than those who are susceptible to neuroticism, and differently again for those who have suffered trauma.

It is beneficial to connect the symbolic with the physically perceptible.

► Example

"When standing, we feel contact with the earth. It reliably carries us and resists us in a positive way. We are able to play around with it and transfer our weight to it, through our right foot and through our left foot, through our heels and through our heels and through the tips of our toes. It is always there for us. Regardless of whether we are dancing, jumping around or we fall over, it's always there to catch us. It is always there and always has time for us, no matter how much we need.

Our body is also a gift from the earth, namely mother nature. It is time to take a moment and be thankful for everything that runs smoothly. Just now, we listened to our feet. Have we every taken the opportunity to show them our gratitude for supporting us on a daily basis for so many years? When you take the time to contemplate it, it really is a great achievement that goes generally unnoticed and unappreciated (at best only first once walking is no longer possible for some reason). Let's sit down and first embrace one foot, then the other with our hands, all at your own pace. Ask each foot internally what it particularly likes: to be tightly embraced and potentially massaged, or to be stroked tenderly, brushed and stretched or relaxed, or rocked and cradled in harmony with your breathing. We'll give it a go and find out what our companions prefer. We enjoy allowing them to feel our appreciation, affection and gratitude. We are both the one who gives and the one who receives, and can internally say: I am capable of giving love, I am worthy of being seen and loved with affection. I am even capable of enchanting my completely ordinary daily life through my heart's loving gaze. ◄

Memories may arise of times when you experience similar feelings, perhaps in completely unremarkable situations, perhaps with people who you randomly crossed paths with, or people who are no longer in your life, or people who are no longer present in the same way as they once were. Yet something continues shine in their trace from back in the day. If I allowed my inner eye to see the light from all of these traces and let them flow together in the sky within, what beautiful kind of illuminating dawn could emerge as a result? The sky would represent me when I am increasingly open to allowing myself to become an amplifying resonance for this light of joy".

The above process provides a waking trance approach for individuals from a rather grey background with a significant number of experiences of deprivation.

The practical part features further approaches with other emphases, including a focus on catching up on structural development, self-care, learning how to handle desire and power in a sound and wise manner, or returning to balance within.

5.8 The Multiplicity of Disorders as the Concurrence of the Primary Tendency Towards Integration and Secondary Tendency Towards Disintegration – A New Approach Towards Classification

As is widely known from old empirical knowledge in developmental psychology-oriented psychotherapy, the same noxa can exert different effects depending on the development stage at which it emerges in a developing individual – aside from his dispositions – and depending on whether or not the individual has already been inflicted in one way or another. (Please refer to the section on development in ► Chap. 3.)

At this point, the 7-stage hierarchy of levels of the pseudo-particle wave tower should only be rudimentarily suggested.

The main pillar for successful development is the biologically embedded social empathy provided by brood care behaviour which has proven to be a required basic experience for the new arrival's subsequent ability to love and form attachments.

A lack of empathy and consistency within relationships appears to cause a more profound and centrally impairing deficit than the common, solely external stresses (although it is also possible to encounter trauma that is no longer experienced as compensatory). Provided this limit is not exceeded, the primarily received and internalised care, love and attachment experiences are typically capable of adequately compensating for the corresponding relationship structures in the social sphere externally as well as in the intrapsychic sphere internally despite damage or trauma. The primarily internalised positive relationship has the potential to become resilient and cushion us.

However, if it is only tenuously developed or missing entirely, this can have a potentiating effect under stress, i.e. more dramatic than under favourable preconditions. This also corresponds to general experience.

We can therefore assume that individuals with strong, primary experiences of acceptance are more likely to feel stamina and go into fight mood than those who have experienced meagre care. Furthermore, it can also be assumed that individuals who feel confident in fight mode for longer energetically partially or entirely remain at the original level of resolve, but that those in flight mode tend to leave the field and move to another level, primarily the psychosomatic level. (For example, after an unsuccessful marital dispute, a wife, who was unable to properly communicate her concerns, develops a headache.)

Specific integration tasks need to be performed at each developmental stage. When the corresponding integration centres differentiate, they are in the vulnerable phase. At this point in time, the organism reacts in a particularly malleable and adaptable way to demands. In unfavourable conditions, the new regulation centres fail to develop as required. As such, significant consequential damage will most likely be inflicted on the structures that are currently developing or further differentiating. Early learning experiences and emotionally charged survival strategies that are capable of ending or toning down crises have a somewhat less significant but nevertheless sufficient impact on our development. In many respects, our organism functions through the interplay of functional units that are assigned to each other in a dual or polarised manner, which are regulated and interrelated through a shared integration centre. These integration centres exist in part prenatally and otherwise develop gradually over the course of development. They are always connected like a complex mobile. Sensitive mothers help their babies to control and dissipate their frustrations through interactions with them. This regulation work is then internalised. Normally, this regulation work in children is further supported through education and in adolescents through a cultivating environment. In the worst-case scenario, the regulation centres fail to reach full functionality.

Stress and trauma are capable of not only hampering the occurrence of the generally progressive integration process, but also reversing it in a structurally regressive manner.

Emotional flooding triggers anxiety and tension, which initially causes the system to be significantly displaced and overridden, consequently weakening its ability to integrate and potentially shifting its target values. The original polarisation may re-emerge once again in addition to the potentially permanent loss of the ability to fine-adjust and centre. This has an effect on the psychosomatic aspect of the symptoms. Through empathy, we are able to differentiate between whether the patient is currently more in fight or flight mode through the clusters of symptoms. Both forms are inherent, useful alarm patterns with a cascade of physiological reactions that are prepared to support a fight or flight response. Both alarm patterns should not be permanently innervated. Instead, the mobilised reserve should be once again released through the corresponding activities. In reality, this has not occurred for quite some time. We no longer need to fight with lions or run away from tigers. We now fight differently, but by no means more successfully and also not in a physiologically adequate manner.

In subjectively hopeless situations, our body effectively calls it quits and shuts down, becoming paralysed, speechless, disorientated and experiences a blackout etc. Sometimes the pursuer then gives up. All of these patterns can be discovered and used as habits. Of course, this comes at a psychological price.

Anxiety is inherently experienced at every stage of development. It is shaped by the respective developmental task in each case. The antidote to anxiety is reliable love, which can be internalised and turned into resilience. Diffuse anxiety is harder to endure than a focused threat. As a result, if often transforms into phobias. Objectified or substantiated anxiety is easier to control and avoid.

Thus far we have addressed potential threats coming from the outside world. Indeed, this used to be the reality. However, internal struggles can also take place according to the same patterns. Conflicts always include two sides that are hostile towards each other. Usually, one of them ends up leaving the field maimed or being completely destroyed. It is interesting to bring together conflict pathology, whose differentiation we can thank psychoanalysis for, and the differentiation from stress research and trauma therapy. The former is consequently embedded in biologically verified findings. Overall, this leads to the formation of a more rounded image.

The present approach assumes that persistent, excessive stress promotes the breakdown of integration, or rather fragmentation, and in the worst-case scenario also leads to structural regression (in terms of loss of function compared with the age-appropriate norm). This disintegration causes "disorders". The organism's resilience to this disintegrative tendency depends on its primary integrative power. Despite a lack of experience at the start of life, it can also be "nourished" by later experiences of supportive relationships, as well as in therapy, enabling the mental fragmentation that has already occurred to be reversed through a re-integration process. –.

Summary of the integration-disintegration concept

1. Existential beginning of life phase:
   - Integrative general factor.

   - Integration tasks: physiological regulatory circuits, coordination of rhythms.
   - + experience: Basic trust, resilience, ordinarily dependable personality structure
     0- neurophys: hardware damage, structural weaknesses, polarisation, lack of balance
- Disintegration (as the result of an activated defence mechanism).
   - Fear of not surviving, fear of being destroyed, fear of falling apart.
   - Fight equivalent: symptoms of aggression (screaming, spitting, biting, vomiting).
   - Flight equivalent: early regulation disorders with the avoidance of contact.
   - Freezing equivalent: anaclictic depression with mortality/catatonia.

2. Stage of tending to basic needs:
   - Integrative general factor.
      - Integrative task: Deficiency + satisfaction, adjusting to the social environment.
      - + experience: Orientation around the protective, social system, formative takeover
      - Risk of addiction and dependency, care as a substitute for love and affection.
   - Disintegration (as the result of an activated defence mechanism).
      - Fear of losing the subjectively vital support of others.
      - Fight equivalent: Issues with demanding care and dependency, addiction.
      - Flight equivalent: Depression with auto-aggressiveness and psychosomatic symptoms.
      - Freezing equivalent: Major depression.
3. Autonomy stage:
   - Integrative individual, social general factor.
      - Integrative task: awakening egotistical power vs. altruistic empathy.
      - experience: Becoming independent, willingness to perform, regulating self-discipline
      - 0- Self-aggrandisement, lack of empathy towards within and the outside world, abuse of power
   - Disintegration (as the result of an activated defence mechanism).
      - Fear of losing power, possessions, control, status and knowledge as a power factor.
      - Fight equivalent: Obsessive-compulsive symptoms within, querulous Personality disorder.
      - Flight equivalent: Depression in the case of self-devaluation or being victim of bullying, outcast.
      - Freezing equivalent: Obsessive-compulsive illness with a pattern for surefire success, powerlessness, victim.
4. Level of social integration of the ego.
   - Integrative supra-individual, social general factor.
      - Integrative task: personal validation vs. holistic ability to love.

- + experience: Social aspects; joy in relationships, ability to empathise and love
- 0- dissocial behavioural aspects and/or withdrawal, lack of empathy

- Disintegration (as the result of an activated defence mechanism).
  - Fear of losing the ability to love, fear or not being loved.
  - Fight equivalent: Dissociality, abuse, jealousy, resentment, devaluing.
  - Flight equivalent: social withdrawal patterns, bitterness, perversions.
  - Freezing equivalent: Burn-out after chronified hatred with self-devaluation.

5. Cortical right/left integration pattern with knowledge as well as self-expression and finding values and meaning.
   - Integrative general factor for differentiating order intra- and interpersonally.
     - Integrative task: Matching digitally conscious and analogue-unconscious.
     - + experience: differentiated interaction with non-verbal nuances/introspection
     - 0- cultural orientation, finding values and meaning underrepresented
   - Disintegration (as the result of an activated defence mechanism).
     - Fear of not reaching full potential and failing the learn the meaning of life.
     - Fight equivalent: one-sided struggles to pursue values, fundamentalism.
     - Flight equivalent: cultural "'phobia", rejection of mistrusted spiritual worlds.
     - Freezing equivalent: Experiences of meaningless, axiological blindness, value chaos.
6. Receptiveness/intuitive "reading", informed energy fields.
   - Integrative task: individual vs. supra-individual, intuitive approach to knowledge.
     - Hardly developed in our culture.
7. The ability to synchronise with extensive reference frameworks.
   - Integrative task: Eliminating polar opposites both internally and externally.
     - Very entirely not developed in our culture.

The last two levels have not been explored in detail as they are rarely experienced by the general population, and if so, then only marginally. The listed pathologies should only be taken as examples. Systematic classification remains to be confirmed.

Bibliography

1. Fuchs T. Das Gehirn- ein Beziehungsorgan. Stuttgart: Kohlhammer; 2008.
2. Rosa H. Resonanz. Suhrkamp, Berlin: Eine Soziologie der Weltbeziehung; 2018.
3. Michael B. Der Arzt, der Patient und die Krankheit. Klett-Cotta, Stuttgart; 1964.
4. Brem-Gräser L. Familie in Tieren – die Familiensituation im Spiegel der Kinderzeichnung. 9th ed. Munich: Reinhardt-Verlag; 1957/2006.

5. Kalff D. war eine Schweizer Psychoanalytikerin nach C.G. Jung, die die symbolisierende Methodik des Sandspiels verbreitete. 1904–1990.
6. Bühler C. Praktische Kinderpsychologie. 1938, Lorenz-Verl., Wien u. Leipzig; 1893–1974.
7. Rorschach H. Der Rorschach-Test. Huber, Bern. – (1927): Rorschach test – Psychodiagnostic plates. Cambridge: Hogrefe; 1921/1927.
8. Wartegg E. Gestaltung und Charakter. Ausdrucksdeutung zeichnerischer Gestaltung und Entwurf einer charakterologischen Typologie. Leipzig: Universität Leipzig; 1939.
9. Bortz J, Döring N. Forschungsmethode und Evaluation in den Sozial- und Humanwissenschaften. Heidelberg, Berlin: Springer; 2016.
10. Lorenz K. Die angeborenen Formen möglichen Verhaltens. Z Tierpsychol. 1943;5:235–409.
11. Harry H The nature of love. In: American Psychologist. Bd.13, 1958. p.673-685; 1905–1981.
12. Dornes M. Der kompetente Säugling. Die präverbale Entwicklung des Menschen. Frankfurt/M: Fischer; 1993.
13. Montagu A. Körperkontakt. 14th ed. Stuttgart: Klett-Cotta; 2019.
14. Wampold BE, Imel ZE, Flückiger C. Die Psychotherapie-Debatte. Was Psychotherapie wirksam macht. Hogrefe, Bern; 2018.

The Practical Approach – Many Paths – Similar Goals: The Diversity and Similarities between Psychotherapeutic Treatment Methods

Contents

L. Hartmann-Kottek, *General Psychotherapy*,
https://doi.org/10.1007/978-3-030-87466-7_6

Overview
The various approaches employed in psychotherapy that have been taken into account in this book, even if not in extensive detail, have contributed to the intellect in the field of psychotherapy over the past century and become significantly widespread – reciprocally copied and modified in part. That being said, the list remains incomplete. The methods are primarily described in publications in the German language and therefore in need of translation into other languages.

6.1 Introduction and Survey

At present, the situation we face in psychotherapy can be described as blind men attempting to find an elephant in the dark. They each have different positions and consequently learn something different from touching the elephant, leading to a debate on the true description of the elephant. Each feel confident that their limited experience has shown them the truth. If they were to share their knowledge, their understanding would likewise expand ◘ Fig. 6.1.

To a certain extent, the multiplicity of psychotherapeutic methods mirrors our stages of development and invites us to determine at which stage and with which topics therapists are most capable of communication – as well as with regard to the heightened responsibility of the patient. In the case of damage that occurred at a very young age or resonances from that time, non-verbal forms of expression are more natural and addressing them can be particularly effective. As such, the integration of aspects from body therapy and creative – artistic elements in the official psychotherapy has proven quite useful.

The extent to which the objectives of the various approaches and clusters are sufficiently similar to convey a sense of shared objectives depends on the criteria applied and is worthy of discussion. A health-conscious approach typically provides the scope for similar objectives.

The chapter on treatment once again touches on the matter of the therapeutic relationship from ► Chap. 2; the medium in which the corrective new experiences should be embedded in order to allow their efficacy to fully develop. As the therapeutic relationship is the top priority in Gestalt therapy, its series of stages shall once again be described as those of the special therapeutic relationship in a natural, temporal sequence. These stages can serve as a guide or safeguard for new therapists.

The optimal approach seemingly revolves around a threefold concept:

1. Around the centre of the "I-Thou" relationship, which provides the patient with a sense of security, appreciation and empathy, and consequently increases their ability to integrate, coherence as well as inner order and creative power.
2. Around the centre between active and passive, i.e. the balance between listening and answering; new impulses may arise from either side of the conversation.
3. In order to build bridges between the conscious and unconscious at all levels; namely: Establishing a link between interpersonal signals, purely mental states,

Fig. 6.1 Elephant

6

imaginary phenomena, body language, psychosomatic events and, first and foremost, the impulses back translated from the symptom.

It is nescassary to identify and bring in contact several aspects:

1. the former impulse that remained concealed within and was contingent on frustration, even if it may have been disfigured beyond recognition, initially emerged as a symptom and needs to be reversed first;
2. its target object, i.e. the adequate counterpart, can be viewed as one of the key links and even as an equivalent to psychological neurorrhaphy. This also requires.
3. the identified divisiving, frustrating or traumatising counter impulse – and
4. the personality of the patient to be uplifting, encouraging and sufficiently self-accepting from the outset (through the therapeutic relationship). In this regard, the task is to establish the tailor-made individual connection for this experience.

This "operation" takes a two-pronged approach: On the one hand, the core of the patient's personality experiences empathetic understanding and appreciative encouragement from the therapeutic relationship, enabling it to encounter traumatic, threatening or frustrating impulses in a different way than beforehand; and on the other hand, when the excluded impulse or part of the experience is retransformed, it may be greeted with acceptance and often viewed as a valuable aspect that can be reintegrated.

This duality corresponds to the notion of integration and disintegration, as explored in the previous chapters, which has been integrated into mainstream therapy.

A conflict can be viewed as a mini form of traumatic struggle which may end in the failure to make a fair compromise in the worst-case scenario, with one side being overpowered by the other and succumbing in a state of paralysis.

The extended standard approach from Gestalt therapy is described in particular detail as an example, with the inclusion of both of its specific viewpoints and the widely accepted ones.

6.2 Many Paths

6.2.1 A Variety of Approaches and Methods as a Historical Treasure

Mindfulness-based therapy approaches (derived from Zen practices).

Analytical psychology (C.G. Jung).

Constellation work: System, organisational, family constellation, large group work.

Autogenic training (J.H. Schultz).

Behaviourism (Pavlov, et al.)

Bioenergetics (Wilhelm Reich, Alexander Lowen, Wolf Büntig, Margit Koemeda, et al.)

Core energetics (John Pierrakos).

Daseinsanalysis (Binswanger).

Ego-state therapy (John and Helen Watkins).

Emotion-focused therapy (Leslie Greenberg).

Existential analysis (Längle).

Existential psychotherapy (Yalom).

Family therapy (psychodynamic, systemic, development-oriented).

Focusing (Gendlin).

Functional relaxation (Fuchs).

Talk therapy/client-centred therapy CCT/supportive therapy (Rogers).

Gestalt theoretical psychotherapy (J. Walter).

Gestalt therapy (F. Perls/L. Perls/Goodman//Buber).

Group analysis (S. Foulkes).

Group dynamics (Kurt Lewin).

Group psychotherapy (Jakob Moreno and others).

Hakomi (Ron Kurtz).

Humanistic psychotherapy (CCT methodically enriched by Gestalt therapy, EFT, TA, logotherapy and existential therapy, humanistic body therapy, integrative therapy).

Hypnosis (classic).

Hypnotherapy (M. Erikson).

Individual psychology (Adler).

Integrative therapy (H. Petzold).

Katathym imaginative psychotherapy (KIP) (Leuner).

Cognitive psychotherapy (D. Meichenbaum, A. Ellis, A.T. Beck).

Cognitive behavioural therapy (CBT).

Concentrative movement therapy (Stolze).

Body therapies (bioenergetics as per W. Reich, G. Boysen, A. Lowen, Pierrakos, Peter Levine).

Expressive arts therapy (Paolo Knill, Hans Prinzhorn, Flora von Spreti, et al.)

Logotherapy (V. Frankl).

Mindfulness-based stress reduction (John Kabat-Zinn).

Music therapy(ies) (Priestley, Gertrud Orff, Nordoff-Robbins, Loos, Schroeder, Schwabe, Boley, Frohne-Hagemann, Tüpker, Weymann).

Narrative therapy approach (Michael White, David Epston, et al.)

Neuro-linguistic programming NLP (Bandler & Grinder).

Neuropsychotherapy [1].

Couples therapy (1) psychodynamic, (2) systemic, (3) developmental.

Positive psychotherapy (Peseschkian).

Positive psychotherapy (Seligman).

Progressive muscle relaxation (Jacobson).

Psychoanalysis (Freud et al.) (1) drive theory (Freud), (2) neo-psychoanalysis (Schultz-Hencke, et al.), (3) objects relation theory (Kernberg), (4) ego psychology (Hartmann) (5) self- psychology (Kohut), (6) relational psychoanalysis (Stolorow, Brandchaft, Atwood).

Pesso Boyden System Psychomotor Therapy (Albert Pesso).

Psychosynthesis (Assagioli).

Psychodrama (Moreno).

Schema therapy (Jeffrey Young).

Somatic experiencing therapy/body and trauma therapy (Peter A. Levine).

Systemic therapy(ies) (Bateson, et al.) & synergetics (Hermann Haken).

Dance/movement therapy (Elsa Gindler, E. Duncan, Trudi Schoop, Ursel Burek, B. Hausmann, S. Trautmann-Voigt, et al.)

Work at the Clay Field (H. Deuser).

Transactional analysis (Eric Berne).

Transpersonal training (Grof).

Trauma therapy (F. Shapiro, I. Olbricht, E. Spangenberg, G. Kahn, P. Levine, van der Kolk, et al.)

Behavioural therapy (1) behavioural therapy, (2) cognitive behavioural therapy, (3) behavioural therapy after the turning point with regard to emotions (see Klaus Grawe), (4) behavioural therapy after the turning point with regard to focusing on mindfulness.

6.2.2 Formation of Clusters throughout Time

Psychodynamic group of methods (= psychoanalysis and depth psychology).

- Psychoanalysis (classic and Neo-Freudian).
- Psychoanalysis-based psychotherapy.
- Psychoanalytic short-term and focal therapy.
- Psychoanalytic interactional psychotherapy (Heigl, Heigl-Evers).
- Psychoanalytic structure-based psychotherapy (Rudolf).

- Individual psychology (Adler), since 1998, (preceded by depth psychology).
- Analytical psychology (Jung), since 1998, (preceded by depth psychology).
- Daseinsanalysis (Binswanger), since 1998, (preceded by depth therapy).
- Psychotherapy based on depth psychology.

Ever since its inception over a century ago, psychoanalysis has gone through a number of stages of development (drive psychology, ego psychology, object relations theory, self-psychology and relational psychoanalysis) and developed a certain level of multiplicity, the representatives of which now exist side by side. This led to both enrichment and tension.

Depth psychology can only be found in Germany. The practice originally pertained to psychotherapists who had received minimal psychoanalytical training and primarily worked in a seated and not lying position, who also typically exhibited more affinity for unconventional methods and were, perhaps as a result, traditionally viewed as less effective by psychoanalysts. In practice, the opposite proved to be true. Studies from depth therapy were urgently required by psychoanalysts to increase the number of studies on effectiveness; as a result of this professional necessity, depth therapy and psychoanalysis were united under the joint term of the "psychodynamic approach", as is commonly employed throughout the world. The primary similarity between the approaches combined under this umbrella term is the emphasis on the topic of the unconscious on which the theory of neurosis is based, and, additionally the incorporation of transference, countertransference, the patient's resistance and interpretation with the objective of obtaining an insight.

6.2.2.1 Behavioural Therapy Group of Methods (BT)

As of 2019, 35 sub-methods can be found listed on the Wikipedia entry for BT.

The Verhaltenstherapiemanual (handbook of behavioural therapy) by Linden & Hautzinger ([2]/sixth ed.) includes 16 treatment programmes and 48 methods + individual processes.

Berking & Rief [3] distinguished between the following methodological groups:

- Cognitive behavioural therapy (including the behavioural approach).
  - Exposure therapy.
  - Operant conditioning.
  - Observational learning.
  - Cognitive therapy.
- Assertiveness training.
  - Social skills training.
  - Emotional skills training.
  - Problem solving training.
- Relaxation techniques.
  - Progressive muscle relaxation (Jacobson).
  - Biofeedback.
  - Imaginative relaxation techniques.
  - Hypnosis.

  - Autogenic training (Schultz).
- Wellness-based intervention techniques.
  - Mindfulness-based stress reduction (MBSR).
  - Mindfulness-based cognitive therapy for depression (MBCT).
  - Mindfulness-based relapse prevention (MBRP).
  - Dialectic behaviour therapy (DBT).
  - Acceptance and commitment therapy (ACT).
  - Vipassana and Zen meditation.
  - Compassion and loving-kindness approaches.

In Germany, schema therapy is considered a behavioural therapy for insurance purposes, but otherwise, it is internationally regarded as a separate hybrid of behavioural therapy and Gestalt therapy, including depth psychology.

Behavioural therapy does not have a common theory or identity. Indeed, in recent decades, it has been particularly "ingestive", attempting to incorporate everything that appears practical and effective. This led to the merging of a collection of fragments of knowledge drawn from various sources. It stands to reason that due to the lack of homogeneity, from a scientific point of view there should not be any common effect sizes, no overall calculations, but rather only calculations on logically compiled healing processes from similar sub-groups. Perhaps the purposefully targeted aim of healing experiences and the promotion of independent initiative can be regarded as a commonality.

Nevertheless, the attempt to achieve practical integration can certainly be considered the positive side of the coin. The identity of behavioural therapy has been subject to several upheavals, incorporations from other schools of thought and transformative manoeuvres.

Prior to now, it has predominantly resided in the belief that efficacy is the result of standardised techniques and methods, not dissimilar to standardised psychological test diagnostics, whereby the goal is to minimise the tester's influence and maximise the consistency of the presentation of the test. This approach is used to determine the statistical significance of the subject's performance on the tested aspect in comparison to the population as a whole (or the reference group). Individual statements are explicitly not made on the evaluation of the performance in consideration of the individual reference systems. To this extent, two identical test results for two different personalities can be of entirely different significance. As such, the type of questionnaire and performance test rarely yields informative results.

The standardised treatment programmes offered by behavioural therapy and the belief in the potential thereof stems from the same inner attitude. Accordingly, it could be deemed sensible to standardise treatment with manuals and demand loyalty thereto. But would this idea truly be realistic? It is reliant on a manageable degree of inner connections, or processing capabilities, similar to the mechanisms of a mechanical clock. If this were the case, for every step in the circuit that goes out of step, there would be a specific correction or treatment programme, which could be applied as prescribed by the manual.

Needless to say, this reductive model does not properly address the complexity of the human brain and its huge potential to connect, which remains incalculable to us.

At long last, "contextual research" (Wampold et al.) has put an end to the belief in standardised and manualised treatment programmes. Psychotherapy that is solely based on statistics, namely psychomechanics, is ineffective. Nevertheless, it can provide supplementary help once the personal aspect in need of correction or support has been identified and activated through personal psychotherapy.

The scientific ideal of behavioural therapy is unilaterally aligned with classic physics by virtue of its statistical calculation of probabilities on large masses. It cannot be denied that these calculations are well suited to large categorical differences, such as those between a gas molecule and a human observer. In this case, the inaccuracy of the average pace, direction and location of the movement of a single gas molecule is relatively irrelevant. When faced with an infinitely large mass, the reaction of the individual is insignificant. Its range of error is extremely small.

However, the same cannot be said for situations where a person faces another one to one, as is the case in psychotherapy. From a scientific perspective, calculations based on statistical probability are out of place if a statement from one person about the other is valid. This constellation ventures into the territory of quantum theory. A certain amount of unpredictability is an unavoidable aspect in this regard, and increasing tolerance of uncertainty is a helpful step. In psychotherapy, certainties are granted through relational structures, i.e. relationships. As luck would have it, the ability to form cohesive relationships is not reliant on one's knowledge of quantum theory.

To date, psychology – particularly behavioural therapy – has failed to adapt to the dualistic mode of existence, the "corpuscle and wave", each of which has its own scientific detection system and acts as a complement to the other, at times in an overlapping and permeating manner. Even if the classic scientific system can provide an adequate understanding of the physical aspects of human beings, it enjoys far more limited success in the psycho-spiritual realm, particularly with regard to its mass calculations, and is almost entirely inapplicable for primarily individualised questions, as is the rule in psychotherapy.

The primacy of the therapeutic relationship signifies a paradigm shift in the field of behavioural therapy.

Ramifications for the future: Whilst undoubtedly costly in terms of training, creating a culture that highlights the importance of the therapeutic relationship is far more effective than a manual-based approach. This expansion, which can be regarded as a turning point for the therapeutic relationship, would significantly change the current identity of behavioural therapy. At present, this approach is advocated by the avant-garde of BT.

This broadening of horizons with the change in the identity of BT could perhaps be eased by denoting the common new beginning "universally integrative." As a result, the notion of expansion in the field would certainly be met with great enthusiasm by the younger generation.

Humanistic Group of Methods.

- Talk therapy/client centred therapy (Rogers).
- Gestalt therapy (F. & L. Perls/Buber).

- Emotion focused therapy (Greenberg).
- Psychodrama (Moreno).
- Humanistic body therapy approaches (Reich, Lowen, Brown, Kurtz, et al.)
- Transactional analysis (Berne).
- Integrative therapy (Petzold).
- Logotherapy (Frankl).
- Existential therapy (Längle).
- Gestalt theoretical psychotherapy (H. J. Walter).

In the USA, the humanistic group of methods comprises: Talk therapy, Gestalt therapy, EFT, psychodrama and humanistic body therapy approaches.

The humanistic-systemic groups of methods emerged halfway through the last century as a response to the gaps left by the two established groups of methods, namely psychoanalysis and behavioural therapy. In particular, these shortcomings could be found in (1) the gap in the relationship with the therapist, (2) to gap in the extended social relationship context, (3) the gap in the here-and-now experience, (4) the gap in the lack of body integration, (5) the gap in the spontaneous, artistic joy in expression and (6) the gap with regard to finding value and meaning.

The common denominator of the humanistic group is a therapeutic relationship with specific acceptance, human sympathy and coherence in addition to a genuine resource orientation, realistic experiences and an enthusiasm for experimenting – with the inclusion of body language, movement, staging and artistic media – alongside a focus on meaning and values.

Systemic group of methods/the bases of systemic family therapy.

- Humanistic systemic family therapy/family sculpturing, inner team, parts party (Virginia Satir, Schultz v.Thun/adopted from Gestalt therapy).
- Intergenerational family therapy (invisible loyalties/Boszormenyi-Nagy)
- Analytical family therapy (Heidelberg school/Stierlin, Sperling, Massing/.
  - Göttingen – multigenerational perspective)
- Systemic family therapy/family pattern disruption, symptom prescription.
  - (Milan model/Selvini Palazzoli, Cecchin, Boscolo, Prata)
- Structural family therapy (outlining subsystems/Salvador Minuchin
- Strategic family therapy (dysfunctional communication/Jay Haley, et al.)
- Constellation work (revised form as per Hellinger/adopted from psychodrama)
- Systematic therapy: narrative approach, construction of realities (White, Epston)
- Solution-focused therapy (Steve de Shazer/Milwaukee school)

Systemic and family therapy place great importance on the social network for the individual, helping to bring this element to the fore and broaden the field of psychotherapy for over 70 years now. A number of sources from different schools of thought have been merged over time. However, a certain level of patience and courage was required to explore new ideas. The range of approaches spanned the extremely resource-oriented, caring approach of Virginia Satir, to the cold, solely strategic methodology of the Palo Alto and Milan groups. In the 1980s, the period of constructive transformation, systemic therapy appeared to be devoted to its linguistic constructs and systemic perspective, which allowed the individual sufferer to be easily overlooked with regard to his personal situation. Since then, new alignments and readjustments have occurred time and again in order to find a meaningful balance between the importance of the individual and the system as a whole.

Unfortunately, even in systemic therapy, the custom is rarely to provide sources and reflect thereon. However, it may be the case that the current generation is unaware of this. Indeed, increasing the awareness of sources could help to emphasise how closely the individual groups of methods are linked. This especially applies to the links between the systemic and humanistic groups.

6.2.3 Depth and Effect Levels

6.2.3.1 General Preliminary Remarks

This section is of particular significance as we are unable to directly reach engrams that have been stored at a young age on the basic needs level through an intervention that addresses a "later" (cognitive) functional layer of our brain. Here we are referring to different, inherent depth levels:

The neurophysiological notion of development and layers is mirrored in the different approaches and methods employed in psychotherapy. First and foremost, in each case, different layers are reached, sometimes primary layers, sometimes secondary, like in a cascade.

Not every therapist or patient feels equally at home and able to resonate on all levels from the outset. As such, the objective is to help the patient to pick up as many means of expression as much as possible. Therapists are required to carefully examine whether their skills can be improved or whether they have a motivated opposition to this (which may be in need of correction); namely, "non-verbal stuff is for kids"; "I want to be taken seriously in my job". Or: "I've never been able to draw, so why start now?" Or: "Isn't physical contact forbidden by our profession? I'd rather not go there!").

This chapter aims to shine a spotlight on pathways to achieve more information and interventions and ventures into extended treatment skills and the doors they open.

Non-verbal approaches tend to be more successful at getting closer to the patient's main concerns and can generally steer them in a more beneficial direction in a more natural manner. The repertoire of communication often expands over the course of developing trust within a therapeutic relationship, unless this is hindered by the therapist's own limitations. In the ideal-case scenario, the therapist is capable of improving his ability to communicate beyond verbal expression to include at least one, and ideally two or three additional non-verbal methods and employing these expressions at the right time to reach new depths. This approach incorporates elements from body therapy, scenic methods and the therapeutic approach to imaginary and artistic media and forms of expression.

6.2.3.2 The Integration of Verbal and Non-verbal Methods

At psychosomatic and psychiatric clinics, non-verbal methods are typically employed as accompanying auxiliary and supplementary therapies alongside the principal verbal therapy and play a role in therapy sessions, to a certain degree. However, most of the information is lost in the process. At many clinics, where the dominating presence of economists is ever-growing, these methods have been cut from the agenda for cost reasons. This constitutes a double loss. In the Gestalt therapy model clinic run by the author, doctors and psychologists who have received Gestalt training are the therapists of choice due to their insight from their diverse backgrounds and proficiency. Others who are only capable of employing verbal forms of psychotherapy, typically depth psychology, are paired with creative therapists to create a therapeutic tandem. The pair conducts all three group sessions per week together with alternate leadership, but with conscious collaboration nevertheless. They are carefully supervised to determine whether they can form an appreciative, mutually supportive model or not. Patients have a fine ear for whether the relationship of the pair in control is in check or not, particularly those who were forced to endure particularly tense situations in their family when growing up.

In addition to his verbal skills, ideally, a fully trained psychotherapist should have received basic training in a variation of body therapy as well as in an imaginative, scenic and/or artistic level of expression and intervention, either in art or music therapy or perhaps in improvisation.

In the future, the latter should not constitute separate auxiliary therapies, but rather integrated components of the principal therapy (as has always been the case in Gestalt therapy). Otherwise, the therapist runs the risk of losing a large amount of valuable mental material, which may have arrived in the conscious in creative sessions. The therapist should address this material at the time at which it arose within the patient, not a few days later once the verbal principal therapist has discussed it with the creative therapist with a very limited outline and summary of the links and observations. In this scenario, by this point, it is usually no longer possible to make a direct link to the patient's current issue.

Example for common separation of methods: In art therapy, free painting emerged as a suitable form of expression, where possible in connection with a dream or daydream. Patient A stared at the blank page in front of her for a while with a sad look on her face, then she reached for the black finger paint and covered almost the entire page with large, heavy, circling black strokes. The only circular clear area could be found at the top right of the page. She sobbed whilst she was painting and quietly remarked that she did not know whether she would ever see the light again. When asked afterwards, she said that her current distress only came to the fore when she was painting and that she is sometimes overwhelmed by anxiety, which is also present in her dreams. She made the connection to her chronically sick husband. However, she did not want to re-enter the mood she was in whilst painting. She was glad to have escaped it. In this case, it would have potentially been possible to bring her latent vital side, including its aggressive elements exhibited in the heavy strokes, into consciousness and to use it as a resource.

Ramifications for training: In order to fulfil the desire for increased qualities of therapists, these skills must be taken into account in the curriculum. If training continues to be unilateral, as it is at present, the patient will be required to transform the idea to the verbalised plane and complete the entire adjustment to the therapist alone to ensure that communication is granted a sufficiently good chance of success. Strictly speaking, the therapist should be the one required to accommodate the patient, not vice versa. Psychosomatic patients, patients from less vocal milieus and those from musically talented backgrounds often benefit from the non-verbal aspects in therapy to a certain extent.

As a psychotherapist is capable of several levels of expression, they are often capable of quickly grasping at which level the patient can be particularly reached. In certain cases, the patient is required to invite the therapist in until the familiarity with an unfamiliar medium has grown to a point that it can be accepted without any issues. This increasing trust leads to greater levels of inner contact and stimulating vitality. Certain aspects need to be taken into account in this regard: At which point is the medium introduced? Does the therapist switch to working with a medium in order to avoid a difficult phase in the relationship? This would, of course, constitute a contraindication. Dealing with resistance and relationship crises forms an essential element of training.

Needless to say, psychotherapy can also be conducted purely verbally or purely guided by others in a psychoeducational manner; in this case, its efficacy can often be improved upon and, from experience, falls short of the potential prospects, causing available opportunities to be passed by.

6.2.3.3 A Focus on Emotions

In this case, due to the fact that emotions are the focal point of psychotherapy, emotion-focused methods should be mentioned first, despite the fact that the body level can be regarded as "deeper." If we are capable of reaching an individual's frustrated basic needs, this offers the patient the chance to "heal" from this level and regenerate completely. Leslie Greenberg (emotion-focused therapy/EFT) can be regarded as a pioneer for this level.

In certain cases, feelings cloak other concealed emotions. Anger, rage and aggression often conceal a disenchanted longing for love, tenderness, understanding, closeness and a hunger to be seen. Anxious vulnerability and sensitivity are sometimes cloaked in courage and outspoken boldness. Overtly forced friendliness can rightly lead to a fear and aversion to rejection.

The objective in this regard is to reach the patient's inner self with understanding and acceptance, and, in the case that he has been subjected to damage, give him a chance to develop and the encouragement to accept himself. As therapists, we are tasked with assuming the role of the parent in relation to the child's ego. Regressive refuelling is entirely permissible during the process of time regression. The relationship with the entire adult personality sometimes involves elements of partnerships on equal terms, comparable to relationships between parents and adult children, who often earn and receive respectful recognition for their achievements and development.

Movement, music, vocal sounds, rhythms, shapes and colours, imagination and improvisational theatre are all suitable media of expression in this regard. Direct access to the unconscious is highly likely. This also applies to the following level:

Body level

Body therapies from the various schools of thought are of great benefit to psychotherapy as a whole. In neurophysiological terms, they are attributed to the lower and middle limbic systems. This level is only partially accessible for the conscious. Body language is an open book for those who are able to read it. As might be anticipated, it can encourage projections. As such, it is also conducive to ask the patient how he finds a certain gesture or posture etc., as he is the one capable of accessing the world of inner experience and the observer is only able to access the world of outer experience. If the observer sensitively imitates a posture or gesture, it is possible that he will likely enter the same state of mind from which the posture or gesture was "born" and subsequently be able to interpret it.

On the one hand, physical responses constitute an ever-present, involuntary event that accompanies mental processes. At times, solely disintegrated aspects are revealed, sensed and addressed on the body level. On the other hand, physical activation can open the door to the journey of discovery to otherwise dormant levels of the mind and old memory traces that would otherwise remain inaccessible. Where there are discrepancies with the conscious level, the body level represents the true angle, namely the vessel of subjective truth in the case of doubt or inconsistency.

The body level is often used as a condensed mirror and enriching information centre to help ascertain issues, despite the fact that they are not primarily expressed on the body level. At its non-verbal level, the body is also capable of bringing incomplete elements and conflict events to a conclusion or concluding a process through an abreaction following trauma. In this regard, it is key that the body's language becomes accessible to the mind to enable the common event to become a whole.

It is also possible to explore and experiment with the body. Which emotions and images are evoked in connection with particular physical responses? When the patient is presented with a topic and allowed to take the lead, his responses should be viewed with interest.

We have previously explored the significant roots of body psychotherapy in the reform movement shaped by art nouveau at the turn of the twentieth century in ► Sect. 2.3.4. At the time, it was referred to as body culture, as the body was now approached with appreciative curiosity and an openness to engaging with its signals, desires and its aesthetics, which were deemed worthy of admiration. Historically, the body level was largely side-lined in psychoanalysis up to this point; its earliest advocate, Wilhelm Reich (1897–1957), was expelled from the International Psychoanalytic Association in 1934. Reich was subsequently celebrated as a pioneer of sexual liberation. The next generation of advocates also faced an upward struggle with achieving recognition and integration of the body level. Physically touching a body causes alarm bells to ring amongst psychoanalysts almost too quickly as if it is always erotic in nature. Unfortunately, by the same token, many sexual assaults occurred in reality. Reich's most well-known student, Alexander Lowen (1910–2008) who focused on chronified patterns of tension, which he distinguished and described as character armour, further emphasised the required grounding and otherwise built new bridges towards psychodynamic thinking with his bioenergetic analyses. In the depth psychology and humanistic fields of training and treatment, the fear of touching was significantly lower, matched with an increased will to engage in experiments. Lowen's colleague in the field of bioenergetics, John Pierrakos (1921–2001) proposed the existence of an energy flow between the body and the mind.

Perhaps the most effective of all variants is that of Gestalt therapy, which works in close proximity to the body. It bears a close, internal proximity to Peter Levine's Somatic Experiencing in that blocked impulses and reactions are carefully looked for, which corresponds to Levine's frozen state. Trembling or shaking phases of autonomous abreaction are also a common passing phenomenon in Gestalt therapy related to the body. The supportive therapeutic relationship addresses the formerly powerless aspect of the self in order to achieve the empathetic "I-Thou" field with it and integrate it.

6.2.3.4 Breathing Techniques and Forms of Meditation

Therapeutic breathing techniques have always constituted a special element of body therapy. They not only facilitate a revealing manner of psychotherapy but also act as an accompanying aid for stress management. The therapeutic use of breathing techniques emerged at the turn of the last century in the form of a secular remnant of yoga-based meditation practices from the Far East (the origins of Pranayama yoga can be traced to the Upanishads; Qiyong yoga is based on ancient Chinese medicine, and similar breathing techniques can be found in Sufism, in early Christian mysticism and amongst Native North Americans.) Many breathing techniques aim to achieve a conscious, harmonising balance between the body and mind. They became the subject of great interest in the 1970s and were enthusiastically adopted into mindfulness-based forms of therapy at the turn of the last century. Stanislav Grof (born 1931) developed a technique with rapid (holotropic) breathing, whereby, astonishingly, perinatal memory engrams are often activated during a trance.

6.2.3.5 Social and Intrapsychic Effect and Access Level

This level encompasses all scenic and spatially symbolic approaches from constellation work and roleplay for the interpersonal and intrapsychic realm. In the case of the latter, the aspect of the relational structure between the parts and their whole overlaps. In this regard, both the forms and levels of communication play a role alongside the change of roles and perspectives, and, last but not least, the quality of the relationship, which may be factual in nature or sustained by empathy and/or compassion. Neurophysiologically, this level is found in the upper third of the limbic system.

6.2.3.6 Means of Access Through Variations of Consciousness

Methods that primarily work with language, narration, imagination, perceptions, realisation, and also with dreaming, trances and daydreams – as well as conscious assessments, will and cognition, neurophysiologically depend on the cooperation between the upper limbic system and the forebrain.

Whether success is achieved through conscious behavioural modification and practice is contingent on the type of motivation behind the patient's will.

Access to the various levels can predominantly be obtained if the therapist takes a natural approach and provides credible encouragement. In the event that a specific change is on the agenda, the engrams will achieve the best success in terms of updating and reconnecting if the trace has been adequately brought into resonance.

6.2.4 Blind Spots Amongst Therapists and Patients

In psychotherapy, the widely accepted initial understanding according to the field's professional identity was that blind sports, namely the suppression of mental substance or functions in the psychotherapeutic partnership can solely be attributed to the patient as part of his issues. However, this exclusivity does not correspond to reality. It solely provides an ideal. This section relies on the underlying conviction that the psychotherapist is the key instrument for providing resonance during the therapeutic process and not that he acts as a conveyor of a standardised manual to which he is entirely subservient, as is the case with the role of a test psychologist. For reasons of scientific comparability, the therapist should primarily be a consistently friendly robot as opposed to an individual being with an open ear for understanding. This normative understanding of the role of a therapist is conducive to test diagnostics, but cannot be transferred to the understanding of the role of a psychotherapist! There is a clear need for an understanding between the various schools of therapy in this regard.

6.2.4.1 Blind Spots in Therapists

In all areas that the therapist is unable to provide resonance, the patient will also fail to reach a conclusion for the aspect in need of clarification. These aspects of the therapist may be.

(a) a result of the structure of his personality, (b) content-related or thematic, or (c) due to training or education that was characterised by certain taboos and prejudices.

(a) The structure of the therapist's personality plays a role if the personality in question exhibits clear imbalances or "accentuations." In a positive case, the therapist can achieve outstanding results for a type of disorder with which he himself has come to grips with as a result of his own issues. However, he may not be able to identify with other issues and fail to get on other wavelengths, lack experience-based understanding or exhibit no ability to resonate. As is widely accepted in the field of addiction, ex-users who have successfully conquered their addiction are generally perceived as more credible, encouraging and consequently successful for patients of addiction than a therapist without this shared experience. This also applies to a certain extent amongst trauma therapists. Indeed, I have come across great examples of this amongst colleagues who have developed impressively successful personalities as therapists despite, or perhaps due to, an initially distressing start in life. Needless to say, there are no lack of examples for how therapists can remain stuck on their own painstaking path to coming to terms with the past whilst nevertheless practising the profession of psychotherapist, often doing more harm than good; the flip side of the coin.

The issue surrounding eligibility criteria and admission to the study and profession of psychotherapy remains unresolved. At the time when psychotherapy training was solely offered by private institutes, admissions interviews, interim assessments and final admissions all formed part of the admissions process.

Those deemed unsuitable or less suitable were, generally speaking, structurally weak people, borderline people, those suffering from personality disorders, addicts, dissocial circles and those in need of therapy, who were subsequently referred to therapy. They were instead required to dedicate their energy and time to further developing and harmonisingly maturing their own personality. Wherever possible, alternative paths for their careers and futures were discussed with them and the potential of illness minimised to the extent deemed possible. Whilst there may have been failings on the part of individual institutes to be adequately uniform and objective in the application of selection criteria and wrong decisions may have been made in certain cases, as a whole, this system did far more to benefit the community than damage it.

An aspect is missing from psychotherapy education on a university level to this day. When no admission criteria are applied for entry into a degree in psychotherapy, the use of a special counselling centre to offer a screened exit from the course or switch to another field of study would be desirable.

On the other side of the coin, therapists with heightened development should be able to work around the fact that they recognise their strengths for their specialist area and similarly are familiar with their limits, and are capable of finding a good way in which to refer patients.

Those who have been required to fight a running battle with the majority of problem areas without permanently collapsing or sustaining significantly damage as a result are ideal for general patient care. Namely, an average level of stress with an average level of resilience as a minimum requirement, and careful conditioning in terms of sufficient self-experience in an individual and group setting, which reaches the required depth, appear to be prognostically favourable. Self-experience simultaneously constitutes "learning by doing". It leads to the emergence of all-rounders. Their resilience can be strengthened by the internalisation of guiding principles and elective affinities.

Supervision and, at a later point, intervision groups are highly recommended for the period following training; a group of colleagues to promote mental hygiene.

(b) Thematic and content-related blind spots can be found both in an individual and collective form. For instance, totalitarian regimes utilise existential fear to create collective taboos. Structures based on hierarchy, like the military, strict and factional ecclesiastical institutions or political parties, as well as patriarchal family clans can also shape individuals through angst-ridden, unwritten rules, creating untouchable areas within. We often encounter this as a normal phenomenon amongst patients, but the notion of explicitly scanning for this among therapists is not part of the job description. A sensitive patient will be able to notice if the therapist fails to provide the expected response and will then learn to avoid unconscious non-responsive topics.

An example of collective suppression: In Germany after the second war, there was initially no opportunity to account for the Nazi past in psychoanalytic training analyses. According to witness statements, the topic was simply not approachable. This would have summoned a response to the atrocity of the holocaust with all of the monstrous horror, which was too extensive for one individual to process. The topic was only first publicly addressed almost one generation after the end of the war at psychotherapy conventions in the 1980/90s, when discussions were organised between representatives and victims from both sides, including certain ones from the resistance, particularly in the first and, above all, the second generation. The emotional involvement of everyone was extremely high and sometimes also publicly cathartic. (This topic also contributed to the former feud between the two psychoanalytic associations, the classic German Psychoanalytical Association (DPV), which was extremely affected by the holocaust but strengthened by re-emigrants, and the German Psychoanalytic Society (DPG), the officials of which came to terms with the politics of the Third Reich.)

Another collective taboo in certain institutions for a long time was the practice of paedophilic abuse. This also often applied to family secrets.

As we are all aware, we all have personal blind spots. In my personal experience, the self-experience groups for prospective psychotherapists at present are more likely to address the previously shame-ridden topic of sexual intimacy, including perverse inclinations and the often-unsuccessful integration thereof into a loving relationship, than the uncomfortable topics of spirituality and death. These topics require stronger familiarity than the former, as the fear of being dismissed and ridiculed is much more controversial. As far as I am aware, this is rather rare elsewhere.

However, patients might well come to therapy with anxiety and sombre question marks surrounding the topic of the end of life and pay close attention to whether this has an impact on their therapist or not, or whether he or she becomes overwhelmed as a result.

Ultimately, it would be unrealistic to expect that our therapists have permanently solved all of their life problems following their training or education if they have been briefly addressed and talked through. Depending on any other hindrances that need to be managed at that point in time, aggravating resonance may arise from problem areas that have subjectively already been processed. This is not an entirely unusual phenomenon and should not be regarded as a shameful event, but rather as an indication of the responsibility to organise accompanying therapy with someone on the basis of trust in the near future. The quicker this occurs, the quicker the ground beneath the therapist's feet will once again be sound.

(c) Delegated prejudices: In this field, as psychotherapists, we unfortunately turn out to be particularly mortal, or perhaps we should say "predisposed?" This is not just a privilege reserved for psychotherapists, but rather a ubiquitous group dynamic phenomenon. Belonging to a group appeals particularly to individuals who automatically feel entitled to feel comparatively superior to others by virtue of belonging to a group. This is a tried and tested means employed by propaganda. Typically, this bestowed boost in self-esteem is warmly welcomed and the accompanying devaluation of the counterpart is only scrutinised by truly critical minds who are not reliant on this addition and want to discover the truth.

We now return to the field of psychotherapy. Which promises do the individual schools make to entice us? Which self-idealised images have they conjured up for themselves with this in mind?

Criticism

With a certain touch of self-irony, G. Rudolf [4], himself a member of the DPG, quoted the English psychoanalyst Sandler, who is a proponent of unrestricted psychoanalysis with 4–5 sessions per week on a self-payment basis, as "a psychoanalyst who believes in the immense value that the discoveries made by Freud and his successors continue to have today", according to her own self-reflection. Rudolf's comment on this was: "It could be suspected that this stance – namely being a classical psychoanalyst – presents a certain allure for many therapists who are, by and large, capable of handling the everyday issues presented by their patients and long to belong to a special community". Followed by another criticism of classical psychoanalysis: "In this day and age, it is not possible for one individual to earnestly claim interpretational sovereignty against another". This reference hopefully adequately summarises the prevailing existence of this attitude: We, psychoanalysts, are special. We represent the "Gold of Psychoanalysis" in the sense of Freud and take care not to be sullied by the "Copper of Suggestion". We are entitled to the ultimate sovereignty of interpretation.

Accordingly, this presumptuous stance does not only give rise to opposition from non-analysts, but also amongst the critical ranks of psychoanalysts.

A change in perspective: The behavioural therapy group offered its scientific tradition, of which BT seems to idolatrously fond, for its self-idealisation. In comparison to other schools of psychotherapy, it had and continues to have the most knowledge in this field. However, "science" can also be used as a weapon in conflict. There is great temptation to abuse scientific methodology if you are relatively sure that your counterpart will most likely not detect the ploy quickly. At times, it is possible to get the distinct impression that BT has tried to bend reality to suit its own method: the psychotherapeutic reality should ideally follow the statistically exact double-blind methodology, including the (randomised controlled trial) RCT concept in such a precise manner that it could be successfully applied in laboratory research in pharmacology. However, this does not accurately portray the reality of psychotherapy. We deal with humans, not molecules, as the unit to be measured. The required "monosymptoms" simply do not exist in reality. They can be regarded as artefacts. It is also possible to turn a blind eye to this.

Students of biased BT professors are often exposed to the prejudice that psychodynamics, particularly psychoanalysis, are hopelessly outdated, ridiculously unscientific, ineffective and belong in a museum at best. Students at psychodynamic strongholds believe that behavioural therapists are encouraged to become superficial couriers of symptoms in addition to reeling off tutorials with limited empathy. Whilst suitable for research, this canned form of psychotherapy is not individualised and, as such, rarely suitable for practice in individual cases.

In any case, the humanist school was already out of fashion (in Germany) and comprised, to some extent, arguably a colourful, creative mix of people, philanthropic dreamers who were not entirely keen on exact science. Many of their methods enjoyed relatively good success in practice. Indeed, it would be conducive to adopt them. All schools have already drawn on humanistic methodology. However, the observation of mutual imitation also applies to all other clusters. Plagiarism may be frowned upon as intellectual theft. Nevertheless, it may also be regarded as an unintentional compliment, provided the imitator makes it clear that he holds this particular method in high esteem and views it as beneficial. In this way, our society of general theft has unintentionally paved the way towards integration.

6.2.4.2 Alliances Born from Blindness
Blind Spots Amongst Therapists and Patients

As mentioned above, a sensitive patient will be able to sense on which topics he resonates with the therapist, perhaps even on a level of defence, anxiety, aversion, anger, experiences of powerlessness, etc. Depending on the relationship constellation between the patient and the therapist, he may register this as a sign to stop, perhaps only at an unconscious level, and put this topic on the back burner. However, in the event that the patient is spoiling for a fight and protest, he may instead prepare to be on the offensive.

It may also be that case that the signals to stop are emitted from the patient himself, as is more often the case. Typically, the therapist is able to consciously register these signals and either respect or address them. However, it may be the case that the therapist unconsciously or preattentively registers the stop signals and is secretly thankful, as they blend in well with the therapist's own blind spot. Following this, the patient and therapist enter into an alliance in joint defence. Whilst this may increase their understanding, it does not yield any results.

Tip

The magic question with regard to blindness in terms of psychotherapy:

- Is it better to be blind than (be forced) to see (everything)?
- Or: What is the benefit to seeing haziness/emptiness/nothing at this point?
- Or: I am the mist. I play an important role for …., namely ….. I've been around since this or that happened ….. If I wasn't there, this would probably happen:…. …. would be the most at risk. I would be superfluous if ….

Blind Spots amongst Patients

This issue is a common one in the field of psychotherapy. The theory of defence is the very core of psychodynamic psychotherapy, or rather conflict pathology. A common functional understanding in relation to modern trauma and polyvagal theory has already been provided in this book.

6.2.5 The Dissonance-Reducing Reconciliation of the Conceptualising Consciousness

6.2.5.1 A Defence-Oriented Principle of Erasure and Blinding

The way in which the conceptualising consciousness works appears to primarily focus on finding approximate solutions with the best possible fit and largest reduction in dissonance in comparison with the stimulus. As long as we are not entirely aware of the reasons behind the almost constant "corriger la fortune" type of functioning of our brains, we remain dependent on hypotheses.

Several generations of psychodynamically minded colleagues have compiled detailed, insightful observations on the fact and process, whereby the contents of experiences that enter into dissonance with the self-image on the fringe of consciousness are repelled, erased, masked and displaced, etc. The observations made by this group focus on the divergence of the current experience from the individual's self-concept, which is strongly oriented around internalised standards. This conformed to the main conflict axis of the Freudian worldview and its social conditions.

However, erasure can also occur amongst individuals with tolerant standards.

It would appear to be a basic principle conducive to survival. Our brain is required to cope with an immense variety of stimuli. It endeavours to (1) present static images and circumstances with memorability through its focus on the relationship between the individual parts to the whole and (2) to increase the level of trust in the world by comparing the current constellations with those that have previously been stored and superimposing them, namely relating to these as templates for interpretation. As such, consistency may enter into a synchronisation process that absorbs, or "enslaves", to use the term coined by Hermann Haken in his laser research, wave formations that do not fit. The familiar accordingly takes precedence in terms of assertion, unless this automated process is consciously switched off and the unfamiliar is prioritised.

As has been exemplified time and again in perception experiments conducted in Gestalt psychology, our perception is influenced by our prior knowledge and familiar concepts. This phenomenon is referred to as hysteresis (◘ Fig. 6.2/▸ Sect. 6.2.5). It is both practical and energy-saving. Our brain strives to present us with things that are familiar to us and minimise uncertainty as a possible source of fear; a sensible survival concept. However, it is not without its flaws: the interpretation is often wide of the mark. The most probable interpretation offered by the brain typically only subsists for around 3 seconds at first, only to be replaced by another. In the case that another interpretation is offered that is a similarly good fit, the

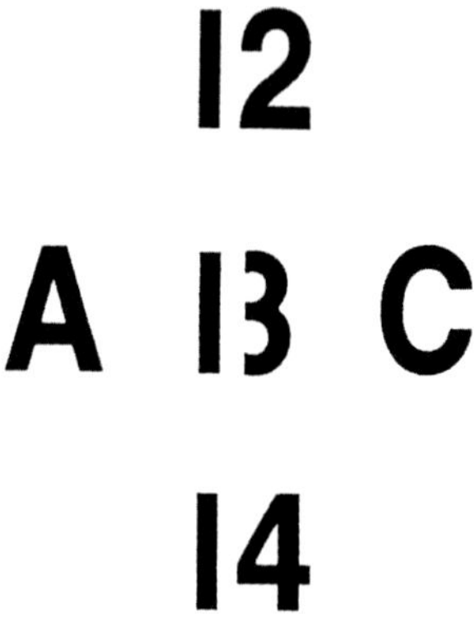

■ **Fig. 6.2** Hysterese

■ **Fig. 6.3** Old or young woman?

former gives way to its competitor. We come across this phenomenon when studying ambiguous images (■ Fig. 6.3/▶ Sect. 6.2.5). Our brain is only ever able to offer one single interpretation in relation to the overall stimuli, not several interpretations at the same time, despite the fact that the world around us could be constantly interpreted (ordered) in a number of ways.

In this chapter, we shall once again delve into Plato's cave, where we deem shadow play to constitute reality. Our fixed concepts delude us into believing that we are aware of the true reality.

If the conscious is provided with two or several suggestions for how to interpret an ambivalent event, depending on the motivation, this may lead to a conflict between which should be energised and which should be erased. Non-accepted suggestions are most likely erased in a central, internal "observed" area, as the information erased there appears to be stored elsewhere and capable of being spontaneously called into consciousness as a result of specific resonances or hypnosis at a later point.

The templates of the conceptualising consciousness help to facilitate recognition and consequently convey a sense of security in an adequately straightforward world, or potentially only convince us thereof. However, templates for interpretation that are too rigid in nature can very quickly lead to discrepancies. Accordingly, it is important that they develop alongside the ever-changing circumstances of life. Every change of template is accompanied by an unsettling crisis. A lack of tolerance for anxiety will cause the change to not occur. As a result, the template becomes fixed. When we reach old age, the enthusiasm for change usually also sinks.

In the case of personality disorders, the originally increased anxiety potential, which is absorbed by the security-giving coping mechanisms for anxiety through the formation of templates, appears to almost entirely curb the modification of templates. The initial stages of freezing set in. The concept of the self largely concurs with the fixed templates. In therapy, the trick lies in reaching an aspect that can be detached from its template and revived.

6.3 Standard Treatment in Chronological Order with the "Unique Therapeutic Relationship"

6.3.1 Conscious Preparation: Agape as the Fundamental Attitude

In this respect, the dimensions of consciousness and the relationship have an effect together, as separately described elsewhere (▶ Chap. 2). They become more potent and are not discussed with the patient. They are instead "given" to him. They provide the prerequisites and background for the future working alliance on the part of the therapist.

This can often require a certain amount of time and energy for newcomers to the field. In contrast, experienced therapists are able to freely, almost unconsciously slip into their practiced therapist demeanour, which appears to operate at a higher structural level and surprisingly is not tiring but rather, on the contrary, brings the therapist a sense of awakening, the subjective feeling of standing in a refreshingly healing current that flows through the therapist with the aim of reaching the patient.

Preliminary phase: **Prior** to each session, the therapist is required to conduct a small amount of calm yet conscious preparatory work, in order to: Reach a centred, balanced state of mind himself, which leaves want to no personal basic needs that may demand something from the patient. The therapist consciously becomes balanced and, to a certain extent, a receptive vessel for his counterpart.

- To provide a sense of willingness in the relationship attuned to compassionate and empathic perception (as per Rogers).
- To focus on a caring, encouraging, potential and resource-oriented viewpoint towards his counterpart. A powerful, meaningful and connective imaginary stream flows from the therapist to the patient (derived from Martin Buber's I-Thou principle for Gestalt therapists and also equitable to unconditional, selfless love).

This demeanour is perhaps most comparable with the religiously neutral fundamental concern (not just as practised in Buddhism), namely the emphasis on: "May all beings everywhere be happy and free!" (The western world of research was taken by storm by the incorporation of this empathetic attitude into meditation by the monk Matthieu Ricard, as it was consequently found to radiate onto his surroundings and counterpart in powerful, high-amplitude, synchronously structured EEG field in this state.) As such, it seems logical to assume that a therapist is capable of radiating empathy onto his patient to a certain degree and providing the atmospherically stabilising basic oscillation of the therapeutic relationship, conveying a sense of trust and security.

Interpreted in terms of the socio-biological level, in this scenario, something is tacitly communicated and subsequently added in terms of holistic resilience, identified as, or at least focused on, the Social Empathy System (SES) by Porges.

6.3.2 The Culture of a Positive Fellow-Beeing in the Therapeutic Relationship

Despite the fact that we all aim to treat all people with the same fundamental approach, it nevertheless remains vital that each person is viewed as an individual and allowed to feel recognition on a personal, individual level.

In this sense, it is beneficial for the therapist to see himself through the patient's eyes. What kind of look would make her feel most understood, appreciated, respected yet not cornered? What kind of inner attitude would this correspond to? The dose needs to be readjusted to suit each individual and every moment.

At first, this may seem like a big ask, but it can turn into an effortless permanent attitude for those who are fundamentally compassionate by nature. If the therapist succeeds in embodying this affirmative positive Fellow-Beeing not just to conform, but rather in a naturally and perhaps even humorous and playful manner, this will provide a strong basis for openness, trust and enthusiasm to become part of the partnership and fill it with joy and amusement, despite the seriousness of the matter at hand. I have found that the dose of this kind of positive affirmative compassion is not always optimal, perhaps due to the therapist being tired, distracted or otherwise aggravated. In this case, it is key that he forgives himself for this human shortcoming and perhaps even addresses it in a suitable manner (such as: "I was somewhat absent-minded last session as there was a lot going on in my head, I'm not sure if you noticed. Nevertheless, I would like to take the opportunity to apologise for this. I just wanted to let you know that my somewhat restricted state of mind had nothing to do with you".)

Many patients have not experienced this form of existential affirmation and long for it. After experiencing it in such a positive light, they often become capable of passing it on to their acquaintances, like the light of a burning candle, and consequently experience positive resonance (for the first time in certain cases).

In this regard, the question is how this positivity inner attitude can "come across" to others in a way that remains relevant to everyday life and natural (ideally not excessively) – and how can I maintain the right distance? In the case of the latter, it can be helpful to consciously fill the role of therapist in a way that is beneficial to both sides, both in maintaining distance and opening up. The patient is or should be aware that the person in this role is simply doing his job and fulfilling a function that temporarily aims to help the patient. Most of the time, patients understand that they can neither adopt nor marry their therapists. Granted, the therapist is required to draw a fine line between solely fulfilling a role and, on the other hand, being a truthful and emotionally credible personality. In the ideal scenario, there are few differences in the fundamental attitudes of the private person and the therapist identity. In this case, the sole difference would lie in the private person's' basic needs, which need to be kept out of therapy and otherwise tended to. The underlying compassionate, appreciative fundamental attitude remains the same in both roles. This way of working can lead to the risk that certain therapists may find their social needs met amongst their patients and, as a result, they may consequently prefer to spend their free time alone, neglecting their circle of friends. An eye must be kept on this tendency, as it may otherwise cause the therapist to drag out therapy and never truly conclude it in order to turn the patients into personal friends without a clear division.

6.3.3 Objective Level of the Therapeutic Relationship

Objective perception: Buber's I-Thou relationship system provides a complementary pole to the objective perceptive ability referred to as I-It. This is comparable to Watzlawick's content and relationship levels. The objective content level tactfully accompanies the establishment of a relationship in parallel and helps us to observe without judgement, classify symptoms and diagnose on the basis of the given criteria. As such, this aspect is a key tool for scientific and insurance classification, but in humanistic methods and all those that recognise the significance of the therapeutic relationship, it explicitly does not have any influence on the personal relationship level characterised by compassion. This level is also not verbalised to patients without particular cause.

As therapists, we learn to act on several channels at the same time.

Needless to say, there are a multitude of considerations in terms of content to take into account. Which classification system are we required to use to decide on a diagnosis? This system will influence our perspective. In the case of an expert opinion, the questions asked attract attention and lead us to not notice certain things or view them as insignificant. As such, it is conducive to get a clearer picture of the objective filter used to observe and to correct it where necessary.

Therapy may make little sense at the time of a pending or ongoing pension application process if the therapy is related to a change that may influence the outcome of the pension approval. The conflict is pre-programmed and not suitable for the patient. These two processes should not occur simultaneously, but rather take place one after another. In any case, the therapist should not simultaneously be required to act as the assessor. This can significantly impair the therapeutic relationship, causing the therapist to violate the medical duty of confidentiality. Unfortunately, it has become common practice in the rehabilitation field to require a detailed, socio-medical assessment from the psychotherapist at the end of treatment, namely a small expert opinion that the patient is aware of in advance. As such, it is perhaps no surprise that the psychotherapeutic outcome is relatively moderate in light of these measures. Nevertheless, the respective institutions are able to save the costs of expert opinions that would otherwise have been incurred. Whether the present amalgamation of therapy and assessment truly saves costs or instead provides a rather naive assessment of the situation through its clear strain on the therapeutic establishment of trust, remains to be analysed.

6.3.4 The Level of Resource Orientation and Salugenetic Resilience

The notion of being consciously open to everything that the patient has achieved, provided this is perceivable for the therapist (even if it is solely the fact that the patient has sought therapy instead of committing suicide), is not referred to as a therapeutic attitude, but instead directly implemented. "I'm glad that you want to take control of your tough situation and have found your way here to improve your situation and the situation of your family. This marks a big step for you and shows that you can and want to take on responsibility. You have most likely managed to pull off things earlier in your life. Would you like to tell me something about that in a few sentences? What are you most proud to have achieved?"

The resource-oriented attitude corresponds to the key requirements of the 1962 humanistic manifesto, written at the time when the humanistic movement in the USA separated from official behavioural therapy and psychoanalysis, instead demanding that the deficit-oriented, scientific and intellectual perspective of the time be supplemented and largely replaced by a resource-oriented view of the patient. This immediately changes the relationship climate, making it more appreciative and welcoming. It reduces the divide between those seeking help and experts, as certain aspects to those seeking help require recognition; the patient feels truly appreciated and receives access to within as well as to his dormant spontaneous healing powers and his own solution know-how. The therapist may symbolically bow down to the patient in recognition. However, this reaction is only conducive if credible, otherwise it can backfire and cause the patient to feel mocked. As therapists, we are constantly walking on a tightrope that only succeeds if authenticity and compassion lead the way.

Aaron Antonovsky's (1923–1994) salutogenesis, based on a mixture of comprehensibility, manageability and subjective meaningfulness, provides resistance-boosting harmony referred to by Antonovsky as a sense of coherence, takes a similar stance in terms of improving health. The world appears to be cohesive, structured and consistent for healthy people, even in the face of challenges posed by tasks; those considered healthy find it logical to accept and confront this world. A positive correlation can also be drawn between health and resilience, and positive social integration.

Psychotherapy promotes the above health-stabilising qualities.

6.3.5 The Compassionate Credible Relationship as Part of the Therapist's Role

6.3.5.1 Namely, a Partnership with Selective Authenticity

In humanistic therapy, the therapeutic identity is not far removed from the identity of a realistic average person, who, despite being a professional, aims to be a positive, compassionate fellow man for those seeking help. In this regard, the goal is to achieve a flat hierarchy between the patient and therapist.

The therapist possesses all of the usual contours of an individual and does not need to assume the role of an empty projection screen, as in classic psychoanalysis. Similarly, he is also not required to enter into a teacher–student relationship, as is typically the case in psycho-educational methods employed in a number of behavioural therapeutic approaches.

The emphasis of the therapist's role is not rigid but instead reacts to what the patient currently needs. This may be different in times of crises than at the end of therapy.

However, in contrast to everyday life, the therapist must not direct any desires related to the satisfaction of his or her own needs towards the patient. Abstinence is key here. The therapist must not respond to the desire for satisfaction from his ward, but rather integrate it into the therapeutic work.

6.3.6 Transference and Countertransference

Transference can be regarded as a distortion of the patient's perception towards an unresolved, previously extremely significant relationship that he continues to hold onto within. It may have negative or positive connotations, or be linked to particular expectations.

Despite their self-experience, which is conditioned by training (often not adequate training at that), therapists can also transfer to their patients. This constitutes countertransference in the strict sense of the term. As perhaps goes without saying, this should not exist at all in the profession and is a taboo subject amongst some colleagues and entire schools. However, this is not a concern of countertransference. It has an unsolicited and prohibited effect. In the ideal-case scenario, the individual therapist is able to recognise it and introduce it in supervision or perhaps even in private individual therapy or Balint groups and work on it there. Therapists are also typically prone to crises, especially owing to the fact that our field attracts more people who have experienced crises than others. (If we achieve in putting all professional and political arrogance to the side, we get results far quicker.)

Countertransference in the broad sense, namely the version commonly referred to, pertains to the completely unconscious, inner extrapolation of all signals received from the counterpart. This may include images, concurring or mirroring responses in the form of body language or psychosomatic responses, atmospheric elements, verbalised ideas, comparisons, sensual resonances or drive qualities etc. It primarily concerns the all-embracing capturing of the counterpart's aura. Countertransference is both used in a diagnostic manner and to form relationships, in addition to providing raw material for future interventions.

Transference and countertransference can be traced back to Freud and are regarded as two of the bedrocks of psychoanalytic treatment, which has since spread far beyond the boundaries of its original school. The way in which they are approached has changed somewhat over time. Transference interpretation is strongly discouraged in the case of structurally weak patients (Rudolf, 2006). Less has been said in recent years on the topic of transference neurosis, which is often, in a manner of speaking, the successful, delayed resistance against resolution in which the patient was previously stuck, which is now addressed by the therapist, as the patient is required to be relatively strong willed to successfully use this concept.

In a group setting, we encounter lateral transferences, whereby the transference is directed to another participant. This generally provides a rich source for clarifying confrontations within the group.

In Gestalt therapy, the therapist also picks up on the transference, but as soon as it becomes concise, it is "resolved" and regarded as a signpost to a problem area that still needs to be processed. Conflicts are resolved in a typical, associatively revived scenario (in psychodrama) or in a fitting spatially symbolic relationship constellation (often encountered in Gestalt therapy), namely at the source of the disturbance, or with the imagined problem person from the time. The therapist takes the pressure off the transference, primarily in the event of negative transference, in order to be able to stand by the patient's side as a guide for his processing and, where necessary, to introduce other stabilising measures to enable this imaginary, symbolic confrontation with the original conflict or problem person in therapy to achieve a successful outcome.

The willingness to resolve the distorting transference is reinforced by the objective to grasp the present reality with as much clarity and distinction as possible.

At all times, the most important role of the therapist is the provision of personal safety! This includes imparting a sense of empathetic solidarity with the despairing or traumatised part of the personality that remains stuck in its former or ongoing crisis. This supportive and trust-promoting relationship function must not be shed in favour of transference distortion. Failing which, only adequately, strong willed personalities will be capable of coping with crises. However, in the strictest sense of the word, psychotherapy is most "necessary" for precisely those who are less strong willed.

6.3.7 Development Promoting Counterpart

In Gestalt therapy, therapists are required to make adjustments to suit their patient's development, namely they should use inner role reversal to sense the counterpart required for this particular patient in order to promote their development. This does not entail playing a role that the therapist simply does not embody, but instead bringing a partial aspect from his own inner range that appears to be most fitting in the above context into the foreground and allowing it to come to life. In this regard, it is tacitly implied that the therapist has developed or sensed all of the essential states of development during his training, at least to some extent, including the old, wise part, despite the fact that he may not have reached this age yet.

To put the above into concrete terms: A distressed patient who is currently experiencing a crisis may temporarily regress to an early infant stage. She becomes receptive to the consolation that is fitting for this age. Generally, we intuitively find the right tone of voice and fitting gestures etc., and intuitively encourage the patient to gradually return to the more stable, age-appropriate state of her identity. The patient determines the tempo, although therapists will most likely advance half a step every so often in order to encourage patients to continue along the stages of development. In any case, it is key to avoid the patient becoming stuck at any intermediate stages solely as we may feel particularly comfortable in the respective supportive role as a therapist. The patient's balance and development are the primary concern. The patient usually requires a variation on parent roles, initially caring but also restrictive and demanding in a good sense, then perhaps also an understanding sibling. However, sometimes he may require a chummy, provocative and challenging friend with whom he can go through thick and thin, at other times an objectively evaluating teacher, and in some instances a grandfather-like old wise man. The aspects that still need to develop vary from person to person. (In schema therapy, a Gestalt therapy manualised by Jeffrey Young, the therapeutic relationship is granted the role of restrictive parenting. This provides a good and feasible simplification of the developmental stages of the potential counterpart situations in Gestalt therapy.)

6.3.8 Experts: Between Liberal Dialogue Partners on the One Hand and Experienced Instructors on the Other

A large number of techniques are employed in psychotherapy. They are most commonly encountered in behavioural therapy. No one should feel pressurised to need, want or be able to use them all. They can be very helpful and beneficial. However, the extent to which they are depends on how they are applied and the context of the relationship. Unfortunately, in certain cases, they can get in the way of a therapeutic relationship and consequently hinder the potentially most effective remedy.

On the one hand, they can be categorised on the basis of their effect level, on which they predominantly pick up on signals – which shall be explored in more detail in its own section below – and, on the other hand, they can be categorised according to the extent of breathing space they grant the therapeutic process. The therapist is required, in a manner of speaking, to act as a contributor and dialogue partner.

(a) Open-Minded Experimentation on All Sensory Channels.

This constitutes the broadest and most unstructured method. It encompasses free play and also artistic creation, even if it only takes place within the scope of limited time and under the inquisitive gaze of the therapist as well as the subject's own registering self-observation. It may involve 10 minutes of exploration around the group room to find a suitable space and structure it. It may include 10 minutes of free play with lumps of clay to see what the patient's hands create. Or perhaps free exploration through improvised movement, confrontation with a large area covered with white paper without any instructions (but with a pen close by). In another scenario, the room may contain several instruments: how does that patient express himself with these? All of the above are natural projective tests, which can subsequently be brought together through self-interpretation in relation to the patient's current state of mind and the unresolved issue. This technique is a substantial expansion of the classic method of free association in the non-verbal realm of expression. In general, intensified introspection induces a waking trance, reminiscent of a creative flow. Contact with the unconscious is very strong throughout.

The following constitutes a productive variant for couple's therapy: The two partners sit facing each other at a small table covered in paper (With a pack of colouring pencils nearby.) "This represents your living space and the next 5 minutes are time that you can both spend as you would like. Please each select a pencil". It can be interesting to observe the interaction, which is also partially reflected in the strokes. There are soft and decisive lines, twisting, obstructive, ornately inviting and hazily dismissive lines etc. Is the space divided? If yes, by who? Are there any overlapping sections? Are any retreat manoeuvres drawn? Invitations to encounter? A shared third area? A joint field or instead a death strip? Is there a rhythmical to and fro that appears to create a link? Or instead an "everyone for himself" mentality in the wilting relationship? Which atmosphere is created? Preverbal qualities may be picked up on and subsequently vocalised.

One variant in this regard is only providing the couple with one pencil and seeing how they interact. However, the disadvantage to this version is that it is not always clear to retroactively differentiate between who drew what.

A short amount of time can be allocated to reflection prior to each experiment to focus attention on the issue that remains unresolved and, if necessary, link this to a question. In this respect, an emphasis is placed on the unconscious from the outset. Sometimes there may not be a specific issue that is capable of entering consciousness. Then the perpetual questions remain: Who am I? What is the next thing that needs to be considered? What needs to be done? What still needs to develop?

This form of structurally free techniques may promote insight, but they are only suitable for average and well-structured individuals. Their breadth can overwhelm structurally weak individuals.

(b) Approaches Based on Processes and Individuals.

Approaches such as classic Gestalt therapy and emotion-focused therapy (EFT) are used to intersperse structured sequences during the flow of the therapy process and at each point when patterns of dissonance or tension arise. This form leads to the emergence of a process-led interweaving of interpersonal and intrapersonal dialogue, which helps to resolve tension and increase coherence. Provided the current therapeutic relationship appears to be resilient, the therapist will address an emerging pattern of dissonance at the opportune moment and ask the patient questions about his self-interpretation thereof. He will attempt to assure himself that it could be interesting to divert the focus of attention in this direction.

▶ Example

Miss XY discusses her younger sister who caused a lot of issues during her childhood due to her princess-like level of entitlement. As the elder sibling, she always wanted to be the sensible one and put everything away. Her little sister took full advantage of this, sometimes painfully pinching her or stealing from her, and before she was able to retaliate, her litter sister would start to cream, causing her mother to punish her instead. No one would listen or believe her pleas. She quickly learned to let it go and endure it. When she was telling her story, the patient was grasping her neck. The therapist tentatively interrupted the flow of speech: "It would appear that your right hand has something to say. Imagine that you are your hand. What does it currently want to do?" "It wants to squeeze. Squeeze the air out". "Nature never intends for us to turn against ourselves. Where else could this impulse fit? It certainly seems to fit this topic". "Ah yes, my sister. That wretched brat! I should have thrown her under the bus when I had the chance!" "Just your sister?" "Well, I suppose my mother could have taken a closer look and not automatically assumed that I was always guilty as the elder sibling. She's the crux of the matter. I haven't been able to look at it from this angle before. That's not great. Perhaps it's too late for things to be fixed between me and her after all this". "I can understand why you are worried. But the real bad thing has already happened. For years. Perhaps talking is the best way to save whatever is left. For you first. What would happen if your mother from back then was to appear somewhere in this room and take a seat in your mind – if that were possible, where would you be most comfortable with her sitting? Over there? Ok. Let's imagine that you were able to use your current abilities to express your feelings from back then. If you're ok with this, I'll remain close to you and by your side. We're in this together. What is long overdue, that is to say what should your mother have heard back then from her eldest daughter? (In any case that difficult side of her mother. She most likely had one side that you liked.) What should she have said at the time that would have been suitable for the child's healthy sensitivity?" "You only ever stuck by my little sister and gave her me to play with as a toy and prey! You never noticed me. You didn't care about my distress. You gave me away! You were so cruel to me! I could have strangled you for how you treated me sometimes! (Patient weeps bitterly.) But I also wanted to be your favourite child from time to time. So I learned to stay silent and become a yes-man. I hate that about myself. Other people feel entitled to walk all over me. I don't want this any more!" "Take a moment to listen to your body; your posture and breathing have changed". "Yes, I feel more honest and free, and just took a deep breath, even though I feel nostalgic and raw inside". "Exactly, you have just spontaneously let go of the yes man. Excellent. You have gone through a lot to get the opportunity to have a loving relationship. This may now be put to the test. It may also provide good motivation for a real talk with your mother today, if you feel brave enough to venture down that route afterwards. But that does not provide a measure of whether you are moving forward within. Perhaps your mother is unable to remember every that happened, it could overwhelm her; Your development will continue independent of this and of course at your own pace. I am curious to know if and how your inner dialogue has developed when we meet next week. If you feel like it, you could always jot down your thoughts. A lot of things may come to the surface. I also have one more request: take special care of yourself today, you've earned it. Maybe treat yourself to something, even if it's just a little downtime dedicated to the secret desires of the girl from back then and the possibility of integrating them into your current everyday life in some way. I look forward to seeing you at our next session". ◀

(c) Semi-Free Exercises.

In a group setting, it can be beneficial to try out an exercise together at the same time and then sit down together to reflect on individual experiences and classify them accordingly. Variations of this are also possible in an individual setting, which may encompass body scan exercises, such as on the topic of proximity and distance or communicating solely through the medium of eye contact, leading and being led, falling vs. being independent, excluding vs. being excluded etc. This form of non-verbal exercises can have an invigorating effect when integrated into the group setting and help to promote group coherence by making the participants connect with each other in a non-verbal manner. It is important the sufficient time remains for individual processing, both in interactions and personal biographical terms. In groups that employ semi-free exercises, more time should be spent on interactions in the group than the amount spent on exercises, or at least the same amount of time. The new experiences must be granted the opportunity to put the previous experiences into perspective or correct them. The opportunities to change on the basis of holistic body-related experiences in anxiety-free (or low-anxiety) interactions are significant. These are tried and tested shortcuts in psychotherapy.

Example from Gestalt therapeutic movement therapy (Ursel Burek, seminar experience) on the topic of changing view and place: This approach can be both intrapsychically imagined or implemented through movement therapy, or also in interpersonal and family therapy settings, and used for various aspects of the therapeutic relationship.

Initial position: – Who or what supports me in a positive way? Does this include the therapist? Who from my life thus far has given me tailwind? What could I rest on in order to avoid becoming fixated on a weak role? Which of my current abilities am I able to use to support my inner child or others? Do I experience any other forces (traditions, a sense of belonging, closeness with nature, material securities, spiritual beliefs etc.) as my tailwind?

Second position: – Who or what is stood opposite me as my counterpart? Is my therapist a counterpart who is capable of freely responding according to the play on words in terms of responsibility (Fritz Perls' concept whereby the ability to answer symbolises the responsibility to do so)? Or does she refuse to answer me and instead let me beat my head against a brick wall, feel at a loss and withdraw like a sphinx-like counterpart? In this scenario, a realistic, mutually affirmative relationship on equal terms is required between both different individuals who are nevertheless potentially interested in each other. Facing each other facilitates encounters at all levels, but does not guarantee it. Contact knows all nuances. Who am I not allowing myself to contact directly? Am I aware of which contact I willingly accept and which I consciously refuse?

Third position: – We, the therapist and patient, stand next to each other and observe the ongoing event as observers from a certain distance as if watching part of a film together. We experience a fraternal and open sense of trust that gives us both security and consequently permits independent perception. There are both agreements and disagreements. Each of us live in our own subjective world on equal footing. We exchange thoughts and find ways to bridge the gaps between us.

Fourth position: – I lead the way as the therapist. When the matter at hand is development, the therapist should provide impulses on a trial basis in order to test whether a new stage can be embarked on, whether traditional objectives can be left behind, whether, for instance, grief has been processed, whether hereunto avoided challenges can be pushed further to the fore, whether the topic of aging and death has become integrable, and perhaps whether it is time to initiate the end of therapy. Who has already led the way for me and unconsciously acted as a signpost and role model? If I lead the way myself: do I know how to meet my own "wise old man" and ask him the way?.

Semi-free exercises can be adapted at will. They provide both patients and therapists breathing space. In addition, they can be adapted to suit different structural levels.

In structurally weak areas, they can be used for planned new experiences aimed at building structure. Example: I practise sending a smile from the inside to all parts of my body, particularly those that are overlooked such as my feet and joints etc. and massaging them from the outside.

(d) A number of set, standardised exercise programmes aim to achieve certain goals through selected, outlined intermediate steps, or at least promise to do so on the basis of findings made thus far. The advantage to these programmes is that the participants are relatively well aware of what they are signing up for. Their need for control is well satisfied at any rate. The same applies to psychotechnically trained therapists. The therapist conveys what is given. He is required to diligently comply with the manual or rules of the game. This loyalty to the manual is regarded as a security for the guarantee of success. However, paradoxically, research [5] has shown that loyalty to the manual may have a negative effect. It can minimise the breathing space in the therapeutic relationship. Originally, the enthusiasm of the manual writer may have resonated despite his specifications and spurred success. It cannot be expected that this factor can be repeated with other therapists. In certain cases, the same programme will have a slightly different effect on the patient than expected due to his individual response. However, the therapist is granted little to no leeway for flexible reactions if his patient's response deviates from the norm.

Therapists with less experience may find precise guidelines to be a beneficial measure that relieves anxiety. Those with more experience will most likely find them restrictive. Patients miss out on the individual resonance chamber and often feel overlooked, treated as a number and consequently like a fish out of water. Nevertheless, they often find relief through the relativisation of seeing others similarly suffer, helping them to learn that others appear to be in the same boat.

Behavioural therapy has provided general psychotherapy with an impressive range of standardised therapy programmes.

6.3.9 Working Alliance and Partnership

To start with, health insurance providers permit up to three preliminary sessions on mutual terms to receive the opportunity to intuitively check whether the notion of a future collaboration is approved by the unconscious and whether an adequate level of trust is established, or whether this is not the case. It can be hard to navigate rejecting a patient in a negative case and to express oneself in such a way that the patient seeking help does not incur any additional damage.

The general working alliance is formed between the part of the patient and therapist that is capable of bearing responsibility. In times of crises, this part may not be accessible to the patient. At best, this may lead to the establishment of a temporary bridging guardianship in which the rules of cooperation are discussed. The primary emphases in this regard are the temporal and spatial setting, potential crisis measures in the case of deterioration, clarification of the insurance circumstances and payment terms, potentially also discussions with relatives, rules on vacations and cancellations and possible also the forms of accompanying cooperation. This may include diary entries or specific observations that may be of importance to the topic at hand. It may also encompass exercises interspersed into everyday life. To me, it appears to be important that the accompanying exercises are agreed upon in a process-oriented and consensual manner. It should be possible to experience them as a consistent continuation of the therapy process. In this regard, it is key that the therapist shows interest in the homework, otherwise the patient may become demotivated. Canned therapy does not motivate patients. In the best-case scenario, the patient is able to determine what needs to be improved of his own accord with additional experiences before the next session.

The objectives of therapy are a key element. They may change intermittently. It may be that at the beginning, the patient is still partially blind to himself or unable to think or hope for anything for himself. Objectives and desires can be readjusted over time. Sometimes overshooting objectives may also need to be scaled back, or the assessment of the situation adjusted. Objectives may become more developed and coherent. This is the desired outcome.

In group therapy, there are a few rules of etiquette depending on the style of intervention, such as "Everyone must only speak about themselves and avoid interpretative 'You-statements'. We say 'I' instead of 'you' to make the respective statement more consciously responsible". "We allow each other to express ourselves". "Interruptions take precedence". "Nobody is allowed to talk about anyone else in the group outside the group, enabling the group to remain an intimate, protected space". Participants are more than welcome to discuss insights that they have obtained to partners to ensure that the partner is not excluded from the development process of others in the group. It must not be possible to identify another group member on the basis of this explanation. The rules specified above are prevalent in interactional and humanistic groups.

A particularly special working alliance is also established, often only non-verbally through nods, if the Gestalt therapeutic work enters into time regression work with time dedicated to the patient's inner child. Regression therapy often takes place in a form of waking trance, during which the therapist is required to have increased sensitivity and responsibility for guiding the patient. She must ensure a continuous rapport through non-verbal signals or a discrete accompanying conversation to guarantee the patient's safety. For the patient, the real-world fades in favour of his subjective one, in which he becomes immersed and changes his tone.

6.4 Variations on the Therapy Setting and Relationships

6.4.1 Couple Setting and Neutrality

When we change roles within, it becomes clear how important it is that both partners need to feel equally understood by the therapist with regard to their concerns and protected to the same extent, provided they sought therapy as a couple. This requires a certain level of practice on the part of the therapist. It is well known that individuals who have only experienced trusting relationships in dyadic constellations but never in triadic, in the case of estranged parents, particularly struggle to navigate the requirement of neutrality in the couple setting. They often feel insecure almost instantly when confronted with a consensual dyad as they fear hostile isolation.

Asymmetry is a natural element of family talks: the patient and his perspective are more trusted than the therapist, as the new arrival in the family. This must be addressed in detail at the start of therapy. "Hello Mr XY, thank you for coming along to help provide clarity on certain issues. You have long been aware of this problem area that is the topic of conversation today and affected by it. Therefore, your viewpoint is particularly important. As I have met with your wife a number of times, the topic on the agenda today is how you experience the issue. In your opinion, what change could be an improvement or perhaps even a solution? What would then be different?"

6.4.2 Multi-Person Settings

- Families can be regarded as hybrid beings; they are a unit that comprises several individuals yet nevertheless a whole in a permanent state of change with various fault lines and breakage points. This poses the therapeutic relationship with the challenge of identifying with a meta-level, changing perspectives and enriching the affected individuals through this change in perspective in addition to putting their own subjective initial positions into perspective. A relatively secure atmosphere can be formed with the assistance of neutrality paired with multipartial empathy.

Questions on the network of relationships within family therapy connect the content and relationship level. We can either elect to focus on the relationship level of those affected and give it space to breathe or emphasise the objective structural aspect of the process as the key element. There have been various phases in the history of family and systemic therapy in this regard.

A spatially symbolic constellation can provide an overview of the emotional network. The key question in this regard is that of whose inner image succeeds (or should) in making it into the scenic constellation and becomes a starting point for the joint discussion. It may be the perspective from the youngest family members, the sickest, the most subjectively responsible, the most excluded....How do the parts view the whole? What is their opinion on the issue? What do they want to change to make belonging to the group – or perhaps their (in)voluntary relative exclusion therefrom at present – better for them? How do they think the others view and, most importantly, feel about the whole and the desires and fears of each individual? Are there any loyalties to the whole, or (only) to individual members? Have any commitments or duties been stated to certain values and concepts of man? How "are we" as a whole? Are there any differences between the external and internal perspectives?

Who is burdened by which differences of opinion between which other members? What is it not permitted to experience, or potentially mention? What is utterly taboo? To satisfy whose need? Are we able to at least rudimentarily address the topic and then respect it as taboo? Is there someone in the group who should be granted the power to decide whether the taboo should continue to apply or be suspended? Who feels bound to it and who doesn't? What does this change for whom? Who or what does which member want to protect?

In which relationships does it make a difference to unpick contradictions to allow the individual to express what they are grateful and happy for and break away from aspects they want to leave behind as they were detrimental? This can help to solve hitches in relationships. If the individual's inner desires had the final say, what would change?

6.4.3 Group Situations

On all accounts, the therapeutic relationship aims to achieve the goal of multipartiality. This also depends on the role the group is required to play. Is it intended primarily as a diagnostic terrain for the therapist's social-communicative skills, or instead as a form of supervised self-help group for the mutual exchange of experiences? Does it primarily provide a space for work and exercises, or a noncommittal creative space for play? Or should it develop into a healing matrix with the potential to solve conflict and change structures? All of these variants may be conducive in the right place at the right time.

The following assertions apply to a form of therapeutic group that regards itself as psychotherapy in terms of the final example, namely as a supportive, healing matrix that enables changes in individuals. The group leader is viewed as responsible for ensuring that nobody is harmed and required to embody these values that provide security. She is responsible for guiding the initial phase of getting to know each other by using structuring, playful techniques in a way that establishes experiences of mutual positive relatedness and networking on a non-verbal, unconscious level and for helping the group to develop a mutually trusting basic identity. Variations of the therapist's fundamental attitude are reflected in the relationships between the participants. The group develops into a "social uterus", which is capable of taking on the (retroactively) comforting function of Porges' social engagement system (SES) where necessary. In this form of group, body therapeutic, creative, staged and imaginative interventions provide the ideal and in-depth ways to reach the basic needs level in imaginative time regression (in a waking trance) and to consolidate or, at any rate, put experiences into perspective there. Patients from a migrant background are encouraged to initially allow verbalisations in their native language.

The basic trust stage is naturally followed by a phase of differentiation, which may be introduced or delayed, where necessary, by the group leader through interventions that address the differences.

6.4.4 Timeframe and Structural Fit

Groups can be understood in a variety of ways, similarly to how there are different types and functions of groups in different contexts. Here, I refer to forms of groups that have proven particularly effective in the context of my psychotherapeutic experience. Considering this, participants need to be arranged in a somewhat uniform manner with regard to their personal structural level with the help of preliminary sessions, and able to commit to 50(−70) hours of sessions in a closed group. In order to get beyond the level of everyday life in groups and explore issues in more depth or clarify personal issues, it would appear that a minimum of 2-hour-long sessions per week with a break is the minimal amount of time required, with 4 hours on a Saturday every 2 weeks or a 2-day group meet once a month being optimal. The longer the intervals between the respective intensive group work sessions, the stronger, more structured and resilient the participants need to be. Those seeking to pursue training should be able to handle block group sessions every 6–8 weeks. Participants are encouraged to write diary entries between sessions to bridge the gap and obtain adequate distance for self-processing. In exceptional cases, a telephone consultation may be organised for crisis situations. Based on experience, this is rarely used, but nevertheless provides a sense of support.

6.5 Methodical Supplementary Knowledge for the Standard Approach as a Process

6.5.1 Recording Traces

6

How can traces be accessed? How may dissonant tensions become apparent? Noticeable changes happen in the here-and-now in the context of spontaneous stories, descriptions of symptoms or memories are best in this regard.

The combination of trustworthy and secure guidance from the calm voice of the therapist with the plea for attention on the recording of traces allows the patient to slip into a waking trance with a restricted focus of consciousness.

Following this, the patient and therapist are required to come to an agreement on one of the jointly observed phenomena that it may be interesting to learn what lies behind it. Typically, there are more traces that it is possible to handle.

6.5.1.1 Unconscious Body Language

Tip

Beforehand: Each confrontation with an unconscious body expression requires appreciation of this additional information and not reprehensive denouncing in the sense of "now I've got you!" If a therapist acts in this way, the patient's scrutiny will be ramped up, leading to a loss of trust in the therapeutic relationship.

Above all, dissonant patterns (dissonant to another non-verbal signal or verbal statement), recurring or irregular patterns are well-suited to recording traces. A sudden change in posture may be noticeable ("What just happened with you? Did you notice anything? All of a sudden you sat completely upright, if I remember correctly" Or (in the case of finger tapping): "Your fingers appear to have something to add. Would you like to repeat it again, perhaps even stronger this time? Yes, thank you. What could they be trying to communicate in their language? …. and with whom is it that they are most likely trying to communicate?")

Recurring patterns usually go back a long way in their genesis. As such, it may be conducive to pose the question: which posture or movement would your childhood and teenage friends most likely recognise you from? (Dr. Hildegund Heinl often commenced her seminars with this question and asked participants to demonstrate, recreate and interpret this gesture or posture. It usually provided an introduction to in-depth individual work in the group). It may be entirely subtle movements, such as biting one's lip instead of saying something, or starting almost every sentence with a "No" and throw one's head back slightly with raised eyebrows.

6.5.1.2 State of Mind, Drive and Temperament

The skill to access lies in separating, centring and potentially personifying the quality from the temperament and drive as an aspect of the remaining personality, enabling it to develop into a counterpart.

▶ Example I

Therapist: – "If you could condense your current temperament first into an attitude and then into a being, what comes to mind?"

Patient: – "He's hanging his head, hunched over, grey, staring vacantly into space and saying nothing as intended – nothing to anyone, and certainly not to my wife".

Therapist: – "And if he could do what he wanted? Why not try it out from his point of view".

Patient: – "He would put his foot down and make everyone jump in fright!"

Therapy: – "Wow, there was a force behind that. It appears you have a bomb inside waiting to be set off. How are you feeling at the moment? Your eyes just lit up for a second".

Patient: – "Well, I think It's clear how I'm feeling right now: like a ticking bomb! That being said, I'd prefer to leave the bombs hidden within, otherwise everything will explode! Where and who would I live with then?"

Therapist: – "And how comfortable do you feel living with a bomb waiting to explore? It already seems to be audible ticking. Why not disarm it here first? That would reduce the risk".

Patient: – That doesn't stand to reason to me, the fact remains the same, along with my anger".

Therapist: – "The fact remains the same. But the perspective can be changed".

Patient: – "Well, I won't let anything be twisted or sugar-coated. My wife was sneaking around behind my back with my best friend, and not just the one time. I didn't realise at first or perhaps didn't view it as possible, or want to acknowledge it, despite the fact that a neighbour alluded to it 3 years beforehand. I finally snapped at her and she realised that this was a serious matter and I had become suspicious. We've all been on a break from each other ever since".

Therapist: – "This break seems to have played an important role. (Patient shows interest) Could you perhaps speak from the point of view of the 'break' and let us know why it has been good and helpful?"

Patient: – "I find what you expect from me to be mind-boggling. But never mind. I am the break for the grandmasters. I prevent everything from breaking that should have lasted an entire lifetime. Things aren't working out for some reason. They're still talking about everyday issues, but not exploring their feelings and their relationship. I've forbidden that. Granted, they've never been very good at that".

Therapist: – "That must be pretty bleak. Now everyone is on the rocks. I'm referring to you again as a whole, the grandmaster, who would still prefer to gain clarity. What do you find to be the worst part of the whole thing?"

Patient: – "I feel betrayed, offended and humiliated, ridiculed in front of friends in groups where I was always the top dog".

Therapist: – "So this is all about being the top dog. What role does your wife play in that regard?"

Patient: (proudly): – "She was and is the most beautiful woman on the planet".

Therapist: – "Did that strengthen your claim to be the best and frame it well?"

Patient: – "Sure, I guess you could see it that way".

Therapist: – "Imagine your wife was listening in on that just now. How would she have felt? How would she have reacted?"

Patient: – "She made the point before about wanting to be more than just a trophy wife. It wasn't enough for her. But she isn't consistent; she also likes having a nice house and attractive husband".

Therapist: – "But he isn't enough in the long term? That would be a tough lesson to learn, which is currently being thwarted by the break".

Patient: – "I hadn't look at it that way before. I don't really know what she feels inside".

Therapist: – "Do you remember how our session began today? Your temperament transformed into an individual with his head hanging low. We could feel your anger. But we were not able to sense the aspect causing your head to hang low. Do you think there is any connection here? Is something missing? The being that represents you, your wife or the relationship?"

Patient: – "In the past I would have claimed nothing is missing. We have regular sex. All is good in that department. We don't have any complaints. But when I see my wife talking to others on the phone or conversing with others, I get the sense that they are all closer to her than I am. I feel powerless, angry and almost impaired in equal measure".

Therapist: – "I'm impressed by your ability to pick up on these differences, namely that certain people resonate better or worse with others, and it concerns me that your being is hanging his head, as though he is sad and has feelings, as though he is able to perceive that he is missing something that connects others. It may appear paradoxical to you: but precisely because you are beginning to suffer from this lack – at least in the background – you may have the chance to make up for it. If your wife was listening in on the last few minutes of our conversation, can you guess how she would have reacted?"

Patient: – "She would be surprised and perhaps even relieved. She never complained that something was missing, only indirectly, long before we became miserable, she would sometimes bring up how she felt the loving, tender connection between us had almost entirely disappeared. She misses this connection, which allowed her to truly enjoy sex. Otherwise, she claimed she felt like a prostitute with a marriage certificate. I thought she was joking. But perhaps not".

Therapist: – "Can you recall any moments when you were moved or affected? Moments of happiness can also lead to overwhelming amazement and goosebumps. In comparisons, eye contact or agreements made via signals are far less spectacular. Are you aware of any non-verbal conversations? You seem a little lost in your thoughts at the moment. Would you like to express what is going on inside?"

Patient: – "I'm currently imagining a scene from back when I was a child with my grandparents, we had everything then. But once I started school and moved in with my parents, who both preferred rational clarity and achievement, I learned to look down on all emotions. This helped me to quickly move on. Now the question begs as to what is really behind this".

Therapist: – "Take the time to consider this. It seems that you are reassessing and reframing your path through life. Has anything changed with regard to your wife?"

Patient: – "Time and again I encountered a feeling of emotional starvation, similar to how I felt as a young child, before I had learned how to tune out emotions and to build up my own armour. My earlier distress had apparently transferred to her. I will ask for her forgiveness, understanding and patience. When I get home this evening, I'll bring her a bouquet of flowers and explain everything". Therapist: "I'm in awe of your incentive to transform bombs into flowers!". ◄

► Example II

Therapist: – "How are you doing? You're nodding your head to and fro a little. Do you want to do that perhaps somewhat more pronounced? What could it represent? Are you able to interpret anything in this regard?"

Patient: – "I'm not entirely sure how I feel; neither fish nor fowl. I don't feel like I am meant to decide here. I see myself walking on a tightrope with a horizontal stick in my hands mumbling to myself: don't fall, don't fall..., odd".

Therapist: – "Coping with tension in the face of an undecided matter? Where could that fit in?"

Patient: – "That's exactly the feeling I get when I think about the fact that my daughter has perhaps married the wrong man, wrong from my point of view, but perhaps good enough to her. It's hard to stand. I keep my complaints to myself. They are perhaps a little too harsh. I wanted someone better for her. She was always the apple of my eye. But I don't want to mess up anything for her that she seems to have her heart set on". (Patient starts to cry.)

Therapist: – "I am touched by your love for your daughter. Whatever the outcome, she will always appreciate your input". ◄

► Example III

Therapist: – "How are you doing? How do you feel?"

Patient: – "Lousy, like I'm being hunted. I was always able to run away, as if there were people after me, as if a storm was closing in. I'm exposed to it. It's just as stormy at home. It often gets loud. From time to time, things hit the wall".

Therapist: – "There seems to be a great deal of dynamics, hunters and the hunted? Your animated voice and flashing eyes would seem to indicate that you could also be the hunter or the storm, or am I wrong there? What is the storm's legitimate concern? What does it really want?"

Patient: – "It wants to make threats, shake things up and scream, it also wants to make tracks, but it can't get anything!"

Therapist: – "A powerless storm that is trying to threaten you? What does it want most of all? What does it desire?"

Patient: – "It wants to curl up and be quiet in a safe place – to simply exist".

Therapist: – "And that's what is missing?" (Patient sadly nods) "My husband has been having an affair for years and is moving out".

Therapist: – "Oh, I'm sorry" Therapist places her arm around the Patient's shoulders, calming her. "Saying goodbye can be extremely beneficial if successful. It can even help you to heal".

Patient takes a deep breath: – "That would be great"…. ◄

6.5.1.3 Daydreams, Images, Imaginations, Dreams About Ambitions and Monsters

Principle: What is propelled into the foreground by emotions as a result of an experience? Or, in the event that a whole sequence of images is recollected in dreams: where does the film stop if I let it play forwards and rewind it a few times? The place where it stops is the point at which there is the most fascination. If I morph into this aspect in the foreground, what does it want, what does it have to say? What does it capable of doing? What does it like? What is it potentially avoiding? What does it view as its task? What is the point in its existence? etc. To whom is its message directed? Which aspects does it simultaneously plunge into the background? Which other details are important? What do they signify? What connects the details to each other? Do they complement each other? Do they conflict? Do they cancel each other out? etc. Can any parallels be drawn to relationships experienced in the patient's everyday life? Are there any gaps in the image? What do they represent? What should not be as opposed to existing? What purpose does anything missing have for the entire situation?

► Example I: Grief and Depression

Patient: – Anna, 43 years old, brings a small drawing to the session. It does not contain much: Darkness and a small light area in the right-hand corner. "When I woke up today", says Anna nostalgically. "I guess that's it", she added. The therapist looks at her inquisitively: Anna: "The light doesn't come towards me, instead it swims across the dark sea away from me and gets ever smaller. Yesterday, at the gynaecologist, she explained to me that the last time I was late on my period was not a sign of pregnancy, as I had hoped, especially due to the fact that I have previously had several abortions, but that it instead marked the first signs of the menopause and a disturbance in hormonal balance".

Therapist: – "You would like to have had another child? Once you enter the middle stage of life, this requires more energy, power and love than when younger. You are worthy of much respect for your commitment!"

Patient: – "Yes, however nothing will come of it now. Praying won't help".

Therapist: – "That is one level of reality. Your mental readiness is another. Recently, you have experienced a longing to love and the confidence to protect and strengthen its light. Perhaps your needed someone like that at the start of your life. You didn't have anyone like that when you were young – and you have nevertheless succeeded in gaining this attitude. In consideration of your difficult journey through life, this represents an extremely valuable, inner life achievement, that I kneel before in awe". Anna looks at the Therapist with wide, somewhat doubtful eyes that appear to be tearing up, and looks within: …

Patient: – "The sea was so frightfully dark – but that has now changed since you were able to interpret it and put it like that. Despite the fact that it is something awful, something inside says that it is also fitting. The sea carries and sways the light, becoming bright in the process. I once had a very similar dream of longing during my childhood".

Therapist: – "How beautiful! If you were to paint it again, what would the picture look like now?"

Patient: – "A sea of light in which the light sways, dances or is calm in the peace. It doesn't really matter, as the light is where it belongs. I am a bright sea, a sea of light. I'll carry it with me in my everyday life throughout the upcoming week. Or, even better, I'll flow through everyday life and make it bright". Therapist: "I look forward to seeing your new picture and our next session". ◄

► Example II: Obsessive-Compulsive Disorder

Therapist: – "What is on your mind? What have you brought with you?"

Patient: – "Even when my eyes aren't shut, I imagine horrifying, recurring scenarios, almost like switching to a second channel of the private cinema in my mind: I take the large, sharp kitchen knife and allow it to fall from above onto my baby's head in an almost playful manner. My baby becomes silent immediately and his head splits apart, just like cutting open a lovely apple. It is so terrifying to imagine, as I love my baby, even when he cries all the time. However, his screams do test my nerves. I'm clearly not a very good mother, otherwise he would be happy from time to time. I did everything to prepare for having a child; I read all the books, got everything I needed, stuck to all the rules and suggestions, but for some reason, it isn't working. I can no longer sleep at night. I have given way to despair and irritation. And now I'm experiencing these obsessive murder fantasies that I seem to be unable to switch off. Something else that has recently also added fuel to the fire is that fact that my baby sleeps perfectly peacefully when lying on my husband's chest. When I see this, my private cinema switches to an image of me dropping the knife onto my husband's stomach, causing fountains of blood to spray everywhere".

Therapist: – "Yes, you are right to find these images disturbing. I can sense the desperation of a young mother who is using all of her energy, perhaps even trying too hard to do everything right and is now at wit's end, in the way you describe your experience. It appears to me that there is a hidden baby crying within yourself. The real baby may be a resonant mirror, a substitute for your internal crying. Ultimately, you have experienced a shared world of experience. You're looking at me with interest and slightly startled. Perhaps if you could try interpreting your own tears – I know you would not allow this in reality, be we could both allow you to imagine it in this setting – what would you, as the baby within, be crying about? What did she need most of all? Did anyone care about how she was doing?"

Patient: – "As far as I am aware, in the first year of my life, there were nurses in the hospital and then in the home. They changed. My mother underwent treatment herself; postpartum psychosis was mentioned. She was always somewhat distant from me. My grandmother was kinder and more natural, she liked me. She helped to make up for a lot later on. However, she would always say: be good so that your mother won't become sick again. I wasn't given much breathing space in this regard. She had previously asked what my crying might represent. Perhaps, I don't want to be at fault if you get sick again! Referring to my mother's 'condition'".

6

Therapist: – "How would you feel about giving your inner child more room to breathe? I think she has some catching up to do! It's lovely to see your eyes gleaming right now!" Patient: "That's a great idea! But how would that work?"

Therapist: – "I think it would be best for us to be guided by your images".

Patient: – "Through that horrible stuff?"

Therapist: – "Exactly, it contains the solution. What does the story of the knife and baby portray?"

Patient: – "Stop screaming! I want to have my own life and be myself!"

Therapist: – "Exactly. If this desire to live is locked away, it will develop into a monster. However, if we acknowledge it and address it, we will come to the realisation that there is something meaningful and beautiful concealed within. What happens for you if you imagine that you were invited to live again and allowed to develop?"

Patient: – "That would make me very happy. I could leap around the room in joy!"

Therapist: – "Exactly. In your image, despite the horror, there was also an aspect of playfulness that wants to live. The murderous impulse clearly comes from the part of you that is trying to defend the rest of your life under desperate circumstances. A remnant of self-defence!"

Patient, looking somewhat wistful: – "Can I really live like other people? Find a new lease of life?"

Therapist: – "Ask the monster what he thinks in his heart. Does he have an opinion?"

Patient: – "Yes, but in doing so no one must be hurt".

Therapist: – "That is an important principle. But it should equally apply to your inner child. I'm sure you remember: "Love your neighbour as yourself". It is a clear statement for equality. I am curious whether the real baby would stop crying and want to cuddle you if your inner baby, your inner child, feels that she is finally looked at and accepted with love, and consequently finds a new joy in life".

Patient: – "Is that possible? Retroactively?"

Therapist: – "If we both take this seriously and fight for it, then yes!"

Patient: – "Can you fight for it? How would that work?"

Therapist: – "You've already being doing so for a while. Just in despair and with little success, but you have already made a great effort. There are two scenarios to deal with. On the one hand, there is a desperate mother who isn't capable of 'reading' or interpreting the needs of her screaming baby and calming him down; on the other hand, there is your former, inner baby, reached via the inner resonance level, who has missed out on a constant, secure relationship experience and also allows herself to be forced into an external control concept in an over-adapted and low-maintenance manner. However, the spontaneity and sensitivity to the primary needs of herself and others fall by the wayside. Ignored aspects transform into a monster whose capabilities are not integrated. Let's help the wrathful monster to transform. What does he primarily desire? Are you able to put yourself in the monster's shoes? That would be helpful!"

Patient: – "Ok, I'm now the monster. When I was young, I wanted to play, dance, dream and live a little haphazardly, just give in to whatever I felt like doing at the time, like over the school holidays. That only happened rarely during happy times with my grandmother. We would also cuddle, which didn't happen later on. Grandmother died soon after".

Therapist: – "Perhaps at the time it wouldn't have occurred to you to speak of a 'monster'?"

Patient: – "No, at the time there were only little kittens there. I liked how they always knew what they wanted. They came in when they wanted a cuddle and left once they were done. Completely independent and confident".

Therapist: – "A helpful creature to have around for those who don't receive affirmation of their right to be themselves. Excellent. Would it be ok with you if we were to symbolically construct your inner situation from the time when you felt similarly constricted as you do now when you were younger? Your later years in school perhaps? You're nodding and looking at me inquisitively. On the one hand, we have your family, who are aware of your mother's borderline psychosis and put you on your best behaviour: be good and well-behaved otherwise you'll make your mother sick. We also have your mother, who was unable to form a special bond with you as a girl. And here, on our side, we have a frightened, well-behaved schoolgirl. If we want to care for the girl, what or who does she need at her side to help her feel better back then?"

Patient: – "My grandmother is already there behind me, then the cats next to me, if it comes down to it, they can also be mature and hiss; my school friend from back then should also be there. I'll also bring in my husband from my current life. I often find joy in his free, easy-going laughter, which I find infectious, despite knowing that it's not entirely mine. I still want to get there. It makes me even more shocked about my horror scenario with the red fountain of blood".

Therapist: – "A striking image. How would you feel about becoming that fountain yourself?"

Patient: – "As the fountain, I would jump into life full of joy and energy, I wouldn't allow anyone or anything to stop me, and I would cross boundaries from time to time. I would push myself".

Therapist: – "How does it feel to hear yourself talk like that? And how does it fit into our constellation of your earlier life situation that we just discussed? Who needs to hear which message? Perhaps also, which indirect order will they get in return?

Patient: – (loud and relatively determined) "To the uncles and aunts who told me I would have been to blame if my mother was to fall sick again, as feared, I return to you this burden that almost crushed me. You should have stood by my side with empathy, not with a raised finger! I denounce you and break off this type of loyal relationship with you. My large kitchen knife would come in handy here! You should have realised that my mother would never entirely succeed in returning to this world with her issues and never again be able to provide love, security and care for her child. (Patient looks sad and starts to tear up.) And now it's too late".

Therapist: – "Too late for empathy towards your inner child? It seems as though she's only now allowing herself to be reached. This is an important development. Thank you. (Patient looks somewhat inquisitive and smiles). Yes, it's a special bond. She's on your wavelength and opening windows and doors. (Patient makes an implied swaying movement.) Let's pretend we have your inner child in our arms, like this cushion here, and rock her a little. You look blissfully relaxed right now. You are free to decide how long we do this for. Similarly to the cats who you were able to connect with so well. In our thoughts we are also rocking your baby at home".

Patient: – "I want to try this out with my husband this evening".

Therapist: – "I'm glad to hear that. All of the 'good spirits' from this session will accompany and support you with this, if you allow it! Until next time!". ◄

6.5.1.4 Recording Traces Through Reflective Resonances, Discrepancies or Gaps that Need to be Filled

Exercise: Go for a walk in nature or spend 15 minutes walking around a garden pondering the inquisitive question: who am I today? Safe in the knowledge that you will recognise everything that symbolises something that currently is significant to you, whether you are conscious thereof or not. What sticks out to you? A bud? A withered stem from last year? A seed pod or cone? A blossomed, beautiful flower? An enchanting perfume? A piece of rough bark? A mound of earth, a weed etc. And when you stop to ask what you have found what it is, what does it respond? Is this element connected to your current state or your everyday life? Does it resonate with your experiences or things you have not experienced within? What led to this detail sticking out to you? What do you associate it with? What does it want to say to you?

The above test is a good example of a projective test, which can just as well make use of other materials, like a toy basket filled with a range of materials such as stones, shells, wood carvings, small instruments, ceramic animals, building blocks, foreign currency, dolls, glass beads etc. and card sets that portray a wide range of different social situations, and coloured cloths.

Everyone introduces their item or colour in the first person and describes the particular aspect that led to their choice.

Variation with multiple-choice answers in the imagination: We'll now transform into a zoo. Everyone is allowed to morph freely two to three times. What is acted out in each case? Are they connected traits or rather diametrically opposed aspects that are difficult to reconcile? How does it feel to be both an elephant and a mouse? Is there perhaps also a symbolic figure for the halfway point between the two, if you can feel this? How does it feel to be both a snake and a foal? How do they respond to each other? Is it all just a confidence-inspiring façade that conceals the danger hidden behind? Or is the naive foal protected as the world is extremely erroneous and dangerous? What is their relationship with each other?

If their relationship was to become more distinct, would they morph into other animals? Do you have a particular animal in mind? What is special about this animal? What does it do and symbolise? Power, speed, charming brazenness, survival skills, dexterity, beauty, elegance, graciousness, slyness?, superiority, independence, wisdom etc. ....How far away from his ideal does the respective person feel in his everyday doing? Does he prevent himself from achieving it?

Another popular notion is transforming into a plant. Which one? What does it look like? Why is it good the way it is? What is it good at doing? What poses a threat towards it? How does it protect itself when necessary? What does it desire? In which environment does it belong? Which other plants does it fit well with? Can any of those be found here in the room? Are there any plants that suit it at all? How do others view it?

Now we shall dive into the world of fairytales. Everyone transforms into a character from a fairytale. They all have naturally determined characteristics, intentions, tasks and fates. What potential of the above characteristics made each individual select their respective character? Are they characteristics experienced by the individual themselves, or rather characteristics the individual wishes to possess? What about the fairy tale character appeals to the individual? What would change in the individual's life if he were to adopt more of this characteristic?

In relation: what would he most like to dress as for the next carnival? What character would he like to try out? What would he then be able to do that would otherwise not be possible, or hardly possible? Who would he like to meet and get to know there?

Each movement is also a statement. The patient should ask himself: Which posture and movement are currently fitting? It is perhaps best when I switch off my mind when trying to establish this and instead allow my body to find a response. I may be surprised by what it offers me in this regard and recognise myself in the reflection of my own physical response. If others are looking at me and following my movement, it is interesting to see how they translate my response through their empathy. Of course, I will balance this to ensure I do not merge into projections from others. However, in certain cases, they are able to see something that had alluded me at first, which can be enlightening. The journey through life can be imagined across the group room. It may also be a winding path. Similarly to how there are different phases with different identities, on this journey through life I may encounter different attitudes, emotional expressions, movements, dances of life etc. if I am in close contact with myself. In the process, the points at which my inner self still requires support and understanding may once again be underlined. How can I communicate this and who in the room will I allow to stand beside me? I may also receive support from myself in the form of another part of myself from another phase in life.

6.5.2 The Spatially-Symbolic Opening of the Conflict Field

6 The unconscious is to be seen as the director of the scene, of the relationship constellations and of the time serties regression.

In this example, the patient starts the session with the following observation, which we can use as the current point of entry:

Patient: – "As I looked out the window into the rainy and stormy weather earlier this morning, I realised that most of the raindrops on the window pane had been blown towards each other by the wind and together found the quickest way to fall, like a fast-flowing stream. Only the drops close to the frame remain scattered and continued to hesitantly seek out their path".

Therapist: – "What did you find most fascinating about this?"

Patient: – "When one of these individual drops didn't want to flow together with its neighbour to form a more powerful force, instead swerving to the side, I wanted to give it a push and say: go that way!"

Therapist: – "It's interesting to hear how engaged your reaction to this observation was. I haven't seen you like this in a long time. Did you perhaps feel that you resonated strongly with this situation?"

Patient: – "Well, I suppose it did hit me how much I have become intentionally distant from the main clique in my office, but I still often notice how the others group together suspiciously and worry that they may have something against me". Therapist:

"You did just mention that you have withdrawn yourself. There must have been a reason for this?"

Patient: – "There may well have been, but I'm not entirely sure what myself. I want to be both part of the group and not. Odd".

Therapist: – "Perhaps it may be good for us to play out the situation going on within in the outside in order to get a better overview. Does that sound ok to you? You're nodding. Good. Which people or groups of people need to be represented in the room? Where and in what distance from each other? To what extent turned towards or away from each other? How appropriate would that be? We can use the empty chairs in the room to this end. What is the real issue at play here? It started with raindrops on the window pane. Then you discussed the situation at your office with ambivalence towards the clique. Stay with this feeling and take a look inwards. Have there been other times and moments that fit this feeling? Your face looks almost startled at the moment".

Patient: – "Yes, I'm imagining myself standing in my childhood living room. My father and three older siblings, my half-sisters from my father's previous relationship, are playing at the living room table. My mother was his second wife. They always got on well, particularly when my father had time for them, which he didn't always have. I was too young to play the games they enjoyed. I was still in school. They would sometimes get annoyed at me because I didn't want to get involved. But I didn't understand the rules. After a certain amount of time, I slipped away and looked for my mother, who was often bedridden with illness at the time. She died 2 or 3 years later. Then I felt truly alone for the first time".

Therapist: – "Your very emotional voice would seem to indicate that this moment from your childhood is more emotionally important to you than the situation at your office. Would you be ok with us stepping into the shoes of the little girl back then? You're nodding. How does everything appear from her perceptive? We're looking at an infant with a longing for contact and relationships. Would you like to invite her to sit down? There in the corner? Ok. And then you described your father with your three older siblings. If you were to give all of them a seat in the room, which arrangement would most correspond to your feelings?" (Patient pushes four chairs very closely together.) "Is anything or anyone missing?"

Patient: – "My mother. She isn't really there any more. She's already looking away in another direction. When I enter the room she notices me and seems happy, she embraces me; however, she looks sad within and appears to worry about what would happen to me if she was no longer around".

Therapist: – "Perhaps it may be beneficial to take another look at your inner constellation, which was of great significance to you, as if you are looking at it from above, like a bird with its wings spread out, slowly circling around a certain area. The bird is able to sense how the little girl feels. Is anything missing that was important to the little girl back then, with positive or negative connotations?"

Patient: – "Yes, our dog. I can't remember which breed he was, it's not important. Above all, we loved each other and he understood me, at least that's how I interpreted it. He comforted me each time I felt rejected. Nero can come and stand beside me".

Therapist: – "Excellent. I'm excited to meet our new family member! Now, let's take the time to once again transform into the little girl from back then, with her favourite dog at her side, surrounded by both of our sympathy, understanding and support. Do you feel any particular sensitivities? Moods? Impulses? If yes, which gestures, concerns or desires most want to be expressed?"

The patient takes a few steps towards her father and half-sisters with open arms, then lowers them, revealing the anger on her face, turns around and returns to cuddle her dog.

6.5.2.1 Brief Interruption of the Example for Directions and Variations

The placeholder does not need to be a chair. It could also be small mats, cloths or boxes with different shapes or colours that are stood on once their respective role has been assigned. The available places or chairs can also be labelled with a particular symbol such as a tiger or shell, etc., or have a coloured sheet placed on top that contains a statement on the relationship structure. However, colours must also be given a new meaning by the patient, as all colours can be experienced with different meanings. The disadvantage to symbolisation is its qualitative determination, which could potentially thwart any impending changes in the relationship.

If individual work is carried out in a group setting, other participants can be selected as the leader. They will receive a small introduction to their role and try to empathise with it. This variant corresponds closely to Moreno's sociogram and psychodrama. The roles allocated by the protagonist should be discussed together, in addition to any impulses that arise within those who are given the role of leader.

6.5.3 Potential Protection and Reinforcement Measures

6.5.3.1 Dialogues and Switching Roles

The question of whether supportive measures need to be taken to encourage the protagonist is always an important one for the scene. These measures are introduced at the point at which the field of conflict develops, but when the impending confrontation remains imminent. How much does the anxiety level rise? It is enough for the protagonist to know the therapist is standing by his side? Various reinforcers are plausible in this regard. On the one hand, it is conducive to address the skills that have developed in the time between the respective crisis and today; the current, mature person does not only lend the person from that time his voice, but also his standing, maturity and insight, his balance and strength etc. However, it can also be helpful to summon an imaginary circle of helpers. This can include all constructive, supportive relationships experienced by the protagonist, as well as those that were only temporary and with people who have since passed away. It can also encompass the individual's experiences of their own ability to love, even if these may not have resulted in the desired success due to reasons for which the individual is not at fault. Furthermore, it may also include everything that the individual is able to recognise and accept in others. These experiences facilitate the establishment of an imaginary supportive group or semicircle behind the individual in a symbolic or personified manner. In the group situation, the group of participants can be motivated easily by asking those who are able to entirely or partially empathise with the protagonist's situation, perhaps as a result of their own analogous experiences, to stand behind the protagonist. Depending on how the protagonist would prefer, the helpers can each put a hand on her shoulder.

During dialogues, which according to the strict form (from Gestalt therapy) should be exclusively led by the protagonist from various perspectives, the therapist should make sure that the protagonist uses direct speech, that she finds a pace suitable for her when the roles are switched to emotionally enter into her role, and that the therapist organically announces the switch in roles at the point when a decisive trail of thought has come to an end and a reaction may have been triggered in a counterpart.

Suitable timing is required for the role change, as it may not be appropriate at times. If the therapist asks the protagonist to switch roles too early, this may cause the primary frustration of the patient or her child ego to not to come to light and be stopped in its by compassion for the counterpart, leading to it being locked away once again. As a result, the all-important deep relief and consolidation stage is skipped. Role changing to introduce a phase of reconciliation during a confrontation is a conscious second step in terms of timing.

▶ Continuation of the Example

Patient (turning once again towards her father and saying in a loud, almost screaming voice): – "You should have been there for me too! Not just for your children that you brought along with you. It's cruel of you to spend most of your time doing things with my elder siblings and having fun with them!" On the one hand, the small girl wants to pummel him, which she symbolically takes out on the table, half in despair, and half because she wants to clutch onto him or ideally sit on his lap and be involved in the group.

Therapist: – "I'm impressed by her strength and vigour. It would seem that her father was very important to her".

Patient: – "Absolutely. He was everyone's favourite. Mine too, although I had somewhat of a guilty conscious about my sick mother. Besides my father, she only had me. My half-sisters were friendly to her, and vice versa, but they always remained somewhat distant".

Therapist: – "You were privileged, you had a mother and father".

Patient: – "I guess you could see it that way. However, I didn't feel like it. I felt like I was too young, my mother was too sick and my father was too absorbed in caring for his other children to feel satisfied, safe and happy. My father wasn't able to grasp this. (Loudly:) Father, why didn't you understand that I wasn't doing well, that I felt rejected by you and your other children at home! I hold that against you!"

Therapist: – "It's time to switch roles. Why not enter the role of your father. What is your situation then?"

Patient as her father: – "Sure, maybe I didn't have enough time for you. However, I often saw you cuddling your mother and got the impression that you had both provided support to one another and you both mutually benefitted from this. Your three elder siblings no longer had a mother, so I was required to be both a father and a mother to them. It was a big shock for them after the accident to discover that their mother had completely gone from one day to the next. For me too, at first. I had just begun to once again see the point in life when I started loving your mother, and the next loss was around the corner. Your mother was diagnosed with cancer. When I was horsing around with the children at the time, it was like whistling in the dark. It represented a desperate attempt to save remnants of a carefree life. I was clearly not sensitive enough towards you. I'm sorry".

Therapist: – "Time to switch back. How do you find your relationship with your father?"

Patient: – "I have pondered this for a long time, but hearing it formulated that way has changed things. My father wanted to convey a sense of security and happiness, and needed to make sure that he wasn't overcome by sadness and existential fear".

Therapist: – "And then the small girl also complained about feeling rejected by her father and her elder siblings at home, from the children who had remained distant to her mother despite her mother's good will. What part did the little girl play in making the others feel like strangers?"

6

Patient: – "I feel like I've been caught red-handed right now. I can't deny that I felt a small sense of victory against the outside children when I was able to emphasize that I was twice as entitled to live where we lived."

Therapist: – "What would you say directly to one of those children?"

Patient: – "If one of us needs to leave, it shouldn't be me. I've been entitled to be here since birth and I'm the principal heir. That's important to me and makes up for everything that is missing".

Therapist: – "Are there any similarities to the situation at your office?"

Patient: – "There are. I'm there longest serving employee there. If one of us had to go, it would be one of them, one of the members of the clique who are a little too garish and cheery at times, who I simultaneously dismiss and envy".

Therapist: – "Do you think there is a need for change? It would appear so in light of the raindrops. If yes, how would it be if you were to address your siblings directly?"

Patient: – "You entered my life as strangers and I haven't let you reach within me. I wanted to retain my privilege and made myself an outsider, expecting too much of my father in the process. We all lost our mothers at a young age. I used my jealousy to subtly turn you into enemies, despite you coming in friendliness". ◄

6.6 Therapeutic Treatment Steps

6.6.1 The Guardrail of the Therapeutic Reanimation Process

» Preliminary remark on the principle of assimilation: The energetically suppressed impulse was referred to by Fritz Perls as "unfinished business" in early Gestalt therapy. Unfinished business aims to arrive at its goal, namely to close the Gestalt. Interestingly, the same principle has turned out to be healing for trauma therapy: the retained impulse, which was unable to be fully resolved in light of the trauma, is consciously and symbolically rectified in therapy. The trembling, muscular storm of movement by the agonists and antagonists provides liberation from the trauma-induced frozen state and constitutes a relieving intermediary stage in the return to natural, goal-oriented vitality. When applied at the right point, this also corresponds to personal causal psychotherapy.

1. Contacting the excluded aspects, or recording traces of acute, recurring, or chronified conspicuous aspects that seep into the foreground, e.g. empty role behaviour, automatic behaviours, micro-movements, changes in body language (contact behaviour, voice, coughing and throat-clearing, sighs or bated breaths, facial expression, posture and gesture, vegetative signals, pain etc.); the forming of an understanding between the patient and therapist that this particular phenomenon is a symptom that they should try to get to the bottom of.

(Caution: The therapist may find a certain reaction conspicuous, but it may be misleading due to the patient's current state, as he is currently stressed, wound-up, tired, cold etc. or because it has been classified differently within by the patient. Just because a patient blows his nose, this does not necessarily have any deeper significance in terms of psychotherapy! On the often hand, patients are often susceptible to not wanting to admit things that are relevant to their symptoms. Consensus is required for cooperation. Perhaps a symptom has not yet been released from the patient's unconscious for deciphering, which may also be the case.)

► Example I

A young man, Karl, 25 years old, a law clerk, loses his ability to speak around older men, particularly those in a superior position. Karl becomes silent. He passed his degree solely through written exams. He finds his loss of voice in certain situations embarrassing, particularly if when asked about his relationship with his father, which he then tends to define as normal and good. Without a voice, he is only granted limited opportunities to practice his career as an administrative clerk, for instance. He worries about the future in this respect. This symptom is the reason why he has come to therapy. ◄

6.6.1.1 Identity Issues Related to the Excluded (Repelled) Part

Who or what is this impulse, this side to me? I (the patient to whose world this side belongs) am trying to identify with it, to completely become this impulse or side, abandoning myself to my sensitivities and associations. I withdraw from myself and make myself available to this other being in order to learn something about his way of being. This is similar to the method of Zen. (If I want to paint bamboo in the traditional Zen manner, I first need to become the bamboo in my meditation in order to perceive something about its being. Only then can I succeed in adequately depicting it on paper to some extent.)

When nothing appears, the following question stands to reason: To what end could it perhaps be good that I am unable to grasp or feel anything, to get any vivid insights?

► Continuation of the Example

What does the silencing of your voice in the presence of older, father-like authority figures symbolise? Attempt to identify: I am the silencer of voices. What would it be if I were not to silence? A risk? Then I wouldn't be a saviour for the threatened person. Not having to listen, not having to think about it, not having to take a position on the emotional turmoil, not having to answer to anyone for this, not having to share my ludicrous fantasies with anyone, not wanting to blame anyone and not wanting to be accused or judged by anyone, being able to remain blissfully unaware: this can provide relief at first. ◄

6.6.1.2 Identifying the Problem Area

6

Which temporal, spatial and relationship context best fits what emerges when the symptom transforms? What is the context behind the resonances and does the transformative chain of associations come to a halt? Who or what represents and embodies the respective impulse? Is there one or several counterparts? Can they be located? Location and direction impulses are spatially symbolically used in this regard. What is the concrete relationship between the players and their opponents? Who is facing whom? How close or far away? Who is turning away from whom? Who turns their back on whom? Who has left the field? Who clings to whom? Who feels trapped by whom? Who is in the centre? Who is in the driver's seat? What kind of position, opinion and power do the key players or the most important person represent? Do abstract or individual names fit their concerns?

The problem area corresponds to an intrapsychic constellation of the current situation in space and time, in which the impulses yet to be excluded have not yet been dissociated, projected onto the outside world.

► Continuation of the Example

Karl's chain of associations remains at a standstill in his childhood. Karl: "I'm between 3 and 4 years old, I'm at home alone with my father. My mother was in hospital at the time. He tried to be loving and caring to me in his own, somewhat unfamiliar way and puts my warm porridge down in front of me to eat, but it's not how my mother usually makes it. I scream and lash out, throwing the plate on the floor and starting another loud tantrum that use to get on my mother's nerves, which I enjoyed at the time as a new source of power. Then my father suddenly left me. At first I didn't realise that he had gone. No one was there for a long time. Crying didn't help. No one came. I fell asleep, most likely from exhaustion. Night-time came and I could hear shouting and footsteps, the door was kicked open. Policemen came into the house. Our neighbour was also with them. My father had been rescued from a car accident and taken to the ICU. The car was totalled. The atmosphere was filled with anxiety. A policeman asked me whether anything notable had happened that day. I was stricken by anxiety. In my childhood mind, I was evil and to blame for everything, including my mother becoming sick. I crawled away under the sofa and don't say a word, only crying, terrified, with my hands in front of my face. My neighbour came to get me and looked after me for the next few weeks. It took several days for me to say another word.

When asked in which scenario something remains unresolved, the two scenarios compete with each other. The fear-ridden scenario of being interrogated by the policeman is repressed by the defiance against my silenced, recoiling father. This scenario can be whittled down to a defiant child and his counterpart, who does not resist but instead slips away, allowing powerful anger to transform into anxiety.

There was also no interaction in the second scenario. In this case, fear blocked any potential interactions. Despite making an effort to converse in a child-friendly way, the policeman ends up enquiring in a way that seems to represent condemnation between heaven and hell". ◄

6.6.1.3 The Meaning behind Mental Blocks

For which purpose is it beneficial, or does appear to be beneficial, that something occurs and remains stuck as it is until now? Are there any suggestions about what could potentially be preventing a stop signal? What are the worst-case scenarios? Has there been any catastrophic ideation? Has this mental block become self-perpetuating and kept, or continue to keep, reality in check with its mystery, despite the fact that it only represents, or previously represented possibilities? For how long has the patient been a hindrance to himself? What motivation or necessity led to the mental block? Who or what was the block required to combat? What is its fundamental purpose? What needs to happen for it to become unnecessary?

► Continuation of the Example

In Karl's childhood mind, he needs protection against the awful possibility that he may have repelled both his father and mother through his defiant screaming. He is unable to bear being this evil. In this case, his parents would never return. Being quiet helped him to avoid incurring any more damage. Driving away his neighbour presented a significant risk. Being quiet and kind provided him with the most security. When he heard the children next door playing, he was often tempted to join in, but at the same time scared that this could lead to another catastrophe. ◄

6.6.1.4 The Dialogue of the Unsaid and Unheard

Is there anything about the unexpressed and unspoken aspects that need to become observable and audible, to enable things to be more coherent between all involved aspects and allow the primary issue to finally reach its goal? In certain cases, nonverbal signals are more suitable in this regard. As has been observed in trauma therapy, the purely physical expression of a hereunto repressed defence mechanism to combat trauma can achieve a new inner balance and liberation. Often, remnants from a relationship can be healed or resolved at deep levels through a handshake and eye contact, or a hug. This can also be achieved in a substitutive manner in therapy groups. As such, the above does not solely pertain to verbal language, but any kind of signal, or also language in the strict sense of the term.

Standard question: What should you or anyone else have told the problem person, or what should they have heard at all times, or at that time they gave their critical opinion?

Preparation: Is the spatially symbolic distance correct? Is it ok to address the problem person face-to-face with eye contact, or would that be too anxiety-inducing? Would a conversation from behind or the side perhaps be more suitable and adequately effective at first? Am I ok with being in the same room as the problem person or would I prefer to notionally kick the problem person out of the door onto the street or banish them to the garden? The imagined conversation will follow nevertheless. It may also be helpful to imaginarily shrink people who induce anxiety, like looking at them through an inverted magnifying glass, in order to gain distance.

Is the confrontation with the person as a whole, or only a side of them? Should one part that needs to be protected against damage carefully be packed away out of harm's way before the awaited confrontation with another part is tackled? In this case, it would be sensible to temporarily use two projection chairs for one person. This can often work well for achieving clarification on the relationship with the patient's parents; the aspect that embodies loyalty and gratitude for vital care is imaginatively placed on one chair. This part cannot be damaged, even if there are heated arguments about other aspects seated on the other projection chair. Without this temporary separation, the fear of destroying the positive aspects of the relationship blocks the required confrontation. Both parts are once again very consciously merged back together afterwards; even if complete integration is not possible, at least now there is the opportunity to say that these aspects both belong alongside each other in this person who is important to me.

Good conditions are required to commence with the dialogue. Is the therapeutic relationship resilient? Can the resources of the current adult be brought into consciousness and incorporated into his imaginary relationship confrontation with the problem person? Which physical posture does the subject adopt to take on this task? Standing is often better than sitting. Could body-centred breathing and mindfulness exercise with the subject's feet firmly on the floor be beneficial before the confrontation? Is an additional, stabilising circle of helpers required? (See ▶ Sect. 6.5.3) Potentially with representatives from the group?

The problem person is always addressed directly. Is the correct tone of voice used? Is it audible? Is it convincing? Confrontations often address reproaches and shortcomings experienced by the subject at a younger age. This represents the aggressive side of the coin. In this regard, it is key that the frustrated needs are also addressed alongside their fears and that they experience consolation as well as caring acceptance. The aim of this process is to allow the individual's various parts, primarily those from their child ego that retreated out of fear and concealed themselves in symptoms, to once again be restored and granted the right to live and develop. This helps to stop processes from being blocked. Furthermore, it also puts the outcome of lost battles into perspective, allowing the individual to start over.

▶ Continuation of the Example

Karl understands that talking to his real parents and his internalised parents, namely the images that he has memorised from the past, are entirely different people. It is logical to consider what can be expected from both the real and imaginative parents, and which chronological order would be most conducive. Initially, Karl wants to face his inner turmoil. This turmoil represents a mixture of earlier and later traces that have been bound together by the same emotional block, consequently resonating as an entire package. He first addresses his father.

In the imaginary scenario, Karl is stood at the edge of a large group room and sees his father fly in through the opposite wall.

Karl (shouting): – "Stop, stay there, come back! We need to talk. It's not ok that you simply ran away just because I was kicking up a fuss. You need to be able to deal with that! You're the father, I'm the child! You act as though I'm some awful monster and guilty for your near-death experience. That was a millstone around my neck for years. And remains so to this day. It prevented me from growing, I became weak, I was the laughing stock of everyone, "he isn't capable of doing anything, he isn't a real boy" etc. Do you know how horrible that was for me? I would have liked to have had a courageous father who I could play around with, like everyone else. A father who would have shouted at other people in my defence when necessary. Maybe then I would have been more able as a person".

Therapist: – "I'm impressed at the voice and posture you used to present your concerns. It's clear that you have been fighting a running battle with this issue for a long time. Kudos to you! How do you feel right now?"

Patient: – "It's a mixture. On the one hand, it's good to encourage this attitude towards fear and to push myself to adopt it. It's an attempt to drop my routine 'flee' behaviour. At least tentatively. On the other hand, I don't know if it's an unreasonable demand and insult towards my father, who I do also like. Then the old, obstructing worry returns".

Therapist: – "You are already spontaneously empathising with your father. This would be a great time to consciously step into his shoes, taking on his role, viewpoint and understanding of the world. Now consciously go to the other side, where he is located in your mind. How does he react to what you just said?". ◄

6.6.1.5 Switching Perspectives

Delving into another person's world can often help to put one's own viewpoint into perspective. It constitutes enrichment, but also the loss of one's own subjective "security" provided by the sole perspective of one's own world. Precisely this can be the counter motivation for avoiding a switch in perspectives. Or, conversely: in the current approach, it makes sense to first suggest switching perspectives once the old, stored, pathogenic resentment has been verbalised. This cleansing process will be blocked if the understanding for another side is deployed too early. It is to be expected that the individual's own (mostly restricted to a childlike) view may need to be retroactively modified and forgivingly retracted, sometimes with an element of shame. When we are children and in times of crisis-related regressions, we are all entitled to have somewhat restricted viewpoints.

► Continuation of the Example

Karl in the role of his father: "I can understand why you want a stronger and more protective father who doesn't run away so quickly. I also don't like these characteristics. I felt similarly about my own father, except much worse. He could never be a role model for me in this regard. His wife used to tell me that he was a completely normal, brave young man in his youth, but once he returned from being taken as a prisoner of war, he was never the same. He was required to participate in truly horrendous things against his will. That took a lot out on him. As a result, the women wore the trousers in our household. But let me make one thing clear, you mother didn't want a strongman or a man hindered by his own strength, which I also didn't want to be.

Back then, when your mother suddenly had to go into hospital because of a ruptured ectopic pregnancy, her life was hanging by a thread. My nerves were exposed. Of course, as a young boy, you sensed the tension and were the final straw with your tantrum. However, we never told you that you were to blame for anything that happened back then".

Therapist: – "Now let's switch back. Ok. How does it feel to be back in your own skin after going through that? Does it feel different in any way?"

Patient: – "I definitely found the comparison between my father and grandfather, who I only really knew for a few years, made a real impression. My father definitely deserves praise. I was also affected in the sense that I was entirely unaware of how dire the situation was for my mother at the time. As a child, I clearly didn't realise. I was only told about it once I was older. I haven't previously been able to merge this with the experience of the young boy in my mind. To my father I saw: If it wouldn't have bothered you at the time, it would have been more than sad. I feel that I don't really know you properly, rather only from a bias perspective. I would like to get to know you better (before it's too late). Perhaps we still have the ability to make up for lost time. How would you feel about going on a hiking trip together? A weekend at first, to try it out, then perhaps something a little grander?"

Therapist: – "I'm impressed by your inquisitiveness, appreciation, willingness to take another look and release yourself from former prejudices, as well as your bold and frank initiative towards your father. You may truly succeed in learning something new. Of course, this is only a roleplay scenario, but do you have any idea how you could perhaps implement aspects of this into your real life?"

Patient: – "Sure, I think my father almost expects this. He enjoys riding his bike. He'd also probably be good at finding a remedy in the case of a glitch. In that regard, a cycling trip would be familiar terrain for him". ◄

6.6.1.6 Integration and Reappraisal

What does the viewpoint and new balance of power look like after the switch in perspectives? The perspectives can be switched several times. Depending on the problem area, a switch in perspectives with various viewpoints, or rather people, maybe on the agenda.

Often it can be helpful to switch to a bird's eye perspective, to the meta level. From there it can be easier to see how everything is going. What is really going on here? What is the real matter at heart? Can this be vocalised? If this notion were given a name, what would a fitting title or caption be?

From the bird's eye perspective, it is possible to get an overview of the behaviour happening in the foreground and also to discover what is going on behind the façade, namely what is happening in the unconscious. This often results in a confrontation between two hurt or unloved beings, who would like their deficiencies to remain hidden from themselves and each other. Once this becomes clear and can finally be admitted, the dynamic changes. There may be common ground for understanding – and also reconciliation. At this point, the former opponents are in the same boat.

► Continuation of the Example

The subsequent distinction of the image of his father, for which the required information was already internally available but not yet integrated, was rectified. Prior to now, the image of the father created by his 4-year-old mind had dominated the relationship in its early stages, alongside his notion of himself as a model. The change in perspective enabled him to be liberated from this fixation. ◄

6.6.1.7 Range of Solutions

What form does the current or interim solution take? Does it constitute a true compromise supported by both sides? Or a somewhat more one-sided compromise? Does the solution consist of an acceptable parallel existence of the opponents in terms of elements of both this and that, both unmixed and perhaps even a chaotic and erratic coexistence, to which they have adjusted.

Or has a creative solution with a new inner structure and evaluation arisen, that is capable of encompassing all of the important aspects? The above case would lead to a change in identity, or a change in the quality of the relationship, primarily through the reidentification with the rediscovered partial aspects that have been granted a place in this newly structured large whole.

► Continuation of the Example

The image of the father and the relationship with him has become more nuanced for the son; conversely, his father has managed to remain open, despite the initial attacks. This can also be regarded as an aspect of the son's fundamental attitude, for which he deserves credit. The symbolic, external change in perspective is actually a change in internal perspectives to aid integration and restructuring. (Certain fundamental attitudes can also be inflexible in nature, instead preferring to persevere with viewpoints that have previously been learned. Any form of change evokes anxiety beyond the normal level.) ◄

6.6.1.8 Trying out and Consolidating the New Identity Using Exercises Deemed Suitable for Everyday Life and Regarded as Realistic, Doable and Sensible by the Patient and Therapist

It can be extremely beneficial if the patient is able to determine the ways in which he can test out the suitability of his new realisation for everyday life. He then takes the lead in making changes to his everyday life and may even explore other avenues that were not on the cards during the therapy session. The more engaged and creative the patient is in this regard, the more his self-confidence and self-esteem will benefit compared to when he is addressed as a student receiving homework to complete. Naturally, in order to encourage the patient, a therapist who conveys a sense of interest and appreciation for the new attempts and successful experiments, and perhaps even manages to draw useful experience from attempts that did not work out, is a prerequisite.

► Continuation of the Example

The patient spontaneously elected to implement his new approach in his everyday life. As a result, he was able to fully identify with this "homework". ◄

6.6.1.9 Reflection, Phase of Peace and Recuperation

This stage is often skipped. However, when something succeeds it provides a source of joy and gratitude that feeds the soul. Time should be set aside for this feeling of joy in therapy at each successful stage. It represents a well-invested 5 minutes spent resonating with joy and confidence. This is important for the patient's identity and key to ensuring the likelihood of success of the next steps.

6.7 Other Proven Therapeutic Methods

6.7.1 Approaches to Establish Contact with Oneself and the Unconscious (Projection Approaches)

Projections facilitate a look into the subject's deeper world, which he is unable to grant the therapist conscious access to as he is not capable of this himself. As such, these approaches aid diagnostics and, depending on the focus of the session, the establishment of resources and areas that still require clarification and healing, yet which continue to represent a source of fear and anxiety.

If a patient is asked to commit to goals too early, before he has been given the opportunity to at least partially assign his inner turmoil, the therapy will not venture into much depth. Instead, a relapse is likely to occur shortly after. Needless to say, it is also possible to end up going to the other extreme and focusing on the fundamental, supra-individual issues in life. In the majority of cases, therapy cannot provide a solution in this regard. It is important to apprehend whether the problem is significant enough for the patient to reach a meaningful goal. In this case, resolution entails true relief.

Projections can be made across all of the sensory channels:

The medium of sound: Simple example with sound generators: If my name was a tone, sound, rhythm or perhaps even a melody, what would it most sound like? Can this impression be transformed into an I-statement? If this question is posed in a group, the others can contemplate it first. Which aspects of the sound does the producer thereof most resonate with? Does this fit his own self-image? Has anything that was previously concealed come for the fore?

The medium of movement: If I am asked "Who are you?" and I am only allowed to answer through movement, a sequence of movements or dance, and if I successfully use my entire body to answer this question, how would it respond? Perhaps it would not only reflect the current situation, but also exhibit a certain amount of range that seems coherent. Perhaps the topic will also transform into: "Good times, bad times, and better times once again"?

The medium of clay (earth): If I take or receive a lump of clay in my hands with my eyes closed and am asked "how are you doing at the moment?", and I were to answer this question using my hands and fingers, I would be astonished at how my movements with the clay would reflect how I (currently) deal with the world around me. There would be a range of variations if I allowed my hands to go on my journey through life, including my various significant encounters. This may also include encounters with animals that were important to me and, above all, provided me with comfort. My hands must be permitted to use the entire range of expression, including conflict and love.

The medium of colour (provided in the form of artist's colours or coloured cloths): If I am asked which colours represent me a) when I am happy with myself, b) when I am going through a rough patch in relation to my most recent crisis, the reason behind why I sought out therapy, c) when I feel how I would like to feel in the future when everything is going well; which colours do I select? And in which combinations? With the assumption that each colour is an entity in its own right and has its own concerns, what would these selections result in? What does the colour from b) need most of all and who is the concern it represents directed at?

The medium of fantasy stories: A small child – a boy or a girl – imagines himself or herself in front of his or her former childhood home. Does he or she want to go in or run away? In this fantasy tale, he or she has the choice to enter the former childhood home or go to another house in a parallel world where things are better than they were in the child's former home, more in line with his or her wishes. The child can also move between the houses. What does he or she want to revisit, take with him or her or consciously leave behind in the old house? What is likely to be found in the new house? How does it feel in one house and the other? Who is here and who is there? Which house feels best? Do they both need to encounter each other?

Medium of imagining a trip to the zoo: A young child has a special ability. She is able to speak with animals, although not to the same degree with all species. Today, she is going on a trip to the zoo. Initially, she starts walking around completely unfettered without a fixed goal in her mind, takes a look around and is inwardly drawn to one of the animals or enclosures. What is in the enclosure? Why does it fascinate her? What being lives there? What does it want? Is there a dialogue? Does it have a plan for the future? Does this story bear any relevance for everyday life?

The medium of painting: When I imagine that there is a world inside me designed similar to the outside world, but constructed in a way that this excerpt symbolically corresponds to my inner self and I can easily feel at one with it and its atmosphere, what would this world look like? Perhaps this requires a combination of several aspects in order to create space to represent the most important elements. Would my inner world be a spring meadow with a stream? A jungle, a vast steppe? A mountainous landscape with cliffs to climb? An ocean shore with sea spray and waves breaking, or with a sandy beach and palm trees? Hilly forests? A metropolitan city with penthouses and roof terraces? An industrial area with wildflower-covered waste heaps.

Where is my safe haven in this world? What does it look like there? How does it feel to be there? Can anything be heard, smelled or seen? Who or what is symbolised there? Are there any other living beings? Does my world also contain a dangerous area when everything feels a little scary? How can it be described? If it had a voice, what would it say? What would its concern be? What does that trigger inside me? Is it (already) time to get involved, or do I first want to return to my safe place and get reinforcement? ….

6

"Improve theatre" as a medium for psychotherapy: The previous inner world approach opened the door to a dialogue featuring allocated roles. If partial aspects of the projection method are personalised and their correlation is revealed in treatment, we then enter the intrapsychic realm, despite the fact that we are acting on the outside. Both Jacob Moreno (psychodrama) and Fritz Perls (Gestalt therapy) were proponents of the healing effect of this form of improvisational theatre. (They met at the end of the 1940s in the USA). The former would ask a protagonist to act out an important real-life situation from the past with selected representatives and help him to explore new opportunities in the process. The latter would ask the subject to set up the inner relationship constellation in the outside world with the space being symbolic and ask him to switch into all of the positions by changing roles one after another. The subject was tasked with defending the respective sub-part's right to exist. The emphasis was subject to change throughout the process. This may or may not lead to good compromises and reconciliation. If successful, the perspective becomes more clearly distinct. However, in the process, the view at the meta-level may find new perspectives. A playful variation on dealing with aspects of the personality within the whole individual was provided by Virginia Satir, who worked together with Fritz Perls in the 1960s, in the form of "The Parts Party". She also helped with the incorporation of variations on improvisational theatre into family and systemic therapy to a certain extent. Bert Hellinger, whose background lay in psychodrama, helped to incorporate constellation work derived from Moreno's sociogram into systemic therapy. This method once again features representatives for the subject's family or social relationship structures, which are arranged in relation to each other by the subject, with the space in between playing a symbolic role. Due to their ability to empathise, the representatives often appear to contribute potentially helpful impulses or thoughts on the hereunto unresolved issue.

6.7.2 Approaches to Determine Hidden Goals

The part-whole approach described above relies on the notion that the subject is latently aware of where the journey of the inner and external development should lead. It is widely understood that the unconscious constantly makes predictions about the probability of upcoming decisions. The conscious mind is unable to tap into this. Though sometimes vague hunches may come to the fore, primarily in the form of a gut feeling.

The dormant potential of inner foreshadowing can also be personified as a partial aspect of the inner self. Depending on the amount of time that lapses from the current development over the course of life, this part may develop into a mature person when the individual reaches middle age, retires or is at the end of his life. In the latter case, the individual is often referred to as an old man or woman. Put in terms of an image: the individual peacefully sits on a bench on the banks of the stream of life at the point just before it flows out into the ocean. He is able to look back over his life with more ease and make connections that are not so clear to the person who is still struggling on his journey through life due to a lack of distance from the current situation. The opportunity to make a wiser decision for the present from the perspective of an old person looking back over his life is hidden in the ability to identify with the chronologically progressive past.

Imagining a white and wise(!) eagle circling above the current issue in wide, calm and effortless curves and yourself sitting on its wings is perhaps somewhat less cumbersome. The distance and calmness physically exuded from this image provide excellent prerequisites for solving the problematic areas.

6.7.3 Approaches to Determine the Potential Meaning to Life or Satisfaction with Life

The above approach provides a personal image in order to establish contact with one's inner wisdom for reaching key decisions in life, which can similarly be applied to the question of the meaning of life. Before commencing, it is important that an inventory is compiled consisting of moments of contentment, the feeling of being at one with yourself, the coherence of being connected and at one with nature as well as peaceful happiness etc. The latter are often unspectacular moments of high intimacy characterised by a lack of words. People may be hesitant to vocalise them, as though this could damage them in some way. In this case, damage would be caused by the unwieldy and rough nature of language and its inability to capture the true essence of the experience.

Women often describe the feeling of pure joy when they have given birth, especially after the first time, and the baby lies on their chest, perhaps even before the umbilical cord has been cut. In my experience, young fathers can also be touched by this. Then, as can be expected, the immense wave of the first great love, including all of its often existentially experienced nosedives, becomes a topic in this context. Love experienced at a later stage can develop a commanding depth and sensitivity that is worthy of praise.

Needless to say, we also experience more outwardly directed moments of joy, such as achieving something and doing well in comparison to others when graduating from school or university, or being on the winner's step in sport, for example. Whilst this may perhaps be more loudly expressed than the events described beforehand, it seems rather modest in comparison. It thrives primarily on the response from the social environment, whereas the other experiences provide inner responses to wider horizons.

The most effective approach includes drawing a river or path of live and marking the stages of joy and contentedness with symbols. Stars, suns, hearts and flowers are the most popular. In order to avoid any bias, the darker elements are also drawn. Whilst they may not be granted as much attention, they must nevertheless not be denied. Dark phases often contain new elements. "If I had continued to ignore the darker periods of life, I wouldn't know how to properly appreciate the good and happy sides to life that I have experienced in spite of everything".

Perhaps the biggest issue faced at the end of life, or when looking back on one's life, is failed reconciliation. Hate is binding. Sometimes it even develops into the central nucleus of the personality, seemingly directing it and giving it a sense of meaning. There are a number of expressive ways of describing things that have not yet been expressed in writing, drawings or speech, or with full use of the body and a loud voice etc. The loudest approach is not always the most effective one. Indeed, the reverse is often true. However, it needs to be capable of reaching the individual's innermost being, the strained core, which is initially strengthened by consolation and understanding. If a role reversal is faced with reluctance, which is often palpable in the case of a relationship characterised by chronic hate, there is still the meta-level, which can be used to imagine what might be going on behind the façade of the problem person. If, in the process, a being emerges that is only meagrely able to mask its shortcomings, needs, old damage or fears, in certain cases, a new balance may be established for the damaged part of the patient, sometimes a subjective balance may even be achieved to the patient's benefit. If it were not for this untold story, pity may arise. It becomes clear that two wounded beings are facing each other, mutually and negatively interlinked. At this point, it can be good to question whether the patient is content with spending the rest of his life running in vengeance, or instead whether he would like to let go of his opponent with good intentions in order to rediscover his love for the world. Needless to say, reconciliation work is dependent on a therapeutic relationship bolstered by true compassion. It serves as the catalyst, in a manner of speaking.

6.7.4 Approaches for a Life that Encounters Fears and the Potential for Life Thereafter

The catalyst function of compassion is all the more required when fears, ultimately also the fear of death, are on the cards. On the one hand, a reliable, supportive relationship in therapy provides the basis. However, it is certainly not capable of replacing experiences of supportive relationships that the individual lacks, instead it is only able to mitigate them, which cannot be maintained forever. That being said, it does provide a model function, which needs to be addressed at a relatively early phase. As soon as it becomes palpable that a relationship has been established, this should be communicated to the patient as an ability and something good and valuable. Notably, that it is important that the patient also gives this opportunity to other people. Relationships with animals can also have a stabilising effect. On the other hand, it can be good to substantiate something that has been achieved or made the patient particularly happy in therapy. Drawings can be hung up or kept at home. They can act as a transitional object, as permanent testimonies of an accepting relationship. A popular and beneficial approach is to imagine a well-rooted tree with the sun and wind playing in its branches. This tree image can be used in body therapy. Security can be alluded to in this process. There are many ways in which someone can be rooted, above all in everything that is physically in check and often overlooked, and also on the social relationship level. It can be highly beneficial if the patient does not remain passive and continue to believe that she is dependent on receiving attention and encouragement from others, but instead sets positive spirals of resonance into motion herself with interest and attention. Experiences of self-efficacy can help to stabilise patients. A playful combat readiness should be encouraged, like in sport or one of the martial arts from the Far East.

I then see whether she is willing to discuss her thoughts on death and welcome any indication that the topic of death is no longer taboo. It can be a relief to learn that the patient has developed a curiosity to learn about how other or previous cultures approach the topic of the death of her own accord. This permits the topic a certain level of breathing space for new thoughts. Further scope has been granted by research on near-death experiences, which has since introduced a number of significant, systematic surveys, the overall conclusion of which tends to convey a sense of optimism as opposed to fear.

I also like to broach the issue of our life cycle from the angle of becoming, passing away and once again becoming, sometimes using the pupation of larvae into butterflies as an example. The intention in this regard is to entrust ourselves to the flow of transformation in which we all participate. The shared fate of being joined together in this process of transformation can help to foster a sense of belonging.

6.8 Degrees of Freedom between Heteronomy and Self-Determination in Psychotherapy

6.8.1 The Scope for Individual Treatment, Including Semi-Structured Approaches vs. Loyalty to the Manual

At present, numerous variants can be found in psychotherapy. The answers provided by therapy reflect the primarily non-reflective basic assumptions about the worldview and concept of man represented in each case. Is it the largely schematising functioning of the individual that is prompted in this regard, or is the emphasis placed on the individual's personal dispositions? Research from the past two to three decades has underscored the significance of the latter. This section addresses the variable of a fixed therapy concept vs. a free, individualised approach. There are many combinations between these two poles. In terms of concept, the purely disorder-specific group treatments are adjusted to suit the shortcomings, potentials, goals and age of the target groups, which should ideally be as monosymptomatic as possible. Of course, in clinical reality, mixed groups of methods and diagnostics tend to prevail purely for organisational reasons. The various training and education courses studied by junior doctors and psychologists at psychiatric and psychosomatic clinics call for different approaches and styles, which can often be confusing for patients due to their multiplicity.

Exercise-based means of reaching a patient: Exercises are fixed approaches that are usually worked on in a group. Depending on the setting, each individual keeps his own experience to himself and may discuss it at his next individual session, or an exchange of experiences may take place directly after the exercise in the group setting. Standardised exercises cannot be tailored to specific individuals; however, personalisation may be possible in the follow-up session.

In this regard, there is also the approach popular in-depth psychology, whereby exercises, especially those based on body and movement therapy, are solely used to trigger deeper responses to significant personal experiences and place an emphasis on the free exchange of experiences.

Exercise-based and psycho-educative approaches fall within the realm of behavioural therapy. Importance is predominantly placed on groups with the same disorders. The original concept of man is the "tabula rasa", the blank page shaped by experiential learning that can be both relearned and forgotten. Work on present problems is preferred, explicitly not issues that happened in the past; a new attitude needs to be developed to cope with the current issues. The ideology of practical feasibility as well as the explicit demand for active cooperation from the patient spreads confidence, hope of a fundamental changeability and high expectations for the personal behavioural correction in terms of the agreed objectives. Standardised guidelines widely dominate in exercises in addition to roleplay guidelines, which have been compiled on the basis of the average experience with the specific disorder. Therapists who know to provide the required leeway to allow important individual experiences that are triggered during exercises to be addressed, largely compensate for the drawbacks of standardisation. They pay a great deal of attention to the transfer of these experiences into everyday life. Largely standardised homework is used to supplement the work; however, it can often cause the patient the feel like a student, which can cause their self-esteem to suffer in certain cases. Nevertheless, this approach has the advantage of encouraging the patient, who will benefit from using the time between sessions for the pursued change, to cooperate in a way that activates personal responsibility.

The therapeutic relationship is considered less important than technical and methodological tools, despite the fact that a friendly and kind approach has been recommended and encouraged as a factor conducive to the collaboration between therapist and patient since the 1990s in behavioural therapy. According to the psycho-educative understanding, the therapeutic relationship leads to the emergence of a slightly vertical teacher-student undertone. This can cause objectionable dissonance to arise, which may have the ability to offend the patient, particularly if the therapist is relatively young and new in the field and attempts to cover up his insecurity, and the patient, as is common, is from an older generation. This constellation is the one most commonly encountered in clinics. If the therapist addresses this constellation, they can gain the trust of the patient.

There are a large range of disorder-specific, standardised and manualised programmes available. These approaches primarily address the patient's readiness to learn, will and understanding. As the programmes emerge with a scientific claim of proven effectiveness, they carry positive connotations; they become beacons of hope with promised success, which often causes a strengthening spiral of positive reactions to be set in motion.

Colleagues in the field of psychotherapy at the start of their career are often particularly grateful to receive clear instructions on how to progress; it helps them to get over the initial hurdle of their natural hesitation due to inexperience more quickly.

Critics are often keen to point out that manualisation and loyalty to manuals may have a negative impact on the final outcome of therapy, as these factors are an obstacle to the development of a therapeutic relationship [6]. They also fail to provide sufficient breathing space for the development of personal impulses, emotions and experiences by the patient, despite the fact that the integration of a resource approach instigated by the therapist has been on the reform agenda of behavioural therapy for over two decades, since the work of Klaus Grawe.

Relaxation exercises also form a component of structured learning approaches. The most traditional is J. H. Schultz's (1884–1970) autogenic training.

In the field of behavioural therapy, Edmund Jacobson's (1885–1967) progressive muscle relaxation (PMR) is the preferred approach and is regarded as easier to master due to its use of the contrast between tension and relaxation, consequently provided easier to perceive differences in muscle tone.

The recent wave of enthusiasm for the healing efficacy of various forms of meditation and yoga that captured the West in the 1970s and 1980s, and experienced a revival in the secular forms at the turn of the last millennium, has firmly anchored a range of mindfulness-based exercises and programmes in psychotherapy. The desired effects are structure-building and structure-stabilising outcomes for those with structurally weak personalities on the one hand, including the establishment of a centring core and an inner observer that is capable of distancing and regulation. On the other hand, mindfulness-based approaches help to strengthen the regulatory capacity of the vegetative nervous system and consequently have a stress-reducing effect. Psychotherapeutic measures are not intended for the sensitisation of spiritual abilities, the purpose for which it was originally developed.

6.9 Characteristics of Structurally Weak Patients

6.9.1 Inner-Outer Boundary

The process of physiologically desired personal shielding from the sea of waves in which we live has received relatively little attention as an objectifiable symptom. However, as a subjective symptom, the experience of being influenced or controlled by others is considered a classic grievance of schizophrenia.

It is perhaps worth exploring how an individual is capable of experience himself as his own subjective world and as a shielded whole dominated by his own wavelength. We have ascertained that the heartbeat pulses through the entire body like a pacemaker and impulse generating organiser, and that the activity of all other living tissue that modifies this basic pulse, transforms it into a unique, individual fingerprint in waves, which cannot be found elsewhere. In a sense, the individual pattern of oscillation can be viewed as a personal mandala.

The notion of symbolically succinct wave patterns relies on a certain level of consistency. It may be difficult to establish an adequately consistent or recurrent stationary wave pattern that provides a formative matrix for internal differentiation in the case of high fluctuation, regardless of whether this is the result of internal or external influences.

6 We now return to the inner-outer boundary: Impressive parallelism for everyday contact behaviour can be found between the capabilities of our consciousness, with its actively shaping forms primed for synchronisation, as well as the pole of active receptiveness, namely the active withdrawal of our own activity to allow us to resonate with the incoming wavelength and simultaneously interpret it on the one hand, and the ability to dismiss or not resonate with this wavelength on the other. This ability to disregard certain stimuli in order to protect the existence of our own oscillation and give it the chance to survive and develop is of vital importance at the beginning of life.

Certain hypotheses have been presented on whether the phenomena of first-rank symptoms (K. Schneider) are exclusively the result of misperceptions due to overexcitation of the body's tissue or whether they (perhaps also) constitute information that was able to penetrate the erasing inner-outer boundary. However, no findings have been made in this regard.

Back to our patients: In terms of intimacy and distance, Mentzos observed an agonising ambivalence between the desire for assimilation and the anxiety of becoming lost in the presence of others (fear of loss of self) on the one hand and the forced withdrawal into isolation, which quickly turns into a subjective prison, on the other in people with borderline symptoms. This phenomenon is dominated by the experience of boundaries breaking down, i.e. the experience of a lack to security, not at an average, interpersonal distance, but instead in a much wider scope.

6.9.2 Centring and Fine Adjustment

In accordance with the hypothesis, structurally weak people, either as a result of genetic or environmental influences, struggled at the perinatal beginning of life to form a presence for normal states of balance between the poles of reward and punishment, namely happiness and misery, inherently found in the interbrain. According to the child's state of mind, these polarised reactions would seek help from those around by crying in the case of an emergency, thus serving survival. The joy exuded from the baby's face once his needs had been satisfied would reward and motivate the helper to continue fulfilling the baby's needs.

R. Hess detected these extremely rewarding and daunting islands in the diencephalon in self-stimulation experiments conducted on apes in the 1950s. The regulation of brain structures, which, when excited, cause the individual to experience reward or dread is also found in the phylogenetically more evolved areas of the brain, and therefore can be regarded as a consistent principle. Creating a balance between these two poles remains the constant task of development.

Caring mothers and caregivers are responsible for helping infants to regulate their feelings. (See ▸ Chap. 2 for more information on this topic). They pick up on the infant's frustration and respond to it in a soft and modified manner through comfort and love. Not all infants are lucky enough to have a consistent caregiver to help with their regulation. Those without a consistent caregiver experience extreme mental states at a young age and are more likely to switch between emotional poles.

Therapy for structurally weak individuals aims to provide subsequent assistance in developing the patient's ability to increasingly adopt a stance located halfway between the poles. This encompasses the poles of intimacy and distance, active and passive, emotionally expressive and restrained, abuse to oneself and others, positive and negative emotions as well as judgements passed on oneself and others. There are many variations on the above. A tendency towards these multipolarised experiences and behaviours is a characteristic of borderline personality disorder. In the case of a somewhat more developed narcissistic disorder, it is primarily the individual's self-esteem axis that lacks a centre. The individual's identity covertly becomes fixed at the pole of inferiority, something that the patient is neither willing to believe nor able to tolerate; he aspires to reach the counterpole, which he is drawn to by notions of greatness and flying high. Being at the extreme end of either pole can endanger relationships.

Despite the fact that the therapeutic relationship is established in here-and-now, it is sensible to pay attention to the ups and downs of an all-embracing arc and keep the flipside of the coin in mind at all times when interpreting polarised reactions. "Yesterday you were more than happy to talk about this relationship, at least that's the impression I got, yet today you seem to have nothing to say on the subject. Maybe we should hold off on reaching a final opinion and instead focus on what causes your assessment of the relationship to go up and down. If you look back on your encounters with this person in the past 2 weeks with the distance obtained from the time that has passed since then, would you say getting to know this person has been worth it thus far, or not? You're giving me a thumbs up. Ok. If the other person was asked, or if you were to step into the shoes of the other person and asked the same thing, what would you answer as this person in response to the question of whether it was worthwhile getting to know xy, namely you? Now you seem to have a puzzled expression on your face. Interesting. What do you think the other person finds difficult about their relationship with you? You're shrugging your shoulders. Ok. Could you do anything to help make the relationship with you more stable and reliable? What has worked well for you in the past?" – "The other day I said that I needed space because my head was spinning. I said it in a rather abrupt manner, but nevertheless in a way that wouldn't annoy anyone". – "An I-statement can often be a helpful tool for regulation". – "In the group we often play around with the hand movements we have learned from body therapy; they signify "come closer", "move back a little", "stay at this distance", but you aren't required to abide by them and can instead continue to pursue what you want, in the case that this is different. Relationships have taken on a whole new meaning for me. It can be fascinating to walk a mile in someone else's shoes, as encouraged by Native Americans".

"What do you do in the meantime when you feel like going crazy, as you often describe your state of mind, before making mincemeat of everything?" – "This used to happen as a response to something which would have previously made me angry of mad. It still does at first, but now I'm quicker to react; I'm a mighty mountain, a volcano ready to explode whenever I want, but instead I now feel the pleasant coolness of the glacier and the distance between myself and the guy whose arrow I allowed to drift through the mountain wind. I keep a pebble in my trouser pocket, which I press. Its firmness reflects the new strength that I associate with this feeling of being a mountain. And once I take three deep breaths, I feel able to respond calmly where I see fit. Any time I encounter situations like this I write about them in the evening. This makes my soul freer. Sometimes even allowing it to laugh in the process".

Fine-adjustment, i.e. the dimming of emotions, is grouped around the 'centre'. The ability to finely adjust is accompanied by simultaneous identification with a delicate mixture of yes and no and of the rejection of the matter at hand with the acceptance of the core personality. This is the magic formula for the internalisation of regulation patterns.

Admiration of our core personality with a certain amount of solidarity (as, by and large, we all struggle to demand something of ourselves) helps us to develop a regulating centre. We are all in the same boat in this regard, even if some of us, such as older siblings, may be somewhat more capable of control. Older siblings can be the ideal role models if they impart a sense of "I know how you feel, but it can be done! Follow my example and you'll do just fine". This approach is often successful with ex-users.

It was the secret formula to success employed by Marsha Linehan. According to her own statements (a speech made at Willigis Jäger's 90th birthday celebration), she was able to gain a strong understanding of the situation going on inside her patients as a result of her own experiences. This allowed a certain sense of partnership to become incorporated in her therapy, which never threatened her position as the leader. At this point, we shall now look at the therapeutic relationship in more detail.

6.9.3 The Therapeutic Relationship for Structurally Weak Patients

The therapeutic relationship has already been addressed to a certain extent in the previous section. However, in this section, we will focus on the striking differences in the therapeutic relationship offered to a) patients in the field of neurosis and b) those who exhibit structural weaknesses. There are also identical accents that require particular attention.

"Meet the patient on the level is he currently at" is a very peculiar task in this area as this often encompasses extremes that the therapist has never encountered in his own life: extreme mistrust, extreme aggressive tension, an extreme hunger for attention and affirmation, an extreme feeling of being along etc. Psychotherapists for psychosis patients are required not only to possess sound theoretical knowledge on the struggles borderline patient's face in trying to relate, but also a resilient inner stability, compassion that is not easily swayed, especially towards those with signs of psychosis, good social skills as a resource and resilience underpinned by their life experience. The journey towards gaining the trust of a patient with psychosis is often long, and once established, maintaining this trust requires tremendous sensitivity. Psychotics are hypersensitive to many different aspects. They are able to tell when someone is lying, cheating and trying to pull the wool over their eyes far quicker than individuals susceptible to neurosis. They are generally transparent in nature and do not attempt to stave off conflicts, as is often the tendency for neurotics. Psychosis psychotherapists are required to be just as transparent. If they are currently emotionally burdened by another issue, it is good to discuss this with the psychosis patient, even if a summary of the matter at hand is not provided. "You may have noticed that my head is a little in the clouds today. I'm dealing with an unresolved issue elsewhere in my life and feel a little under pressure. It has nothing to do with you. My energy and attention are a little overwhelmed and divided due to my current situation with the additional stress. If I am unable to change this, I would still like for you to be able to relate to my current state of mind".

Special attention needs to be paid to the regulation between intimacy and distance. It always remains a balancing act. In individual sessions, you could suggest that the patient moves his chair as close or as far as he feels comfortable at that point in time. His escape route to the door must remain unobstructed at all times.

Navigating the polarising experience and the corresponding actions of the psychotic can often be the most strenuous aspect. The therapist needs to be capable of realising how much it means (or can mean) to the psychotic when he is able to grasp something as completely unspectacular and normal and act accordingly unobtrusively. A reaction from a place of balance often constitutes a great achievement for many patients and deserves special recognition: e.g. someone new to the group sits in the patient's usual seat. He is kindly informed of this and moves to another chair without a fuss. This is certainly something to be celebrated. Previously, this event would have triggered a hostile reaction. Conversely: If the patient relapses into polarised interpreting and behaviours, the therapist should acknowledge the incident as an unfortunate relapse and not grant it any further attention. This requires the therapist to reverse their inner rewarding attitude. Love and appreciation are yet again the most effective factors in luring development in this respect.

I found the sensitive relationship developed by Benedetti towards his numerous patients of psychosis particularly impressive, which led him to describe finding his patients to be "beautiful," whether children or adults, as soon as he was able to reach them with no doubt as to his objective perspectives. His feeling of beauty arose in relation to an internal process of mutual understanding, perhaps due to a connecting, synchronous field.

Bibliography

1. Grawe K. Neuropsychotherapie. Hogrefe, Göttingen; 2004.
2. Linden M, Hautzinger M. Verhaltenstherapiemanual. Heidelberg: Springer; 1996, 2000, 2005, 2008.
3. Berking M, Rief W. Klinische Psychologie und Psychotherapie. Band 1 u. 2. Heidelberg: Springer; 2012.
4. Rudolf G. Psychodynamisch denken – tiefenpsychologisch handeln. Praxis der tiefenpsychologisch fundierten Psychotherapie: Schattauer-Verlag, Stuttgart; 2019.
5. Wampold BE, Imel ZE, Flückiger C. Die Psychotherapie-Debatte. Was Psychotherapie wirksam macht. Bern: Hogrefe; 2018.
6. Norcross JC, Wampold BE. Evidenced-based therapist responsiveness. New York: Oxford University Press; 2019.

The Spirit of the Times and Cultural Epochs as Reference Systems Along with Their Transformation of the Ideal Level of Human Development – A Historical Journey Into the Past to Learn for the Future

Contents

L. Hartmann-Kottek, *General Psychotherapy*,
https://doi.org/10.1007/978-3-030-87466-7_7

Overview

The culture and spirit of the time, including its ideological and religious background, shape our inner mindset, our understanding of social normalcy and of the sense and objectives for how we interact with ourselves, as well as social cohabitation. The spirit of our ethnic origin and the time in which we live provides a formative reference system for individuals and shapes their development potential. Comparison facilitates a larger scope of development for the here and now.

This extremely concise overview aims to put the apparent normality of our current understanding into perspective. Values, worldviews and concepts of man are all relevant to psychotherapy, particularly if and as long as we do not reflect them.

However, comparisons and overviews do not account for the impression that we are currently living in a time that represents the peak of development. With this in mind, we shall take a step-by-step look back in time, perhaps also in spirals, in order to finally secure a future in which the development of human potential is more adequately challenged and promoted.

7.1 Psychotherapeutic Trends After 1800, the Birth Century of Current Psychotherapy, up to Today

7.1.1 The Analytical and Reductive Perspective as the Mainstream

True to the spirit of the enlightenment that swept through Europe in the aftermath of the French Revolution, the new-found appreciation of reason and precise observation which harked back to Aristotelian teachings introduced elementary psychology to the intellectual minds of the nineteenth century against the backdrop of Newtonian physics and the technological advancement it spurred.

In terms of philosophy, the advent of German idealism in the eighteenth century with thinkers such as Immanuel Kant (1724–1804), Johann Gottlieb Fichte (1762–1814), Friedrich Wilhelm Schelling (1775–1854) and Georg Wilhelm Hegel (1770–1831) marked an intellectual current intent on restoring the Western values of truth, ethics and religion in addition to a unified raison d'être in the face of growing social ills through the advent of realism and materialism, providing inspiration for Ludwig Feuerbach (1804–1872), Karl Marx (1818–1883) and Friedrich Engels (1820–1895). This movement was joined by waves of positivism and empiricism from across the pond in England, led by minds such as John Stuart Mill (1806–1873) and Herbert Spencer (1820–1903), the primary focus of which was the inductive, logical processing of material with utilitarian objectives in terms of development and progress. Transsubjectivity was frowned upon.

Wilhelm Wundt (1832–1920), widely regarded as one of the pioneers in elementary psychology, established the first academic university institute for experimental psychology in Leipzig (Germany), which quickly amassed international acclaim.

Interested scholars travelled from all corners of the globe to Leipzig to adopt and disseminate the academic approach cultivated at the institute, which required the will to make precise, objective observations without any interplay of subjective thoughts or emotions. In essence, it should be possible to get the same results from repeat experiments. Detailed observation led to a clearer image of the smallest units, or rather elements, and initiated the search for the ultimate cause there.

W.M. Bechterew (1857–1927) travelled from Russia to study under Wundt, later going on to innovatively research conditioned reflexes in St. Petersburg with his student I.P. Pavlov. His work provided the foundations for the rudimentary understanding of reflexology in learning theory and behavioural therapy.

One of the strongest proponents of Wundt's work, his student Hugo Münsterberg (1863–1916) emigrated to North America, bringing with him the spirit of experimental psychology and the approach of unerring, objective, psycho-technical observation to Harvard, where he was a lecturer, and to universities throughout the USA; he also passed it on to the field of economics in an innovative manner.

The observational focus on the "object" and its inner substance, particularly its inherent, logical laws, led to the emergence of phenomenology, of which Edmund Husserl (1859–1938) was a key pioneer. Husserl also studied for a period under Wundt, later shifting the focus of his work from the natural sciences to the humanities, eventually becoming a professor at the University of Freiburg.

Sigmund Freud (1856–1939) was borne of the same Enlightenment spirit with its reductionist worldview, which inspired and shaped his work. He exchanged letters with Wilhelm Wundt. Freud applied the scientific stance nurtured by Wundt to his "talking cure", taking it as a model for the therapeutic relationship in his psychoanalysis.

It would be extremely far-fetched to assert that it would be possible for research in the field to recognise that the lack of consideration towards the interpersonal relationship constituted the most significant shortcoming in early psychotherapeutic approaches after barely a century as a result of the contemporary scientific understanding. This shortcoming had a profound effect on the initial development of psychoanalysis and behaviouralism, namely behavioural therapy, which prevails to this day. It continues to shape the trends of both psychotherapeutic schools of thought, which remain plagued by tradition.

7.1.2 An Integrative, Holistic Stance as the Countermovement

The end of the nineteenth century marked a turning point with the emergence of a corrective, intellectual resistance to the notion of the intellect and its logic being the sole ways to access individuals. Wilhelm Dilthey (1833–1911) saw understanding as key in psychology, accompanied by the vitalism advocated by Henri Bergson (1859–1941) and Maurice Blondet (1861–1949). Vitalism came to the fore in the work of Hans Driesch (1867–1941), noted for his work in the field of zoology. His experiments on sea urchin regeneration demonstrated that the whole could be found represented in the parts of the sea urchin and that these parts had the power and inner structure to become a whole. Each part acquires its gestalt, purpose and being from the whole.

This countermovement represented a departure from the mechanized, schematic and statistically calculated way of thinking. It enabled unique, individual and inner emotional experiences and inner freedom to be regarded as virtues in their own right. Above all, it liberated the creative potential of the generation at the turn of the century, transforming it into a new expression, culminating in Art Nouveau. As a new form of lifestyle, it marked a positive awakening in terms of the approach towards nature and physicality, including sexuality.

In the latter half of the nineteenth century, the discovery of holistic phenomenon emerged as a psychological countermovement to the fragmented, analytical understanding of science at the time. Holistic psychology gained increasing recognition through experiments conducted in the field of perceptual psychology, providing impetus for Gestalt psychologists. As a technical term, "gestalt" is synonymous with "wholeness" and can be regarded as a relational structure within and figure in the foreground that stands out from the background on the outside. Ever since Kurt Lewin, one of the most renowned field theorists who was far ahead of his time, a gestalt has been understood as a field and structure.

The first cited intellectual founder of Gestalt psychology was the Austrian Christian von Ehrenfels (1859–1932). He was the first to use the melody to explain a "gestalt", namely that a gestalt is defined by the inner relational structure of its notes. If transposed into another key, each individual note will change; yet, the melody will remain the same. Another example: The length of a poem can be calculated in terms of lines, words or individual letters, but its meaning cannot be deduced from the individual parts. Fragmentation and destruction of the mutual relationships render it unintelligible. In other words, as the ancient saying goes: The whole is greater than the sum of its parts.

The turn of the century and the first half of the twentieth century marked a period of great interest in Gestalt therapy at a number of institutions and amongst holders of professorships (Graz, Leipzig, Würzburg, Frankfurt, Berlin, Münster, Tübingen, amongst others), research groups, publications, etc. It would not be possible to mention this period without referring to the following personalities: Alexius Meinong (1853–1920), Oswald Külpe (1862–1915), Narziß Ach (1871–1916), Karl Bühler (1879–1963), Felix Krüger (1874–1948), Max Wertheimer (1880–1943), Otto Klemm (1884–1939), David Katz (1884–1953), Kurt Koffka (1886–1941), Wolfgang Köhler (1887–1967), Friedrich Sander (1889–1971), Kurt Lewin (1890–1947), Richard Meili (1900–1991), Wolfgang Metzger (1899–1979) and Wilhelm Witte (1915–1985).

The numerous holistic phenomena were empirically studied and categorised. Wolfgang Metzger combined them under the phrase "Laws of Seeing". Nevertheless, the dynamic background to their occurrence was not explained during his lifetime. Kurt Goldstein (1878–1965) the neurologist and Gestalt psychologist, Adhémar Gelb (1887–1936) and other contemporaries held the strong conviction that the information processing of our central nervous system follows the laws of Gestalt therapy. As it happens, following a period of latency lasting around half a century, these notions were revisited in lectures on neurophysiology (Gerhard Roth) when complex, cerebral information processing was on the agenda.

Psychological phenomena found in Gestalt psychology were also observed by quantum physicists such as Hans-Peter Dürr, Heisenberg-Schüler and Hartmann Römer, the professor of quantum physics at the University of Freiburg. Should the holistic gestalt phenomena correspond to wave-like manifestations similarly to all dualistic modes of being, which superimpose and help to form matter? In this context, the dualistic mode of being refers to the particulate nature of matter on the one hand and the wave nature on the other.

This notion has been somewhat simplified for the current generation in comparison to our colleagues over a century ago. In the time since, humanity has come on leaps and bounds in terms of embracing Einstein's findings and quantum physics, despite the fact that this can still be regarded as "going against the grain" and hard to comprehend for naive minds. At any rate, we are now able to broach and address issues from a scientific standpoint.

We have the ability to understand the analytical and integrating approaches as meaningful complementary poles. In terms of science, this corresponds to the processing of classical physics on the one hand (in the case of calculations of large numbers with large categorical distances between the calculated and the calculator, for whom an average value is precise enough) and, on the other hand, the application of the laws of "New Physics" (particularly in the case that there are only small numbers and no or only a small categorical distance in size between the observer and the observed). By and large, we need both in science. For psychology in the strictest sense of the world, we primarily require the processing and general approach of quantum theory. First and foremost, psychotherapy constitutes a 1:1 relationship between spontaneously reacting individuals (or even an 8:1 relationship, the latter does not make any categorical difference).

If an insurance company would like to find out how often the average German male suffers from depression, how many children he has, how much money he spends, how much beer he drinks, how many cars he totals and at what age he dies, classical calculation methods and the resulting statistics are most likely to provide the best answers. An analogous approach is taken in diagnostic survey questionnaires to find out whether a certain patient can be regarded as in the average range for his age group in terms of the extent of his reactions of withdrawal or fear, for instance. Unfortunately, this does not provide the patient's sitting opposite with any access to his reference systems, which have been shaped by his life thus far, his learning processes, the way in which he processes conflict and experiences, or his resilience and potential for his own personal understanding.

When the emphasis is placed on the individual in psychotherapy, namely when authentic, free and compassionate behaviour that is modified by the unconscious and its associations in addition to external factors is involved, we require other scientific methods that exceed classical statistics. These methods must be capable of adapting to the various courses a process may take. Furthermore, we need to shift the focus from psychotechnics (canned interventions), which should not be regarded as equal to psychotherapy (particularly in view of the fact that psychotechnics is only capable of explaining less than 10% of variance in findings), to factors that are relevant to therapy, namely the therapeutic relationship and the therapist's personality.

At present, psychotherapy is rather inconsistent. On the whole, it is currently in a phase of upheaval in terms of its concept of man, identity, methods, objectives and its research methodology. This does not entail a dramatic period in the history of psychotherapy, but rather a challenge that, in the best-case scenario, may result in a leap forward with regard to quality.

7.2 European Cultural History of Ideal Human Development

7.2.1 The Remainder of the Modern Era

The Renaissance, the Baroque era, Classicism, Rationalism, Empiricism, Materialism, the Enlightenment, Sturm und Drang, Weimar Classicism, Romanticism, Biedermeier, Lebensphilosophie, Impressionism, Art Nouveau, Existentialism – all regarded as the pinnacles of art history and philosophy after the 1500s that helped to shape the spirit of the respective period and its transformation.

The previous section touched on the conflicting ideals that emerged in the eighteenth and nineteenth centuries: the tedious emancipation and liberation from religious and secular-societal tutelage over the course of generations in the external world, the roots of which can be traced back to the Renaissance and even the Middle Ages in part, which was coupled with an inner counterpart, in which the force of reason, logical persuasion and personal reconciliation of experiences were allowed to prevail both over externally controlled convictions and over emotions, drives and the openness to faith on the whole. In terms of society, the characteristics of trusting good faith and emotionality were abused repeatedly in various ways by religious and secular players. The great collective effort to gain protection from external threats of abuse, to either shrug them off or at least mitigate them, led to a change in the value placed on inner sources of perception, responses, ways of processing and access to inner potential. Notably, at the time, personal beliefs based on inner veracity could result in dangerous conflict with the external authorities.

Only a small number of privileged individuals were able to risk having their own opinion and expressing their views without worrying about the consequences. The church maintained its right, in terms of its understanding of the Bible, to interpret reality and impose its way of thinking. In the seventeenth century, Germany endured significant suffering as a result of the Thirty Years' War (1618–1648), which also spread across the borders of its European neighbouring states. Protestants and Catholics wreaked havoc across the territory and killed each other, purportedly in the name of their true faith. In reality, the war was merely a power play between advocates for the church and state. The average individual very quickly ended up being caught in a conflict between wanting to survive and betraying their beliefs, usually opting for the latter: "Ubi bene, ibi patria" (where I feel good, there is my homeland). As such, the baroque period (approx. 1600–1750) commenced with the looming fear of death and culminated in a lust for life, perhaps understandably gaining a reputation as a period of great contradictions. The bridges between the emerging rivalry in splendour between the powerful and the renouncing transient spirit of other contemporaries could not be wider. In a similar vein, the trauma of anarchically turbulent experiences on a daily basis, on the one hand, contrasted with the propensity for almost over-regulated artistic design on the other. The latter could conceivably be viewed as an attempt to cope with the former.

Starting at the end of the Middle Ages, spanning the Renaissance, the Baroque Era, Rationalism (Descartes), Empiricism (Hobbes, Locke, Hume), the Enlightenment (Voltaire, Montesquieu, Rousseau), Idealism (Kant, Fichte, Schelling, Hegel), Existentialism (Kierkegaard, Sartre, Marcel, Heidegger, Jaspers) and the various intermediary forms of these philosophies – a struggle for the understanding of reality emerged over time, with intellect and its logical and analysing potential on one pole in addition to a materialistic grasp of reality led by the senses and adjusted by emotions, which is vital for survival in a distressing and insecure world. At the other end of the spectrum lies the pole that makes people susceptible to abuse and which, despite various innovative attempts to revisit this topic in the modern era, has lost its significance and been pushed into the background of general consciousness. This pole represents an intuitive, more holistic grasp of reality shaped by emotions.

From a modern perspective, the latter could be assigned the characteristics of attachment security and inner coherence, with its cognitively and materially accented counterpole of incoherency characterised. The notion that the analytical and dissecting perception down to the smallest detail fits the bias preference of the mind and its logic, and is potentially also capable of promoting a materialistic, mechanical worldview, has been explored in the previous section. The materialistic perspective from the eighteenth and nineteenth centuries remains popular amongst the mainstream to this day, particularly amongst individuals who have not encountered "new physics" in their lives and especially in view of the fact that the materialistic, reductionist concept of "common sense" appears to fit the worldview taken in by the limited sensory channels.

■ Partial Perspectives Are Also Capable of Flourishing

We shall now return to the matter at hand: Kant's (1724–1804) deontology and self-disciplinary ideal guided several generations like an intellectual beacon. It even sculpted the conceptual notion of man adopted by nations, particularly in Prussia, and, in a somewhat milder form, the notion of the traditional "German" abroad. Friedrich Schiller (1759–1805), a profoundly loyal devotee to Kant, erected an intellectual monument to the latter's moral concept with his poetry.

When scaled down to the level of the bourgeoisie, deontology flourished amongst the people alongside other customs from virtues adopted from Christianity, namely in the modest variants of the bourgeois virtues: punctuality, orderliness, diligence, cleanliness, frugality, honesty and reliability. The women in the background were expected to make a considerable large contribution without making a fuss. For men in the Prussian civil service, these virtues shaped their careers.

Let us take one step back: from a social point of view, the gap between the concepts of virtue and man significantly increased in the eighteenth-century feudal system. The notion of simply accepting the place in society into which one was born without questioning it, as was the case in the Middle Ages, came increasingly under scrutiny in this period. Nevertheless, the ethical and moral standards upheld at court were worlds apart from those of their subordinates. In other words, Quod licet Jovi, non licet bovi: What is permissible for Jupiter (a godlike ruler) may not be permissible for a bull (a commoner). The upper and lower classes were so far removed from one another that empathetic identification with one another would be considered inappropriate, or perhaps even abhorrent. This phenomenon can be referred to as social decoherence. Decoherence leads to a lack of understanding and, in tense situations, can be regarded as a prerequisite or perhaps even an invitation to war or rebellion, both externally and internally.

Throughout Europe, an outcry against the inequality faced in comparison to the privileged class gathered pace amongst the underprivileged. Pent-up aggression is capable of poisoning from within, making us sick or deceitful and distorted in our character. A certain amount of courage is required to rebel in an external, unprotected sphere, as this could entail the existential end. The people faced the risk of losing their heads or entire families. An understanding of solidarity with their comrades seemingly provided the sole element of subjective protection.

During this period, waves of emigration reached the shores of North America. The American Declaration of Independence (1776) is often regarded as the first milestone in the overthrow of absolutist rule and the transformation of society. However, on average, the once-excluded and underprivileged members of society were only capable of showing limited compassionate solidarity and marginalised themselves. The ideals of the new-found camaraderie were put to test with racial unrest.

In Europe, the people unleashed their rage in response to a lifetime of existential despair in the form of the French Revolution (1789 ff) with shockwaves felt around the world. The momentum and cruel excesses of the revolution were ultimately only quelled with great difficultly and at the cost of many lives. At long last, rational forces had successfully established the human ideals of "liberty, equality and fraternity", laying the foundation for the establishment of human rights. This ideal is not only based in reason, but also postulates empathy, respect for others and a willingness to communicate on equal terms.

In a similar vein, the Russian Revolutions (1905, 1917 ff) also reflected the above momentum amongst the socially divided classes, their alienation, their pent-up destructive rage and their yearned-for notion of equality. However, in this case, these ideals were only achieved sporadically and temporarily.

The search for adequate and liveable ideals of social coexistence continues to this day, with no definitive answer in sight. Two world wars have caused us to come to this bitter conclusion. Time and again individuals succumb to the temptation of self-glorification, the amplified glorification of their reference groups, the hostile isolation at everything outside the outer border of their recognised identity and the abuse of power, regardless of the size of the interpersonal cohesion, whether it occurs in politics, in more limited relationships (such as in the professional politics amongst psychotherapists) or in an individual's relationship with themselves. Hostility and wars are waged first in our minds, or hearts, when we fail to empathetically identify with others. A certain level of awareness in this regard and responsibility is beneficial. Only a small number of exceptional people appear to have cultivated the greatness, power and maturity to achieve the overarching ability to identify with others.

Our search for ideals and principles throughout history and cultures now returns to the eighteenth century. In the visual arts, the rationalist attitude towards classical ideals as exemplified by J. J. Winckelmann's (1717–1768) leading concept of "noble simplicity and quiet grandeur" was primarily expressed through the clear and calm alignment of classicism. Winckelmann was an advocate of the revival of the understanding of art from Ancient Greece, even the imitation thereof, to counter the grandiose, exuberant lifestyle and excessive design that characterised the Baroque and Rococo period. Whilst his appeal was largely met with a sympathetic ear, it was not free from critique. An interesting example can be found in

J.W. Goethe's (1749–1832) "Ipheginia in Taurus", for which the content was derived from antiquity; yet, the characters have the same experiences, struggles and inner battles as those of his time.

The perspective on life from the antiquity was previously mirrored at a much earlier period, in the well-proportioned perception of form from the Renaissance (approx. 1500–1600). Men of the Renaissance developed their potential, experienced life following the model from antiquity as the benchmark for every aspect of life, and revived the divine ideals from the antiquity to the extent possible. Nevertheless, the men of the Renaissance remained divided in terms of whether to adhere to the holistic perspective of Plato or instead to follow the more detailed, nature-observing view promoted by Aristotle. Be that as it may, the relative balance between the intellectual urge to discover and the sensual affirmation of existence shaped the Renaissance man and distinguished him from the later ideals advocated by the predominantly rational men of the Classical era and the Enlightenment.

In the eighteenth and nineteenth centuries, the concept of rationality was glorified as the rational, clearly structured and light field to which the Greek god Apollo was assigned. At the time, people unintentionally found themselves in a field of tension as a result of this bias: the darker aspects scorned by the rationalist cult of the Enlightenment entered the general consciousness with Romanticism, at the same time to a certain extent and partially a little later in a reinforced manner to compensate, thus shaping the lifestyle of the countermovement. The repressed elements of the human psyche forged ahead, in part in the form of the culture of sensitivity, and in part as the Dionysian principle that dominated life at the time, namely the longing for the emotional aspects of human nature and archaic creative power. On the one hand, this was experienced as vigorously fascinating, and on the other as the highly ambivalent aspects of human existence, which was linked to the god Dionysus. Friedrich Schlegel (1772–1829), Friedrich W.J. Schelling (1775–1864) and Friedrich Nietzsche (1844–1900) are often cited as the intellectual forces behind this form of polarisation, which appeared to characterise the contemporary attitude towards life.

In terms of science, during the romantic period, understanding and empathy were granted more importance than sober observation. Gaining an overall view was preferred to an observation focused on the details.

Romanticism often substantively revolved around the glorification of medieval chivalry, whilst turning a blind eye to its dark sides.

This complementary pole of Romanticism allowed feelings, longings for a breakdown of boundaries, conflations, subjectivity, a sense of meaning, intuition, ecstasy, myths, a connection to nature, sensitivity, a cult of transience, enigmatic searches for transcendence and breaking free from the restrictive shackles and structures of all forms to flourish. The preoccupation with the unconscious, hypnotic phenomenon, emotions, but also with the processing of emotional struggles came to the fore.

This was the viewpoint substantively adopted by Sigmund Freud with the emotionally multifaceted pole associated with Romanticism. However, in terms of the scientific-methodological understanding, he preferred rational, sober observation.

The Romantic Era began at the end of the eighteenth century and spanned the entire nineteenth century, ending at the start of the twentieth century. It oversaw the revival of Art Nouveau, providing impulses to help the longing embodied by the blue flower to finally be satisfied.

It is not possible to overlook the formative influence of the following personalities during the Romantic Era: the Brothers Grimm, Novalis, E.T.A. Hoffmann, J.W. v. Goethe, Ludwig Tieck, Bettina von Arnim, Joseph von Eichendorff, Clemens Brentano, Horace Walpole, Joseph Addison, Charles Dickens, etc.

The start of the twentieth century marked a major societal, philosophical, scientific and cultural shift in almost all facets of life, which called the previous perspectives and system of values into question. As a result of its substantial influence on the early approaches of the humanistic and systematic group of methods, whilst the older approaches, namely those derived from learning theory and psychoanalysis, remained relatively untouched, this phase of reorientation is also relevant to psychotherapy.

In spite of this interesting perspective, we shall now continue our search for development guidelines further back in time.

7.2.2 The Middle Ages

This period is largely accepted as the time between approx. 500 and 1500 AD. A certain amount of fluctuation can be found throughout various regions, with the period starting earlier in southern Europe and somewhat later in the north of Europe. The Early Middle Ages, initially characterised by the turmoil of the Barbarian invasions, lasted until approx. 900 AD. The High Middle Ages then lasted until 1250 AD, which marked the start of the Late Middle Ages.

Following the crusades, which saw the import of valuable knowledge from the Middle East to the shores of Europe, Central Europe, particularly France, experienced a boom in the impressively vigorous and enthusiastic construction of churches. This collaborative effort helped to build social bridges between the classes at the time, something which had previously never occurred, and has since remained an isolated incident. In the late High Middle Ages, the predominately earth-bound "romantic" architecture transformed into Gothic architecture, which reached to the heavens with a focus on the vertical, allowing light to pass through the man-made stone walls and flow through the stained-glass windows, glistening like gemstones, reminiscent of the heavenly Jerusalem as described in The Apocalypse of John. It seems logical to assume that the sense of identification amongst the social group that constructed a cathedral with their work must have been colossal. Indeed, the voluntary collective effort based on spiritual motivation remains a marvel to this day. This miraculous occurrence happened all over Europe at the same time.

Which principles were the people led by? The general populace had no access to education, meaning that they were illiterate and had no knowledge of Latin, the dominant language of the educated at the time in Europe. Instead, their education was ruled by the Christian church, based on the illustrative images and sculptures on the walls of churches and the teachings of the clergy.

Seven Christian virtues can be traced back to the High Middle Ages alongside the Ten Commandments and the divine virtues of faith (fides), hope (spes) and love (caritas) as the benchmarks for the focus and development of a person: (1) humility (humilitas), (2) love (caritas), (3) morality (castitas), (4) patience (patientia), (5) temperance (temperantia), (6) humanity (humanitas) and (7) diligence (industria).

Each virtue is faced with its deadly sin counterpart: (1) pride (superbia), (2) greed (avaritia), (3) lust (luxuria), (4) wrath (ira), (5) gluttony (gula), (6) envy (invidia) and (7) sloth (acedia).

The canon of virtues is reminiscent of the monastic vows of poverty, obedience and chastity. In turn, the distorted cardinal vices evoke images of the tales of monastic excesses, which repeatedly led to the emergence of reform movements, not solely during periods of dissolution. A shining example in this regard is the wave of western Benedictine monasticism that spread throughout almost all of Europe following the example of Cluny.

Those who were born into nobility and became knights were required to develop Christian virtues, above all humility, generosity, kindness and moderation in addition to other special moral values: adherence to the rules of decency, steadfastness, courtesy, courtly manners, a chivalrous reputation, emotional high spirits, courage, loyalty, devoted love and dignity. These ideals can be found in poetry of the time.

In the Late Middle Ages, following the eradication of the function and purpose of chivalric combat as the result of the advent of gunpowder, the culture surrounding chivalry gradually decayed and was lost in its original form.

The Investiture Controversy marked another bitter conflict between church and state, namely the emperor and pope, in the High Middle Ages. In principle, the main controversy surrounded who should be permitted to hold positions in the church. For an extended period, a vexing and unresolved atmosphere of tension persevered. The leading figures in the church, who made a number of enemies, failed to abide by Christian virtues, leading the way to their downfall.

The advent of the plague throughout Europe in the fourteenth century marked a dark and sombre period for the entire populace. Disease swept through the land, from the Crimea to north-western Europe, claiming the lives of a third of the inhabitants in its path. With death now an unavoidable and large part of everyday life, the people started to view life on earth as a short preparation for eternal life. Memento mori was portrayed in numerous Danse Macabre paintings.

To this day, the Inquisition remains viewed as a particularly morally dark chapter in the history of the church. The church asserted its power through fear, laying claim to the sole, bible-compatible interpretation of reality with no regard for scientific knowledge and mercilessly, if not diabolically, asserting this interpretation against independent thinkers. The darkest depths of the human mind rose behind hypocritical façades. Blazing funeral pyres were faced with many an upright backbone. The church dug the grave for its own credibility.

By the end of the Late Middle Ages, the church was in a comparably dark predicament as a result of its profiteering from the sale of ministries, selling of indulgences and the reprehensible lifestyles pursued by many of its representatives. Several reformers (Huss, Zwingli, Calvin, Luther, et al.) were willing to put their life on the line, sometimes losing it, in rebellion. These uprisings were accompanied by the uprisings led by the secularly oppressed people, which joined the outrage at church representatives, who had since lost all credibility. In addition, the Thirty Year's War (1618–1648) claimed the lives of an immeasurable number of people, adding further fuel to the fire. Ultimately, for the most part, solace was provided by the corrective reflection on the virtues and a more veritable new beginning in all camps spurred by spiritual and intellectual reform.

7.2.3 The Antiquity – Greek Prosperity with a Focus on the Foreground and Background

- **Life in Greece**

The antiquity is the common term for the period spanning the emergence and flourishing of the Greek and Roman civilisations, namely from approx. 800 BC to 500/600 AD.

Cultural development in Europe has been shaped on a number of levels by influences from the antiquity: (a) socio-political, (b) art and (c) philosophy. At this point, it is conducive to ask ourselves how these ancient civilisations perceived themselves. Which ideals did they most aspire to?

First, we shall take a look at Athens and the "golden age" of Pericles. The best thing would be to go to the forum and listen to the articulate Pericles (490–429 BC), who was instated in command of the army multiple times (Thucydides XXXVIII).

Note

Pericles was a proud advocate of democracy in Athens, referred to as popular sovereignty. All citizens were granted the same rights for their concerns. Each person was respected by virtue of their competence and ability to adhere to the law and customs, not on the basis of whether they were rich or poor. Athenians allowed free access to the city for everyone, relying on their ability to act decisively in the case of an emergency. Knowledge and education were met with similar open-mindedness, in addition to a strong affinity for neat simplicity. Reflection and consultation had proven worthwhile prior to taking bold actions.

So much for the notions from Pericles' Funeral Oration on the points related to the understanding of the self and coping with life. Granted, this oration was not targeted at the entire populace, but rather solely at males born in Athens, not women, immigrants or slaves. Whilst an impressive feat at the time, democracy still remained relatively limited in terms of those it served. Nevertheless, this agenda was based on the ideal of a free, fair, social, relationship-forming, generous, art-oriented, assertive, operational, level-headed, intellectual and responsible individual, an ideal that has time and again provided the basis for the development of Western intellect at times when it was necessary to shake off arbitrary rule and stand up in defiance as a dignified human being.

We now return to Ancient Greece. A chapter that is often overlooked in school and history books, in spite of the fact that it may have had a stronger, lasting impact on the lives and identity of the entire populace more than the ideals mentioned above, is that of the Eleusinian Mysteries. It is estimated that this festival, together with its predecessors, was held for almost two thousand years, certainly since 1350 BC until its ban in the fourth century AD by the newly Christianised Roman rulers (Emperor Valentinian). The proconsul at the time, himself an initiate of Eleusis, attempted to prevent this ban as, according to his understanding, the Eleusinian Mysteries were inseparably connected to the spiritual existence of the Greeks.

Note

Initiates faced the penalty of death for divulging what they had experienced in the secretive festivals. As such, we are forced to rely on a puzzle constructed from various sources. The epitaph on one Eleusinian grave writes: "Beautiful indeed is the Mystery given us by the blessed gods: death is for mortals no longer an evil, but a blessing". Pindar (522–446 BC), goes on to say: "Happy is he who, having seen these rites, goes below the hollow earth; for he knows the end of life and he knows its god-sent beginning". Sophocles (497–406 BC, Athens) summarises his interpretation in a similar way: "Thrice happy are those of mortals, who having seen those rites depart for Hades; for to them alone is it granted to have true life there; to rest all there is evil". And finally the words of another Roman, Cicero (106–43 BC), to his friend Titus from Athens: "For among the many excellent and indeed divine institutions which your Athens has brought forth and contributed to human life, none, in my opinion, is better than those mysteries. For by their means, we have been brought out of our barbarous and savage mode of life and educated and refined to a state of civilisation; and as the rites are called "initiations", so in very truth we have learned from them the beginnings of life, and have gained the power not only to live happily, but also to die with a better hope" (quotes from Pindar, Sophocles and Cicero taken from Nedden, p. 12 ff).

In contrast to the democratic political activities taking place in Athens, in Agora, women, men, immigrants and slaves were all allowed to partake in the Eleusinian Mysteries. All that was requited was two guarantors who had already been initiated in order to become a candidate. Those who had committed murder, desecrated a shrine or who did not speak Greek were not permitted to partake.

The path to initiation was split into the preparatory, small Mysteries in February and the large, nine-day festival in September with advanced entry in the subsequent year. Purification, fasting, teaching and self-reflection aspects were alternated with chants, communal activities, tests of courage, confessions and pilgrim rituals. Candidates for initiation were required to learn to endure mockery, teasing and perhaps even confrontation with justifiable accusations without the ability to defend themselves, on a special bridge. Reports tell of up to 30,000 participants per event, even citing a panhellenistic movement. People travelled from all over in the latter centuries, primarily from Rome. Truces were called for battles during the Mysteries. In addition, special arbitration bodies were established at the festival to handle any clashes that occurred during the Mysteries.

Despite the primary goal of purification, for a time, appalling excesses appear to have become standard practice, casting a shadow over the festivities.

In terms of content, the Mysteries initially and primarily revolved around the tale of Demeter and Persephone. This tale was based on a Homeric hymn from the eighth century BC, which may be traced back to ancient Mediterranean, matriarchal philosophy. The "Great Mother" Demeter, the goddess of harvest and agriculture, was a central figure of the Eleusinian Mysteries. Her daughter, Persephone (Kore), was abducted from her mother by Hades, god of the underworld. Following extensive negotiations, the gods reached a compromise: Persephone would return to her mother in the world of the living for two-thirds of the year, allowing nature to flourish and grow, only descending to her husband Hades for the winter third of

the year.

The Eleusinian Mysteries appear to convey that the dark time corresponds to the time spent in our bodies here on earth, with the brighter and more joyous time corresponding to a brighter plane, which we are required to depart from our bodies to enter. As such, the Mysteries were primarily concerned with the understanding of birth and death, the underworld, the earth as the realm of transience, resurrection, the consciousness of immortality and reincarnation.

Over the centuries, further influences were incorporated: a strong, orphic influence from Persia via Thrace (with the gods Dionysus and Orpheus as its priests), an Egyptian line of influence, various Bacchus legends, in addition to one variation from the cult of Mithras with the worship of purifying light and fire, which similarly came from Persia, and finally the myth of Hieros Gamos, purportedly traceable to the Middle East. Dionysiaca (Book 6, cited from Greiner), the principal work of Nonnos, addresses the notion of birth in heavenly fire, a clearly deeply Eleusinian mystery: Light as the seed of life?

For the average Greek citizen, this notion may have inspired a relatively anxiety-free sense of being embedded in nature and the cosmos in addition to an acceptance of love, sexuality and other sensory pleasures of life, an appreciation of beauty (aesthetics), exoneration from guilt as a result of the cleansing rituals and an ethical orientation with courage, loyalty and sincerity held in high esteem.

In terms of the history of philosophy, Pericles was alive during the period of pre-Socratic philosophy. The thinkers of the time had largely outlined important fundamental approaches that would run like a common thread through the subsequent two and a half millennia in various forms.

To summarise, Parmenides (540–470 BC), and his school of Eleatics in Southern Italy, directed his inner and spiritual gaze towards the bigger picture of god and his quality of his everlasting immortality. In addition to grasping the "way of truth", he also claimed to know the "way of opinion" (doxa) derived from sensory perception, which was viewed as delusion (Reminiscent of the Vedic notion of maya).

Heraclitus (544–484 BC) focused on the dynamic becoming and passing in a world full of conflicting complementary poles, including those of life and death.

The constant change, which is not lamented but positively accepted as a result of insight into higher connections, appeared to be controlled by a distant yet nevertheless wise divine Logos principle.

The mechanical worldview held by Empedocles (approx. 492–432 BC), Leucippus and Democritus (approx. 460–370 BC) aligned with that of materialism. Empedocles proposed the theory of the four classical elements (fire, water, earth and air), which constantly mix in new combinations of one another and are indestructible as a basic substance. Conversely, Democritus held that everything is composed of one type of matter, the atom, which, as its name entails, is inseparable, impenetrable, heavy and indestructible. The atom is moved and mixed by mechanical forces. Democritus did not believe in the existence of a connective whole as purported by Parmenides. Nothing occurs without a plan; everything is determined by causation. This marked the first step in the quantitative, mechanistic view of the natural world. Both souls and thoughts exist in the form of atoms. However, as Aristotle later wrote, it remains to be seen where their movement originates.

The golden age of philosophy in the antiquity was the passage of teacher-student relationships between Socrates, Plato and Aristotle. Each of these individuals has been previously mentioned in ► Sects. 3.6.1, 3.6.4, and 3.7.2 as key shapers of the thinking traditions and cultures of the following millennia. In the Middle Ages and the Renaissance, several schools of thought were set against each other on the basis of the difference between Platonic and Aristotelian teachings, namely with regard to Plato's holistic approach and Aristotle's detail-oriented, sensually observable and describable worldview. In addition to others, this key difference is one of the fundamental aspects addressed in this work with the aim of helping to form a bridge of understanding.

Greek philosophy as a whole was taken as a blueprint for the Roman Empire, which itself had a lasting impact on the culture of Europe in its contribution of the concept of law and state-military organisational strategies, which were considered progressive at the time.

7.3 Psychological Understanding of Development in Non-European Cultures

7.3.1 Varied Paths to Realisation in Terms of Place and Time and Across Space-Time

Despite the fact that a wealth of detailed explanations of the worldviews and concept of man held by highly developed, lost cultures would be conducive at this point, we shall refrain from this level of detail as it would simply extend beyond the scope of this book. As such, instead we will make do with a number of concise depictions.

The logic in this approach may lie in visualising for which alternative focuses of development or aspect assigned particular importance experiences can already be found in the functional processing of the human mind as a spontaneous, cultural large-scale experiment. Could we benefit from one or the other field of experimentation at present? Could today's field of psychotherapy profit from this form of broader view? Knowledge of the various worldviews and concepts of man is a prerequisite in this regard. They are closely linked with both religious beliefs and expectations pertaining to considerate and social interactions. It is often interesting to see which special skills can be developed with the assistance of a human group and the impact thereof on survival, i.e. the potential of an additional specialisation for the development of the entire group, its concept of man, its resilience and perhaps also its vulnerability. Each topic is worthy of its own book, including the unwritten aspects that have not been mentioned due to the conciseness of this summary.

7.3.2 Africa, the Cradle of Humankind

Phylogenetically speaking, Africa takes priority over all other continents. This applies to both early forms of modern humans and humankind itself. The modern form of humans evolved about 160,000 years ago. Following this, mankind reached the far corners of the world in several waves, initially towards the Middle East, then along the South East Asian coast towards Australia. As perhaps goes without saying, we have no knowledge of how humans at the time viewed themselves and their world over time. Our current understanding is solely based on conjecture.

Any attempts to get an overview of the various current spiritual trends found in the African continent on the internet would quickly end in confusion on account of the sheer variety. It is also advisable to practise a certain level of caution towards reports written with the filter of culturally different groups of people.

In the present day, the north of Africa is largely Islamic, partially Christian and home to a small minority of followers of the Yoruba religion, which developed in West Africa in the last 2500 years, experiencing its heyday around 1300 AD. (A later section is dedicated to Yoruba.) The south is home to a range of long-standing, animistic religious beliefs featuring high deities and intermediary demi-gods featuring distinctive ancestor worship and variety between tribes.

The history of the settlements of the remaining indigenous tribes, which can most frequently be encountered in modern-day South Africa, appears to stretch back at least another 20,000 years. These tribes include, amongst others, the San people and the Khoehoe. Their habitat is under constant threat from the outside world with the construction of dams, tourist hotels and national parks, etc. Human rights organisations have only recently started to make a significant push for the indigenous population's right to stay in their original habitat.

Based on the available reports at this time, it would appear that a standard religion that has remained independent throughout time cannot be found in Africa. Nevertheless, a common belief in a supreme god, a supreme being afar, who appears to be divine for everyday concerns is a standard feature of the religions practice in Africa. This god cannot be contacted by mere mortals. Instead, there are demi-god who are prayed to for advice for almost every everyday concern. All of nature, including humans, is "animated" by a spirit (animism). The belief in the almost healing power in living people is cultivated, often in combination with its affinity to good or evil spirits. In certain areas, animals are sacrificed to appease and win the grace of spirits. On the other hand, certain animals are regarded as representatives of demi-gods and worshipped, or they may serve as totems. Ancestor worship is also widely practised. Almost every tribal community follows traditional rites to mark each phase of life such as birth, initiation and death.

Beyond the main religions, the belief in supreme beings has also spread throughout southern Africa. Perhaps the most well-known supreme being in East Africa is Mulungu: the Earth God, responsible for rain and fertility, but also viewed by some as a divine ancestor; in West southern Africa, Kalunga, both an ancestor god and in certain places the god of the underworld who is primarily responsible for fertility. Central southern Africa is the realm of Leza, the sky god responsible for thunder, lightning, rain and rainbows, who united with the Earth God Bulongo to enable crops to grow, amongst other powers. The Benin Sun and Sky God "Lisa" join with his wife Mawu, the Moon Goddess when there are solar or lunar eclipses. Subordinate to the supreme gods are a number of more accessible demi-gods, who do the groundwork for the former.

However, above all, the ancestors are accessible with their creative powers and also as protective gods. They are said to have the creative ability to call forth something or someone, i.e. to make contact with them.

Individual tribe members gain vital security and orientation from the social integration provided by the tribe, its rituals and beliefs. The hierarchical structure of the tribal organisation, including its breathing space for each individual member, appears to be reflected in the transcendental world of imagination.

7.3.3 Indigenous Australians

According to recent genetic findings, the indigenous people reached Australia in stages, starting with the first waves of emigration heading eastwards from Africa across the coast of South East Asia approx. 62,000–75,000 years ago. The final stage of emigration was most likely at the time when the land bridge was still in existence between the main continent and New Zealand. Excavations indicate that people reached the southern edge of Australia around 50,000 years ago. Aboriginal Australians are widely regarded as the earliest representatives of modern humans found outside of Africa.

Due to a somewhat bemusing translation of the Central Australian word "alcheringa", the Aborigines are often referred to as "dreaming", or the term "dreamtime" is used with regard to both an individual and collective concentrated, trance-like contact, and a form of meditative reflection with the world of their ancestors. This contact also relates to their understanding of the cosmos, including its natural, moral and social order. Stories of creation and tribal development are recited at their meetings. Notably, the Aborigines believe in perpetual creation with its formless and timeless source of existence in the background. The past only exists in this world, not on the Alcheringa side. Living and unliving beings of this world have counterparts in the transcendent energetic Alcheringa world, and their actions have an effect there. The beings found there are in equal contact with all beings, including the ancestors. A human soul is immortal, a spark from the Alcheringa level of the ancestors that remains in constant connection with it and returns to it after death.

The Rainbow Serpent, or the rainbow, represents a combination of two key principles: forces of spirit and matter ([1], p. 35).

Songlines, also called dreaming tracks, record the history of the tribe and the transcendent creative plane located in the geography of this world in a spiritual map. The stories associated with this plane are told in the form of songs and only passed on verbally. These stories form a bond between the people and their land. They serve as an open book due to this connection to history.

For the Aborigines, the difference between women and men is highlighted by the sometimes extremely rigorous initiation rites for young men (circumcision, scarification). Thirteen-year-old boys undergo a journey entitled a "walkabout", during which they are left alone, not allowed to speak and tasked with establishing and developing contact with the Alcheringa level. (Not dissimilar to the "vision quest" found in numerous Native American cultures.) These rights of passage involve a certain level of trauma within the protection of the supportive group as a development stage that requires and promotes frustration tolerance and resilience. Accordingly, this method can be a balancing act and require great intuition and experience from the older generation.

7.3.4 Polynesian Culture

There are numerous different theories and postulations behind the settlement of the Polynesian islands. It is widely accepted that several waves of migration came from South East Asia in the years between 4000 BC to 1000 AD. Rather exceptionally, it is also postulated that people from South America may have reached Easter Island and other East Polynesian islands.

A small side note on Ancient Japan: The prehistoric, ethnically varied settlement of Japan may have been partially integrated in these early waves of migration from the Asian mainland to the islands to the east, particularly via Thailand and Korea, but also in the North from Siberia, and potentially from Europe. Nevertheless, the early prehistoric area of Japan did create a few local forms which led to the emergence of Shinto as a nature religion that shared an inner similarly to the beliefs held by early Polynesians, above all with regard to the central role of ancestor worship in addition to the worshipping of natural phenomena (mountains, cliffs, waterfalls, springs) and natural forces as gods, kami. (After 600 AD, Shinto, the primary religion in Japan, spread and mixed with Buddhism in a partially harmonious and partially strained combination.)

Back to the prehistoric opening to this section: Impressively, despite their geographic dispersal, the Polynesian island chains developed a relatively similar cultural area with regard to their languages, social structures, religious beliefs, navigation systems based on astronomy, skilled craftsmanship in ship and house construction, trade and agriculture.

The key common feature in this cultural area is ancestor worship. The dead are symbolised in towers featuring caved figures crouching on top of each other, which are consulted for questions related to everyday life. In addition to totem poles, Polynesians are also protected by protective gods and spiritual beings. The term "manahune", used to refer to the indigenous people, means expert in "mana", the powerful, spiritual energy that flows through everything that exists, including inanimate nature. "Mana" means more than just strength, power, influence, purity and wisdom: it also represents the connection of all beings to one another. As a result, the burden and relief of interpersonal relationships are of particular importance. Ritualistic blessings in all shapes and forms are laced throughout everyday Polynesian life. "Mana" corresponds to the mainland Asian notion of "Chi", which can be focused on and influenced. Those with a particularly high level of mana exude self-confidence and are regarded as predestined for leadership roles.

The entire world is experienced as animated. There is no separation between this world and the other side. Each body, whilst animated, is inhabited by a soul that continues to exist after death and continues to seek and maintain contact with the living.

Taboos play a significant role in this region. In Hawaii, it is believed that breaking taboos and violating spiritual and religious laws can cause illness. The resulting illness can only be healed by the sick person, in the presence of the imagined ancestors and higher deities, by doing penance, possibly with the assistance of a prayer expert (kahuna pule). A herbal practitioner (kahuna la'au lapa'au) can also provide assistance in this regard. In the traditional rite, the people and family members involved are present. The gods or the person with whom the dispute was held or who was hurt as a result are asked for forgiveness. This ritual is intended to restore a strong relationship between family members. According to traditional belief, illnesses are caused by concealed resentment, stress, feelings of guilt, recriminations and a lack of forgiveness. Healing takes the form of spiritual purification through truth and honesty ('oia'i'o) and open, repentant confession, which should be followed by forgiveness.

The Hawaiian healer Morrnah Simeona (1913–1992), who received a Christian upbringing, updated ho'oponopono into a psychospiritual self-help therapy as a 14-step praying process, which no longer requires the presence of the traditionally present family members. Prayers are solely addressed to god. The ritual also resolves karmic ties. (The Hawaiian people have always believed in reincarnation, a belief that has survived the conversion of the island to Christianity.) During the ritual, the three elements of the self, namely the waking conscious, the subconscious and the unconscious, are brought into mutual balance and united with the divine creator. In addition, prayers are also made for the liberation of other beings in an altruistic manner. The aim is to transform the painful parts of memories held by those prayed for into pure light.

The traditional forgiveness ritual that is widespread throughout Polynesia, which is reminiscent of family therapy, provides an impressive template for psychotherapy and widely compatible with general approaches.

7.3.5 The Culture of the Ancient Far East

As is widely accepted, the European and Asian region was first settled following a large wave of emigration from Africa at a later date, around 25,000 years ago. In later stages, people reached Europe and the Far East, initially from the Middle East and Central Asia, and finally made their way to the American continent in several waves around 12,000 years ago. Millennia then passed before the emergence of advanced civilisations.

7

■ India in the Vedic Period: The Early Seers

In the Indus valley, evidence has been found of a farming civilisation dating back to 3300 BC. The mature form of the Indus civilisation lasted from around 2500 to 2100 BC. Around 1800 BC, the Indus and Saraswathi Rivers dried up, causing large settlements to be abandoned. The population moved to the banks of the river Ganges. Around 2000 BC, nomadic groups referred to as the Aryan (Sanskrit: noble) migrated from the Eurasian Steps and conquered the subcontinent for the most part. The Aryans worshipped Agni, the god of fire and Indra, the "personification of the thunderstorm". They were a settled people who merged themselves, their customs and their beliefs with those of the indigenous people, leading to a cultural renaissance.

Between the mid-second and mid-first century BC, the Vedic scriptures, which had previously only been passed on verbally and may have come a long way from the original tales, were recorded in Sanskrit. At a later point, the Vedic scriptures were incorporated by Hinduism, mixed with early Buddhist philosophy and carried up the rivers into the Himalayas to Tibet, where they could still be found until recently. According to Bede Griffiths (p. 22/23), the Vedas were put into writing through the insights of meditating "seers" ("Vid", the Indo-Germanic language root of "Veda" means to see, 'knowledge'). They are "apauruṣeyā" (Sanskrit), i.e. not of man. The seer of the oldest text of the Vedas, the Rigveda, states: "ekam sat vipra bahudha vadanti" (Sanskrit), i.e. that all gods can be understood as names and forms of a single reality. (Former polytheism transformed into monotheism through holistic perception.) This one reality was understood to be a tumescent, growing energy (Sanskrit root: bṛh) and referred to as "Brahman". Brahman is the indivisible yet creative principle realised in all that exists. Atman, the individual self, is a part of Brahman and can be found in every individual. In meditation, the unity of both is found in the experience of participation. The doctrines of karma and rebirth (in the Upamishads) provide the foundation for the constant coming and passing of souls. In a predating tradition, this primal energy of Brahman is realised in the form of Agni, the god of fire, to whom thought, knowledge, wisdom and spirit are also attributed. Agni (later known as Indra) was offered fire sacrifices. His counterpart, Vrata, embodies the energy than keeping the soul locked in the hard rocks of matter, ensuring humanity remains imprisoned. According to the tale, with help from Agni (Indra) and the Vajra (Sanskrit: thunderbolt), light, consciousness, soul and life (symbolised by water and cows) were freed from the rocks, i.e. matter. This clearly entails where the true plane of being is. In this ancient myth, the "sacred cow" symbolises life free from concepts, namely structure-free energy. (Einstein may have rejoiced to read this ancient parable of the Vedic seers and relate it to his thoughts on the transformation between matter and energy.)

We have previously explored the attempts made in Vedic tradition to reach the plane of knowledge related to the whole time and again and not become overwhelmingly fixated on conceptual notions that may be compatible with everyday life but nevertheless perplexing in ▶ Chap. 3. The ability to differentiate between the two is not something that we learn and develop in this day and age, or rather, at most it is cultivated in contemplative niches, therefore only existing as a peripheral and exceptional phenomenon.

It is perhaps notable that the focus of ancient Vedic philosophy was placed on the releasing transformation of energy from its material and structural solidification. This marks a reversal of the otherwise conventional stories of creation, which focus on the creation of all kinds of new structures, and celebrates their newly created existence.

Ancient India – The Longing for Reform

By the sixth century BC at the latest, monastic reform movements were growing in popularity and distancing themselves from the social abuses of the ruling classes. A number of monastic movements longed for self-liberation and strived to achieve purity, frugality and an end to violence. At the same time, criticism of the Brahman caste system and its ritual sacrifices was slowly brewing in the background, alongside the lack of social relevance of Hindu teachings and practices for fellow man. The need to devalue other (lower) castes and women became rife. The reform movements of the period gave rise to Jainism and Buddhism, which continue to exist to this day.

Separate to the social upheaval, during this period interest shifted towards sensually perceivable, concrete structures: the structure-preserving force was worshipped and personified in the form of the god Vishnu, with the structure-destroying force represented by the god Shiva, who was originally revered as the god of death. Energy flowed to and from Brahma, who rested in the background. In one variant established at a later point primarily in southern India, the notion of cosmic energy was centred around Shiva who danced life into and out of forms with the cosmic dance. Shiva danced the cycle of life and death, and rebirth.

Note

Gautama Buddha (officially determined to have lived from around 563 BC to 483 BC, potentially up to 100 years prior) left the aristocratic family into which he was born at 29 years of age in order to meditate, practise asceticism and learn philosophical insights. He began to teach following his spiritual enlightenment. In doing so, he avoided the entire edifice of religious Brahman activities, including their sacrificial rituals, that had started to wane during his lifetime and, above all, rejected the caste system imposed as a result thereof. His teachings focused on the liberation from the adherence to illusory, earthly pleasure and motivated his followers to pursue blissful devotion to Nirvana. To him, truth, peace, wisdom and compassion were the ultimate goal.

He proclaimed that life on this plane is afflicted to suffering due to birth, sickness, age and death; that we are easily won over by greed, delusion and hatred; that we can find happiness in avoiding these root causes; and that we are capable of practising the noble eightfold path, namely with right view, right resolve, right speech, right conduct, right livelihood, right effort, right mindfulness and right meditative absorption. "Right" in this regard pertains to the above viewpoints on avoiding suffering for ethical reasons. This process should also lead to an increase in the desirable virtues, meditation, compassion and wisdom.

The existence of a constant "self" is denied, similarly to the constant of the same souls during the transmigration of souls. (Our categorical thinking may delude us here; a mixed form of both identical and non-identical would perhaps be more suitable in this regard.)

In Buddhism, Nirvana is regarded as the ultimate state of consciousness which transcends all objects, a state that can be experienced with the conscious yet for which words do not suffice. On the face of it, Buddhism is a religion without a god or creator. Be that as it may, he appears to exist concealed and tangible in the potential of Nirvana. Buddha discovered that the mind exists independently of its senses. The essence of reality can be found in unstructured, undifferentiated, universal suchness. This covers the philosophical aspects.

The other elements of his teachings were concerned with human sympathy: Buddhism fulfilled the deep longing of people for solace, empowering compassion and loving solidarity.

Buddhism is the fourth largest religion in the world. It has spread to the far corners of the globe and split into many variations. Beginning in the seventh and eighth centuries, it travelled along the river valleys in the Himalayas towards Tibet; however, it was most well received in China, where the practice of Zen Buddhism flourished and enjoyed much popularity in South East Asia.

Ancient Persia: Zoroaster – The Guardian of Fire and Social Responsibility

Historical records on Zoroaster make it difficult to pinpoint precise dates for his lifetime. However, key findings would suggest that he lived as a historical figure between 640 and 553 BC. (Other sources cite approximations of around 800, 1000 and even 1800 BC.) He is said to have been born in northwest Iran and travelled to the mountains of northeast Iran to study the scriptures (?) and meditate in his twenties. Researchers are still confounded by the parallels between certain verses penned by Zoroaster and ancient Indian texts, which would seem to imply that he had established direct contact with the Indian cultural region. Zoroaster then moved westwards in various stages as a prophet and social reformer with hymns (gathas), encountering several nomadic peoples before finally arriving at the court of Cyris II (585–530 BC) and his son Darius the Great in Persepolis. Darius recorded the gathas and made Zoroaster's doctrine the state religion. This marked the first mention of "Ahura Mazda", the god of Zoroastrianism.

Note

In terms of appearance, the Zoroastrian cult of fire, the symbol of the purifying, healing flame, is somewhat reminiscent of the early-Indian god of fire, Agni. In contrast to the Hindu practices at the time, Zoroastrianism did not feature sacrifices or a caste system and only worshipped one god, Ahura Mazda, the all-wise.

The world was understood as the opposing interaction between good and evil, an open-ended process, namely a duel on equal terms. Man is called upon to voluntarily and verily stand up for the victory of good in this world as an ally of the good side with his "good thoughts, good words and good deeds" by his own volition and thus support Ahura Mazda in his battle. Man is at liberty to choose between hate and forgiveness. From this time on, man is challenged to listen to his own inner voice and mind, be truthful, recognise situations that require a decision and to make a positive decision in this regard. Those who struggle to reach a decision should seek assistance in their community, but, first and foremost, from Ahura Mazda.

In the event of good deeds, his imaginary, spiritually symbolic figure "Farvahar" will grow feathers on his wings and conversely, in the event of bad events, feathers will grow on his underbody. The ring held by the symbol represents loyalty, the direction in which the right hand is pointing symbolises the devotion to the divine and the circle in the centre portrays the eternal spirit without a beginning or end, of which man is part.

In the cosmos there are forces that are unnecessarily in conflict and seek balance. In this situation, the unwise are susceptible to creating their own hell borne of their selfish behaviour (represented by Ahriman); however, the wise, who exhibit more adaptable behaviour, are capable of turning this conflict into a paradise. The analogous term to Tao in this regard is "asha" ("that which adapts").

Note

Ahura Mazda created universal order, which man is to inhabit in harmony, love, modesty and truthfulness, serving it with his working capacity. Asha is the force that harmoniously connects everything in the universe, including man with god, and can primarily be accessed through meditation and intuition.

According to the teachings of Zoroaster, at the end of life the individual is judged based on his or her actions, which are balanced against each other. This concept includes all people with whom the individual came into contact throughout his or her life on truthful and equal terms, even subordinates and women! It underscores the merit in working for the community and developing a keen commitment to the benefit of all. It forms the path of development based on practical deeds.

Note

The notion of social equality and the assumption of personal responsibility contain undeniable revolutionary and social reform elements. It appears to represent a contrast to the caste system. Even women were widely granted the right to self-determination. First and foremost, this concept is a call for active participation, for the development of social skills whilst accounting for all of the unwritten rules of the reference group, for the development of a personal conscience and for the assumption of individual responsibility for the group.

- **Ancient China: The I Ching Oracle Game Amidst Polar Forces**

I Ching (Yi Jing) is widely regarded as the oldest collection of texts from ancient China on the basis of oracle bones (as grave furnishings) dated back to 3000 BC. It is often referred to as the "Book of Changes" and consists of 64 hexagrams, each containing six stacked horizontal lines, which are either broken or unbroken.

The symbols were used for the ancient Chinese practice of divination. The complex commentaries that have emerged over the course of the centuries reflect problem-coping mechanisms relevant to everyday life, intrapsychic conflicts, aspects significant to development psychology, interpersonal issues, socio-political upheavals as well as philosophical and cosmological approaches.

Despite that fact that I Ching significantly predates the teachings of Yin and Yang, over time, parallelisation has emerged between these concepts with the following denotations:

Note

Unbroken line: Yang (the creative principle, masculine in nature, sky, light, day, life, odd numbers, expansion, penetration, solid substance, mountains; Denotation in India: Lingam; symbolic figure: The dragon)

Broken line: Yin (receptive principle, feminine in nature, earth, darkness, night, death, even numbers, constriction, soft substance, streams; Denotation in India: Yoni; symbolic figure: The tiger)

> **Note**
> I Ching attempts to grasp the permanent transformation of the two polar elemental forces as a whole united on the basis of certain laws.

An emphasis is placed on the notion that the two polar, complementary or "entangled" forces form a common system – a concept reminiscent of the complementarity principle in quantum theory.

Arguably the most highly regarded translation of I Ching (into German) was published by the sinologist Richard Wilhelm in 1924. It provided particular impetus for C.G. Jung, who incorporated aspects from I Ching into his psychotherapy. He highlighted that it was not based on the causality principle, but rather, similarly to the unconscious, on a synchronised principle based on the wave nature of reality.

■ Tao Te Ching and Laozi – Trust in the Wisdom of Tao

Richard Wilhelm also translated Tao Te Ching, a collection of cosmological findings, suggestions on how to approach the polar elemental forces and dualisms, guidelines for personal development and the development of a peaceful state, a collection that was most likely compiled by several "old masters" around 400 BC. Tao Te Ching is the holy book of Taoism.

> **Note**
> With regard to interacting with the elementary forces, it is recommended that one remains centred at all times, or rather pays attention to the complementary and balancing nature of the poles.

The central terms in Tao Te Ching are Tao and Te, which were initially translated as "The Way and Virtue" or "The Way and Power", subject to much critic and later by R. Wilhelm as "Meaning and Life".

> **Note**
> It is not possible to precisely define this all-encompassing highest and deepest principle of reality that eternally infuses, transforms and shapes everything with the bounds of our language, which is merely a product of our constructivist thinking. Nevertheless, we are able to trace around the notion and define what it isn't.

Tao Te Ching employs numerous parables. The scriptures advise the reader that those who abide by its commands can hope for an increase in energy (in terms of Chi).

Note

The wise man sets his self aside, desiring nothing for himself from his accomplishments; he practises doing nothing; yet, nothing remains unaccomplished (Wu Wei), as the positive appears of its own accord when man does not disturb the natural order. The weak and pliable overcomes the strong and hard, as often cited from ▶ Chap. 36. He who conquers himself is truly strong. He who knows when he has got enough is rich (▶ Chap. 33). The able man is humble. The sage opens his heart to the world and lives in eternal peace and harmony. The sage's Tao is sharp but not cutting (81). He who is filled with Tao requires no moral commandments duties or laws. He allows himself to be guided by love, facing even his enemies with childlike veracity and love.

At the time it was recorded, this ancient philosophy was out of step with the political turmoil and various power struggles throughout China, though it clearly struck a chord amongst a deep-rooted belief held by the people.

Needless to say, the above belief in relinquishing rules and regulations marked a glaring contrast to the outlook and strict set of rules that emerged from the teachings of Confucius. It constitutes another attempt to handle the unrest throughout the land.

▪ Confucius – Enforcer of Tradition and Society

Confucius (551–479 BC) himself appears to have handled the demands placed on himself in a somewhat more flexible manner than his later followers.

Note

Confucius was a pragmatist of interpersonal order. According to him, this could be found in harmonious balance with the universe. A noble and moral man is guided by this all-encompassing order, something viewed as irrefutable by Confucius. The path to attainment could be found through education, which Confucius valued highly and believed should be accessible to all. First and foremost, education serves moral self-cultivation and not egoistic purposes.

Confucius stated the following characteristics as fundamental virtues: (1) benevolence, (2) righteousness, (3) filial piety and (4) proper conduct in terms of polite etiquette with appropriate respect.

When asked what wisdom is, Confucius responded: "When working for what is right for the common people, to show reverence for the ghosts and spirits while maintaining one's distance may be deemed wisdom". For Confucius, the rules of the external social system provided the initial basis for clearly defined free will. He viewed social hierarchies as a natural phenomenon, despite that fact that he viewed all people as inherently similar – only separated from each other by education. Nevertheless, he accepted differences that emerge in a pragmatic manner and did not refute them.

This approach supports and underpins existing familial, social and political structures, whilst also simultaneously providing an orientation towards the ideals of benevolence and righteousness.

- **Tai Chi – Improving the Balance of Energy**

Tai Chi, traditionally referred to as "shadow boxing" and literally translated as "Supreme Ultimate First", emerged as an internal martial art to cope with individual and collective inner tensions, to remain in balance and develop inner chi in the process. Tai Chi was first recorded in China in 206 BC. At the time, it would appear that there was great demand for its balancing effect. Since the sixteenth century, Tai Chi has primarily developed into a martial and movement art. Its flowing movements enable the expression of the goal of harmonious balance. Other variations also exist with a stronger emphasis on the combative elements. These have taken on special forms and are particularly distinct, as with those developed by the Japanese.

- **Japanese Martial Arts – The Domain of Yang**

Japanese martial arts are deeply rooted in the hardships of everyday life. In studying them, we encounter a concise Yang side to this culture. The collective term "Budo", translated as "the way of war", is the Japanese term used to describe Japanese martial arts. These include: Jiujitsu, Judo, Karate, Suijutsu, Aikido, Kendo, Naginatajutsu, Kobudo, Kyudo and many others (approx. 60 in total). All of the modern martial arts can be traced back to the original martial art, Kalaripayattu, which originated from India. Japan was under the rule of the Samurai from 1185 AD for 800 years. Bujutsu, along with its preoccupation on killing and survival, reigned supreme; this form of the martial art first required modification before its transformation into a fair sport. The first Jiujitsu school was established in 1532.

In the background, martial arts convey an inner teaching very similar to Zen. Their objective is the harmony between body and mind, contacting inner truth, the targeted use of energy resources, channelling energy and increasing responsiveness. Meditation aids the mental state free from distracting thoughts (Mushin) through mindful meditation, becoming one with the bow (in archery), or with the cultivated motor processes of the respective martial art. However, the true art lies in the concentration and guidance of Chi energy (Japanese: Ki).

In the internal martial arts, special breathing techniques are employed to build up Ki in Hara, the abdomen (approx. 4 cm below the navel), where it is stored and propagated with the assistance of the power of imagination via the energy gates (meridians) by the body, or, in the case of external martial arts such as karate and kung fu, transported to the arms and legs. In karate, it can often help to imagine one arm extended by a sword. "Kime", instantaneous tensing, directs the concentrated Chi to the obstacle just before the collision in karate. "Kaia", the shouting whilst fighting, supports this energetic balance process. It is essential that the mind, trained by meditation, is capable of binding the energy at the intended point in a matter of seconds in order to shatter walls, or other objects, with bare hands without inflicting any injury. According to hypothesis, the mind, the concentrated consciousness, may be the force that enables the mobilisable quanta to deploy laser-like characteristics.

As is perhaps clear, this martial art variant indicates that we (potentially everyone) are capable of transforming energy into a Yang form, or informing it with our mind in such a way that it is capable of penetrating and cutting through hard matter.

- **Japanese Aesthetics as the Domain of Yin: In the Field of Intellectually Stimulating Subjectivity – As Artistic Play with Emptiness and Nothingness**

The sensitive treatment of emptiness is widely regarded as an influence from Zen Buddhism, which reached the shores of Japan through China, where it merged with Shinto. Negative space plays almost as significant a role in Japanese paintings as the subject, balancing its importance paradoxically with its "nothingness".

7

A particularly striking example of the aesthetic spurred by this concept can be found in the form of Zen gardens. The beauty carefully selected to appeal to our senses, usually a plant, statue or perhaps even a rock appears to be engaged in a dialogue with the apparent "nothingness", namely a grey pebble surface. Raked surfaces are used to indicate that this nothingness does not represent chaos, but rather a covert, subtle order. Each pole highlights the distinctiveness and beauty of its counterpart, appearing to bow down to it.

The way in which Japanese poetry employs the medium of language to express the unspeakable beyond the meaning of words is astonishing. In this regard, the top virtue is the self-aware, self-illuminating, spiritual subjectivity of the moment, referred to as "kokoro". This mental state intuitively searches for an adequate inner language (omoi) in addition to an expression for the emotional and mental state (yo), which both need to be transposed into external language as adequately and simply as possible. The renowned poet Teika regarded kokoro as "a living and genuine subjectivity, a state of subjective equilibrium which transcends the transiency of the psychological sphere of phenomenal commotion" ([2], p. 16/17). This understanding is derived from practice of contemplation established by the master Chigi (Chinese: Zhiyi), 538–597 AD, in which enlightenment experiences provide an outlet for kokoro. Chigi's kokoro plane of experience is derived from transcendence (or an early form thereof), yet not from the level of constructive or conceptualising thinking.

Traditional measures in Japanese poetry are the 31-syllable waka and the 17-syllable haiku.

In terms of content, these poems depict an atmospheric, insightful and emotionally moving situation as the here-and-now plane. They convey the content of an experience with a transcending realisation for which language does not suffice. However the images of nature and sentiments may better fit to condense and symbolize the content. The associations of words used, such as clouds or the upper notes of an instrument, resonate and colour the language, which moves in the realm of content. A high level of cultivated receptivity, Yin energy, is required to understand this form of aesthetics.

Note

In the background we encounter a metaphysical approach which is familiar to us from the Far East. The nothingness (mushin), namely that which cannot be attributed a term or articulated, is defined as the greater whole, the sole reality. That which can be defined is referred to as "yu". This concept pertains solely to the detail that the human consciousness brings to the fore through its naming and becomes an empirical aspect of life. In contemplation, it is possible to reverse this process of emphasising a certain element until it once again sinks back into the lack of clarity found in the non-articulated whole. Man gravitates towards the whole and as such, this interpretation of nothingness is granted the utmost value and significance. It is regarded as immaculate, pure and supremely beautiful. Those who engage with it will find that its darkness ultimately transforms into radiant light.

Thinking can be split into: (1) when a subject focuses on a cognitive focal point, which emerges when the subject uses his logical and linguistic capabilities to focus on an object in a linear and casual manner and sets it apart from the rest and (2) in the contemplative field, thinking tends to take on more associative and visual aspects than linguistic. This second form is the one designated for cultivation on metaphysical grounds.

In terms of its structure, the difference between types of thinking can be regarded as a parallel to the Western psychoanalytical categories of the conscious and unconscious.

In Japanese art, an emphasis is placed on the contemplative plane, which develops in stages: the perception of the plane is followed by the selective awareness of its details and finally by their rich, associative connections to each other, from which higher dimensions can emerge. Then the consciousness of awareness merges with what is perceived and creates a new balance between the two, enabling colours and shapes to harmoniously blur, fade, grow dark and suddenly transform into radiant life. These stages can be regarded as the steps leading up to contemplative experience, which is substantiated by the metaphysical and depicted in a modified form in art. In Japan, this is referred to as the structure of "The Way of Craft" (geido). The primary emphasis is always the inner experience.

Traditional Noh theatre subtly maps out the gradual development of the stages of inner maturity. The treatise of Zeami (1427) has become particularly well-known in this regard. His nine levels of development are modifications of styles, mental states, aesthetic values and value systems; there are three lower, three mediocre, and three higher steps. According to Zeami, the highest level of development is transcendental non-duality of the inner landscape (Izutsu, p. 65).

In reflection, this chapter highlights several key areas where the field character of mental activity is emphasised in addition to the search for balance, the esteem in which self-organisation is held and the objective of non-duality. To link this with the sciences, it could be stated that the patterns of thought from the East are comparable in structure to quantum theory, as the Western version thereof, which exhibits a creative power in conceptual theory.

7.3.6 Native Americans

The Native Americans most likely reached North America between 15,000 and 12,000 BC by walking across the land bridge that linked North East Asia to North America at the time. Once they set foot in America, they split into over 2000 groups and migrated in various stages primarily along the West Coast towards Central America, finally reaching the southern tip, yet also settling in the interior. Findings
7 have confirmed the genetic match between the Asian peoples, predominantly those from North East Asia, and Native Americans.

As might be expected, the tribal beliefs about the visible and invisible world vary to a great extent. Nevertheless, almost all tribes seem to share the same basic approach: the spirit and its spiritual energy flow throughout all of nature, including people, and connect everything. Everything is animated, even lifeless nature. Accordingly, all beings are worthy of respect. This secures cosmic harmony, for which man is responsible in addition to the survival of the community at large. The earth, largely referred to as "Mother Earth", spawns all beings, nourishes them and receives them once again in death. She plays host to humanity. The Great Spirit is the father, a supreme, formless energy. (Man is explicitly not destined to rule over nature.) Living men and the spirits of their forefathers look over and (for the most part) care for one another. Animals are also worshipped. They also possess a spirit. The Great Spirit is the supreme power that rules over all souls, uniting everything. In certain regions, special forces such as wind, streams, particular animals and plants are worshipped as independent subordinate deities. People attempt to establish contact with the spirits and ancestors in a number of ways. The most common methods feature a combination of ritualistic dances, vision quests in solitude and concentration on healing and soothsaying skills supported with rituals in addition to the special interpretation of dreams. Importance is attached to external and internal purity, which is promoted by fasting and sweat ceremonies. Overall, a fundamental respect of others is key. The older generation is viewed as significantly wiser the younger. As such, the elders are responsible for raising the tribe's young. The history of an entire village or tribe is showcased on totem poles, which guided the Native Americans through their special abilities.

Secret societies were established with other tribes to mutually strengthen and protect each other. Special secret societies also existed for women. In general, older women were considered to possess stronger spiritual powers than younger women.

The Native American cultures perceived time as circular: only what is, what always has been and what always will be. Similarly to time, the souls of people are also recurrent. Man possesses two types of souls, one embodied in the breath that dies with the body in death and the other that is released upon death, passing through an intermediate realm before returning to the realm of the living. This latter soul is also capable of leaving the body during sleep or otherwise flying away.

To summarise: the individual man learns to become an empathetic member at one with nature as part of the reverent integration into the social uterus of his group. Respect towards the living and ancestors is the most sacred of virtues.

7.3.7 Ancient Egypt – Mystery and State Morality

The history of modern-day Egypt has been traced back to around 10,000 BC. In Ancient Egypt, the state was governed by the pharaoh, or his vizier on his behalf. The state functioned due to the control of the civil services. District governors were responsible for laying down law and order and collecting taxes in addition to their role as priests of the local deities. At the time, the profession of the scribe was held in high esteem. The majority of Egyptians were farmers, craftsmen and tradesmen. Prisoners of war from other territories were kept as slaves.

Note

The kingdom of Osiris was ruled over by the goddess ma'at for almost 6000 years. Ma'at is the daughter of Ra. She is the personification of law, rules, governance, morals, order, balance, truth and justice. The state and ritualistic religious set of rules appeared to align with the cosmic order. Priests of ma'at were judges. Compassion was exhibited and human action was judged according to ma'at. The hearts of the deceased were weighed against the feather of the goddess ma'at in the weighing of the heart rite. Her counterpart is chaos and evil.

The pharaoh, the son of the sun, the mediator between the gods and the living, was tasked with bringing prosperity, wealth and abundance to his kingdom. It was essential that he remained in close contact with the gods and was assured of their protection and guidance.

Both the multifarious realm of the gods and the state featured hierarchical structures. The pharaoh, including his family, were at the heart of state religion. The belief in one god only first emerged at the time of Akhenaten. Priests benefitted directly from the power held by the pharaohs. Their Mystery Schools cultivated secretive, occult knowledge which was guarded with strict initiation rituals and oaths of secrecy. For rulers and the upper echelons of society, the highest concern was the immortality of the soul and its possible return to the body after death. This strong conviction led to the emergence of the renowned art of embalming and burial culture that flourished in Ancient Egypt.

The soul: Ba, the principal aspect of the soul which appears in bird form, would live after the body died and join the body in the afterlife. Ka is the aspect of the soul that is said to be the protective life force. After death, Ka was capable of granting some of its vital essence to statues erected in its honour. Akh is the intellect as an ancestor spirit in the form of a light, symbolised on earth by a crested ibis. It would only come into being if the judgement of the deceased was positive regarding Akh.

In order to be resurrected in the afterlife, Akh requires a dignified burial in this world beforehand and needs to remain in social memory for a long time. Every human being is first required to acquire Akh over the course of life; Akh is not present from birth. Under favourable circumstances, Ba and Ka unite to become Akh after death. Akh belongs to heaven, the corpse to earth. Akh becomes a star in the sky. In the early history of Ancient Egypt, it was believed that only the souls of the kind were capable and predestined to be resurrected. However, following the end of the Old Kingdom, Egyptians believed that normal mortals could also hope to become immortal by fulfilling the commandments of virtue and righteousness, as imposed by ma'at. As a result, each individual was able to seal his own fate through the way in which he led his life. Death was understood as a prerequisite for the start of a new life. Those judged unworthy upon death would cease to exist entirely.

Egyptians celebrated the return of fertile land and its harvest in the rhythm of the recurring flooding that occurred along the banks of the Nile. Individuals were accustomed to viewing themselves from the perspective of the community at large and joining their compatriots for community building projects. The needs of the individual came second to over-arching societal issues. It was beneficial to learn this at a young age.

In this sense, the timeless, inspiring work rendered by the general population, namely the centring of power, cast long shadows on the individual Egyptians.

Note

People aspired to possess the virtues of obedience, gentleness, self-discipline, verity and tireless diligence in addition to sharing with others in the spirit of fraternity and showing kindness towards those who are disadvantaged.

7.3.8 Yoruba, the West African Approach to Asian Spirituality (100 BC)

- **Aye l'ajo, orun n'ile – The Earth Is a Marketplace; Heaven Is Home**

This phrase is understood to have originated from the Yoruba (Baba Ifa Karade, p. 123). It causes the listener to listen attentively. The West African Yoruba ethnic group is one of the most widespread peoples found in the northern African continent. Yoruba stands out from the customs and beliefs held by the many other ethnic groups and tribes found in the region. The religious customs of Central and North America were shaped to a certain extent by descendants of the Yoruba people as a result of the devastating slave trade, albeit in vastly different and with greater variation to the original form practised in Africa. Yoruba people also adopted local Christianity and Islam in the process.

Historically speaking, the period between 2000 and 500 BC, a point at which the Sahara was still relatively luscious, witnessed the migration of people from the Middle East, Egypt and East Africa along a route that started in the Middle Nile and ended to the west in the Middle Niger region. Civilisations dating back around 65,000 years have been found in modern-day Niger. The civilisation established there in the period spanning 2000 BC to 500 BC is referred to as Nok culture. The people of this civilisation met and joined with the emigrated Yoruba group, marking a settlement for the latter and leading to the establishment of Ife (archaeologically dated to after the fourth century BC) with the Yoruba king Oduduwa. This merging of ethnic groups led to a cultural golden age lasting centuries, which reached its pinnacle in the thirteenth century BC. Ife became the holy city of the Yoruba.

Today, it is widely accepted that several waves of east–west migration occurred during this period. One legend tells of emigrants who fled from Mecca to escape the pressure to convert to Abrahamic Islamic monotheism.

According to Dierk Lange, the neighbouring city of Oyo contains records of the key ancient oriental dates and names of kings, which would suggest that refugees also reached the Niger region on the east–west route at the time of the collapse of the Assyrian empire (612–605 BC).

Worldview. Ashe is the Yoruba concept of life force, energy and inner essence which forms the basis of all matter and nature. Ashe is inherent in every creation and manifests itself in various stages. The worship of elements of nature such as trees, rocks or streams is not the worship of the tangible form, but rather of the inherent Ashe life force found therein. Ashe also manifests itself in the realm of gods and goddesses. Olodumare is a creator god. His sons and daughters, the mediators between him and man, are referred to as Orisha, angelic beings with various characteristics. Orunmila, the Orisha of wisdom, transcendence and rituals, is the patron of the Ifa oracle. He knows the fate of all things and serves the balance and harmony of world order. Orunmila is considered a prophet, a highly spiritual person, who provided the origin for meditation in the Yoruba faith and the rites thereof, a faith that has existed before him, since time immemorial. Orunmila is said to have lived on earth during the period when the Yoruba arrived in Ife. He was a well-travelled person, having travelled through Ancient Egypt and the principal Jewish civilisations in the pre-Christian period at any rate. His goal was to dedicate his life to learning and the development of man to allow his earthly consciousness, Ori, to unite with the spiritual consciousness known as Iponri. Whilst the path to achieving this may have been tedious, lengthy and require a pure heart, the reward at the end proved worthwhile. In the Ifa scriptures (Isha = Sanskrit term for God), which were originally only passed on through oral tradition, harmony with the divine is defined as the central pursuit.

Orunmila introduced the oracle ritual. This ritual is intended to make it easier to determine destiny, namely what man should devote his life to, and to engage with it. In the end, these rituals served to enable man to become aware of his own divine nature and reflect it in his own earthly life in order to return home to the gods. On earth, in addition to oracle enquiries and rituals, this could be achieved by abiding to the ethical and moral duties placed on man, which require him to help defeat negative forces. An emphasis is placed on his own effort. It is said that Orunmila obtained his wisdom from 16 divine prophets (Odu), who came down to earth for a period and were tasked with elevating man to a higher plane. There are an almost infinite number of spiritual verses, moral teachings and parables in the Yoruba faith.

The polarity principle can also be found in the oracle, who is thrown into a form of trance: cowrie shells are interpreted on the basis of how they fall on the table, either with their closed or open side facing upwards. This is reminiscent of the coin toss method in I Ching readings. A closed side is interpreted as darkness, with the open side representing light, analogous to Yin and Yang. Eight shells are thrown to produce a certain constellation, which is then interpreted on the basis of the distance between the shells.

These constellations or destinies are referred to as "paths", an expression also used to refer to and interpret the undefinable "Tao". Harmony with the divine and the desired modesty of the sage emerge time and again as the objectives.

Astonishingly, Yoruba also features a chakra system with seven planes, comparable to the structure of the Ancient Indian system. Each chakra has a corresponding Orisha, an angelic being considered responsible for the development of the special functions and capability of this plane. Human consciousness is required to work its way up the ladder of development. If the individual makes an effort, he may receive support from the corresponding Orisha. He is rewarded in becoming a balanced being in consciousness and conduct.

Followers of Yoruba have a somewhat more holistic understanding of prayer than other religions, involving the entire body and its movements. Dance often turns into prayer, allowing the individual to enter into an ardent state of trance supported by rhythmic music. This provides powerful tailwind for the matter being prayed for. Fervent African body consciousness meets Asian spirituality.

The divine essence permeated into every creation by Olodumare is the life force "Ashe" ("Asha" in Zoroastrianism and Brahma in Hinduism). Ashe takes the form of nature. In the Yoruba faith, it is Ashe that is worshipped, not the creation itself but the force inside, which can also be found in man.

In moral conflicts, the divine side requires our willing support. The Yorubas also believe in a moral struggle between the Orishas that ascend into the heavens and their counterparts, the Ajogun (demonic beings) in addition to the duty of man to side with the former. This belief is not dissimilar to the struggle between good and evil forces in Zoroastrianism, even if duality is not explicitly addressed in this case.

For the Yoruba, one support system was provided by their ancestors. The notion of reincarnation is a common attribute of many religions; however, in this case, the ancestors do not return from their sentence, but instead support their family and tribe. In between, the soul of a deceased person will travel to a "good" or "bad" heaven. However, similarly to the second death following judgement after death in Ancient Egypt, the souls of those who were cruel in life are sent to another place after death, from where they can no longer be resurrected.

A number of souls reside in each person and desired or undesired, short-term, long-term and permanent states of possession may be encountered. This is a common occurrence.

Those who succeed in living their lives according to ancestral tradition will develop the following virtues throughout their lives: calmness, serenity, respect, patience, prudence, knowledge, wisdom and understanding. In order to allow individuals to move along this path, the Yoruba ancestors (Egun) compiled guidelines for harmonious self-development. They are as follows (as abbreviated in Karade, p. 88):

1. There is to be no practice of wickedness.
2. There is to be no stealing. (Even if Man does not see, Olodumare sees.)
3. There is to be no selfishness.
4. There is to be no covenant breaking nor falsehood.
5. There is to be no hypocrisy.
6. There is to be no acts of atrocity committed against one's neighbours.
7. There is to be honour and respect to the elders.
8. There is to be protection of the women. (Women are the flowers of the garden; Men are the fence around them.)
9. There is to be truthfulness and uprighteousness.
10. There is to be kindness and generosity.
11. There is to be sensitivity in respect to person-to-person relationships.
12. There is to be chastity in respect to vows of mates. (Man, do not secure another man's wife. Woman, do not seduce another woman's husband.)
13. There is to be hospitable directives.

Orunmila is said to have spent time in Jewish communities before the advent of Christianity, so he most likely was aware of the Mosaic covenant. Although Orunmila is referred to as a creator god, the Yoruba faith does not stipulate the belief in a monotheist one-god system. The Orisha are granted an external half-way status, yet still remain in close proximity to man. Orunmila, the former man and visionary prophet, was elevated to the position of son of god after his death, without detracting from Olodumare's status.

Yoruba commandments and bans appear to be somewhat more mildly formulated than those found in Abrahamic laws. They are not threatening in nature. (It is important to note that this assumption is based on the linguistic properties of the translated version of the original text.) There is seemingly not a commanding tone. Those addressed retain their dignity despite the restrictions. Notably, truthfulness and fairness are mentioned several times in different forms, as though a lack thereof is the worst imaginable notion. Emphasis is placed on a cultivated approach of respect towards others. Interestingly, there is no commandment that calls for the protection of children, perhaps as it was simply not necessary (?) Nevertheless, an outright ban on killing is also missing from the guidelines.

Perhaps due to the fact that the prophet Orunmila seemingly condensed the best approaches from his day and age into a common doctrine, drawing on his interest in trends in religion and identification with the balancing of forces. It is widely accepted that he experienced close encounters with both traditional Chinese and Daoist culture in addition to Indian tradition and Zoroastrian teachings. We now find ourselves at the second half of the last millennium before Christ, which marked the fall of the Egyptian empire. It is also possible that Orunmila was not required to travel to each of the lands in which these beliefs were held, but rather met knowledgeable representatives from the above cultures in Egypt, in Alexandria, for example. His genuine interest opened the doors to tolerance and intellectual incorporation. Arguably, Orunmila should have been included in the pivotal figures in Karl Jasper's Axial Age.

The amount of cultural material from the golden age of the Yoruba people in the Middle Ages that remains in its original form following several centuries of the slave trade and spreading abroad is simply incalculable.

7.3.9 The Understanding of Development in the Large Sphere of Abrahamic Cultures

▪ Judaism – Islam – Christianity

A number of influences from neighbouring kingdoms came into play on the topic of development. By virtue of its power and ma'at-oriented identity, Egypt cast a long shadow over its neighbours – or, put simply, it exerted its influence on them as a role model.

> **Note**
>
> In Judaism, i.e. the Old Testament, abiding by the law and obedience are also considered the highest virtues.

This pressure had to be endured. What could have assisted in this regard?

The Old Testament also contains ideas that may have escaped the Egyptian Mystery Schools with knowledge on Hermes Trismegistus passed on in secret despite the hermetic oaths of secrecy. Ecclesiastes 3: "For everything there is a season" (Luther text: There is a time for everything, and a season for every activity under the heavens ... etc). Furthermore, the Kybalion, a book that records the philosophy of each common school of thought in Ancient Egypt, contains the following principle of rhythm: Everything flows, out and in; everything has its tides; all things rise and fall; the pendulum-swing manifests in everything; the measure of the swing to the right is the measure of the swing to the left; rhythm compensates.

This notably resigned attitude may have been required to counter the strictness imposed by the law, both in Egypt and amongst the Israelis.

The Sabbath, a day of abstinence from work in the name of God not observed by the Egyptians, may have constituted a reward and been regarded in a particularly caring light. It also aimed to strengthen man's relationship with god.

Another adoption of customs can be found in the Old Testament, which is often viewed as a relief from the burden of the strict commandments, in the form of a cultural nucleus from Mesopotamia: the Song of Songs. The song is purported to have its roots in Ancient Sumerian fertility worship.

Of the three monotheistic religions that recognise Abraham as the first prophet alongside Moses, Judaism is the oldest, followed by Christianity (the virtues of which have been previously addressed in the section on the Middle Ages in Europe) and Islam as the youngest.

Judaism

The Jewish understanding of the relationship between man and god has many facets; yet, its principal notion appears to be rooted in its early beginnings. The history of Judaism begins with Abraham, who made a covenant with god for the people of Israel. This formed the basis for the establishment of the first monotheist religion with an invisible god in the last millennium before Christ. The oldest religious texts in Judaism are the Five Books of Moses (Tanakh), the Torah (law) and rabbinic writings thereon.

The Tanakh contains the Ten Commandments outlined by Moses. They take the form of a contract, as was customary at the time among the Hittites between the ruling king and a subordinate liege. The commandments can be regarded as a decree. It can be assumed that the social commandments included therein already existed and were rooted in those practised by the semi-nomadic Jewish tribes. The first three commandments appear to first have been placed first following the Babylonian captivity (586–529 BC). Two versions of the Ten Commandments exist, with little variation between them. The following commandments are taken from the Book of Exodus.

Note

"I am the Lord your God, who brought you out of Egypt, out of the land of slavery.

You shall have no other gods before me.

You shall not make for yourself an image in the form of anything in heaven above or on the earth beneath or in the waters below.

You shall not bow down to them or worship them; for I, the Lord your God, am a jealous God, punishing the children for the sin of the parents to the third and fourth generation of those who hate me, but showing love to a thousand generations of those who love me and keep my commandments.

You shall not misuse the name of the Lord your God, for the Lord will not hold anyone guiltless who misuses his name.

Remember the Sabbath day by keeping it holy

Six days you shall labour and do all your work, but the seventh day is a Sabbath to the Lord your God.

On it you shall not do any work, neither you, nor your son or daughter, nor your male or female servant, nor your animals, nor any foreigner residing in your towns.

For in six days the Lord made the heavens and the earth, the sea, and all that is in them, but he rested on the seventh day.

Therefore, the Lord blessed the Sabbath day and made it holy.

Honour your father and your mother, so that you may live long in the land the Lord your God is giving you.

You shall not murder.

You shall not commit adultery.

You shall not steal.

You shall not give false testimony against your neighbour.

You shall not covet your neighbour's house. You shall not covet your neighbour's wife, or his male or female servant, his ox or donkey, or anything that belongs to your neighbour".

In the cultural history of the region prior to this point, it was not customary to forbid the worship of foreign gods or make an image of god or the gods, at least not an image representing god. However, figures of speech can also be employed for descriptions.

In spite of the fact that Judaism is based on a monotheistic image of God, it is depicted with characteristics that restrict humans and incite fear, similarly to those found in an earthly ruler who demands complete subservience from his subjects. The commandments mention a "jealous God", who threatens to exact revenge if he is abandoned by man. This is an unsettling notion and reduces his magnitude to the scale of a man who wants to rule. By contrast, the social commandments appear to be pragmatic and mutually beneficial. Granted, owing to its age of 2000–3000 years old, they need to be read in the context of the history and region at the time in order to put them into perspective.

The Halakha (part of the Talmud) contains 613 commandments and legal principles. Its primary goal is conduct in line with the commandments of the Torah. In the Middle Ages, the writings of Maimonides, a legal and religious scholar who reformulated the belief in an incorporeal God, were met with high acclaim.

The preservation of traditions is the primary concern of modern-day rabbinic Judaism. There is no universal definition for the Jewish principles of faith, not even on the existence of God (unlike Islam and Christianity) and they remain open to interpretation. Nevertheless, the circumcision of boys continues to be considered obligatory.

According to a restrictive, time-honoured tradition, a Jew is anyone born to a Jewish mother. In modern times, this has been reinterpreted to include anyone with a Jewish parent. Yet more liberal is the interpretation that anyone who identifies as a Jew or converts to Judaism can be regarded as a Jew. The Jews do not have their own version of the missionary tradition upheld by Muslims and Christians.

Judaism can be split into the following sects: Orthodox, Reform and Reconstructionist Judaism. Furthermore, Jews may be split into the following groups according to the ethnic heritage of their ancestors: Ashkenazi, Sephardi, Mizrahi, Yemeni, Sabra, amongst other groups. As illustrated here, there is a vast amount of diversity to bridge within the religion.

Hasidism, which has existed since the antiquity in various places at various points in time, is characterised by its high moral standards and the cultivation of a closeness to God experienced through mysticism in prayer. The sheltering, comforting and encouraging potential of God can be found in the relationship experienced with him. (In Hebrew, Hasid means "the pious one" who stands in grace.) A particular Hasidic movement emerged in Eastern Europe in the eighteenth century in response to the pogroms against Jews in the region. Martin Buber (1878–1965) was a prominent German-speaking representative of the Hasidic movement in the twentieth century.

▪ Islam

The Prophet Muhammad (said to have been born in 570 or 573 AD in Mecca and died in 632 AD in Medina) was the founder of Islam. Since then, Islam has grown to become the second largest religion in the world. There are several sects of Islam which can primarily be found in Pakistan, India, the Middle East, North Africa and Southeast Asia.

Note

In order to be considered a Muslim, the following prerequisites must be fulfilled at a minimum: (1) The declaration of the Shahada in front of two witnesses: "There is no god but God, and Muhammad is the Messenger of God"; (2) the ritual of daily prayer five times a day, including the performance of ablution; (3) the donation of a fixed portion of income to the poor; (4) fasting during the daylight hours of Ramadan; (5) making at least one pilgrimage to Mecca. However, in the Quran, Surah An-Nisa 4:136 states: "Believers, hold on to your faith in God and His Messenger, as well as the Book that He has gradually sent down to His Messenger, as well as the Book that He sent down before. He who denies God, His angels, His Books, His Messengers, and the Last Day has gone far astray".

In pre-Islamic Arabia, there was a history of meteorite worship in Mecca prior to contact with representatives from Judaism and Christianity. In addition to Muhammed's revelations, Islam is also based on the teachings of Moses and considers Jesus Christ another prophet. Similarly to the two other monotheistic religions based on scripture and revelations, Islam can also trace its roots back to Abraham. Various similarities and differences can be found in the three main religions: Islam and Judaism are strict proponents of the law, whilst Christians and Muslims strongly engage in missionary work (including all of its regrettable downsides known to al).

In Muhammad's eyes, after their failure to accept his Islam, Judaism and Christianity were to be regarded as the failed continuation of the common religion that needed to be defeated. Muhammad had demanded that Jews and Christians recognise him as the messenger of God and his Quran as the word of God. Conversely, the Christians sought recognition for their doctrine of the Trinity and original sin, for which they remained the sole proponents. At the time and to this day, there remains a dispute between the Jews and Muslims concerning the recognition of Jesus as a prophet. Against this backdrop, there has been no further rapprochement between the three monotheistic religions.

The ninth century saw the emergence of Sufism as a sect of Islam in Iraq. Sufism continues to exist to this day beyond the borders of Iraq. Sufi practice focuses on the renunciation of worldly drives and needs in addition to the glorification of God, resulting in experiences of mystical unity with God. Those who experience intense, mystical experiences are highly revered in Sufi culture. Practices of Sufi are organised into Sufi orders. Sufism has come to be known for its trance-inducing circle dances and its great poets.

Today's political Islam does not pursue a religious agenda; instead, it is primarily concerned with extending the influence and cultural reach of Islam. This can be distinguished from Jihad, the holy war, in the strict sense of the term, which, depending on the interpretation of the Quran, can be considered to call for the eradication of all other faiths. According to the Quran, Jihad is one of the religious duties of a Muslim. Jihad can be translated as "striving, struggling, endeavour and commitment". The Quran does not clearly define whether this battle is defensive in nature or offensive, against other faiths. However, in the centuries following the death of Muhammad, Islamic legal doctrine openly pursued the defence and expansion of Islamic territory to secure Islam's place as the leading religion.

The Quran states (9:33, according to Paret): "It is He who has sent His Messenger with guidance and the religion of truth to manifest it over all religion, although they who associate others with Allah dislike it".

The following text can also be found in the Quran (9:5, according to Khoury): But once the Sacred Months have passed, kill the polytheists "who violated their treaties" wherever you find them, capture them, besiege them, and lie in wait for them on every way. But if they repent, perform prayers, and pay alms-tax, then set them free. Indeed, Allah is All-Forgiving, Most Merciful.

7

There verses and similar ones in the Quran are referred to as Sword Verses, namely a call to arms.

However, modern-day interpretations are restricted to a non-military approach to the term Jihad. There have been attempts to invoke a form of inner Jihad, a struggle against moral demerit, debauchery and ignorance.

A major similarity between the Abrahamic religions can be found in the Ten Commandments. There is clear, if not entirely complete, consensus between the Mosaic and Islamic laws, which can be found below in abbreviated form:

Note

1. Do not ascribe anything as a partner to Him.
2. Be good to your parents.
3. Do not kill your children in fear of poverty.
   We will provide for you and for them.
4. Stay well away from committing obscenities, whether openly or in secret.
5. Do not take the life God has made sacred, except by right.
6. Stay well away from the property of orphans, except with the best [intentions], until they come of age.
7. Give full measure and weight, according to justice. We do not burden any soul with more than it can bear.
8. When you speak, be just, even if it concerns a relative.
9. Keep any promises you make in God's name. This is what He commands you to do, so that you may take heed. This is My path, leading straight.
10. So follow it, and do not follow other ways: they will lead you away from it – This is what He commands you to do, so that you may refrain from wrongdoing.

In spite of the clear similarities to the Biblical Decalogue, there are also a number of differences, particularly those pertaining to the well-being of children. Muhammad is said to have been an orphan himself.

If we now return to the question of the relationship of man to God, or the worldview and concept of man, we discover a striking discrepancy between the universal ban on killing others and intolerance to those from other faiths for whom the inhibition against killing, which is inherent in all of us, is seemingly overruled. "Those who fail to recognise our God are worthy of death!" "Those who think differently to us are our enemy!" This form of indoctrination through seemingly divine commandments leads to the creation of religiously motivated criminals and a perversion of the values. By deactivating the conscience of the believer, these convictions can cause him to lose any sense of empathy for his counterpart, to kill others with no emotional involvement and even view his acts as ordained by God – a phenomenon widespread around the globe and not solely found within the realms of Islam. Atrocities can also be committed in the struggle against political conformity, for example. Many are sat in the same boat as far as this is concerned. Those most susceptible of committing acts of violence in the name of religion or political freedom are people who grew up in totalitarian regimes in fear under a high level of pressure to conform.

The good news is that the interpretation of Jihad has posed a significant issue throughout the history of Islam, even if certain marginal groups are unable to accept that a non-military interpretation is approach strived for. As such, the question on whether and under what circumstances the combative approach can be replaced by internal and external peace remains unresolved.

■ Christianity

The order of contributions made to Christianity is intentionally not listed in chronological order.

With over two billion followers, Christianity is the largest religious community in the world. However, it is split into several confessions (Roman Catholic, Orthodox, Protestant and Anglican, to name but a few).

The central figure in Christianity is Jesus of Nazareth, now referred to a Jesus Christ, a Jewish itinerant preacher who possessed healing and prophetic powers and spread the word that he was the son of God and the Messiah prophesised by the Jews. Jesus was condemned and crucified in Jerusalem in 30 AD. The testaments of his disciples and others are recorded in the New Testament, including the story of his resurrection. The Old Testament corresponds to the Hebrew scriptures of the Tanakh. Christianity emerged from Judaism and is based on the same history and Abrahamic roots.

The Christian faith is monotheistic. Its concept of the Trinity, namely the harmonious unity of the Father, the Son and the Holy Spirit, is somewhat difficult for the basic human power of imagination to comprehend.

(Potentially a comparable challenge for our conceptualising Either-Or way of thinking as the notion that a quantum, the smallest energetic unit of our reality, is both a pseudo-material particle and an interweavable wave, which predominately presents itself as a hybrid of the two in various degrees of order. It can take on a certain form, or not. On the other hand, our brain function has not evolved on the basis of its ability to adequately depict reality and to adapt to what it learns from this reality, but instead on the basis of its ability to survive this aspect.)

Note

The revolutionary aspect to the Christian message is the promise of God's unconditional love towards man, which does not discriminate on the basis of nationality, race, gender or social status.

Jesus of Nazareth treated the principle of loving one's neighbour from the Ten Commandments as equal to the love of God. Furthermore, he provided his followers with the guiding principle for interpersonal balance, namely "Love your neighbour as yourself". He was explicitly against any temptation to abuse power.

Man is usually incapable of resisting these temptations. The commandment requiring universal, compassionate love appears to overwhelm the majority of people.

7

The concept and interpretation of original sin, which is not found in Judaism or Islam, are often contested within the Christian church. According to one interpretation, original sin denotes the distancing of man from God, which necessitates death. The death of Christ, who died for our sins, is then understood or interpreted as the redemption of our souls from (spiritual) death. In recent years, theology has avoided the personal attribution of original sin, instead preferring to refer to it as universal solidarity in sin. When we abide by Augustine's commandment "love and do what you will", then the fundamental sin is a lack of love. This notion has many faces. Particularly in psychotherapy.

Christianity officially has the most followers in the world. Yet, despite this, the world is not in good shape. Even if much good has been done in its name, which often goes unnoticed and occurs quietly in the background, the darker elements of Christianity in the past and present cannot be overlooked (the crusades, the Inquisition, the ransom of the soul, the abuse of power, missionary work against one's will, sexual assaults, Pharisaism, etc.). It seems that the message of compassionate love has only been grasped by a minority. This also applies to a certain extent to the representative of the Church itself.

It may be the case that the bold commandment of love, including love of one's enemies, simply transcends the average state of development achieved by man, meaning that we are only able to experience it in its isolated forms at best. In the case, we should celebrate the existence of each of these individual forms to bring them into resonance with one another and allow them to mutually encourage each other in their existence.

7.3.10 The Belief in Matter as the Sole Form of Being

This perspective has already been addressed at the start of this chapter. The understanding of reality consisting of accumulations of the smallest, solid particles that are mechanically connected to one another first emerged in Ancient Greece (Empedocles, Democritus, Anaxagora, et al.) and, notably, later during the periods of Enlightenment and Empiricism, Positivism and Materialism in the nineteenth century (not lease due to the limited scientific methodology and knowledge). This perspective appears well-suited to our sensuous experience and consequently our sound judgement. It suggests the notion that everything is constructed into larger systems, comparable to the innumerable and diverse bricks of Lego. This worldview has persisted among the general population and is widely accepted in applied natural sciences such as medicine and psychology, for instance. As such, the human body and its mind are considered to be a kind of Lego construction. The impulse for change always comes from outside, according to the assumption.

The reputable side to this approach lies in its self-discipline, which is carried forward with the conviction of refraining the acceptance of unproven hypotheses. Proponents of this perspective do not accept anything that they are unable to convince themselves of or comprehend with their own senses. They are not misled by wishful thinking or fantasies about other hypotheses if they are unable to produce any tangible evidence to support them. Vague assertions are not regarded as adequate. Illusions are avoided. Those who follow this approach often have a somewhat sober attitude towards life. They regard death as the end of everything. Adults are required to accept this stoically. People must not delude themselves. They feel that others may simply not be capable of dealing with this sober realisation. Perhaps they are unable to handle the notion that they are not immortal, but limited, like everything else. Whether life has meaning remains an unanswered question. People often like to attribute a meaning to life because that is more satisfying than reality. Everything exists by chance. Evolution could have taken an entirely different turn. Yet, in the reality we find ourselves, everyone is and dies alone. Social projects provide a way to find balance during life, potentially also communal families in various forms. On the other hand, certain individuals may alternatively elect to structure their life through an isolated existence or a position of power. Values cannot be defined and meaning cannot be found beyond the scope of materialistic life. With this in mind, it is entirely reasonable to make the most of one's freedom, for as long as permitted, and cover one's own back well in terms of the material.

The basic assumption that matter is the only existing form is posited belief. It is based on a scientific partial truth: matter, a form of energy, exists, which creates the illusion of solid matter and allows us to inherently regard it as such.

7.3.11 Collectively Encouraged and Trained Cerebral Development Conditioned by the Zeitgeist

Concepts from the Far East have been addressed multiple times throughout this book (► Sects. 3.2.4, 3.2.5, 3.6.4, and 3.7.3). The ancient tradition in the Far East appears to have developed a significant complementary pole for the understanding of our brain, which could help the Western world to find a new balance in this day and age. The incorporation of mindfulness-based approaches into therapy could be viewed as the initial steps in this direction. The goal in this regard is to obtain balance, a sense of calm, reduced stress levels, the ability to establish distance within, increased resilience, enhanced presence with a heightened ability to experience things and, first and foremost, the freedom to keep automatic, ego-centred judgements at bay. These attributes are not inherent virtues.

In comparison to the aspirational virtues handed down to us by the Vedas and their Tibetan scriptures, as representatives of modern Western culture (or perhaps it is best to instead modestly refer to a civilisation in this regard), on average, we are at a comparable stage of development. The mainstream focuses on the struggle for power and satisfying their needs. Only a small minority of the population is predominantly rooted in higher levels. The bias of Western civilisation has hindered the development of the mainstream, relative to its potential, causing opportunities to be ignored.

However, it is able to benefit from significant developments in the field of the conceptual mind, the technology it has developed and the applied sciences. These factors were not developed in the culture of the Far East, but instead hindered by the cultural zeitgeist.

7.3.12 Current Collective Social Development Conditioned by the Zeitgeist

The Current Situation

It would be rash to attempt to make a general statement on the average state of development with a view to the current social situation. As such, for the sake of putting the following assertions into perspective and the limited applicability thereof, I would like to state from the outset that this book, published in 2020, is the work of a psychotherapist who was alive during the Second World War and has borne witness to world events through the lens of a defeated Germany, including the subsequent confrontation with its sinister contributions to world history, but who was and remains – like many of her generation with similar experiences – committed to learning from these events. Whilst it may have initially appeared that man was keen to move towards increased humanity and common social responsibility in the immediate aftermath of the war, as famously exemplified by the establishment of German Basic Law, for example, this harmony has waned in the 75 years since the war, and can no longer be found in Germany, Europe or around the globe. It's almost as though the pendulum of development is now once again swinging in the direction of fragmentation, along with all of the negative consequences thereof.

Are we capable of detecting a disaster waiting to happen in advance? Society often ponders how totalitarian leaders were able to rise to power through fair elections and supported, or at least tolerated, by the populace, as has occurred time and again throughout history, not just in Germany. Unfortunately, this issue prevails to the present day, indicating that whilst certain regions may have learned something from the last century, the world continues to turn a blind eye.

Fragmentation requires an initial starting point of the potential for dissatisfaction, distress, grievances and existential fear to be incited amongst the population. Aspiring rulers then focus on these problem areas, addressing them speaking in the language of those affected and hitting all the right notes in a show of feigned solidarity. As a result, those in need view the ruler as one of them, someone who understands them, appreciates them, promises them help and appears to be their knight in shining armour who will be able to make all of their worries disappear in return for their support. His promises, i.e. illusions, must be realistic enough that the target group are able to take them seriously. The greater their distress and grievances, the more desperately they will grasp at straws. In general, those seeking to rule will then attribute the issues of the populace to external aspects, ideally generalising and attributing it to a foreign enemy or minority group, which then takes on the stigma of "black sheep". This provides the negative feelings of the target group with a way to let off steam. The need for causation is satisfied at first, provided no critical aspects are registered by the target group. This attitude generally leads to increased radicalisation and social fragmentation throughout the general population. However, it can also lead to the emergence of a supportive social "we", formed as a result of the inner relief and heightened collective self-esteem amongst the (typically national minority) group that feels addressed by the claims of the politician. The aspiring ruler draws them in with claims of an idealised future, the antithesis to the current situation, whilst simultaneously highlighting the terrifying scenarios that would unfold were he not to be elected. If an individual is incapable of seeing any alternatives or has reason the fear reprisals, he will support the politician as if in a walking trance, suppressing all critical thinking for as long as possible and regressing his own inherent autonomy. This marks a step backwards in terms of development.

How can a regime that abuses power and leads to the regression of its uncritical followers be prevented? How can the inherent balance between the needs of the individual and the collective be upheld? This level of consciousness requires the participation of everyone. If it succeeds, we will be in true balance.

The following is hardly a new notion; yet, it remains a hot topic: we require an adequately large number of people, critical thinkers, who first manage to find a balance for themselves between individual balance on the one hand and the constructive interplay between themselves and the social reference group in their region, and beyond where possible, on the other. This provides the basis for establishing strong forms of resilient and functioning democracies. The ability and willingness to identify with all other members of the reference group(s) also needs to be developed in order to ensure that their place in society is guaranteed just as we would like our own to be. Each individual should finally ask themself: how far does my compassionate and appreciative sense of belonging stretch; how extensive is not just my tolerance, but also my ability to love? Does it only extend to those around me? Only those from the same area, social class, political party, religion, nation, race or culture? Does my empathy extend to animals? Plants? An appreciative and caring approach to nature? Am I capable of listening to my surroundings and perceiving myself and my environment from a higher perspective, and subsequently letting this flow into my actions?

Depending on the level of consensus reached on the notion of allowing all living beings to live as they please, and whether individuals are willing to make compromises to achieve this harmony, we may find ourselves able to leave the permanent state of the abuse of power and enter a state in which power is used to govern in a wiser manner on the basis of humanity and social responsibility. This would represent the development step currently strived for by the mainstream. Although our civilisation may be highly advanced in terms of technological development, we continue to climb the ladder of development at a snail's pace.

If the next stage of development brings an increase in the creative potential of human beings, their awareness of meanings and values, their intuitive receptivity and their advanced ability to form relationships, this would constitute a desirable bonus and motivation for the qualitative further development of man as a whole.

7.3.13 Personal Oscillatory Activity – Fragmenting Chaos or Resonant Ordered Structures Capable of Synchronisation?

We will now address the question concerning which stage of human development calls the shots. If, at the end of our journey through several periods in our cultural history, we take a moment to recall the "towers" of development stages at the end of ▶ Chap. 3 and ask ourselves: "Which of the stages were particularly resonated and were addressed in which era, by which zeitgeist, or by which worldview and concept of man?" We will arrive at different answers.

In other words, which parts of an individual were particularly addressed, demanded or affirmed in which religious, secular or other worldview and consequently gained significance within the system as a whole and its identity?

It is not surprising that the notion of the divine plane can, but is not always required to, amount to a screen for projection. Tales of gods with human characteristics, with the petty, worldly grievances of an averagely chaotic family tend to put a smile on our faces, whether they concern the gods of Olympia, Germania or Egypt, etc. In certain cases, it would seem a few sessions of family therapy would not go a miss. In comparison, monotheistic faiths seem to be more manageable. But are these faiths truly all-encompassing and in harmony? Is any missing? Is there a lack of life-enhancing love? Is there a lack of encouragement to assume responsibility for one's own actions, even in light of the risk of making the wrong decision? What kind of formative model does it provide for man?

▪ Chaos or Order/Solidarity and Fragmentation

It is worth determining the degree of order and relationship quality within one's own definable identity: chaos or a holistically structured reflection (as internal individual gestalt). This question can be applied to both one's own frame of mind and the imagined nature of our transcendence, regardless of whether it is a projection or not. We shall first explore the individual: How harmonious or coherently do we oscillate within ourselves? Are we able to imagine that we are in our core and able to resonate, as if on our own personal mandala? This can only be answered with a certain stretch of the imagination. Nevertheless, the belief that it is possible to bestow an attitude upon oneself and the entire organism which causes harmoniously structured wave patterns to emerge of their own accord through interactions provides a valuable aid and motivation for centring All forms of meditation are beneficial in this regard.

These patterns are receptors for analogous wavelengths that are potentially capable of resonating. However, this property is also simultaneously responsible for the failure to produce a reaction due to the lack of an analogous, suitable wavelength nearby. It can cause the individual to feel alienated and uninformed, or lead to resonance with chaotic energies. Certain individuals may intentionally turn a blind eye to any resonances that cannot be categorically assigned on ideological grounds in order to avoid the acceptance of any consequences that could require the individual to expand their worldview (which could lead to mockery by the individual's social circle and put them at a disadvantage), for instance.

We will now take a look at the variation we have encountered on our journey through the wide range of transcendental and religious concepts. As an individual, do I find it logical and plausible that

(a) The relationship between the divine beings, forces and ancestors as the players on the invisible plane is unconnected and somewhat random, and that those who reside on this earth are required to ensure that these unpredictable entities are kept satisfied with sacrifices in order to adequately stay in their good graces? This often paints a very needs-oriented and self-centred picture of the divine, making them seem human in nature. Failure to meet their needs or grant them enough attention will lead to fear and destruction. At the same time, a lot of support within the social group and practical knowledge is required to survive under these circumstances. Man focuses on the search for protection and allies amongst the gods, in addition to appeasing incalculable forces.

Or, (b) as an individual, do I believe that the divine, who take on the role of gods, experience similar suboptimal difficulties in their relationships with each other as humans, with similar preferred sympathies, weaknesses, jealous competitions, blinding passions, affairs, envy, betrayal and quarrels up to mortal combat with the annihilation of entire generations? The clan of gods seemingly reflects a large human family with all of its intrapsychic infighting. Deities can be regarded as aspects of life. A certain dose of sound knowledge of human nature is required to accept these gods, who are ultimately no better off, which provides us with a certain amount of comfort. The gods are focused on their struggle for power, attractiveness and influence amongst their ranks. Man continues to search for allies amongst the dubious and often all-to-human gods.

Or, (c) do I find it plausible and likely that if I believe in solely one god, I am required to dedicate a great level of care towards this one god (with whom I identify) to ensure that nothing else compares to him and could threaten him with the loss of his power and splendour? Does this entail that a state of war must automatically occur if two or more beings appear on the transcendent plane? Is battle mode always engaged when another being emerges on the horizon and needs to be brought to heel or eradicated? (This totalitarian principle can also be established around an absolutist monarch or party leader in place of a god). Identification with an autocrat can often provide an uplifting release for those with low self-esteem; being at his mercy can trigger feelings of powerlessness. This is reminiscent of the biologically anchored power struggles observed in the animal kingdom: It's him or me, either-or. There are only winners and losers, a concept which triggers anxiety and potentially aggression in some. Followers of a ruler, the enslaved people, fully relinquish their power, instead participating in their minds by identifying with the ruler and assuming that they are always on the right side of his orders. Their imagined (borrowed) power conceals their weaknesses.

In more positive terms, a balanced person who is aware of his limitations will feel a great level of security and comfort in acceptance by the sole god. However, if this relationship is not adequately tangible for the individual and solely conveyed in the theoretical form of a mental concept, the connection to god will collapse and fail to last. With this in mind, fragmented individuals run the risk of chipping away at themselves, functionalising themselves and spreading disintegration.

Version (d) It is possible to imagine a scenario whereby individual personified forces with limited autonomy to express their own special function can also be addressed as subpersonalities and brought to the fore yet nevertheless remain a cooperative element of the group of deities as a whole and never become detached from the mutual, amicable solidarity, under the umbrella of a superior deity. This concept assures support in spite of range. It focuses on integration, cohesion, reconciling relationships with a whole and, in the best-case scenario, love.

Version (e) The all-encompassing or divine mode of being encompasses the possibility of both structured and unstructured existence, including all forms in between and Both-And variations. As a result, the relationship between life and death is also changed. The matter and formless state of being, which is regarded as nothingness by our senses, can be perceived as the radiant, more vibrant, attractive pole. Experiences of occasionally tipping from apparent emptiness into overwhelmingly vibrant and omniscient abundance have led man to search for the conditions and methods required to induce this tipping phenomenon and even the ability to induce it of our own accord (Zen culture).

Version (f) In the Western world, it is not uncommon to grow up without any transcendental ideas and instead focus solely on interpersonal relationships and the ethics thereof, provided the latter are exemplified by role models. This blueprint is combined with a purely material understanding of reality. Concepts of interconnectness, field interactions and resonance processes fall upon deaf ears.

The parts and the whole. A Both-And principle also applies to the awareness between the whole and its parts. This pertains to both the intrapsychic and intrapersonal relationships experienced by an individual within and, above all, the individual's relationship to his all-encompassing reference framework, including its subjective pyramid of values.

Note

Generalising partial perspectives only ever lead to an impasse.

The interconnected whole, in which all opposites are neutralised, is simultaneously in a state of peace and balance, and possesses the highest degree of freedom for its potential transformation. The whole is mirrored in its details, which provide it with affirmation and recognition, and are capable of reshaping it, causing it to subtly change. The details exist in participation with the whole, allowing them to receive recognition and dignity. The whole is reflected in every individual, even if only usually in a hazy manner.

The primary emphasis of this version is placed on receptivity, inquisitive perception, a playful readiness to react and an openness to higher insight. Contact with the all-encompassing dimension is neither abstract nor rational, but instead a positively staggering, meaningful experience for which words do not suffice.

■ Existentialism and Its Ties

Mircea Eliade, often referred to as the "Einstein of the history of religion" by his peers, is perhaps more renowned for his work in the field of existentialism than history. He laments the loss of spiritualism in modern, irreligious society, which he was clearly still capable of observing on an archaic level. This assertion was made with forms of spiritualism with a strong focus on experiences such as Siberian shamanism in mind. The balancing of vital emotionality and its connection to a conscious outlook is the secret behind the grounded and effective guidance and transformation of emotional strength into spiritual power. This potential has been developed into various shamanistic rituals on every continent. In order to address this notion in a responsible and ethical manner, each form requires its own critical acknowledgement.

It may be the case that shamanistic rituals were also widely practised within the Egyptian Mystery Schools and by the Celtic druids, albeit in different forms. This potential is encountered amongst almost all indigenous people. It mobilises abilities which are regarded as miracles, but which clearly belong to the inherent mental, spiritual and physical healing and transformative potential of man, which has neither been promoted nor academically studied or categorised in the present-day field of psychotherapy. This aspect has been almost entirely excluded and sidelined by today's book and high forms of religion, resulting in a loss of potential and the removal of this aspect from official culture, which has led to the emergence of grey zones.

Two elements are encountered once again in this grey zone, namely content and structure. Content rides on the waves of structure. It can be regarded as an exertion of influence on the substantive ideas and assessment filter verified by the individual himself or others. This also touches on the endeavours of cognitive psychotherapy. It is notoriously employed in hypothesis, marketing, propaganda and similar measures of persuasion. In this regard, all conceivable crossovers between conscious and unconscious deception to the detriment of the targeted individual can be observed on the one hand, and encouraging, life-enhancing, suggestive encouragement is experienced on the other, including the individual's own stimulation of his self-healing powers. The latter is the primary focus of resource work. A clear ethical attitude is required in this regard.

The other level is that of the interconnectable, structural side of this potential, which is experience-oriented in nature. It permits engagement in the present through conscious, mindful, sensual experience, usually in the form of receptive awareness in a timeless mode. Cognitions have a tendency to interfere with this and are best left in the background. This resonating aspect also provides the individual with the opportunity to engage in improvisations, intuition, unconscious, self-organised processes and waking trances, experiences of creative flow, centring movements and rhythmic dance. In certain cases, endurance sports can also induce energy-enriching intense experiences, which are often accompanied by a sense of meaningful satisfaction. They can also increase the differentiating perception of others and facilitate a more holistic grasp on reality.

In the best-case scenario, the individual is able to clarify which of these modes of contact with reality is preferential – not just externally, but also within. Are both fully developed? Are they in conflict with one another or do they work alongside each other? One observes and substantiates, whilst the other perceives in resonance. The former defines and arranges, whilst the latter lives in the here-and-now.

Remnants of this general shamanism, or, in more prosaic terms, the subtle, perceptive, energy-forming and guiding potential that can inherently be found in all of us, live on in the placebo and nocebo effects of spontaneous healing, above all in relation to the immune system, in the martial arts with conscious energy guidance, and also covertly in medical compliance and psychotherapy, with the strong influence of the therapeutic relationship and personality.

When Eliade writes about the loss of the spiritual, this can be understood as a loss of man's relationship to wholeness, which constitutes the loss of the resonant form of existence that allows us to experience a sense of belonging and involvement in the whole. In any case, in etymological terms, the terms holy and whole are part of a series of identity transformations derived from the same meaning. The loss of common resonance between man and the partially tangible and partially transcendental level that embraces him is typically experienced as a sense of being excluded and lost bereft of meaning. It constitutes the underlying feeling of nihilistic existentialism.

As this book is dedicated to the topic of psychotherapy, it seems only logical to draw a parallel to the situation of an individual who loses the support of his empathetic and supportive social system and falls into a state of isolating fear, followed by a state of shock. In this regard, it is also beneficial to apply Buber's and Watzlawick's (indirectly even Einstein's) distinction between an existentially affirmative, resonating ability to form relationships on the one hand, and a rational, observing and distancing (resonance-free) objective relationship on the other hand. The former connects, whereas the latter fragments the overall situation. The ability to experience the energetically expanded, transcendental surrounding world is blocked by the individual, above all when this is socially sanctioned; it also loses its ability to differentiate when it is not trained and is quickly outshone by the seemingly clear outlines of conceptualised thinking. In terms of the transcendental level, these outlines are provided in the form of established beliefs and myths, institutional commandments and bans on the subject of the higher dimension, dogmas, revelations and stories from others. However, this yet again leaves us as prisoners in Plato's cave engaging in shadow play. Revelations from others can be a helpful and encouraging addition, provided one's own connection to transcendence is promoted and developed. Calls to repress others who are already in a state of emotional shock as a result of the threat of punishment solely condemn them to internal alienation and denaturing. In psychology, we often have our work cut out to first reach these frozen inner beings. In the worst-case scenario, a combination of empathetic compassion and the pragmatism of Kant's imperative can have a beneficial effect (namely, act in such a way that the principles of your conduct could be applied to all). Walk a mile in another's shoes and identify with his situation or distress. This will change your perspective.

The opportunity and risk in this regard lie in the relationship between us, as personified individual beings, and our speculated superordinate referential wholes. There remains a lot of uncertainty, which is overshadowed by the significant efforts made to bridge the unknown through dogmatic pseudo-knowledge. The question of tolerance could be a key development criterion for the reference group: As an individual, am I capable of allowing myself the freedom to explore the possibilities, viewpoints and concepts on the kind of relationships I experience in my faith or political associations in order to freely elect whether I am for or against these? Or would this spiritual exploration condemn me to death?

A community is both a helpful and necessary factor of life, which, on the one hand, promotes this form of relationship building that transcends the individual and, on the other hand, simultaneously provides binding ethics for its members in mutual, caring respect. This also corresponds to the spiritual legacy of the Dalai Lama XIV. The good news is that the ethical rules of social cohesion are very similar throughout the various cultures in the world. A certain number of these have already been explored in this book. The ethical rules of social cohesion are also usually underpinned by biology.

One element that the Dalai Lama struggles to tolerate is war launched in the name of a religion and, furthermore, when religion is abused to promote economic and political interests. At their core, all religions are primarily devoted to the promotion of love, even though they may feature different mental frameworks. Love and inner virtues may be developed within and outside of religions, as claimed by the Dalai Lama, the Buddhist spiritual leader of the Tibetan people, to the great astonishment of many shortly before his retirement on his 80th birthday. This represented the opening of the spiritually restrictive barriers of religion – at least on his part.

This show of tolerance even exceeds that found in "Nathan the Wise" by E. Lessing [3], in which he pleaded for tolerance and humanity in light of the centuries-long conflicts between Judaism, Islam and Christianity. Whilst none of these religions lost sight of their purpose entirely, they were also not without their aberrations, which violated humanity.

We now return to Eliade and his concerns. As has possibly become clear, the loss of the "sacred" lamented by Eliade in terms of being moved by it cannot primarily be reversed through shamanism, but instead through the characteristic of a relationship and natural order that can be experienced by individuals in sufficient harmony – a characteristic that can coordinate with others or even be synchronised on all levels: the existential (including wave-corpuscle identity), the pseudo-somatic, the intrapsychic, the interpersonal, social, societal and transcendental. According to his daughter, Christine Mann [4], Werner Heisenberg primarily regarded god as a perceptible, central order. Tangibly reaching this sublime order, feeling at one and in the right place, and contributing to its flexible stabilisation with mindfulness are a great and rewarding feat.

Conclusion

Monotheism is not automatically the optimum form of faith. The personal relationship between man and god is the decisive factor behind whether a supportive, life and development enhancing "field" emerges. Furthermore, the concept of intermediary angelic deities that work together with the primary deity can be an encouraging trendsetter in certain cases, or may instead hinder further development of the faith. When perceived orthodoxy is used to justify the elimination of others, it purely conceals the activated archaic level in its destructive form.

Conclusion

Both the individual and collective worldview and concept of man, alongside all of their religious and otherwise guiding concepts and value systems, are as equally parts of the human psyche as the reflections of mental events on the social and psychosomatic level. Constructive and destructive impulses may originate in worldview level in addition to harmonising and fragmenting, supportive and threatening, development hindering or promoting impulses, which shape the motivations at the level of our basic needs and even have the power to override the biologically anchored protective reflexes. It is precisely due to the fact that each individual views their own nonreflective worldview as normal that the inclusion of superordinate beliefs in psychotherapy is of such great importance.

The goal at this level is also that of loving, supportive integration to counter fragmentation, which usually increases the likelihood of anxiety.

Bibliography

1. Cowan J. Offenbarungen aus der Traumzeit. Das spirituelle Wissen der Aborigines. Stuttgart: Lüchow-Verlag; 2004.
2. Toshihiko I, Toyo I. Die Theorie des Schönen in Japan. Beiträge zur klassischen japanischen Ästhetik. DuMont, Köln (Original in Englisch, 1981: The Theory of Beauty in the classical Aesthetics of Japan. Nijnhoff, The Hague/Boston, London); 1988.
3. Lessing GE. Nathan der Weise. Stuttgart: Ein dramatisches Gedicht in fünf Aufzügen. Reclam; 1779/2000.
4. Mann F, Mann C. Es werde Licht. Die Einheit von Geist und Materie in der Quantenphysik. Frankfurt/M: Fischer-Verlag; 2017.

Research and Research Crises in Psychotherapy

Contents

L. Hartmann-Kottek, *General Psychotherapy*,
https://doi.org/10.1007/978-3-030-87466-7_8

Overview

The good news first: around the globe, there are still faithful scientists who are primarily concerned with knowledge and not the political struggles between the interests of associations. I would like to extend my deepest gratitude to these scientists for their prudent research and sense of justice.

One of these groups decided over half a century ago to contribute regular updates (every 8–10 years) by publishing a handbook on psychotherapy research in which all schools are invited to publish their peer-reviewed findings. In the beginning, the editors were Bergin and Garfield; in more recent years this position was taken over by M. Lambert (Publisher: Wiley, New York, 6th edition 2013). Accordingly, Lambert's summary cites that all methods are able to produce at least sufficiently good results and that, from a scientific standpoint, the exclusion of none of the methods published in the handbook is justified.

Scientists and their teams who have dedicated their work to high-quality, true to life (contextual) psychotherapy research and fill us with the hope that the reality of life can merge with the world of academia are also worthy of praise.

Last but not least, I would also like to express my gratitude to every scientist who helps us to uncover the misuse of science through "fake science", open our eyes to it and help us resist it.

As is clear from the above, gratitude is primarily directed at the individual beacons of scientific ethics, both individuals and research teams. Unfortunately, the notion of general ethics is not always guaranteed in the scientific field. Increasingly, over the past 50 years, there has been an increased need for an unbribable arbitration board for research issues amongst the scientific community, not only in the field of psychotherapy. Scientific methods are an expert field that does not only pertain to general academic knowledge. It can be easy to pull the wool over the eyes of laymen such as lawyers, political decision-makers, insurance representatives, etc. with misinformation and scientifically veiled, interest-driven expert opinions. To counter this, we require a globally recognised, truly neutral platform with legal competence.

Those who focus on differences, exaggerate the importance thereof and thereby highlight their own superiority for competitive reasons. As perhaps goes without saying, higher efficacy in the general school ranking grants the respective approach a market advantage, provided the population is aware of the ranking. Naturally, the representatives of this group are keen to put the "dodo bird" message into perspective, namely, the belief that all methods are adequately effective and should therefore coexist. This group argues that the differences are not significant enough to require an existential "exit" for any of the subgroups. The diversity found amongst those who enter the field of psychotherapy increases the chance that more adjustments may be required between both the patient and therapist, and the budding therapist and his psychotherapeutic approach in light of the diversity of man. Based on experience, therapists are often only convincing and helpful in their work if their style matches their personality. However, style is subject to change over the course of a lifetime.

Research provides scientific data for both "adversaries" [1]: (a) for the assumption that there are differences in effectiveness between the individual methods. And (b) there are also data to support that the differences between efficacy are not significant enough to justify officially merging psychotherapy into a standard approach to the benefit of the seemingly more effective group and abandoning multiplicity. Differences in efficacy are a quantitative, linear threshold. However, the differences between approaches are always linked to somewhat different characteristics that may be key to the responsiveness and affinity of the individuals seeking help.

So far, we have addressed the truthful handling of scientific data. We will now take a look at the opposite side of the coin, arriving at the bad-news part of this section. Psychotherapy research is also subject to the risk of misuse and has succumbed to it in certain fields. The attempt to cover up, falsely interpret or downright falsify occurs amongst individuals or collectives when good results lead to advantages, whether immaterial or material in nature. This chapter is not concerned with individual methods, but rather with the real-life application of research methodology to gain knowledge in addition to the use thereof as a weapon with the aim of deception, concealment or cheating.

The effect sizes of all age groups of psychotherapy and all of the various approaches are an average of ES 0.80 to 0.83 around the world.

"General psychotherapy", when intentionally geared towards the scientifically proven effect factors, would likely achieve a significantly

above-average effect size. As the present approach, even if in extended form, sprang up from the same dimensional consolidation as Gestalt therapy and provided it also keeps pace with the diversity of approaches that have since developed, "general psychotherapy" is likely to follow its predecessors and ranking among the most effective methods.

8.1 History – With or Without the Dodo

Competitive constellations happen. This is belonging to the archaic nature of man. And this can also be the basis of the nature of human-induced social entities that rely on shared beliefs. In order to belong to one or the other camps, worldviews and the impact of the zeitgeist are just as important as the individual character that develops through differing amounts of awareness towards social, socio-political, economic, gender-related, racial, religious or spiritual issues.

Competition and rivalry spur development, provided they remain fair. Acting in a fair manner requires the ability to walk a mile in your rival's shoes and grant him the same right to life that you grant yourself. This is a key sign of human maturity. To what extent can the rivalry between the schools of psychotherapy be regarded as fair? It can to some extent, but unfortunately, this is not always the case. We are not spared the task of looking into each individual case.

According to which criteria was and does the rivalry between the schools of psychotherapy continue to be supported and what is the primary issue? Which is the most persuasive school and has gained a following, both of patients and prospective therapists? Which school holds the positions of power in socio-political associations and professional bodies? Which camp has made a name for itself in publications? Which camp has gained decisive influence to advise insurance providers, hospitals or the government? Which camp claims (1) to be particularly effective in terms of treatment? (2) to work in a particularly economic manner? and/or (3) to be based on "science" in a unique way? The criteria can be listed as follows. They are not of equal importance when we ask: who or what should psychotherapy serve? The issue of patient well-being is often neglected and drowned out by the battle for survival between advocacy groups and lobbyists, provided there is no corrective influence from a higher meta-level. In Germany, the country home to the author of this book, there is no competent, neutral mediation body.

How can a "scientifically recognised" method be defined? In the field of medicine and other medical sciences, there is no legal basis for these criteria. As a rule, lawyers generally make do with a concept that corresponds to the level of awareness and acceptance within the community of experts. However, psychotherapy is primarily concerned with obtaining evidence for the effectiveness of a method. The way in which the evidence is provided can vary significantly. For over 30 years, the rivalry in the world of psychotherapy research has been led by two fronts: the contextual research approach, which attaches great importance to effect sizes and the inclusion of naturalistic relationships, on the one hand, and efficacy research on the other, which transfers the proven criteria and notions of purity from pharmaceutical laboratory research to psychotherapy. The transfer of this principle is an irritating category error.

Initially, however, the primary focus is the "healing" of an individual, without the ability to objectively, subjectively or broadly define the term. The "scientific" definitions of psychotherapy vary widely and, at present, largely depend on the worldview and concept of man held by the respective examiner or his school. In this respect, we are currently in a process of radical change, or rather an intermediary state, in which we are required to determine which scientific approach is best able to encourage, recognise and comprehend the transformation desired by the individual towards his personal "healing". Comparative, mass-statistical calculations on the data of individuals, as are often undertaken in survey tests, rarely provide any meaningful knowledge on the individual, solely in relation to his reference group. The constellation of the queried characteristics in the individual's inner relationships and his experience-related, individual connections are not recorded. Whilst rarely used in the comparison, this void is beginning to be filled by process research and time-series analyses.

"New physics", namely quantum theory, is yet to be incorporated into research methodology. It remains to be seen whether this will be achieved. The need to take the freedom of the individual alongside his positive incalculability, prospects for development and ability to change into consideration in psychotherapy is palpable. This presents a challenge for scientists.

For over four decades, there has been a constant debate on whether the efficacy of all methods is statistically similar, making it justifiable to "award them each a prize" as the Dodo suggests in Alice's Adventures in Wonderland, or whether the focus should instead be on the differences between the methods that exist in subtle tones, in a competitive sense. The dispute primarily pertains to their assessment and consistency across the board. In their book, The Great Psychotherapy Debate (2015, p. 179), psychotherapist researchers Bruce E. Wampold and Imel clarify that the differences between methods used around the globe are too minimal, and as such, the Dodo's suggestion of awarding everyone should receive special attention.

Wampold ([2], p. 336) summarises that the thousands of studies conducted do not indicate that, on average, any special treatment for a particular disorder was clinically or systematically superior to any other psychotherapy. None of the examined psychotherapists were assessed as substandard or even detrimental with any true certainty. All methods can make a useful contribution which is sufficient. Some will, of course, be more effective than others.

This is the basis used to justify the notion that "general psychotherapy" can be expected to produce above-average success in the outcomes of therapy in terms of the current state of psychotherapy, particularly when it focuses on the aspects of the therapeutic relationship and the therapist's personality.

We now return to the history of psychotherapy: In the USA, the commandeering approach is that of the scientific style of the analytical, cognitive approach shaped by the Enlightenment since the arrival of Hugo Münsterberg, who immigrated to America from the school of Wilhelm Wundt (Leipzig, Germany). His work is also linked to the eminence of the learning and behavioural theories from the Pavlovian school of psychology. (Pavlov's teacher Bechterev was also a student of Wundt's in Leipzig.) His emigration helped to spurn the dissemination of the learning-theory approach in the USA at an earlier date and to far more acclaim than in Europe. In German-speaking countries and throughout Europe, psychoanalysis was particularly popular in the first 30 years of the previous century. Despite the fact that both schools of thought show similarities in terms of the scientific ideal of the neutral, analytical observer, in certain aspects they developed into rivals. One banished the elusive aspect of emotions and relationships into a black box for an extended period, whilst the other used hypotheses to tackle these issues, particularly the activity of the unconscious. Hypothesising not only condenses timeless findings, but also viewpoints conditioned by the respective zeitgeist and experiences of the observer, and always requires readjustment.

In 1933, psychoanalysis reached the shores of the USA from Germany with the significant and tragic emigration that occurred during this period. This helped to reinforce the psychoanalytic colleagues located in the USA.

Some of the German-Jewish emigrants returned to Germany in the 1950s after the Second World War. They then set out on their task to create a network of psychoanalytical institutes and initially constituted the sole psychological healing method against mental disorders. As such, they held the monopoly. In 1967, psychoanalysis was first included in the services offered by the statutory health insurance system on the basis of an efficacy study conducted by A. Dührssen, marking a respectable step forward for the school. Psychoanalysts could be found working as consultants in medical associations, the health insurance system and government-related bodies. At the time, Pavlovian animal experimentation had long been practised at German universities and academic institutes; however, behavioural therapy was not yet on the radar of the population as an effective therapy for individuals. It was instead primarily associated with the admirable learning abilities of trainable rats.

The 1950s in the USA saw a rise in discontent with academic psychotherapy, namely cognitive behavioural therapy and psychoanalysis, which both appeared to fall short of the requirements of the day. This revelation led to the emergence of the humanistic wave, or, perhaps more aptly put, its revolution. The humanistic manifesto was published in 1962. Founders of the Association for Humanistic Psychology included: Maslow, Ch. Bühler, Goldstein, Rogers, Köstler, May, Huxley, Mumford, Sutich, Bugental, et al. Virginia Satir and Fritz Perls were also closely affiliated with the work of the humanistic movement and worked alongside them for a period. Virginia Satir always referred to them as the "Human Potential Movement". The humanistic movement took the individual with his entire range of needs seriously, including his longings and desires for dignity, subjective values and sense of meaning. Furthermore, this movement was the first to recognise the importance of a resource-oriented perspective. Prior to this point, psychotherapy, particularly psychoanalysis, had solely been focussed on the sick, unsuccessful and broken. The new methods introduced by the humanistic wave worked on the basis of scenarios, creativity, emotions, physical activity and, above all, relationships. The therapist was viewed as a selectively authentic counterpart, who was capable of interacting with the patient. This aspect of a real relationship with the therapist opened the door to encouraging and inclusive new experiences. "Transference", one of the key aspects of psychoanalysis, was relativised; it was used (particularly in Gestalt therapy) as a projection to distort the patient's perception and as a path to the fixated source of emotional imbalance. However, the working plane moved away from the objective of abreaction through transference neurosis on the substitute object towards the healing understanding of the frustrated basic needs of the patient and empathetic acceptance for their subjective distress.

The humanistic wave hit Europe in the 1970s, initially still closely aligned and based on the same sources as couple's and family therapy in terms of methodology, which gradually went on its own way with the objective of establishing a system initially synonymous with the technical expression "Gestalt/whole/relationship structure". The focus shifted from the initially context-related individual to his reference framework, namely, on the whole as a system and its connective structure. For a time, this "system" was regarded as more significant and fascinating than the individuals that made it. Ultimately, the pendulum once again swung back, with the implications of a particular constellation on the individual member of the observed system placed at the fore. By virtue of its nature, couple's and family therapy is concerned with the tension between the individual and his reference frameworks and simultaneously attempts to find solutions from its understanding of this tension.

In the 1960s/1970s, the humanistic wave marked another anti-establishment, peace-seeking and spiritual awakening movement driven by the youth. Meditation was a daily part of life for many. Others instead felt alienated by it. As such, mainstream society and psychoanalysis viewed this movement as an enfant terrible; a threat to their world order. It was nevertheless inspirational to the youth and enjoyed immense popularity. Certain psychoanalysts even undertook training in fields such as Gestalt therapy, psychodrama, talk therapy, bioenergetics and other body therapy approaches, logotherapy, transactional analysis etc., learning new aspects that they had previously overlooked. This invoked fear amongst mainstream circles, who feared that their dominant position may be at risk and consequently united against the new, positive, spiritual wave.

In certain cases, some colleagues discovered "science" as a means of professional warfare in their subjective distress and injury. Others were predominately drawn by materialistic factors in light of the considerable economic success to be found in educating future psychotherapists. At this point, we are broaching on the abusive dark side of science.

As research methodology is a highly specialised, theoretical subject and cannot be found in general education, indeed not even as part of the general expertise of practice-based psychotherapists, it is particularly suitable for influencing, misleading and manipulating not just colleagues in the field, but first and foremost ministerial, legal and political decision-makers. It is not a difficult feat to convince and bluff committees with data that are purportedly backed by science yet cannot be verified by laymen, particularly if the presenter has a strong academic background and confident demeanour.

8.2 Fake Science and Its Alternative Facts

It is certainly troubling that even psychotherapy, a field dedicated to mental health and human development, is affected by this epoch of the misuse of science, primarily motivated by competition, power, money, influence and appearance etc. Individual nations have experienced different levels of misuse. Unfortunately, Germany is one of the biggest perpetrators. Nevertheless, it remains important to distinguish between the small group of career politicians who, despite criticism from within their ranks, have proven themselves to be quite capable of the above, from the majority of those in the field who instead benefited from this misuse but were not behind it and would have made different ethical decisions had they been asked.

The following variants provide a limited snapshot. They include some of the crassly and regularly used data distortion and points of critique.

1. Divide and conquer: Not solely a strategy utilised by the rulers of Rome, but a strategy employed by the "alternative scientists" in terms of Klaus Grawe's data pool.

Grawe compiled several groups of methods according to leading characteristics and inner similarities for his evaluation (1984–1986) of studies (from 1934 to 1984 around the globe): psychodynamic, behavioural therapeutic and humanistic interpersonal (systemic) groups, relaxation techniques and hybrids. In this respectable meta-analysis, the efficacy of the humanistic and interpersonal groups was tied in the first position. Conversely, behavioural therapy, which placed a strong emphasis on behavioural aspects at the time, was ranked midway in terms of quality, with the contemporary psychoanalysis/psychodynamic groups in the last place.

Now, when one group is able to lay claim to being the best with the backing of a "research report commissioned by the Ministry of Health" that claims to be the largest meta-analysis conducted in the world in the latter half of the last century, laypeople are easily won over, given the power to provide their own statements and convinced that only their own two groups of methods form homogeneous clusters. Accordingly, all other groups of methods should be viewed as fragments due to their heterogeneity. However, statistical fragments are too small to be scientifically evaluated. This form of fragmentation can lead to the statistical eradication of clusters, as though they never existed in the first place. (In Grawe's meticulous meta-analysis, all groups of methods were viewed as similarly homogeneous and heterogeneous by his team. The source materials and their clustering were only published subsequently and can still be accessed by those who are interested) ([3], Hogrefe).

This scientifically unjustified, arbitrary perspective creates a division between what exists and what does not. In this regard, it was key that those denied the right to exist were not privy to the manipulation taking place to ensure that they could not launch targeted objections. It was only in 1991 that they were informed that the evaluation of the extensive meta-analysis would result in them losing the right to treat patients by law. Studies conducted in subsequent years to highlight how non-mainstream approaches were just as successful as those permitted by law were not recognised by representatives of the latter.

Representatives from the scientifically recognised, legally valid approaches mastered their ruthless game of persuasion and invented "alternative facts" on the political stage, with the backing of their association lawyers. Fictitious scientific data were employed against political decision-makers as a smokescreen for lobbyist power grabs. Targeted false information was used to the benefit of a select few and the detriment of others. Perhaps most deceitfully, this initiative was proliferated by those at the bottom of the ranking, namely the psychodynamic group, which had been working towards this for over 20 years with the assistance of the average-ranking group in terms of quality, contemporary behavioural therapy, which successfully broke the statutory monopoly on patient care held by psychoanalysis in 1987. The top-ranking groups of methods (from Grawe's meta-analysis) were still in their early days and at the stage of establishing themselves. It would appear that they were legally liquidated, de facto through the employment of a statistical trick.

2. Bias presentation of findings, interest-driven changes in the reference system: The data from the aforementioned meta-analysis were not viewed in terms of its overall context by its misusers, but instead in carefully selected comparisons from which an advantage could be obtained. For all intents and purposes, a tour de force.

The misusers emphasised, always with a certain level of self-praise, that the numerical data were proof that psychoanalysis is effective and that the null hypothesis for psychoanalysis can be rejected. This is entirely correct. (The findings were indeed more than simply chance and beyond the placebo effect.) The scientific misleaders failed to mention that the psychoanalytical studies were not found to be more effective in comparison with another method, even in comparison with relaxation techniques, in one single case. They concealed that as a group, they were at the very bottom of the efficacy ranking and that the newer methods (humanistic and interpersonal clusters) were at the top of the table. The authors of the misleading research report (1989) successfully covered up their knowledge of true reality from the public and German lawmakers. It is key to emphasise in current times that development has progressed in all approaches and that current psychoanalysis and behavioural therapy are able to present more favourable data. Furthermore, the ruthless approach adopted by certain career politicians must not be generalised to pertain to all or the majority of colleagues in the legally recognised approaches. This would certainly not paint a picture of the situation how they view it. Scheming has been well concealed behind apparent legality.

However, the above approach was and remains acceptable to the majority of directors and professors at German institutes and has not yet been stopped in its tracks by a veto from a sufficiently large group of colleagues on the basis of their scientifically truthful knowledge. The collective body of those responsible supports the division between all colleagues in the field and the social elimination of their rivals. From a psychodynamic standpoint, this could perhaps be regarded as a very primitive form of collective defence against humiliation and existential fear.

3. In Germany, behavioural therapists have also largely lost their "scientific innocence". For example, in 1987, during their official battle for approval by the health insurance system as part of their comparative battle with the monopolising psychoanalysis, they decided to not account for the systematic skewing of results through statistical adjustments, a known occurrence at the time, in their comparison of studies. This pertained to allegiance bias (namely, the positive distortion of results as a result of the researcher's allegiance to a specific school of thought) and the popular use of symptom scales that react quicker than other mental characteristics in behavioural therapy, and other factors. Accordingly, in 1987, behavioural therapeutic studies appear to have performed much better than those conducted in the field of psychoanalysis, which, needless to say, helped to persuade the representatives of health insurance providers to include behavioural therapy in the services offered under statutory health insurance. Following a required correction, the values would have remained adequate, but not quite as sensational as they were presented. (personal correspondence with U. Strümpfel).
4. A very popular yet regrettable custom around the globe is the failure to publish dissertations or studies that only feature moderate or average findings. This appears to indirectly serve the reputation of the respective school. However, it certainly does not represent a gain in general knowledge.
5. The restriction of the subjects to monosymptomatic diagnoses automatically leads to the creation of a younger, less severely ill, less chronified patient base in comparison to an average patient group, which inherently leads to higher healing potential with restitutio ad integrum, including the chance of spontaneous healing. In general, this homogeneous group produces more similar measured data, namely a small distribution around the average, than heterogeneous groups and consequently automatically achieves a higher effect size value purely on the basis of mathematics.
6. Comparative group dross: the trick of working with a trained pseudo-therapist or a recently qualified assistant in an unfamiliar therapy concept, or with ther-

apists from the "true" schools of psychotherapy who attempt to imitate another therapy approach for leadership of the control group that they are not convinced of, to ensure that one's own approach can contrast well in comparison; unfortunately, it is a very common approach in the initial decades of research. In order to protect the comparative therapies, attention must be paid to the equivalence of the competence and experience of the therapist in serious comparisons.

7. Elevating the scale: The creation of the methods paper published by the German Scientific Advisory Board for Psychotherapy (WPB) can also be considered among the list of factors behind the proliferation of alternative facts. The paper is an instrument that requires methodological ideals on the basis of approximately 40 criteria from efficacy or laboratory research. It could be employed as didactic teaching material to account for the various criteria that can play a role in the assessment of the quality of a study produced in the school of behavioural therapy. This is the beneficial aspect of the publication. However, it lacks criteria for proportionality. In reality, ideals are not fulfilled, or only fulfilled in part, especially since no perfectionist pressure similar to the pressure applied by the WBP has been established on a global scale in order to invalidate individual studies that it regards as anything but perfect. Statistical relativisation is automatically already carried out in all individual studies with a smaller number of test subjects (e.g. less than 60 test subjects), meaning the study is consequently not disqualified. With a smaller number of patients, it naturally produces higher correlations but is not invalid. As a rule, individual studies are usually subjected to a meta-analysis, which, due to an extremely large amount of patient data, leads to smaller random samples and minor defects or even incidental incorrect data that can be regarded as interpolated, being balanced out through statistics.
8. Transgressive paternalism in the selection of criteria for one approach compared to another: The methods paper, namely the criteria scale for the WBP, is also concerned with the transfer of one research design to all other approaches, such as a learning programme defined in small steps that appears to have proved successful for behavioural therapy. This includes the standardisation and manualisation of diagnostics and therapy, an inadmissible transgression and paternalism of other approaches. Behavioural therapy is not the standard for all variants of psychotherapy! Its categories do not cover those addressed by other approaches and do not automatically occur everywhere. The approach towards therapeutically useful stages of uncertainty, creativity, decision-making scope and the dialogue with the unconscious is not suitable for strict manualisation, yet nevertheless of particular importance to psychotherapy. Somewhat absurdly, the German WBP requires loyalty to the manual as a criterion for a scientifically acceptable study, despite the fact that it has long been known that this characteristic is intertwined with a negative therapy outcome. As such, it is pure nonsense to require this. It solely proves that the methods paper is not aimed at scientific knowledge, but rather at other interests.

9. Unscientific misinterpretations: Once the key significance of the "common factors", primarily the therapeutic relationship, was officially recognised and could no longer be argued away following the decades of research produced by contextual psychotherapy, and once the relative insignificance of the technical aspects of psychotherapy had been proven (only around 10% if the final outcome), all studies, programmes, techniques and methods that had previously attributed their good results to their "tools" and the enthusiasm of their founder had remained discounted, were re-evaluation. Founders of programmes are generally fans of their work and suggestively infect them with their enthusiasm and high expectations when the efficacy thereof is revalidated. Later users of this programme may be surprised to learn that the success achieved is far more modest.
10. RCTs (randomised controlled trials) are a proven research design from pharmaceutical research. During these trials, it is possible to control all conditions and keep them at a constant, except for the certain aspects being tested. They were originally intended as a pharmakon. Several participants can be blind: the patient, the investigator and the evaluator are all kept in the dark about who has received the "verum" and who has received the empty, pseudo-pharmakon. The control group and treatment group are drawn by lots. Needless to say, in psychotherapy, there is no such blinding. The therapist is aware of the therapy form he uses and the patient knows in which sort of group he participates, whether he wants to or not. He is selected to join the group. This can reduce motivation if he was hoping to be placed elsewhere. Patients often exhibit preferences to certain therapies and tend to perform better when placed in the desired groups ([2], p. 342). Wampold (p. 336) summarises that the introduction of RCTs may have advanced the general establishment of psychotherapy as a science by producing certitude that psychotherapy is indeed effective (and with a respectable average effect size of 0.80), yet RCTs did not fulfil what was and remained required of them in the psychotherapy field for the distinction of methods: The blinding required to answer this question is not possible in this case. The knowledge gained from RCTs is often inflated. Efficacy studies in naturalistic settings are required to answer this question (Wampold, p. 139).
11. Suppression of average or bad results: Most associations, institutes and pharmaceutical companies appear to have made a secret pact that only good or outstanding results will be published. Anything else is swept under the carpet. This is part of the ongoing rivalry between the schools of psychotherapy, where the Dodo bird has not yet been heard.

From the opposite standpoint, it is important to remember: Credible research and data that can be trusted still exist around the world. It is difficult for the everyday layman to distinguish between fact and fiction. Our field is simply too specialised. Consequence: We required an independent research ethics commission to turn to in cases of suspicion.

Conclusion

The research results provided by groups engaged in competition with others can only be taken with a pinch of salt at this point in time. The example of Germany, where the misuse of science and political power joined together to create an alternative system based on deceptive misinformation that serves neither the development of science nor the patient's well-being but instead works in the favour of a handful of long-established associations, has caused alarm bells to ring.

8.3 Collective Insult-Retort Sequences as a Problematic Motivational Focus

This chapter is not solely concerned with science, but also with the questionable art of using scientific data that aid concealment to improve professional positions of power, set other schools aside and compensate for wrongdoings.

Certain sequences of flourishing and withering of certain schools of psychotherapy can be regarded as a chain of (r)evolutionary replacements and subsequent fixations, which, in turn, require (r)evolutionary replacement. The new repels the old approach, often only in certain aspects. Over the course of generations, these sequences can occur peacefully or laden with insult. The more appreciated, differentiated and inclusive the old generation acts towards the impulses of the new generation, the more inclusive and attentive the new generation is likely to treat the intellectual legacy, without hindering its own development.

Part of Freud's unresolved legacy was his ability to reject, review and renew his own thoughts; however, he struggled when colleagues presented their own thoughts and concepts from their different experiences and wanted to introduce them. He seemed unable to appreciate new and foreign intellectual concepts and allow them to exist alongside his approach. This intolerance may be linked to Sigmund Freud's own socialisation, who was thrust into the role of head of the family as the first-born child of a Jewish family. Furthermore, he was the only one of his siblings sent to grammar school at great cost to his almost penniless family. As such, he was touted as a genius on whom all the family's hopes were pinned from a young age. Great things were expected of him, which, in turn, may have given strength to the group supporting him. The significant hierarchical gap between him and his siblings did not provide him with the basis for a partner-like understanding of relationships. His deep relationship with his mother, Amalie, who idolised him as her first-born child and provided the template for the Oedipus complex, further underlined his prominent position, which no one was permitted to question. Perhaps it is no wonder that Freud came not to expect intellectual exchanges, but rather a confirmation from his followers. As soon as one of his followers introduced their own ideas, he was quickly viewed as a hostile dissident who needed to be ostracised in order to avoid putting the idolised Freud into perspective. The list of dissidents from the first and second generation is particularly long. From the meta-level, psychotherapy owes its diverse and wide understanding of the human mind to its work.

In terms of professional politics in the subsequent generation, a certain number of colleagues identified with the seemingly superior gold standard of psychoanalysis and consequently believed that they too were superior in comparison to others, which led to insults being hurled and the establishment of rifts between schools. As it became clear in the 1980s in Germany that the results of psychoanalysis, as practised prior to this point, performed embarrassingly badly and worse than those that had remained overlooked until this point by the meta-analysis, many colleagues were gravely insulted and denied these new findings as much as possible to collectively ward off non-fundamentalist approaches according to the motto: anything that relativises us must be destroyed. All means were justified in this regard, even dissocial ones! There were two main camps at the time, the "orthodox" victors and the separatist, defeated camp. Division is a primitive form of defence. In other countries, various approaches have established an increasingly appreciative coexistence that incorporates healthy competition alongside mutual enrichment and cooperation.

In the 1960s–1990s in Germany, psychoanalysis looked down dismissively on behavioural therapy and all other approaches. Remnants from this time can still be found in the corresponding institutional strongholds. There were also assertions of abusive excesses of power in the internal relations of a number of psychoanalytic training institutes by brave individuals at national conventions. Insulting devaluation had become increasingly noticeable both within and outside the school of psychoanalysis. If the psychoanalytic community wanted to touch on approaches from other schools such as the humanistic/Gestalt therapy camp, the source was veiled beyond recognition, such as in the case of interactional therapy, which adopted ideas from the humanistic group dynamics in 1968 (Heigl/Heigl-Evers) or structure-oriented therapy (Rudolf), which can trace its roots to Gestalt therapy (as is the case with schema therapy).

In the early 1980s, the interpretation of behavioural therapy as veterinary therapy was a topic of ridicule at conventions; the speaker could be sure that he would generate laughter in the hall.

This laughter came to a quick standstill when behavioural therapy succeeded in breaking the monopoly held by psychoanalysis in 1987 and joined the services offered under the statutory health insurance system with its persuasive (and inflated) studies. It was a bitter pill for the school of psychoanalysis to swallow. Both approaches viewed themselves as complete opposites of one another. Advocates from the school of behavioural therapy began to ridicule psychoanalysis as old-fashioned, hypothesis-oriented, ineffective and unscientific. This rhetoric has been relayed to psychology students by professors of behavioural therapy at universities throughout Germany for 20 years, leading to a bias amongst the majority. Mutual prejudices impoverish the understanding of psychotherapy and its practical approaches.

The career politicians from the forced marriage between behavioural therapy and psychoanalysis agreed that no other group should be included in statutory health insurance. Redirecting anger towards an external enemy is always a successful way to relieve inner tensions within a group. Behavioural therapists offered to raise a pseudo-scientific fence around the "feeding trough" for the statutory health insurance system' with large-scale schemes, drawing on the example set by pharmaceutical research, to protect behavioural therapy and psychoanalysis, despite the fact that they were unable to fulfil these criteria, through the preservation of their vested rights, irrespective of the proof of efficacy. The blueprints provided by laboratory research were not viewed as scientifically uncontroversial as a means for psychotherapy around the globe.

Colleagues around the world agreed on the differentiable measurement of effect sizes (according to Cohen). This was rejected by the WBP in Germany, as their findings proved that the notion of the division was a lie.

I believe that work with integrity and unjust practise takes place in all camps, and that it would be conducive to remember the equality of the rule of law, particularly the basic law in Germany. This could assist in motivating colleagues who want to press the reset button on the Psychotherapist Act (PTG) introduced over 20 years ago, in order to put a stop to surreptitious pseudo-legalities.

Future psychotherapists are entitled to learn the variety of possibilities in order to decide their own path or opt to pursue several approaches. They should be allowed to listen to their inner voice and react accordingly. Only then will they be capable of developing variants in harmony with their personality.

With the opinionated attitudes that have since become tradition, what is lost in the respective approach from others remains timelessly valid and endures out of deference to innovative, individual, creative achievement within the scope of a particular life situation and in line with the respective zeitgeist. The latter needs to be qualified by the subsequent generation.

First and foremost, relativisation is required if a charismatic founder of a school feels entitled to generalise his personal findings. Humanity is far too multidimensional for this. Nevertheless, If we lose sight of the value concealed within, it would constitute our own loss of the shared intellectual legacy and the experiential harvest from the lives of an entire generation.

Of course, this does not mean that old knowledge should be accepted without evaluation, yet the contrary: It is key to carefully distinguish between what appears to still be helpful and integrable, and what should be disregarded as too specialist, clearly outdated or incoherent.

There is also another form of cancellation relevant to psychotherapy. It is remarkable how long emotional factors such as the therapeutic relationship and the impact of the therapist's personality were blind spots in the field and remain partially overlooked to a certain extent. When good results are achieved in the field of behavioural therapy, especially in the work of the enthusiastic pioneers of new learning programmes, the success is traditionally still attributed to the techniques and methods applied, despite the fact that it has been known for many years that the technical influence on the clarification of the overall outcome is 10% at most [4]. Accordingly, the often impressive initial results achieved by the founders of subsequent therapies cannot be replicated.

A special aspect and hurdle in behavioural therapy can be found in the inhomogeneity of the entire group of methods. Colleagues in the field of behavioural therapy have been struggling to find a uniform approach to no avail since the introduction of the Psychotherapist Act. They have always been open about this to the public. Good. Nevertheless, the unresolved issue surrounding the extent to which results from studies within the behavioural therapy conglomerate can be generalised prevails. It is not scientifically possible and can solely be applied to internally related branches that are linked by a similar approach. Mindfulness-based approaches appear to constitute one group, cognitive approaches another and behavioural approaches yet another. In essence, schema therapy is a combination of Gestalt therapy (including psychodynamics) and behavioural therapy. In light of the conglomerate structure of behavioural therapy, the humanistic group of methods was met with insult in 2018 through the non-recognition of their intellectual thread that runs throughout by the WBP. Once again proving that the WBP is a committee solely run by politics and not science.

The flexibility and willingness of behavioural therapy to change are worthy of its own chapter. Behavioural therapy has two faces; one from within and one directed outwards. From within, there was likely always the feeling that a certain improvement in competence needs to be achieved and developed. The original foundation for learning theory was meagre and lacked any emotion. The will to expand the field appeared significant, rewarding and conveyed a sense of an enriching turning point for cognitive, emotional and mindfulness-based therapies. From an outside perspective, many colleagues were familiar with individual aspects of techniques or methods from other therapies (preferably a light version of the methods suppressed by behavioural therapy). These aspects were impetuously proven without the corresponding inner attitude. The resentment of some is understandable, as is the inquisitive, practical will to try out new approaches exhibited by others.

For a while, the notion of plagiarism forced its way to the fore amongst those who felt at a disadvantage and vulnerable. A move towards general, common, yet still diverse, psychotherapy has since gained traction around the globe. This trend towards integration provides us all, and our successors, with a great opportunity: The opportunity for a transformation of the emotional climate in psychotherapy. Mutual respect would be beneficial to us all.

What about the scientific ideals harboured by the group of humanistic methods? (This includes the following in the USA: talk therapy, Gestalt therapy, psychodrama, logotherapy and existential analysis; in Germany, it also encompasses transactional analysis and humanistic body therapy methods.)

In the 1950s/1960s, and at the start of the 1970s, nobody, particularly amongst the Gestalt therapists at the time, believed that the highly complex, emotional dynamics of an experience process could be adequately scientifically documented. One individual who attempted this in the 1980s, Les Greenberg, was ridiculed for his efforts by his colleagues in the field of Gestalt therapy. He, therefore, turned to talk therapists, who had already put forward an effort to evaluate the success of their therapy according to shared criteria since the 1950s and maintained a more open mindset towards scientific studies.

A number of early scientific contributions were also made in this uncharted territory by J. Moreno, the father of psychodrama and the sociogram, who began to systematically document relationship constellations in group and social fields.

Unfortunately, the largely excellent work conducted by the initial generation of pioneers amongst various humanistic psychotherapists has been lost in part due to disinterest and blindness towards the significance of evidence of efficacy, but primarily due to their excessive scepticism towards science. Nevertheless, they honed their expertise in a self-critical manner, adjusted the method to suit the patient and directed a significant amount of energy towards training their successors. A method is not equatable to its scientific output!

Fortunately, since the 1980s, the scepticism initially held by humanists, particularly those from Gestalt therapy, has become a thing of the past and scientific documentation has caught up. In the current international efficacy ranking, measured in terms of effect size as coined by Cohen, the humanistic group of methods and behavioural therapy is ranked equally (ES 0.93) in the above-average field, with Gestalt therapy (including Greenberg's emotion-focused therapy) at the top with an effect size of 1.12–1.42. As a pioneer in process research, Greenberg was able to focus on microanalysis and move forward with a common concern in Gestalt therapy. The time-series analyses of single-case experimental design (Herrera, Chile) had a similar purpose. Process research is compatible with the system of humanistic methods. Indignation is aroused when behavioural therapists attempt to manualise humanistic methods, (a systemic requirement for ready-made programmes in behavioural therapy).

We now return to the problematic area characterised by the WBP.

The WBP (German Scientific Advisory Board for Psychotherapy), which comprises 12 representatives from standard approaches entitled to a voice (namely psychoanalysts and behavioural therapists), exercises its political power granted by virtue of ministerial self-government to abandon the internationally recognised measurements of science and effect sizes, in favour of new proposals. It simply does not grant a response to the effect size outcomes of large meta-analyses that prove the lack of scientific grounds behind the division within the German field of psychotherapy into standard and non-standard approaches, but that both groups are essentially equally effective as a whole, that indeed suppressed ones often tend to achieve slightly better results than those behind the suppression. The willingness to ignore international effect size calculations is comparable to the denial of reality. In light of the political power and economic profit gained for certain associations on the basis of "fake science", this practice is particularly ill-founded, unapologetic and/or consciously dissocial.

Systemic therapists, who, according to my observations, initially developed in concurrence with the common humanistic wave and then started to head in a different direction around a decade later with a focus on couple's and family therapy, were required to overcome initial difficulties in terms of science. Nevertheless, in 2008, (contrary to the expectations of the WBP), they succeeded in providing scientific evidence in the required style sheet by consulting closely with affiliated institutions abroad, only to be kept waiting for 10 years between initial scientific recognition and approval for patient care for adult psychotherapy, even longer for child and adolescent therapy, without cause.

This is a perfect example of intolerable arbitrary rule. This group was long excluded from inclusion in the services covered by statutory health insurance, despite the fact that it had fulfilled the rigorous conditions with studies conducted for precisely this purpose. At the same time, psychoanalytic child and adolescent psychotherapy remained approved as statutory privileged patient care, despite the fact that it had never presented adequate efficacy studies and abandoned all efforts in this regard in 2007, postponing them indefinitely and discontinuing with this pursuit in 2019.

The scheming employed by the guideline procedures to fight their way into their privileged position and the existentially threatening way in which they have asserted their political power against the rest, namely underprivileged colleagues, has left behind a trail of renewed resentment in wide circles throughout Germany with long-term repercussions. In the rest of Europe, there may be rivalries between the various schools with different victors, though there is not any official exclusion that threatens the existence of certain approaches.

If we were to jointly succeed in bridging the unjustified division between the approaches and forming a whole from the various schools of psychotherapy, our successors may find it easier to see how the various approaches complement one another and that it is good and helpful to be familiar with and practice a wide range of interdisciplinary treatments as a therapist.

8.4 The Changing Trends of the Past Century and Possibilities for Integration

Complete exclusion of emotions by the therapist in the therapeutic relationship is no longer regarded as conducive. Broadly speaking, there is no ideal for therapists in terms of maximum neutrality (as intended by Freud), but rather a shift towards selective authenticity; neutrality is solely limited to the emotions of the therapist that are purely private and do not aid the therapeutic process.

Complete exclusion of the patient's emotions (which were previously condemned to the behavioural black box as emotions could not be charted at the time and this behaviour corresponded to the scientific ideal at the time) is also no longer required. Instead, an increasingly secure and appreciative attitude towards emotions has been adopted by recent behavioural therapists for the therapeutic relationship.

From a psychodynamic viewpoint, the notion of stories and role-play is no longer excluded and frowned upon as "acting out", instead stories largely developed into a welcome, symbolising consolidation that can help to diagnose the patient's symptoms and provide clarifying interpretations on the relationship level, perhaps also a cathartic process or involuntary dialogue with the unconscious.

Body and corporeality are also generally no longer excluded from psychotherapy, instead leading to an openness to learn to understand and decode the specific language of each body as a valuable, additional source of information.

The originally somewhat rigid definition of the "there-and-then" (in psychoanalysis) as the key reason behind disorders has been relativised in favour of a flexible approach with the temporal allocation of the disturbing, mobilised, symptom-forming resonance at present.

The original monopoly of the libido as the sole cause behind mental illness was already viewed with scepticism within the field of psychoanalysis during Freud's lifetime and later replaced with a range of childhood desires (Schultz-Hencke) and later by motivation and drive hierarchies (self-psychology).

Until recently, the official, psychodynamic understanding still required commitment to the notion of the monocausality of disorders. However, this trend has since pursued the expanded view of the inner and external multicausal network of the patient in his past and current situation with traditional monocausality solely used for didactic simplicity. As such, it is more conducive for peer reviewers, enabling them to recognise their familiar, theoretical structures, than the understanding of reality with which practitioners are faced.

A gradual decline in the use of external interpretations has been observed among the new generation of therapists. Scepticism is increasing in terms of their efficacy, as they come from the therapist's own experiences and can be perceived as an encroaching abuse of the power of interpretation; instead, the interest in the patient's self-interpretations appears to have increased.

The experience that structurally weak patients tend to gain little from long, unstructured sequences in individual and group settings, but instead achieve better results with shorter, semi-structured approaches, appears to be a general experience in the field; likewise, the therapist should not treat structurally weak patients as a sphinx, but instead should be emotionally transparent, or clearly "readable",

in order to give the patient a sense of security and orientation.

Conversely, manualised learning programmes have proven to be moderately suitable, particularly in the case of close loyalty to the manual; the more rigid the programme, the less the studies correlate with success. A high level of loyalty to the manual can also have negative effects. Psychotherapists have since learned that the technique applied only provides a minor counterweight compared to the quality of the therapeutic relationship.

Compared to the early beginnings of psychotherapy, value orientations, cognitions, meaningfulness and convictions, including misguided convictions and delegations have been awarded a key place in psychotherapy.

In comparison to the early days of the field, which were shaped by the analytical spirit of the Enlightenment, the Gestalt psychological, i.e. holistic systemic view has helped with the perception of the existence of entities that come into being through relationships: Dyads, triads, couples, families, small and large groups etc., but also experience-related, confrontational connections in the mind and physiological patterns of sure-fire success to remove tension in cases of overexcitation and during inhibitive or freezing reactions.

Integrated psychotherapy can first and foremost be characterised by the notion of balance:

1. Balance on the relationship level between the patient and therapist, namely the insertion of impulses, active perception and a receptive approach towards the counterpart.
2. Balance in the therapeutic perception between a resource-oriented viewpoint and a problem and shortcoming-based approach.
3. Balance between engaging with and establishing contact with the unconscious and its reflexive processing and conscious allocation.
4. Balance between awareness of inner needs and the requirements and inviting opportunities of the environment or context.

5. Balance between self-control inside and out, and largely conflict-free consent to being controlled by others.
6. Balance between interpersonal proximity, affiliation and commitment and the need for autonomy, self-realisation, boundaries and self-determination.
7. Balance between the individual's self-image, namely the self-referential perception of oneself, and the experience of participating with different, potentially changing fields and reference frameworks.

Bibliography

1. Wampold Bruce E, Imel Zac E, Christoph F. Die Psychotherapie-Debatte. Was Psychotherapie wirksam macht. Bern: Hogrefe-Verlag; 2018.
2. Wampold BE, Imel ZE. What do we know about psychotherapy? And what is there left to debate? 2015. http://www.societyforpsychotherapie.org/what-do-we-know-about-psychotherapy-and-what-is-there-left-to-debate?24.jan2019.
3. Grawe K. Psychotherapie im Wandel. Von der Konfession zur Profession. Göttingen: Hogrefe; 1994.
4. Wampold BE, Imel ZE, Flückiger C. Die Psychotherapie-Debatte. Was Psychotherapie wirksam macht. Bern: Hogrefe; 2018.

Conclusion

Contents

L. Hartmann-Kottek, *General Psychotherapy*,
https://doi.org/10.1007/978-3-030-87466-7_9

Overview

The paramount importance of the therapeutic relationship for a positive therapy outcome supports the theory that healing is primarily a holistic process from within, in which nature restores its regulatory capacities in the protective space of insightful, loving affection. Disorder-specific therapy typically plays a supportive and stabilising role in this regard. If a symptom is viewed as a consolidation of the relevant, troubled problem area and accordingly re-interpreted, the meaningful part of the excluded aspect can help to find the path to a solution. The development of a therapist's personality and the ability to facilitate emotionally corrective experiences are two additional cornerstones.

Psychotherapy is a field with encounters on multiple levels: not just encounters between people who come to enter a more coherent and healing state of organisation with the help of psychotherapy. In the background, the natural sciences also encounter the humanities, social sciences and cultural sciences; a highly meaningful match concerned with the essence of human existence.

9.1 Foreground: The Practical Approach

9.1.1 Mindfulness and Personality Training

Prospective psychotherapists should be granted the opportunity to process their own mental stress, whether traumatic, conflict-based or otherwise, in the protective framework of a sufficiently generous amount of time allotted to self-awareness, whereby a higher level of introspection and reflection may be required in comparison to the common therapies. In the case of severe mental stress, the candidate in question will be required to reassess pursuing this profession with the assessor. Candidates who succeed in processing their stress often make excellent psychotherapists, particularly in their field of experience. As a minimum requirement, all candidates need to learn how to distinguish the resonances that come from their own unresolved conflict and a reflection of an external issue.

Prospective psychotherapists require basic mindfulness training in order to focus on sensitive receiving and focused sending of mindful energies, which can be acquired in differing ways through various meditation and self-awareness exercises. This training enables the psychotherapist to (a) become receptive towards himself and his counterpart on the various levels and capable of differentiation, (b) to have concentrated mindful energy at his disposal in order to be precise and create new structures, and furthermore, (c) in order to deliberately construct new fields, for which both of the above are required. These abilities, namely subtle, resonating perception and creative structuring are a prerequisite for nuanced therapeutic contact behaviour. They require tangible acceptance through common fields of resonance on the one hand, and the simultaneously clear, confrontational absence of communication, often in sub-areas, on the other hand. The latter may be experienced as rejection or being erased. An absence of communication can be an especially useful protective mechanism and must be constantly on hand as an option. However, saying no as a sure-fire success can also become a prison. It prevents resonance. Each individual appears to be existentially reliant on affirmative resonance at their core. (Babies who do not receive it die.)

Mindfulness training also touches on the difference between empathy and the unexpressed subtext of unconditional love, i.e. agape therapy. Both are important aspects of therapy. The latter concerns the sending of a strong, active, yet simultaneously emancipatory characteristic of existentially supportive acceptance in line with Martin Buber's I-Thou concept (or empathy, as described by Matthieu Ricard) to the existential core of the individual in need, especially if it is palpable that the patient has not adequately experienced this at an early age, for example, or if the individual currently appears to be distressed or even traumatised.

9.1.2 The Many Facets of the Therapeutic Relationship

The unexpressed, positive existential message is conveyed through the therapist's inner attitude, as mentioned above. It establishes seemingly paradoxical equal footing between the therapist and the patient on the superordinate, holistic level, meant in the most profound sense of compassion: "We both inhabit the same, often tough world and want to get by. In this sense, we share the same fate". In the case of an existing rudimentarily spiritual background: "We are each a reflection of the greater whole and consequently worthy of profound respect". In contrast, the difference between the roles, which is not denied, is secondary.

Resource perception: This aspect of appreciation can be vocalised. What has this person achieved despite difficult circumstances? What when well? What can be credibly acknowledged?

From which level does the patient send relationship signals and from which level are they responded to by the patient? Does this match, or is there dissonance? (Often-encountered negative example: older patients inherently find it tough, or perhaps even embarrassing, to work with younger therapists, particularly when the therapist attempts to cover up his insecurities, hides behind his role and lacks the courage to address this issue. This combination is a challenge for both parties.)

At which age of mental development does the patient appear to be according to my inner eye? Can I sense his aura? Which counterpart does he require to reach his next stage of development? Should I act as a nursing, protective mother for him, or instead an encouraging, cheering and challenging brother, a loyal colleague or a wise old man, etc.? Which impulse or experience of a shortcoming is he maintaining at an earlier stage if it does not correspond to the stage of development he has reached? Which facets of the relationship are easy for me to handle on the basis of my personal experience and which are difficult? Should I try to continue expanding my range of relationships?

Is there a strong projective perceptual distortion of the therapist in terms of transference, particularly negative transference? If so, I separate myself from the transference by differentiating between myself and the projected version of me, as otherwise I am neither able to assist the patient in critical situations, nor enter into a therapeutic alliance with him. The issues that would otherwise be handled in the case of transference neurosis focused on the role of the therapist are clarified in an imagined dialogue with the genuine version in this case. In order to clarify the relationship, we address the true source behind the distress and consequently promote a realistic perception of the therapist in his semi-active and emotionally semi-authentic role behaviour from the outset.

The objectifying viewpoint (Buber's I-It component) registers the symptoms, or pathological anomalies, at the same time without compromising the I-Thou relationship field. This is most likely to succeed if the inner observer refrains from making any emotional judgements.

As a therapist, can I fend a level to establish a working alliance? If yes, we can then get to work on forming a team. Which level of my counterpart am I most likely to resonate with? Is this level adequately capable of forming an alliance? If the answer is no, crisis intervention is required. It is then important that contact and relative security are established from the outset and consequently developed.

Forming an interpretative team between the therapist and patient for their dives into the unconscious: this pertains to a cooperation in the sense that the patient does not feel ashamed to report or be made aware of changes in his body language, but instead feels secure and valued enough within the confines of the therapeutic relationship. Only once this has been established is the patient able to confront and calmly discuss the phenomenon, which is generally a supplementary or corrective dissonance experience to the one reported. This experiential confrontation brings defence mechanisms to light in their nascent state in the presence of a witness. The evidence thereof can lead to distress. It can help to spur change, but requires a strong sense of security in the trust towards the therapist and his humorous, supportive and appreciative compassion. This approach is typical in Gestalt therapy at its full potential.

Revitalising relationships in the intrapsychic realm, directed towards all symptoms, "numb", abandoned, blocked, armoured and tabooed aspects in addition to all inadvertent dissonances and tension. This stage aids individual diagnosis. Last but not least, the therapeutic relationship aims to uncover the original impulse that developed into psychosomatic symptoms, amongst others, and invite it to be reinterpreted; the focus in this regard is typically on a sympathicontonic, functional overreaction, or perhaps excessive counter-regulation, muscular dysinnervation, an implied defence mechanism or abortive freezing.

The therapist should invite the patient to engage in inquisitive rediscovery, to make associations and self-interpretations. In a similar vein to Socrates, the therapist does not know the patient better than himself but is nevertheless aware that the patient understands more than he is currently aware of in the depths of his unconscious and tries to encourage him to listen inwards by showing interest. The revival of excluded aspects with the assistance of resonant association is both useful and insightful. We use our abilities from the stage of magical thinking, which has mostly fallen into latency, at which point it felt easy to empathise with animals and plants. Everything was capable of speaking and interacting with everything else. A somewhat playful, anxiety-free, perhaps even cheeky atmosphere is required to obtain the unconscious as a helpful partner in therapy. The primary aim is to re-establish contact with scattered, submerged, distorted or masked aspects, exploring them to determine their identity and concerns in order to associatively use the resonance principle to trace them back to their source and arrive at the interpersonal unresolved core conflict.

- **Relationship Mirroring Between the Individual and Multiperson Level**

Mirroring takes place in both directions: (a) the individual internalises the relationship constellations in his environment, such as the happy or perhaps the estranged relationship between his parents, and is shaped by these relationships in terms of his fundamental emotional attitude; or he is currently the victim in a bullying situation and has developed a depressive, paranoid reaction. (b) An individual can be particularly strongly influenced by his environment if he is in a position of leadership or a role model. This influence may encourage empathy or perhaps head in the opposite direction towards the devaluation of others. In companies, the sickness absence rate of employees is taken as a measure of the working climate by management.

Dialogical relationship cosmos: from chaos to reorganisation: The therapeutic re-enactment of unresolved, often structurally formative conflict constellations corresponds to relationship-clarifying parts and whole work.

In practice, the problem area is symbolically represented with empty chairs or symbolic items in the room. The therapist is required to assume a more active, stimulating role whilst ensuring that the protagonist remains the director of the scene at all times, to render a true depiction of his inner reality. It is crucial that she remains in close contact with what is going on within as the patient can be described as hypnoidal and relies on her guidance depending on the situation whilst diving into his past and attempting to clarify relationships, something he had previously avoided, within the safe confines of the therapeutic relationship. She ensures that he re-emerges from the focused regression with healing and emotionally corrective experiences.

Risking relationships for seeming nothingness: In certain instances, the patient may struggle to dismiss a difficult yet almost cherished or at least familiar part of his identity, almost as though he would no longer feel like himself or recognise himself without it, equatable to leaving a job that he became over-identified with, or another identity crisis. He feels as though he is nothing, standing in the fog with his self-awareness. He runs the risk of reconnecting with the old familiar aspects. At precisely this point, it is important that he can feel and hear the support of the therapist, who tells him that this state of not knowing is important for his transformation inner reorganisation and that over time he will be able to sense the outlines of his new, more coherent identity.

At the end of the therapy cycle, it is then time to stabilise the new aspects. In one case, the patient succeeded in telling his father from the viewpoint of his 4-year-old inner child in his mind how much he had missed him and felt guilty when his father left home after he had had another tantrum the day before, as he had done many times before. This inner child scenario made it clear to the patient that despite his tantrums, he was not the reason behind his parent's divorce. He had rarely dared to show any resistance before. Now, he has finally experienced showing resistance towards his father in the therapy exercise. As a result, he said that he wanted to look for opportunities to express his wishes or opinions to men such as colleagues, neighbours or perhaps even the postman in everyday life. Furthermore, he also intends to call his father, who is still alive, and explain the therapy he has experienced with the hope of restoring their relationship, but not as it was before, instead in a more grown-up manner.

Consensus-based experimentation in everyday life is preferential and suggestions from the patient are gladly taken on. This helps to avoid the establishment of a disempowering teacher–student relationship. The key aim of the therapeutic relationship is interpersonal teamwork with varied distribution of tasks, despite the clear gap in professional aptitude.

9.1.3 Openness to Expertise, Experience and Experimentation

▪ **Substantial Alternative Methods**

Emotionally corrective experiences spurred by therapy and accompanied by trust are seemingly one of the key factors that can be found to differing extents in all approaches. In this regard, they can be regarded as the core component of general psychotherapy.

9 The pragmatic help provided by the multitude of exercises and learning programmes is well suited to the large majority of the population, especially since the associated active engagement in participation constitutes a healing antidote to experiences of powerlessness. The student–teacher attitude from behavioural therapy may not be suitable for all client structures and constellations, but it does nevertheless appear to be adequately acceptable as an interim solution for the majority. Variations on homework have since become established in depth psychology and humanistic methods, even though they are primarily used as a flexible exercise instead of a rule in this regard.

The various methods may be derived from the field of body or movement therapy, from art or music therapy, they may use dreams, daydreams or the imagination, they may initiate spontaneous or semi-structured re-enactments or document the symbolic content of constellations and sociograms. A wealth of experience has been obtained, particularly in experience-based methods. This knowledge has been widely incorporated in one variation or another in all other methods. It is key that adequately large, experimental leeway is granted for projections of the unconscious alongside scope for the patient's self-construal and additional space for therapeutic dialogue.

The high effect size of Gestalt therapy is most likely due to its consciously differentiated therapeutic relationship culture and, to a less extent, on the techniques employed. Parts-whole work (i.e. chair work) and inner-child work are methods from Gestalt therapy regarded as formulas for success, which have already been adopted by several other approaches and contributed to the merging of the field of psychotherapy for several decades.

Exploration diagnostic methods are used to detect patterns. They also have the potential in terms of the therapeutic confrontation: their reflection causes the patient to question whether they can see this aspect in themselves and identify with it, or whether they immediately experience a mismatch and the spontaneous need to change.

For immature individuals, in the sense of development, the motivation to change is often the result of pressure to socially adjust in order to belong to a certain group. However, for developed individuals, this motivation can generally be found in the unconscious desire to attain a higher degree of personal coherence.

Exercises that put unclear issues into concrete terms are often particularly helpful. This pertains to both creative and representative forms and, above all, spatially symbolic constellations. Perhaps most widespread are the substantiations of inner circumstances, problem group or family relationships that initially appeared vague or confrontational. Whilst aspects are often presented in a somewhat simplified manner, this experience rewards the patient with relative clarity that paves the way towards solution-oriented discussions or other actions. The efficacy of this approach is contingent on the directness with which the dialogue takes place and the depth of the experience. This is a feature of waking trances, which occur when the patients engage with their imagination and truly imagine themselves in contact with their imagined counterparts and behave accordingly. Their brain interprets the imagined scenario as reality and adapts the neurophysiological memory traces within (engrams).

It is helpful to determine at which point in the patient's development the fight, flight or freeze pattern mobilised as a result of trauma or conflict and became a sure-fire success for certain circumstances.

Trauma therapy has proven to be an extremely helpful method as a link between all of the approaches employed at present and is now available in variants based on psychodynamics and behavioural therapy. Trauma therapy appears to have been influenced by a number of sources.

At the start of the 1970s, the author (L. HK) became familiar with crisis intervention from Gestalt therapy, the predecessor to trauma therapy, which (1) combined "grounding" from bioenergetics and mindful, body therapeutic mindfulness exercises in the style of Elsa Gindler as stabilisation exercises with the imaginative approaches from art therapy such as the "safe place", namely the home within, and the friend and helper team. (2) Theoretically, distancing techniques were regarded as a useful aspect of the disintegration pole, namely the avoidance of contact. This is the counterpole to the usually aspired pole in psychptherapy, the pole of integration. The rejected aspect may be locked away in a vault by the patient or sunk to the deepest depths of the pacific, frozen in an ice cave, or exiled to the dark side of the moon etc. Solely the patient should be granted the ability to determine the location and degree of exile, who will then return from powerlessness to self-efficacy. (3) For subsequent phased confrontations, the patient is offered an imaginary TV that can be moved further away if desired, with a screen that can be reduced in size and colours switched to black and white, which can be muted or switched off completely using a remote control. (4) In addition to this screening technique, in order to confront trauma, the patient may be invited to address the traumatising scenario from a standing position with contralateral tapping and eye contact with the therapist, who is inwardly focused on conveying a life-enhancing, compassionate and dignifying I-Thou field towards the patient, but all spontaneous defensive movements, emotions or vocal expressions, which are most easily expressed from a standing position, are permitted. (5) If required, following the emotional catharsis, the therapist could take on the role of a nonverbally comforting and embracing counterpart. Peter Levine, who primarily works in the field of biology, developed a similar way to resolve trauma and its frozen state with regard to non-verbal, physical aspects and the therapeutic attitude independently of this tradition, which has been met with approval by the trauma therapist Bessel van der Kolk. This convergence constitutes mutual confirmation and appreciation for the various pathways followed.

The addition of EMDR (eye movement desensitization and reprocessing) to trauma therapy in the mid-1980s by Francis Shapiro is also of great importance. EMDR pertains to the bilateral stimulation of both sides of the brain through eye movements, sounds or short touches. The approach postulates that this will mobilise blocked, stressful memories and subsequently process them in an integrating manner. EMDR has enjoyed widespread success, directing interest towards the notion of healing, neural pathways. It can be sufficient as the sole method used to heal simple traumatisation. Indeed, it has proven to be a stable brief therapy by catamnesis. More severe and complex traumatisation requires more in-depth trauma therapy.

The charisma of the therapist's personality plays a decisive role in all interventions in terms of whether the opportunity to change provided by the intervention can be used by the patient. Openness to change requires trust.

Emotionally corrective experiences accompanied and induced by therapy as disorder-specific interventions: It is easier to bypass counterimpulses such as the fear of stroking a dog when the patient has a phobia of dogs, in the protective confines of a supportive relationship. If the invitation to engage in this experience is not sufficient to jump over the anxiety barrier, it is important that the counterimpulse and the reason behind it are first put on centre stage and processed. In the ideal scenario, the therapist is able to combine a range of proven approaches from behavioural therapy with extensive experience in psychodynamic conflict resolution.

Therapeutically guided experiences with previously unexperienced potentials, longings and compensatory fantasies enable the effects of traumatising shortcomings to be less pronounced and open the door to new, healing aspects of the patient's identity. These approaches are often particularly effective when tailored to the specific patient.

Collaborative exercises in groups on the topic of power and powerlessness in relationships, for instance, can help the patient to address their memory traces and establish a dialogue with previously repelled memory traces within the protective setting of the group, to feel uplifted and comforted in the general consternation and benefit from the solutions suggested by other participants.

This can become problematic if no space is granted to strongly affected participants for the personalisation of group exercises.

Accordingly, exercised-based approaches should be used process-oriented. They are always useful in psychotherapy if they help to stabilise newly acquired ego-syntonic experiences. It is important that these experiences do not get reversed or lost through the power of familiar routine processes. These exercises need to be individually tailored to the event that has just occurred. It is also conducive to establish suitable "homework" during therapy sessions.

9.2 Background: Overriding Principles

Psychotherapy cannot be viewed as isolated, neither in terms of the cultural zeitgeist nor in the context of the natural sciences or humanities. It appears to provide a point of intersection for the various aspects of defining existence. One of the key questions in this regard pertains to the primary substance of all life, to mind and/or matter. This question has split worldviews and concepts of man into two camps for millennia, and this division remains significant for the field of psychotherapy, not only division in this sense.

Psychotherapy provides a number of practical examples for our existence at the shoreline between the wave field and structuredness (matter). However, this situation has not yet been categorically addressed by the mainstream.

The either-or fallacy was proven as inadmissible over a century ago in light of the discovery that matter is a structured special form of energy and both can be transformed into one another. This spontaneous transformation potential, which exhibits its greatest level of freedom in the smallest energetic unit, the quantum, appears to exist consistently across all levels of complexity, including the human brain and mind in addition to social relationship structures and organisations. We constantly deal with an interplay between structurally determine and free, self-organising energies.

As such, in psychotherapy, on the one hand, two scientific approaches collide with one another, namely the "classical" approach, which creates rules on the basis of the on-off logic of thinking with the mind as the activity of the dominant hemisphere of the brain with all of its flows of information via the sensory channels and the ability to focus consciousness, and, on the other hand, the approach of "new (quantum) physics", which, corresponding to the characteristics of the subdominant hemisphere of the brain, engages in the formation of waves, holistic pattern recognition, relational structures, transformations, the tolerance for uncertainty and, subjectively, higher cognition of the mind. Accordingly, the vibrational and relationship level can be regarded as the original, analogue form of existence and the structural level as existence given form, i.e. instructed existence, the digital form. All communicative symbolic systems such as writing and speaking in addition to all mathematical structures are digital forms of existence. Our consciousness is capable of creating digital structures with impressive brilliance. However, it is also capable of consensually surrendering itself to the current of life free from judgement and being guided by the self-organisation of the unconscious waves, in line with our analogue potential. We require both possibilities in psychotherapy.

As humans, we would benefit from consciously feeling at home on this shoreline in this eternal process of change between the structural and wavefield for the duration of our time on earth, and remain full of curiosity with regard to the future stages of development.